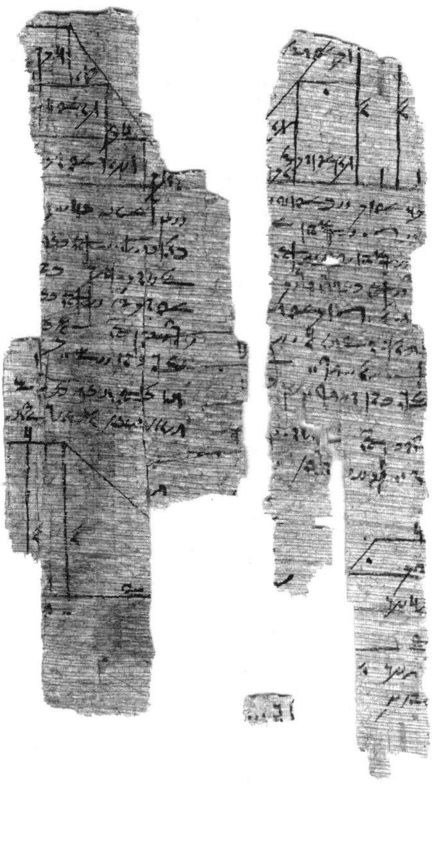

AMAZING TRACES OF A BABYLONIAN ORIGIN IN GREEK MATHEMATICS

AMAZING TRACES OF A BABYLONIAN ORIGIN IN GREEK MATHEMATICS

Jöran Friberg

Chalmers University of Technology, Gothenburg, Sweden

World Scientific

NEW JERSEY • LONDON • SINGAPORE • BEIJING • SHANGHAI • HONG KONG • TAIPEI • CHENNAI

Published by

World Scientific Publishing Co. Pte. Ltd.
5 Toh Tuck Link, Singapore 596224
USA office: 27 Warren Street, Suite 401-402, Hackensack, NJ 07601
UK office: 57 Shelton Street, Covent Garden, London WC2H 9HE

British Library Cataloguing-in-Publication Data
A catalogue record for this book is available from the British Library.

Credit line for the photo on the front cover and the spine:

VAT 8393, Vorderasiatisches Museum, Staatliche Museen zu Berlin.
Photo: Olaf M. Teßmer.

Credit line for the drawing of a pentagram on the back cover and the full title page:

VA 5953, Vorderasiatisches Museum, Staatliche Museen zu Berlin.
Hand copy: Tomas Wahlberg, Chalmers University of Technology, Gothenburg.

Credit line for the papyrus fragment on the half title page:

P. Heidelberg Inv. Dem. 663, by permission of the Institute for Papyrology, University of Heidelberg.

AMAZING TRACES OF A BABYLONIAN ORIGIN IN GREEK MATHEMATICS

ISBN-13 978-981-270-452-8
ISBN-10 981-270-452-3

Printed in Singapore by World Scientific Printers (S) Pte Ltd

Preface

A discussion of the ways in which Greek commentators in the (late) antiquity tried to explain the origin of Greek mathematics, as well as a historical survey of early cultural contacts between the Greeks and the Near East can be found in, for instance, van der Waerden, *Science Awakening 1* (1975 (1954)), 83 *ff*.

In the present book, a sequel to the author's *Unexpected Links Between Egyptian and Babylonian Mathematics*, Singapore: World Scientific (2005), Greek and Babylonian mathematical texts will be allowed to speak for themselves.

The following passage, of interest in the present connection, can be cited from the Preface to my *Unexpected Links*:

> "My observation that there seems to exist clear links between Egyptian and Babylonian mathematics is in conflict with the prevailing opinion in formerly published works on Egyptian mathematics, namely that practically no such links exist. However, in view of the dynamic character of the (writing of the) history of Mesopotamian mathematics, not least in the last couple of decades, it appeared to me to be *high time to take a renewed look at Egyptian mathematics against an up-to-date background in the history of Mesopotamian mathematics!* ....
> The detailed comparison in this book of a large number of known Egyptian and Mesopotamian mathematical texts from all periods has led me to the conclusion that the level and extent of mathematical knowledge must have been comparable in Egypt and in Mesopotamia in the earlier part of the second millennium BCE, and that there are also unexpectedly close connections between demotic and "non-Euclidean" Greek-Egyptian mathematical texts from the Ptolemaic and Roman periods on one hand and Old or Late Babylonian mathematical texts on the other."

Also of relevance in the present connection are the following words from the summing-up in the last few lines of *Unexpected Links*:

> "The observation that Greek ostraca and papyri with Euclidean style mathematics existed side by side with demotic and Greek papyri with Babylonian style mathe-

> matics is important for the reason that this surprising circumstance is an indication that when the Greeks themselves claimed that they got their mathematics from Egypt, they can really have meant that they got their mathematical inspiration from Egyptian texts with mathematics of the Babylonian type. To make this thought much more explicit would be a natural continuation of the present investigation."

The following deliberation is in agreement with the cited passages:

The simplest way of explaining the many parallels found in this book between (certain parts of) Greek mathematics and Old or Late Babylonian mathematics is to assume that *in ancient Greece elementary education in mathematics for young students (not necessarily intending to become mathematicians) was conducted in terms of metric algebra in the Babylonian style*. Here "metric algebra" is a convenient name for the very special kind of mathematics, with *an elaborate combination of geometry, metrology, and linear or quadratic equations,* which is first documented in proto-Sumerian texts from the end of the fourth millennium BC, and which prevailed in Mesopotamia without much change to the Seleucid period close to the end of the first millennium BC. During the 2500 years of its existence already *before* the dawn of Greek mathematics, this kind of mathematics ought to have had ample opportunity to spread to more or less distant neighbors of Mesopotamia itself. That this hypothesis is correct in the case of Egypt was demonstrated in *Unexpected Links*. To show that it may be correct also in the case of ancient Greece is the object of the discussion below.

It is important to understand that one of the obstacles in the way for a better understanding of possible relations between Greek and Babylonian mathematics is the circumstance that Greek mathematics is documented mainly through copies of copies of important manuscripts with advanced mathematics, while Old Babylonian mathematics is documented mainly through clay tablets with relatively low level mathematics, written by mediocre scribe school students, and Late Babylonian/Seleucid mathematics is documented only through a small number of texts, for the simple reason that in the second half of the first millennium BC clay tablets had been replaced by more easily perishable materials as the preferred medium for writing. For these reasons, it is difficult to know what Greek mathematics at a lower level was like, and equally difficult to find out how advanced Old and Late Babylonian mathematics at a higher level may have been.

It is also important to understand that since the heated but inconclusive debate about Greek "geometric algebra" in the late 1970's, much has happened in the study of Babylonian mathematics. Thus many new mathematical cuneiform mathematical texts have been published since then, several of them with unexpected and astonishing revelations about the scope of Babylonian and pre-Babylonian mathematics, and many of the earlier published mathematical cuneiform texts have been explained in new, and much more satisfactory ways. Therefore, it is now obvious that the mentioned debate was conducted against a background of regrettably insufficient knowledge about the true nature of Babylonian mathematics.

More or less accidentally, the dedicated search in this book for parallels between Greek and Babylonian mathematics has, in addition, resulted in a rather extensive survey of certain important parts of Greek mathematics, as well as in new answers to a number of open problems in the history of Greek mathematics.

Here follows a brief survey of the contents of the book:

Chapter 1 is a continuation and more or less definite conclusion of the debate about what has been known as the "geometric algebra" in Euclid's *Elements* II. In this chapter it is shown that far from being Greek reformulations in geometric terms of Babylonian (non-geometric) algebra, the propositions in *Elements* II are *abstract, non-metric reformulations of a well defined set of basic equations or systems of equations in Babylonian metric algebra*, that is of *quadratic and linear equations or systems of equations for the lengths and areas of geometric figures*.

Strictly speaking, *Elements* II is not about "geometry" at all, in the literal sense of the word, which is 'land-measuring'.

Characteristically, as a consequence of the different Greek and Babylonian approaches to geometry, diagrams illustrating non-metric propositions in the *Elements* are what may be called "lettered diagrams", while diagrams illustrating Babylonian metric algebra problems are "metric algebra diagrams" with explicit indications of relevant lengths and areas.

As a whole, *Elements* II is a well organized "theme text" of the same kind as similarly well organized Babylonian mathematical theme texts.

Chapter 2 begins with a presentation of Euclid's proof of *El.* I.47, and of Pappus' proof of a generalization of *El.* I.47. Then follows a discussion of the OB (Old Babylonian) forerunner of *El.* I.47, "the OB diagonal rule (for rectangles)". It is suggested that the rule may have been discovered accidentally in connection with the study of "rings of four rectangles (or four right triangles)".[1] The argument is supported by the recent discovery of an OB "hand tablet" with a picture of a *ring of three trapezoids*. The hand tablet is published in the author's *A Remarkable Collection of Babylonian Mathematical Texts*, New York: Springer (2007).

Chapter 3 is a confrontation of Greek rules for the generation of pairs of numbers (integers) such that the sum of their squares is also a square with OB rules for the generation of "diagonal triples", rational sides of right triangles. The Greek rules are attributed to Euclid (lemma *El.* X.28/29), Pythagoras, and Plato, while the OB rule is manifested in a number of OB "igi-igi.bi problems", as well as in the famous OB table text Plimpton 322.

Chapter 4 begins with a discussion of Euclid's important lemma *El.* X.32/33, which says, essentially, that a right triangle is divided into two right sub-triangles similar to the whole triangle by the height against the hypothenuse. That this result was known also in Babylonian mathematics is demonstrated by an OB problem for a right triangle divided by use of a recursive procedure into a "chain of similar right sub-triangles".

Chapter 5 contains a completely new approach to the study of the notoriously difficult tenth book of the *Elements*. It is shown that the theory of inexpressible straight lines in *El.* X is based on a number of fundamental lemmas and propositions such as the lemmas X.28/29, X.32/33, X.41/42, and the propositions X.17-18, X.30, X.33, X.54, X.57, X.60, all of which can best be explained by use of Babylonian metric algebra. As a matter of fact, a particularly great role is played in *El.* X by "quadratic-rectangular systems of equations of type B5", by which is meant problems where both the sum of the squares of two unknowns and the product of the unknowns are given. Such problems appear as well in Babylonian mathematics.

Also discussed in this chapter is the relation between Euclid's "parabolic application of areas" in *El.* I.44 and Babylonian "metric division".

---

1. Note that, since angles was a relatively unknown concept in OB mathematics, it is less anachronistic to speak of OB "right triangles" than of OB "right-angled triangles".

Chapter 6 is devoted to a discussion of *Elements* IV, a well organized theme text concerned mainly with "figures within figures". It is shown, through a great number of examples, that figures within figures was a popular subject also in Babylonian mathematics.

Chapter 7 explains in terms of metric algebra the cutting of a straight line *in extreme and mean ratio* in *El.* VI.30, as well as the theory of the regular pentagon and the equilateral triangle in *El.* XIII.1-12. It is pointed out that the propositions *El.* XIII.1-11 can be interpreted as a "metric analysis" of the regular pentagon *relative to the radius of the circumscribed circle*, while a (hypothetical) corresponding Babylonian metric analysis of the regular pentagon necessarily would have operated *relative to the side of the pentagon.*

The relation of such a metric analysis of the regular pentagon (alternatively the regular octagon) to the theory of inexpressible straight lines in *El.* X is investigated.

The chapter ends with a survey of examples of regular polygons and related objects in Babylonian mathematics.

Chapter 8 is an account in terms of metric algebra of the construction of regular polyhedra inscribed in spheres in *El.* XIII.13-18. The account highlights the role played in some of these constructions by the diagonal rule in three dimensions.

Then follows the presentation of a Kassite (post-OB) text with the computation of the interior diagonal of a gate by use of the diagonal rule in three dimensions, and of another Kassite text with the computation of the weight of a colossal 'horn-figure' (icosahedron), constructed by use of 20 equilateral triangles with sides measuring 3 cubits and made of copper sheets 1 inch thick. Both texts are published in the author's *Remarkable Collection* (2007).

Chapter 9 begins with Euclid's demonstration in *El.* XII.3-7 of (essentially) the fact that every triangular prism can be cut into three triangular pyramids, each one of which has a volume equal to one third of the volume of the prism. Then follows a discussion of texts showing that OB mathematicians could compute correctly the volumes of various kinds of whole and truncated pyramids, as well as of whole and truncated cones. The manner of computation of the volume of a "ridge pyramid" in an OB mathematical text is compared with the dissections used in *El.* XII.3-7 and with similar dissections used by the famous Chinese mathematician Liu Hui in his commentary to problems in the Chinese mathematical classic *Nine Chapters*. It is pointed out that there are indications that also Babylonian mathematicians knew about similar dissections of prisms and pyramids.

Chapter 10 contains a detailed discussion in terms of Babylonian metric algebra of Euclid's parabolic, elliptic, and hyperbolic "application of areas" propositions *El.* I.43-44, *El.* VI.24-29 and *Data* 57-59, 84-85. In addition, a completely new explanation is given of Euclid's intriguing proposition *Data* 86, which is here shown to give the detailed solution to a complicated "quadratic-rectangular system of equations of type B6", related to the already mentioned quadratic-rectangular systems of equations of type B5 in the proofs of *El.* X.54 and X.57.

Chapter 11 begins with an account of some of the most interesting propositions in Euclid's lost book *On Divisions*, known mainly from an abstract published by a 10th century Persian geometer. Particular attention is given in this account to problems where triangles or trapezoids are divided by lines parallel to the base, and to an appealing proposition where the problem of dividing a triangle in two parts in a certain ratio by a line through a given point in the interior of the triangle is reduced to the problem of solving a certain quadratic equation.

Then follows a detailed discussion of numerous OB parallels in the form of problems for triangles or trapezoids divided in certain ratios by one or several transversals parallel to or orthogonal to the base. Among these problems are several of the most interesting and sophisticated of all known Babylonian mathematical problems. In particular, a completely new explanation is given here of an OB quite sophisticated "boundary value problem", where a trapezoid with known base and top is divided into a chain of three rational bisected sub-trapezoids.

The "confluent trapezoid bisections" in a couple of OB mathematical texts show that OB mathematicians knew how to combine a solution to an indeterminate quadratic equation of the form sq. s_{a} + sq. $s_{\mathrm{k}} = 2 \cdot$ sq. d with a solution to the indeterminate quadratic equation sq. a + sq. $b = 1$ in such a way that the result is a new solution to the first equation.

An interesting observation is that the famous "Bloom of Thymaridas" is a generalization of a system of equations connected with an OB method for the construction of solutions to trapezoid bisection problems.

Chapter 12 compares Hippocrates' *quadrature of lunes* with various Old and Late Babylonian computations of the areas and diameters of certain figures with curved boundaries, in particular certain double circle segments, but also "concave squares" and "concave triangles".

Chapter 13 contains a discussion of a large number of examples of parallels to Babylonian metric algebra in Diophantus' *Arithmetica*. Thus, for instance, *Arithmetica* I is organized precisely like an OB theme text with equations or systems of equations for one or two unknowns. Particularly interesting here is the appearance of the word *plasmatikón*, the meaning of which has been debated. However, it is likely that when a problem is called *plasmatikón*, that means that it is 'representable', namely by a metric algebra diagram. It is also interesting that the *diorisms* appearing in certain problems are conditions for the existence of solutions which seem to have been derived from the study of such diagrams.

In *Arithmetica* II, some "basic examples" which are usually explained by reference to the "chord method", can just as well be explained by reference to metric algebra problems for triangles or trapezoids inscribed in circles, or by reference to trapezoids divided into parallel stripes. Similarly, the interesting and well known method of "approximation to limits" in *Ar.* "V".9, which can be explained by a variant of the chord method, can just as easily be explained with reference to the OB method of "confluent trapezoid bisections".

Diophantus' extremely interesting but obscure construction in *Ar.* III.19 of *a square number equal to a sum of two squares in four different ways* can with advantage be explained in terms of metric algebra with reference to a "birectangle" (a quadrilateral with two opposite right angles). This construction, too, seems to be intimately connected with the OB method of confluent trapezoid bisections and with the OB rule for the composition of a solution to an indeterminate quadratic equation of the form sq. s_{a} + sq. $s_{\mathrm{k}} = 2 \cdot$ sq. d with a solution to the indeterminate quadratic equation sq. a + sq. $b = 1$.

An indeterminate "price and number problem" which appears totally out of context in *Ar.* "V".30, is closely related to similar OB problems leading to systems of linear equations, but it is interesting also because it is solved by use of solutions to "quadratic inequalities" obtained through "completion of the square".

Arithmetica "VI" which is concerned with indeterminate equations for right triangles, has, like *Arithmetica* I, precisely the same form as an OB theme text. The construction problem *Ar.* "VI".16: *To find a right-angled triangle in which the bisector of an acute angle is rational*, appears in *Ar.* "VI" totally out of context, and is solved by what looks like metric algebra.

One of the few occurrences in *Babylonian* mathematics of indeterminate equations is particularly interesting because it equates (in a totally artificial way) the interest on a loan with a square, a cube, or a "cube-minus-1", the latter term meaning a "quasi-cube" of the form "cube n – square n". It is interesting to note in this connection that in *Ar.* "VI" all the undetermined right hand sides of equations are, likewise, either a square, a cube, a "quasi-square", or a "quasi-cube".

Heron's well known *area rule for triangles*, the Indian mathematician Brahmagupta's closely related *area rule for cyclic quadrilaterals*, and Ptolemy's and Brahmagupta's *diagonal rules for cyclic quadrilaterals* are treated together in **Chapter 14**. It is shown that all these rules can be derived in simple and straightforward ways by use of metric algebra, as long as no other cyclic quadrilaterals are considered than *triangles, rectangles, symmetric trapezoids, and birectangles or "cyclic orthodiagonals"*.

In **Chapter 15**, Theon of Smyrna's "side and diagonal numbers algorithm" is explained in terms of an "ascending chain of birectangles". It is shown that a similar construction works just as well when the equation sq. $d = 2$ is replaced by more general equations of the forms sq. p = sq. $q \cdot D - 1$ or sq. p = sq. $q \cdot D + 1$, where D = sq. d.

In this connection is discussed also a previously never clearly understood OB mathematical table text which may be related to an "ascending spiral chain of trapezoids". An OB "ascending and descending chain of trapezoids with fixed diagonals" is considered in Appendix 1.

Chapter 16 is devoted to a detailed discussion of two methods for the approximation of "square sides" (square roots) used in Heron's collected works. One method, which is essentially the same as a Babylonian "square side rule" is used in the great majority of cases. A second, more accurate method is explained here in terms of "third approximations", by which is meant approximations obtained through a kind of repeated composition of an initial approximation with itself, resulting in a "formal third power".

Interestingly, the use of third approximations can explain not only Heron's accurate square side approximations, but also the well known and much debated Archimedian accurate estimates for sqs. 3, as well as the accurate square side approximations in Ptolemy's *Syntaxis* I.10.

The chapter ends with a discussion of Babylonian square side approximations and of examples of an elegant OB method of eliminating square factors from an area number before the computation of its square side.

In **Chapter 17** it is suggested that Theodorus of Cyrene's famous irrationality proof for square sides of non-square numbers, mentioned in *Thaetetus 147 C-D*, can have been carried out by use of a "descending chain of birectangles", of the same form as the *ascending* chain of birectangles used in Chapter15 for the explanation of Theon's side and diagonal numbers algorithm. The irrationality proof by use of such a descending chain of birectangles works only as long as a solution (in integers) is known to the equation sq. p = sq. $q \cdot D \pm 1$, where D is the given non-square number. If the pair p, q is a solution to an equation of this kind, it is convenient to call p/q an "optimal approximation" to sqs. D. As it turns out, it is easy to find such optimal approximations for all non-square numbers (integers) D from 2 to 17, when $D = 13$ by use of a "third approximation" of the kind discussed in Chapter 16, but not for $D = 19$. This circumstance may explain why Theodorus stopped his demonstration after reaching the case $D = 17$. (The case $D = 18$ can be neglected, since sqs. 18 = 3 · sqs. 2.)

There is an interesting connection between the explanation above of Theodorus' irrationality proof and Brahmagupta's well known observation that he could find a solution to the equation sq. p = sq. $q \cdot D + 1$ in every case when he already knew a solution to the equation sq. p = sq. $q \cdot D + r$, with $r = -1, \pm 2$, or ± 4. As a matter of fact, the method used by Brahmagupta in the non-trivial cases $r = \pm 4$ can be explained in terms of "formal third powers".

In **Chapter 18** it is observed that the Heronic *Metrica* is a typically Greek (Euclidean) mathematical hand book, while the "pseudo-Heronic" *Geometrica* is a compilation of various sources, some of them clearly influenced by Babylonian mathematics. The chapter contains, among other things, surprisingly simple new explanations of the solution procedures in *Geom.* 24.1-2 for a couple of indeterminate problems for the areas and perimeters of a pair of rectangles. Another interesting problem discussed in this chapter, with an obvious relation to a number of Babylonian mathematical problems, is concerned with the sides of a right triangle at a distance of 2 feet from a right triangle with given sides. The chapter is concluded with an explanation of an intricate *division of figures problem* in *Metrica* 3.4, which can be reduced to a rectangular-linear system of equations for two segments of one side of a triangle.

In **Appendix 1**, a new OB mathematical problem text of extraordinary interest is published jointly with J. Marzahn, curator of the collections of clay tablets at the Vorderasiatisches Museum, Berlin. The text begins with a diagram showing a chain of five trapezoids, all with the diagonal 3. The explicit computation of the various sides and transversals of this chain of trapezoids demonstrates that OB mathematicians were familiar with Ptolemy's diagonal rule in the case of symmetric trapezoids, and that they had found, in addition, an elegant rule for the construction of a linked pair of symmetric trapezoids with diagonals of the same length. The recursive procedure used for the computation of the sides and transversals in the chain of five trapezoids starts with the central trapezoid and continues with ascending and descending chains of trapezoids, much like the ascending and descending chains of birectangles discussed in Chs. 15 and 17 above.

The book ends with **Appendix 2**, which is a catalog of all plane and solid geometric figures appearing, in one way or another, in Mesopotamian mathematical texts. There are also an index of texts, an index of subjects, a bibliography, and a comparative set of Mesopotamian, Egyptian, and Greek timelines showing periods of documented mathematical activities.

The work with the manuscript for this book has been supported by
The Royal Society of Arts and Sciences in Gothenburg,
Wilhelm och Martina Lundgrens Vetenskapsfond 1, and
Gunvor och Josef Anérs Stiftelse.

September 1, 2006 Jöran Friberg

Contents

Chapter 1

Elements II and Babylonian Metric Algebra

The enigmatic nature of Euclid's *Elements* II and the related propositions *El.* VI.28-29[1] (Heath, *TBE I-III* (1956); *HGM 1* (1981), 379-380; Christianidis (*ed.*), *CHGM* (2004), Part 6) has given rise to a heated debate among historians of mathematics, summarized by Artmann (*Apeiron* 24 (1991)) in the following words:

> "Traditionally VI.28 and 29 have been considered under the rubric 'geometrical algebra', a concept introduced by Zeuthen (1896), 7, following Tannery (1882). Subsequently Neugebauer (1936), van der Waerden (1954), Freudenthal (1977) and Weil (1978) adapted and extended Tannery's and Zeuthen's position. Heath followed Tannery in his comments on II.5 and 6, which he interpreted as solutions to quadratic equations. This traditional position was attacked by Szabó (1969), Unguru (1975) and Unguru and Rowe (1981), (1982). Van der Waerden (1954), 118-126 gives a clear statement of the position of the proponents of 'geometrical algebra'. His main claims are:
>
> (i) The real content of VI.28 and 29 is algebraic (as solutions of quadratic equations); geometry is only a mode of expression.
>
> (ii) Geometrical algebra originated with the Pythagoreans, who took it (somehow) from the Babylonians.
>
> (iii) The Greeks had to use a geometrical formulation of the theory of quadratic equations because they had no other way to deal with incommensurable magnitudes."

Since those words were written, one of the basic premises for the whole controversy has been shown to be invalid. Thus, it has been demonstrated by Høyrup, through a detailed analysis of the technical vocabulary in mathematical cuneiform texts, that Old Babylonian (OB) mathematicians understood quadratic equations in terms of the dimensions and areas of rectangles and other *measurable geometric magnitudes*, and not primarily in terms of anything like our school algebra. (See, for instance, Høyrup, *LWS* (2001).) Subsequently, it has been shown by Friberg (*BaM* 28 (1997),

---

1. A useful survey of the contents of all the thirteen books of Euclid's *Elements* is given online by D. E. Joyce, <http://aleph0.clarku.edu/~djoyce/java/elements/elements.html>.

Ch. 1) that also Late Babylonian mathematicians used a similar "metric algebra" in order to visualize and solve quadratic equations. Intriguingly, the roots of the Old and Late Babylonian metric algebra can be traced back to examples of "metric squaring" and "metric division" in Old Akkadian and Early Dynastic mathematical texts, half a millennium older than the better known Old Babylonian mathematical texts (Friberg, *CDLJ* 2005/2; *RC* (2007), Apps. 6-7)), and perhaps even to the surprising "field expansion procedure" in proto-cuneiform texts from the end of the 4th millennium BCE (Friberg, *AfO* 44/45 (1997/98); *RC*, Sec. 8.1 b).

The changed premises will make it possible to resolve the mentioned controversy by showing, in this chapter, that the alleged "geometrical algebra" in Euclid's ***Elements* II** is of the same nature as closely related results in Old and Late Babylonian metric algebra, and that therefore the assumption that the Greeks had to use a geometric reformulation of an originally purely *algebraic* theory of quadratic equations "because they had no other way to deal with incommensurable magnitudes" must be false.[2]

1.1. Greek Lettered Diagrams *vs.* OB Metric Algebra Diagrams

The style of Euclid's exposition in Book II of his *Elements* is shown by the following analysis of the text of one of the propositions in Book II:

***El.* II.5** (Heath, *TBE 1*(1956)) begins with *a statement in general terms*:

If a straight line is cut into equal and unequal segments,
the rectangle contained by the unequal segments of the whole
together with the square on the straight line between the points of section
is equal to the square on the half.

Then follows *a more comprehensible reformulation of the statement in terms of a suitable diagram*:

For let a straight line *AB* be cut into equal segments at *C*
and into unequal segments at *D*;
I say that the rectangle contained by *AD*, *DB* together with the square on *CD*
is equal to the square on *CB*.

2. The ideas discussed in this chapter were presented at the *Zeuthen-Heiberg Centenary Symposium on Current Studies in Ancient Greek Mathematics*, Copenhagen, August, 1994.

In a careful construction of the following complete diagram, step by step, this initial diagram is then extended into a combination of rectangles and squares, where lettered vertices are introduced in alphabetic order:

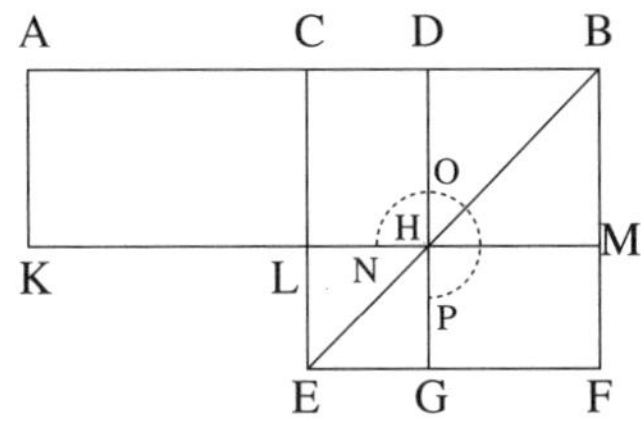

For let the square *CEFB* be described on *CB*, and let *BE* be joined;
through *D* let *DG* be drawn parallel to either *CE* or *BF* (cutting BE in H),
through *H* again let *KM* be drawn parallel to either *AB* or *EF*,
and again through *A* let *AK* be drawn parallel to either *CL* or *BM*.

Since the diagonal *BE* has been drawn, the proof of the statement can begin with an application of the "diagonal complements rule" in *El.* I.43:

Then, since the complement *CH* is equal to the complement *HF*,
let (the square) *DM* be added to each;
therefore the whole (rectangle) *CM* is equal to the whole (rectangle) *DF*.

Next, by a transitivity argument,

But (the rectangle) *CM* is equal to (the rectangle) *AL*,
since (the segment) *AL* is also equal to (the segment) *CB*;
therefore (the rectangle) *AL* is also equal to (the rectangle) *DF*.

Hence the following intermediate result:

Let (the rectangle) *CH* be added to each;
therefore the whole (rectangle) *AH* is equal to the gnomon *NOP*.

This intermediate result is rephrased in terms of the initial diagram:

But *AH* is (equal to) the rectangle (contained by) *AD*, *DB*, for *DH* is equal to *DB*,
therefore the gnomon *NOP* is also equal to the rectangle (contained by) *AD*, *DB*.

The last step of the procedure is the completion of the gnomon to a square:

Let (the square) *LG*, which is equal to the square on *CD*, be added to each;
therefore the gnomon *NOP* and (the square) *LG*
are equal to the rectangle contained by *AD*, *DB* and the square on *CD*.
But the gnomon *NOP* and (the square) *LG* are the whole square *CEFB*,
which is described on *CB*;
therefore the rectangle contained by *AD*, *DB* together with the square on *CD*
is equal to the square on *CB*. Therefore etc.

The consequent use of lettered vertices in all geometric diagrams is perhaps the most visually striking feature of Greek mathematics of the kind that one meets in Euclid's *Elements*. The lettered vertices are used not only in the diagrams themselves but also in the text, in all references to the diagrams. In the example above, straight lines are named after their end-points, as in *AB*, *CD*, *etc.*, rectangles or squares after their vertices, as in 'the square on *CB*', or 'the square *CEFB*', or simply '(the square) *DM*', and 'the rectangle contained by *AD*, *DB*', or simply '(the rectangle) *AL*', and so on. There are never any metrological or numerical specifications for given plane or solid figures or their parts, such as their lengths, angles, areas, or volumes. The device that is used, perhaps a bit too cleverly, in order to avoid any mention of lengths, areas, *etc.*, is to say that one straight line is 'equal to' another straight line, or that one plane figure is 'equal to' another plane figure, *etc.* In the statement in the example above, for instance, a rectangle and a square are said to be equal to another square.

The situation is completely different in Babylonian mathematical cuneiform texts, where in all diagrams showing plane or solid figures, straight lines are denoted by their lengths and/or suitable names such as 'the upper length', 'the middle length', 'the lower length', 'the first length', 'the second length', *etc.*, and where similarly areas or volumes are denoted by numbers and/or suitable names. (A good example is IM 55357. See Sec. 4.3 below.) The numbers or names for the lengths are normally placed alongside the figures in their proper places, while the numbers for the areas or volumes are placed inside the figures. The situation is similar in Egyptian hieratic or demotic mathematical papyri, and even in Greek-Egyptian mathematical papyri from the Ptolemaic and Roman periods. (See the many examples in Friberg, *UL* (2005).)

There is another obvious fundamental difference between the example above and a typical Babylonian mathematical text. In *El.* II.5, the object of the text is to *prove* that two geometric figures 'are equal'. The object of a Babylonian mathematical text is nearly always to *compute* something. So, how can there be any kind of relation between a Greek text like *El.* II.5 and Babylonian mathematics? To begin to see why, one has to see what becomes of the lettered diagram in *El.* II.5 *if the letters are removed and instead lengths and areas with their numerical values are explicitly indicated* in the Babylonian style. In Fig. 1.1.1 below, a (hypothetical) example of such a diagram in the Babylonian style is shown to the left, and a modernized version in the same style to the right.

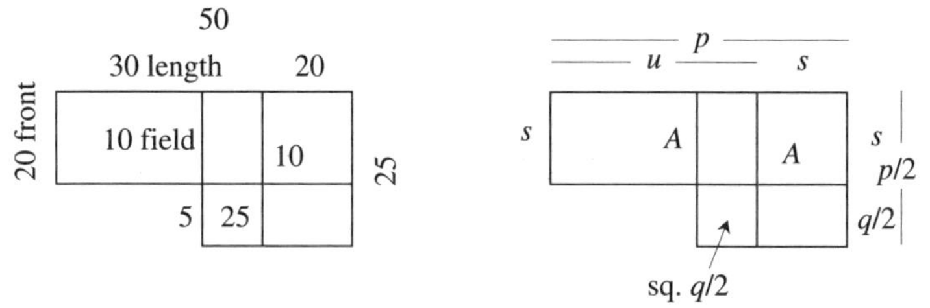

Fig. 1.1.1. A diagram in the Babylonian style (left), and a modernized version (right).

The names used for the long and short sides of a rectangle in OB mathematical texts were normally the Sumerian terms uš 'length' and sag 'front'. The most commonly chosen values for the length and the front were 30 and 20 length units (Sum. ninda = c. 6 meters, or 1/60 ninda = 1 dm). In the diagram above, to the left, the length is 30, the front 20, the sum of the length and the front 50, and half that sum 25. The area of the rectangle is 30 · 20 = 10 (· 60), the area of the small square is sq. 5 = 25, and the area of the large square is sq. 25 = 10 25 = 10 (· 60) + 25.

The numerical example shows how Babylonian mathematicians could arrive at interesting results through experimentation with numerical values for the parameters of a geometric figure. Another way in which they could find new insights was through shrewd observation. Thus, for instance, it is known that OB mathematicians were familiar with what they called a.šà *dalbani* 'the field between' two plane geometric figures.

In the example in Fig. 1.1.2, *the field between two concentric and parallel squares* is what may be called a "square band". Now, if you want to divide the square band equally into four simple pieces, you can do it in several ways. In particular, you can divide the square band into four equal rectangles, as in Fig. 1.1.2, left, or into four "square corners" (what the Greeks called "gnomons"), as in Fig. 1.1.2, right. Evidently, the area of any one of the four square corners is then equal to the area A of any one of the four rectangles. It is also clear from the figure that if p is the side of nígin *kīditum* 'the outer square' and q the side of nígin *qerbitum* 'the inner square', then the area of the whole square band is sq. p – sq. q, while the area of one of the square corners is sq. $p/2$ – sq. $q/2$. In Fig. 1.1.2, right, the notations p and q have been chosen for the sum $u + s$ and the difference $u - s$, respectively, where u = uš, the 'length', and s = sag, the 'front'.

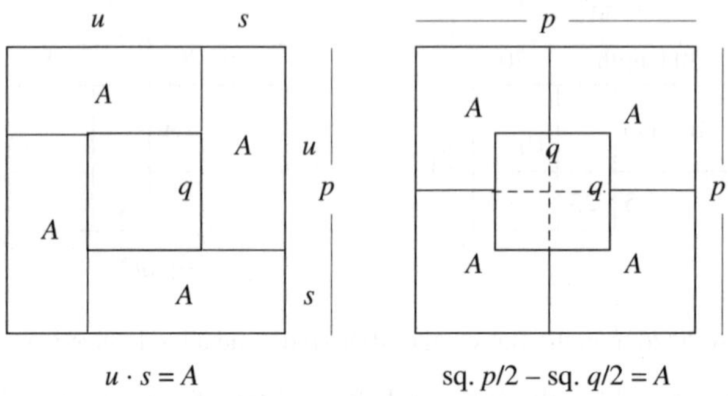

Fig. 1.1.2. Two simple ways of dividing a square band into four equal pieces.

Rectangles, squares, and square corners played a dominant role in OB metric algebra. Often, the first step in the solution of a given metric algebra problem was a transformation of the problem into one of a small number of OB "basic" metric algebra problems (Friberg, *RlA 7* (1990), Sec. 5.7 c):

Two basic *rectangular-linear systems of equations*:

B1a: $u \cdot s = A,\quad u + s = p$

B1b: $u \cdot s = A,\quad u - s = q$

Two basic *additive quadratic-linear systems of equations*:

B2a: sq. u + sq. $s = S,\quad u + s = p$

B2b: sq. u + sq. $s = S,\quad u - s = q$

Two basic *subtractive quadratic-linear systems of equations*:

B3a: sq. u – sq. $s = D,\quad u + s = p$

B3b: sq. u – sq. $s = D,\quad u - s = q$

Three basic *quadratic equations*:

B4a: sq. $s + q \cdot s = A$

B4b: sq. $u - q \cdot u = A$

B4c: $p \cdot u$ – sq. $u = A$

The important thing to remember is that all these types of rectangular-linear, quadratic-linear, or simply quadratic metric algebra problems were actually *visualized as problems for rectangles and squares*.

Below, *the thirteen propositions El.* II.2-14 *will be compared with this list of nine OB basic metric algebra problems.*

1.2. *El.* II.2-3 and the Three Basic Quadratic Equations

The proposition ***El.* II.1** states that if two straight lines are given, and if one of them is divided into a number of segments, then the rectangle contained by the given lines is 'equal to' the (sum of) the rectangles contained by the second line and the segments of the first. The purpose of this proposition is not at all clear, although it is likely that the proposition is meant as a reminder of *the additivity of areas*. In this sense, it paves the way for the following two propositions, *El.* II.2 and *El.* II.3.[3]

El. II.2

If a straight line is cut at random,
the rectangles[4] contained by the whole and both of the segments
are equal to the square on the whole.

El. II.3

If a straight line is cut at random,
the rectangle contained by the whole and one of the segments
is equal to the rectangle contained by the segments,
and the square on the mentioned segment.

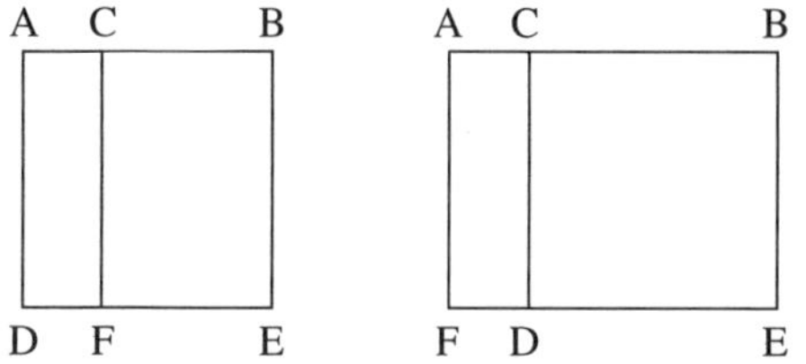

Fig. 1.2.1. Diagrams in *El.* II.2 (left), and *El.* II.3 (right).

The diagram in ***El.* II.2** (Fig. 1.2.1, left) is replaced in Fig. 1.2.2 below by a diagram in the (modernized) Babylonian style, which shows that for any triple of straight lines (of length) u, s, and q, with $u - s = q$, the statement in *El.* II.2 saying, essentially, that, by the additivity of areas,

$$u \cdot s + u \cdot (u - s) = \text{sq. } u$$

can be reformulated[5] as *a quadratic equation of type B4b*:

$$\text{sq. } u - q \cdot u = A, \quad \text{where} \quad A = u \cdot s.$$

3. All translations of propositions in the *Elements* are borrowed from Heath, *TBE* (1956).
4. Note the plural. *Cf.* the remark in Vitrac, *EA* (1990), I: 328, fn. 3.
5. Contrary to Euclid who avoids talking about one plane figure *subtracted* from another.

Alternatively, the same statement can be reformulated as *a rectangular-linear system of equations of type B1b*:

$u \cdot s = A, \quad u - s = q.$

See again the diagram in Fig. 1.2.2.

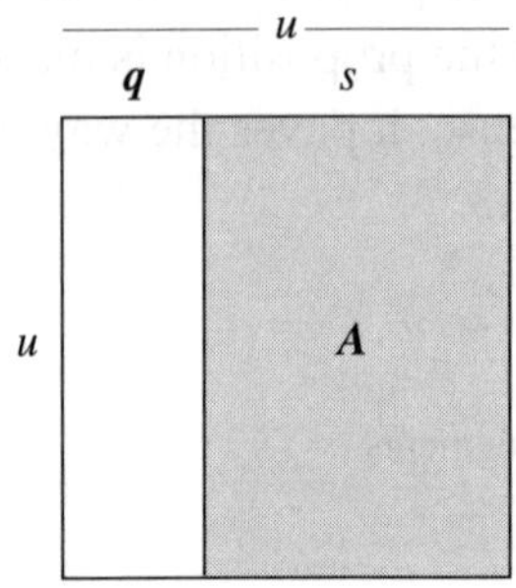

***El.* II.2 :** $u \cdot s + u \cdot (u - s) =$ sq. u
B4b : sq. $u - \boldsymbol{q} \cdot u = \boldsymbol{A}$
B1b : $u \cdot s = \boldsymbol{A}, \quad u - s = \boldsymbol{q}$
(Here u, s, q are *straight lines*, sq. u a *square* with the side u, and $u \cdot s$ a *rectangle* with the sides u, s. Simultaneously, u, s, q denote the *lengths* of the straight lines with these names, while sq. u and $u \cdot s$ denote the *areas* of the square and the rectangle with these names.)

Fig. 1.2.2. The diagram in *El.* II.2 replaced by a diagram in the Babylonian style.

Therefore, the purpose of *El.* II.2 may have been, essentially, to demonstrate that any *quadratic equation of type B4b*:[6]

sq. $u - q \cdot u = A$

is equivalent to a *rectangular-linear system of equations of type B1b*:

$u \cdot s = A, \quad u - s = q.$[7]

Fig. 1.2.3 below shows that there are *two* ways of similarly replacing the diagram in ***El.* II.3** with a diagram in the Babylonian style. According to the interpretation in Fig. 1.2.3, left, the statement in *El.* II.3, saying, essentially, that

$u \cdot s = (u - s) \cdot s +$ sq. s

can be reformulated as a *quadratic equation of type B4a*:

sq. $s + q \cdot s = A$, where $A = u \cdot s$.

Alternatively, the same statement can be reformulated as *a rectangular-linear system of equations of type B1b*:

$u \cdot s = A, \quad u - s = q.$

6. Necessarily with q and A positive, if u and q are interpreted as lengths and A as an area.
7. Necessarily with s positive and s less than u.

Therefore, one purpose of *El.* II.3 may have been to demonstrate that any *quadratic equation of type B4a:*[8]

$$\text{sq. } s + q \cdot s = A$$

is equivalent to a *rectangular-linear system of equations of type B1b:*[9]

$$u \cdot s = A, \quad u - s = q.$$

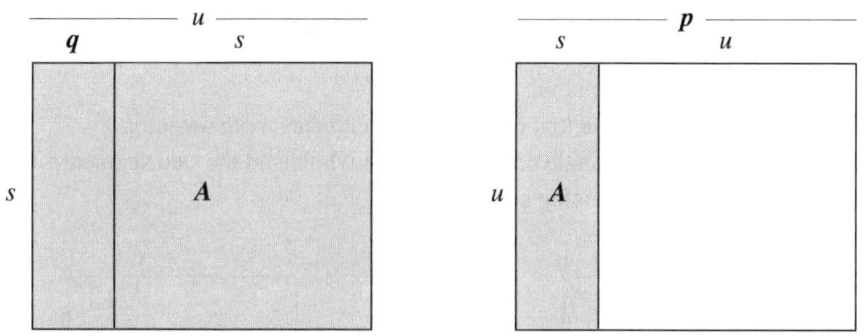

***El.* II.3 :** $u \cdot s = (u - s) \cdot s + \text{sq. } s$

B4a : $\text{sq. } s + \boldsymbol{q} \cdot s = \boldsymbol{A}$

B1b : $u \cdot s = \boldsymbol{A}, \quad u - s = \boldsymbol{q}$

***El.* II.3 :** $(u + s) \cdot u = u \cdot s + \text{sq. } u$

B4c : $\boldsymbol{p} \cdot u - \text{sq. } u = \boldsymbol{A}$

B1a : $u \cdot s = \boldsymbol{A}, \quad u + s = \boldsymbol{p}$

Fig. 1.2.3. Two possible interpretations of the diagram in *El.* II.3.

According to the interpretation in Fig. 1.2.3, right, the statement in *El.* II.3 can be reformulated as a *quadratic equation of type B4c*:

$$p \cdot u - \text{sq. } u = A, \quad \text{where} \quad A = u \cdot s.$$

Alternatively, the same statement can be reformulated as *a rectangular-linear system of equations of type B1a*:

$$u \cdot s = A, \quad u + s = p.$$

Therefore, another purpose of *El.* II.3 may have been to demonstrate that any *quadratic equation of type B4c*:

$$p \cdot u - \text{sq. } u = A$$

is equivalent to a *rectangular-linear system of equations of type B1a*:

$$u \cdot s = A, \quad u + s = p.$$

8. With some obvious restrictions because of the geometric interpretation.
9. With some obvious restrictions because of the geometric interpretation.

1.3. *El.* II.4, II.7 and the Two Basic Additive Quadratic-Linear Systems of Equations

El. II.4

If a straight line is cut at random,
the square on the whole is equal to the squares on the segments,
and twice the rectangle contained by the segments.

El. II.7

If a straight line is cut at random,
the square on the whole and that on one of the segments, both together,
are equal to twice the rectangle contained by the whole and the said segment,
and the square on the remaining segment.

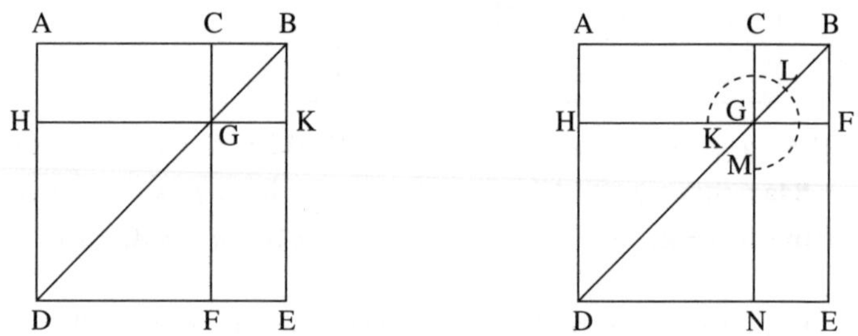

Fig. 1.3.1. Diagrams in *El.* II.4 (left), and *El.* II.7 (right).

In Fig. 1.3.2, left, below, the line AB is called p, its segments u and s. The statement in ***El.* II.4** can then be interpreted as saying that

$$\text{sq.}\,(u+s) = \text{sq.}\,u + \text{sq.}\,s + 2\,u\cdot s.$$

This equation, in its turn, can be reformulated in the following way:

$$\text{sq.}\,p = S + 2\,A \quad \text{where} \quad p = u+s, \quad S = \text{sq.}\,u + \text{sq.}\,s, \quad \text{and} \quad A = u\cdot s.$$

Therefore, the purpose of *El.* II.4 may have been, essentially, to demonstrate that any *quadratic-linear system of equations of type B2a*:

$$\text{sq.}\,u + \text{sq.}\,s = S, \quad u+s = p$$

is equivalent to a *rectangular-linear system of equations of type B1a*:

$$u\cdot s = A, \quad u+s = p \quad \text{where} \quad A = (\text{sq.}\,p - S)/2.$$

The interpretation of ***El.* II.7** in Fig. 1.3.2, right, is not quite as straightforward, since in order to get an interpretation where *El.* II.4 and *El.* II.7 are closely related, one has to assume that the diagram in *El.* II.7 is only the upper right corner of a larger diagram, based on two concentric and parallel squares. If this assumption is allowed, the given straight line AB in *El.* II.7 can be called u, and its arbitrary segments s and q, where q is the side of the inner square. The statement in *El.* II.7 can then be interpreted as saying that

$$\text{sq. } u + \text{sq. } s = 2\, u \cdot s + \text{sq. } (u - s).$$

This equation, in its turn, can be reformulated in the following way:

$$S = 2\, A + \text{sq. } q \quad \text{where} \quad q = u - s, \quad S = \text{sq. } u + \text{sq. } s, \quad \text{and} \quad A = u \cdot s.$$

Therefore, the purpose of *El.* II.7 may have been, essentially, to demonstrate that any *quadratic-linear system of equations of type B2b*:

$$\text{sq. } u + \text{sq. } s = S, \quad u - s = p$$

is equivalent to a *rectangular-linear system of equations of type B1b*:

$$u \cdot s = A, \quad u - s = q \quad \text{where} \quad A = (S - \text{sq. } q)/2.$$

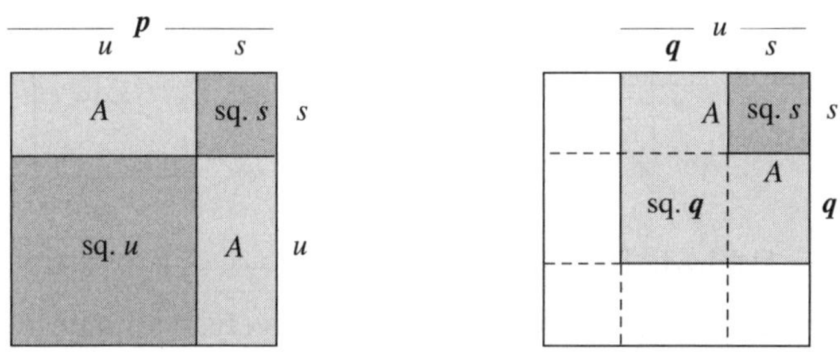

El. II.4 :

$\text{sq. } (u + s) = \text{sq. } u + \text{sq. } s + 2\, u \cdot s$

B2a :

$\text{sq. } u + \text{sq. } s = \mathbf{S}, \quad u + s = \boldsymbol{p}$

$\Rightarrow \quad u \cdot s = A = (\text{sq. } \boldsymbol{p} - \mathbf{S})/2$

El. II.7 :

$\text{sq. } u + \text{sq. } s = 2\, u \cdot s + \text{sq. } (u - s)$

B2b :

$\text{sq. } u + \text{sq. } s = \mathbf{S}, \quad u - s = \boldsymbol{q}$

$\Rightarrow \quad u \cdot s = A = (\mathbf{S} - \text{sq. } \boldsymbol{q})/2$

Fig. 1.3.2. Interpretations of the diagrams in *El.* II. 4 and *El.* II.7.

In Sec. 1.4 below it will be shown how systems of equations of type B2a (or B2b) can be solved by use of *El.* II.4 in combination with *El.* II. 5 (or by use of *El.* II.7 in combination with *El.* II.6).

Similarly, it will be shown how quadratic equations of type B4a (or B4b or B4c) can be solved by use of *El.* II. 3 in combination with *El.* II.6 (or *El.* II.2 in combination with *El.* II.6, or *El.* II.3 in combination with *El.* II.5). See Figs. 1.2.2 and 1.2.3 above.

1.4. *El.* II.5-6 and the Two Basic Rectangular-Linear Systems of Equations

El. II. 5

If a straight line is cut into equal and unequal segments,
the rectangle contained by the unequal segments of the whole,
together with the square on the straight line between the points of section,
is equal to the square on the half.

El. II. 6

If a straight line is bisected and a straight line is added to it in a straight line,
the rectangle contained by the whole with the added straight line, and the added straight line, together with the square on the half, is equal to
the square on the straight line made up of the half and the added straight line.

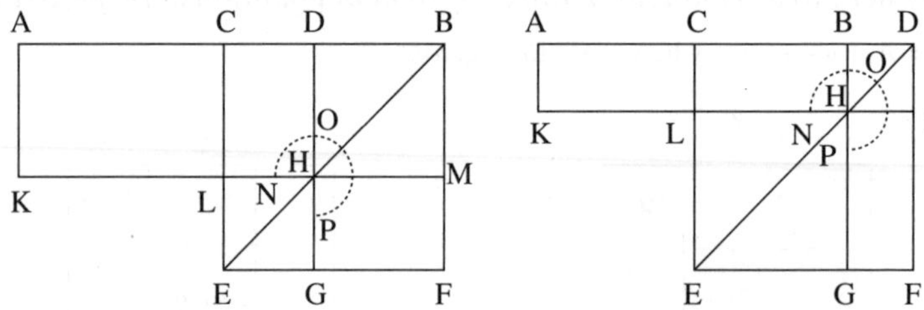

Fig. 1.4.1. The diagrams in *El.* II.5 (left), and *El.* II.6 (right).

The proofs of *El.* II. 5 and *El.* II. 6, respectively, both start by assuming that the straight line AB in the associated diagram is the given line.

In Fig. 1.4.2, left, the given straight line AB is called p, and so on, as above. Then, the statement in ***El.* II.5** can be interpreted as saying that

$$(p - s) \cdot s + \text{sq.}\ (p/2 - s) = \text{sq.}\ p/2.$$

This equation, in its turn, can be reformulated in the following way:

$$A + \text{sq.}\ q/2 = \text{sq.}\ p/2 \quad \text{where} \quad A = u \cdot s, \quad p = u + s, \quad \text{and} \quad q = u - s.$$

Therefore, the purpose of *El.* II.5 may have been, essentially, to demonstrate that any *rectangular-linear system of equations of type B1a*:

$$u \cdot s = A, \quad u + s = p$$

can be solved as follows (with sqs. meaning “the square-side of”):[10]

$$(u - s)/2 = q/2 = \text{sqs.}\ (\text{sq.}\ p/2 - A),$$

10. With some obvious restrictions because of the geometric interpretation.

$u = p/2 + q/2 = p/2 + \text{sqs. (sq. } p/2 - A),$
$s = p/2 - q/2 = p/2 - \text{sqs. (sq. } p/2 - A).$

Here sqs. (short for "square side") stands for the side of a given square. Note that when both p and q are known, u and s can be found as the "half-sum" and "half-difference", respectively, of p and q.

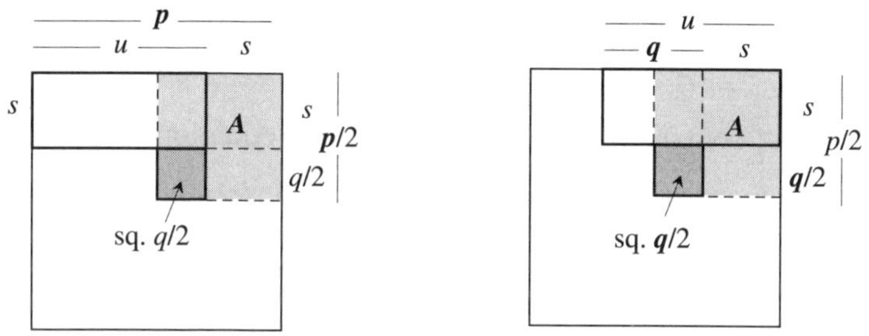

***El.*. II.5 :**

$(p - s) \cdot s + \text{sq. } (p/2 - s) = \text{sq. } p/2$

B1a : $u \cdot s = \mathbf{A}, \quad u + s = \mathbf{p}$

$\Rightarrow \text{sq. } q/2 = \text{sq. } \mathbf{p}/2 - \mathbf{A},$ *etc.*

***El.* II.6 :**

$(q + s) \cdot s + \text{sq. } q/2 = \text{sq. } (q/2 + s)$

B1b : $u \cdot s = \mathbf{A}, \quad u - s = \mathbf{q}$

$\Rightarrow \text{sq. } p/2 = \mathbf{A} + \text{sq. } \mathbf{q}/2,$ *etc.*

Fig. 1.4.2. Interpretations of the diagrams in *El.* II. 5 and *El.* II.6.

In Fig. 1.4.2, right, the given straight line AB is called q, and so on. Then, the statement in ***El.* II.6** can be interpreted as saying that

$(q + s) \cdot s + \text{sq. } q/2 = \text{sq. } (q/2 + s).$

This equation, too, can be reformulated in the following way:

$\text{sq. } p/2 = A + \text{sq. } q/2$ where $A = u \cdot s$, $p = u + s$, and $q = u - s$.

Therefore, the purpose of *El.* II.6 may have been, essentially, to demonstrate that any *rectangular-linear system of equations of type B1b*:

$u \cdot s = A, \quad u - s = q$

can be solved as follows:

$(u + s)/2 = p/2 = \text{sqs. } (A + \text{sq. } q/2),$
$u = p/2 + q/2 = \text{sqs. } (A + \text{sq. } q/2) + q/2,$
$s = p/2 - q/2 = \text{sqs. } (A + \text{sq. } q/2) - q/2.$

As mentioned above, the solution to *a quadratic-linear system of equations of type B2a* can be obtained by use of *El.* II.4 in combination with *El.* II.5. Indeed, suppose that

$\text{sq. } u + \text{sq. } s = S, \quad u + s = p.$

Then *El.* II.4 can be used to show that

$$u \cdot s = A, \quad u + s = p \quad \text{where} \quad A = (\text{sq. } p - S)/2.$$

In combination with *El.* II.5, this shows that

$$(u - s)/2 = q/2 = \text{sqs. } (\text{sq. } p/2 - A) = \text{sqs. } (S/2 - \text{sq. } p/2).$$

Consequently,

$$u = p/2 + q/2 = p/2 + \text{sqs. } (S/2 - \text{sq. } p/2),$$
$$s = p/2 - q/2 = p/2 - \text{sqs. } (S/2 - \text{sq. } p/2).$$

Similarly, of course, in the case of a system of equations of type B2b.

In the same way, *a quadratic equation of, for instance, type B4a* can be solved by use of *El.* II. 3 in combination with *El.* II.6. Indeed, if

$$\text{sq. } s + q \cdot s = A,$$

then it can be shown by use of *El.* II.3 that if $u = s + q$, then

$$u \cdot s = A, \quad u - s = q.$$

Therefore, in view of *El.* II. 6,

$$s = \text{sqs. } (A + \text{sq. } q/2) - q/2.$$

1.5. *El.* II.8 and the Two Basic Subtractive Quadratic-Linear Systems of Equations

El. II.8

If a straight line is cut at random,
four times the rectangle contained by the whole and one of the segments,
together with the square on the remaining segment, is equal to
the square described on the whole and the mentioned segment as on one straight line.

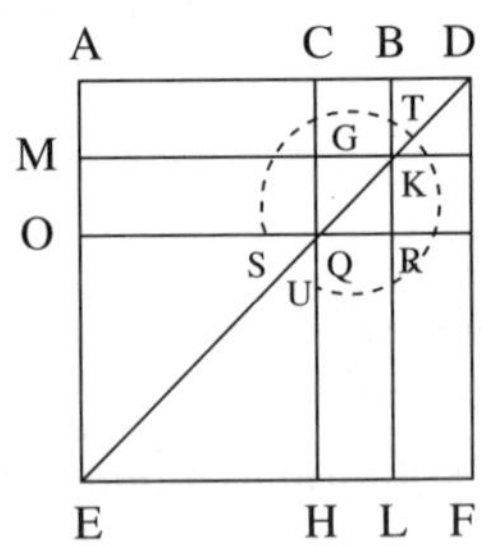

Fig. 1.5.1. The diagram in *El.* II.8.

In Fig. 1.5.2 below, the given straight line AB is called u and the two segments into which it is cut are called s and q. If also, as usual, $u + s$ is called p, then the statement in ***El.* II.8** can be interpreted as saying that

$4\,u \cdot s + \text{sq.}\,(u - s) = \text{sq.}\,(u + s)$ (*cf.* Fig. 1.1.2!)

This equation, in its turn, can be reformulated in the following way:

$4\,A = D$ where $A = u \cdot s,\;\; p = u + s,\;\; q = u - s,$ and $D = \text{sq.}\,p - \text{sq.}\,q.$

In other words, if

$\text{sq.}\,p - \text{sq.}\,q = D,$

then

$D = 4\,A = 4\,u \cdot s = 2\,u \cdot 2\,s.$

(Note that $2\,u$ and $2\,s$ can be interpreted as *the length of the mid-line* and *the width*, respectively, of the square corner formed by removing a square of side p from a square of side p, as in Fig. 1.5.2.)

Therefore, *any rectangular-linear system of equations of type B3a*:

$\text{sq.}\,p - \text{sq.}\,q = D,\;\; p + q = 2\,u$

can be solved by use of *El.* II.8 as follows:

$p - q = 2\,s = D/2u,$
$p = u + s = u + D/(4\,u),$
$q = u - s = u - D/(4\,u).$

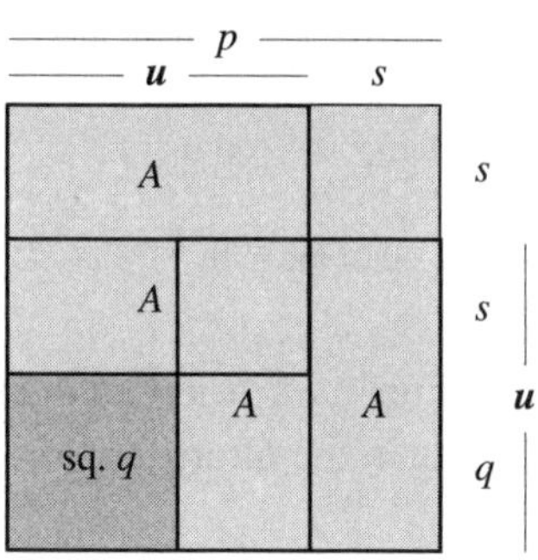

***El.* II.8 :**

$4\,u \cdot s + \text{sq.}\,(u - s) = \text{sq.}\,(u + s)$

B3a-b :

$\text{sq.}\,p - \text{sq.}\,q = \boldsymbol{D}$

$p + q = \boldsymbol{2\,u}$ (or $p - q = \boldsymbol{2\,s}$)

$\Rightarrow \quad 4\,A = 2\,u \cdot 2\,s = \boldsymbol{D},$

$p - q = 2\,s = \boldsymbol{D/2\,u}$

(or $p + q = 2\,u = \boldsymbol{D/2\,s}$), *etc.*

Fig. 1.5.2. Interpretation of the diagram in *El.* II. 8.

One would now expect a further proposition related to the case of a rectangular-linear system of equations of *type B3b*. However, this additional proposition was omitted by the author of *El.* II, obviously because this case, too, can be taken care of by use of *El.* II.8.

1.6. *El.* II.9-10, Constructive Counterparts to *El.* II.4 and II.7

It Secs. 1.2-1.5 above, it was demonstrated that *the first half of Elements II, comprising the seven propositions El. II.2-8, can be interpreted as a catalog of various steps in the geometric solution procedures for the nine basic problems of OB metric algebra, six kinds of quadratic-linear or rectangular-linear systems of equations, and three kinds of quadratic equations.* In this first half of *El.* II, all the proofs are based on *manipulations with squares and rectangles.*

It will be shown below that *the second half of Elements II, comprising the six propositions El. II.9-14, can be interpreted as a parallel catalog of various steps in geometric solution procedures for six of the nine basic problems of OB metric algebra, namely the six kinds of quadratic-linear or rectangular-linear systems of equations.* In this second half of *El.* II, all the proofs are based on *manipulations with right triangles and circles.*

El. II.9

If a straight line is cut into equal and unequal segments,
the squares on the unequal segments of the whole
are double of the square on the half and of the square on the straight line
between the points of section.

El. II.10

If a straight line is bisected, and a straight line is added to it in a straight line,
the square on the whole with the added straight line
and the square on the added straight line, both together,
are double of the square on the half,
and of the square described on the straight line made up of the half
and the added straight line as on one straight line.

In Fig. 1.6.1 left, below, the given straight line in *El.* II.9 is called p, and the unequal parts of p are called u and s, just as in the interpretation of the diagram in *El.* II.4, in Fig. 1.3.2 above.

A considerable part of the proof of *El.* II.9 is devoted to a careful construction of the various parts of the plane figure shown in the diagram. The most essential part of that plane figure consists of two right triangles with the sides u, s and a, b, respectively, joined along a common diagonal of length d. A plane figure of this kind can be called a "birectangle", because it has two right angles.

The most essential part of the plane figure appearing in the diagram for *El.* II.10 consists of two *partly overlapping* right triangles with the sides u, s and a, b, respectively, joined along a common diagonal of length d. A plane figure of this kind can be called an "overlapping birectangle".

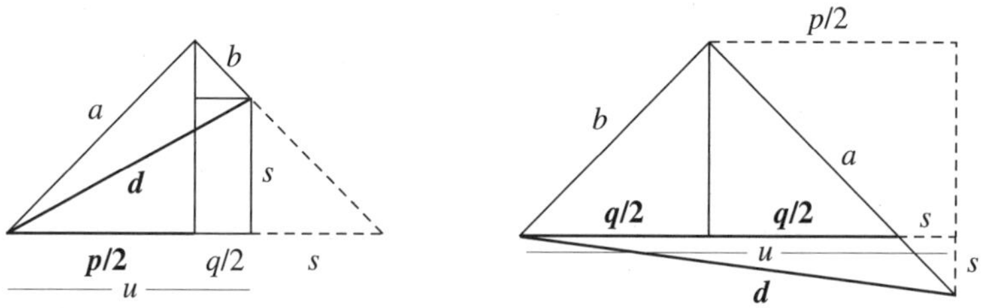

***El.* II.9 :**

sq. u + sq. s = 2 (sq. $p/2$ + sq. $q/2$)

B2a :

sq. u + sq. s = $\boldsymbol{S}$ = **sq.** $\boldsymbol{d}$, $u + s = \boldsymbol{p}$

$\Rightarrow$ $q/2$ = sqs. ($\boldsymbol{S}/2$ – sq. $\boldsymbol{p}/2$), *etc.*

***El.* II.10 :**

sq. u + sq. s = 2 (sq. $p/2$ + sq. $q/2$)

B2b :

sq. u + sq. s = $\boldsymbol{S}$ = **sq.** $\boldsymbol{d}$, $u - s = \boldsymbol{q}$

$\Rightarrow$ $p/2$ = sqs. ($\boldsymbol{S}/2$ – sq. $\boldsymbol{q}/2$), *etc.*

Fig. 1.6.1. Interpretations of the diagrams in *El.* II.9, and *El.* II.10.

The simple proof of the proposition in *El.* II.9 is based on repeated applications of the "diagonal rule" in *El.* I.47. On one hand,

sq. d = sq. a + sq. b = 2 sq. $p/2$ + 2 sq. $q/2$,

since a and b are the diagonals of two *half-squares* with the sides $p/2$ and $q/2$. On the other hand,

sq. d = sq. u + sq. s.

Therefore,

sq. u + sq. s = 2 (sq. $p/2$ + sq. $q/2$).

The proof of the similar proposition in *El.* II.10 is similar.

The purpose of ***El.* II.9** may have been to show that any *quadratic-linear system of equations of type B2a*:

sq. u + sq. s = S, $u + s = p$, with S and p given,

can be solved as follows: The diagram in Fig. 1.6.1, left, is constructed, with d = sqs. S. Then it can be shown, as in the proof of *El.* II. 9, that

S = sq. u + sq. s = 2 (sq. $p/2$ + sq. $q/2$).

Consequently, u and s can be computed in the following way:

$(u - s)/2 = q/2 = \text{sqs.}\ (S/2 - \text{sq.}\ p/2),$
$u = p/2 + q/2 = p/2 + \text{sqs.}\ (S/2 - \text{sq.}\ p/2),$
$s = p/2 - q/2 = p/2 - \text{sqs.}\ (S/2 - \text{sq.}\ p/2).$

Similarly, of course, ***El.* II.10** can be interpreted as a geometric solution procedure for *a quadratic-linear system of equations of type B2b*:

sq. u + sq. $s = S$, $u - s = q$, with S and q given.

The reason why *El.* II. 9 and 10 can be understood as "constructive counterparts" to *El.* II.4 and 7 will be disclosed below, in Sec. 1.9.

1.7. *El.* II.11* and II.14*, Constructive Counterparts to *El.* II.5-6

El. II.11

To cut a given straight line so that
the rectangle contained by the whole and one of the segments
is equal to the square on the remaining segment.

El. II.14

To construct a square equal to a given rectilineal figure.

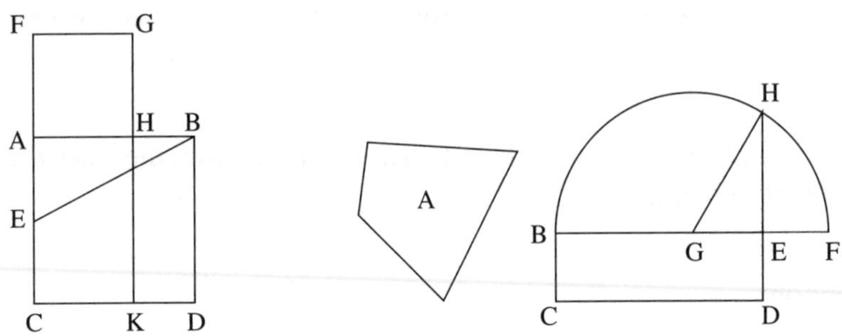

Fig. 1.7.1. The diagrams in *El.* II.11 and *El.* II.14.

In these two propositions, the author of *Elements* II has chosen to consider *two particularly important constructions* related to two propositions that would have been the "constructive counterparts" to *El.* II. 5, 6.

In the diagram for ***El.* II.11** (Fig. 1.7.1, left), AB is the given straight line. The first step of the solution to the stated construction problem is to construct the square ABDC with sides of length h. AC is bisected at E, and the diagonal BE is drawn. A point F on the extension of AC is found such that EF = BE. (This can be done most easily by finding the intersection of an extension of AC with a circle through B with center E.) The square AFGH is drawn on AF, and the side GH is extended to K. With this, the construction is completed, and it remains to prove that the given line AB is cut by H in the desired way.

The construction in *El.* II.11 can be explained as follows: In Fig. 1.7.1, left, let h be the length of AB = AC, let s be the length of AF, let $u = s + h$ be the length of CF, and let $p/2 = s + h/2$ be the length of EF = BE. Then,

$s + h/2 = \text{EF}$ and $h/2 = \text{EA}$.

Now, according to *El.* II.6,

$\text{CF} \cdot \text{AF} + \text{sq. EA} = \text{sq. EF} = \text{sq. EB}$.

In other words,

$u \cdot s + \text{sq. } h/2 = \text{sq. } p/2$.

An application of the diagonal rule in *El.* I.47 then shows that

$u \cdot s = \text{sq. } h$ where h is the given length of AB.

This means that *the rectangle FGKC is equal to the square ABDC*. Hence, if the rectangle AHKC is subtracted from both, it follows that also the rectangle HBDK is equal to the square FGHA. Therefore the point H divides AB in the desired way.

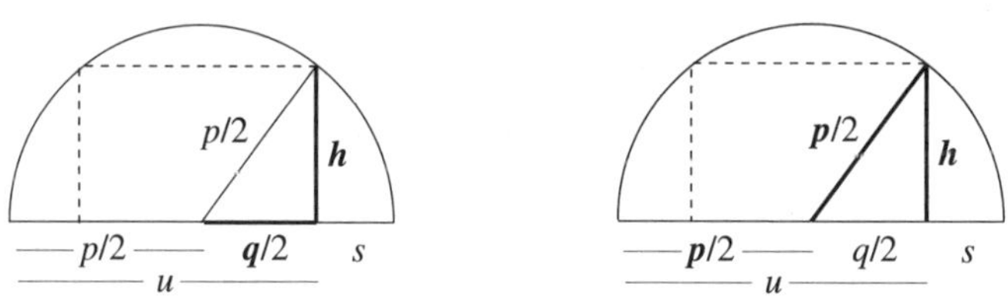

***El.* II.11* :** $\text{sq. } h + \text{sq. } q/2 = \text{sq. } p/2$
B1b : $u \cdot s = \boldsymbol{A} = \textbf{sq. } \boldsymbol{h}, \quad u - s = \boldsymbol{q}$
$\Rightarrow \quad p/2 = \text{sqs. } (\boldsymbol{A} + \text{sq. } \boldsymbol{q}/2),$ *etc.*

***El.* II.14*:** $\text{sq. } h + \text{sq. } q/2 = \text{sq. } p/2$
B1a: $u \cdot s = \boldsymbol{A} = \textbf{sq. } \boldsymbol{h}, \quad u + s = \boldsymbol{p}$
$\Rightarrow \quad q/2 = \text{sqs. } (\text{sq. } \boldsymbol{p}/2 - \boldsymbol{A}),$ *etc.*

Fig. 1.7.2. The general ideas behind *El.* II.11 and *El.* II.14.

What is going on here is revealed in ***El.* VI.30**, where it is shown that if a straight line AB is divided in the way described in *El.* II.11, then it is "divided in extreme and mean ratio". Note that the diagram in *El.* II.11 is nearly identical with the diagram in *El.* VI.30.

Now, consider the construction of the diagram in Fig. 1.7.2, left. Begin by assuming that q is a given length and A a given area, and construct a right triangle with the sides $q/2$ and h = sqs. A. Then, according to the diagonal rule in *El.* I.47, the diagonal of the right triangle is also known. If it is called $p/2$, then

$$A + \text{sq. } q/2 = \text{sq. } h + \text{sq. } q/2 = \text{sq. } p/2.$$

Next, construct a semicircle with the radius $p/2$ and with its center at the lower left vertex of the right triangle. The result is the diagram shown in Fig. 1.7.2, left. Let

$$u = p/2 + q/2 = \text{sqs. } (A + \text{sq. } q/2) + q/2,$$
$$s = p/2 - q/2 = \text{sqs. } (A + \text{sq. } q/2) - q/2.$$

Then

$$u + s = p, \quad u - s = q,$$

and it can be shown geometrically, as in Fig.1.1.2 above, that

$$u \cdot s = \text{sq. } p/2 - \text{sq. } q/2 \quad \text{so that} \quad u \cdot s = \text{sq. } h = A.$$

Therefore, the lengths u and s constructed in this way with departure from the given quantities q and A are solutions to the following *rectangular-linear system of equations of type B1b*:

$$u \cdot s = A = \text{sq. } h, \quad u - s = q.$$

It is important to realize that *proposition El. II.11 in the form that Euclid gave to it is, essentially, the special case when h = q of the more general proposition El. II.11*, illustrated by the diagram in Fig. 1.7.2, left.*

Now, consider instead the diagram in Fig. 1.7.2, right, related to the diagram in Fig. 1.7.2, left. Begin by assuming that p is a given length and A a given area, and construct a right triangle with the diagonal $p/2$ and the upright h = sqs. A. Then, according to the diagonal rule in *El.* I.47, the length of the base of the right triangle is also known. Call it $q/2$. Then

$$\text{sq. } p/2 - A = \text{sq. } p/2 - \text{sq. } h = \text{sq. } q/2.$$

Next, construct a semicircle with the radius $p/2$ and with its center at the lower left vertex of the right triangle. The result is the diagram shown in Fig. 1.7.2, right. Let

$$u = p/2 + q/2 = p/2 + \text{sqs. } (\text{sq. } p/2 - A),$$
$$s = p/2 - q/2 = p/2 - \text{sqs. } (\text{sq. } p/2 - A).$$

Then

$$u + s = p, \quad u - s = q,$$

and it can be proved as above that the lengths u and s constructed in this way with departure from the given quantities p and A are solutions to *a rectangular-linear system of equations of type B1a*:

$$u \cdot s = A = \text{sq. } h, \quad u + s = p.$$

What does this result have to do with ***El.* II.14**, where Euclid shows how to “construct a square equal to a given rectilineal figure”? The proposition is illustrated by the diagram in Fig. 1.7.1, right. Euclid begins by constructing a rectangle equal to the given figure (which is a paraphrase for *a rectangle of given area* A) by use of *El.* I.45. How he then continues can be explained as follows: He lets u (BE) and s (ED) be the sides of the rectangle with the given area A, and constructs a semicircle with the diameter $p = u + s$ (BF). Next, he constructs a perpendicular, whose length may be called h, in the semicircle from the point (E) where the diameter of the semicircle is divided into two segments of lengths u and s, and draws a right triangle with the given upright side h (EH), the given diagonal $p/2$ (HG), and the base $p/2 - s = q/2$ (GE). This is, essentially, the same construction as in Fig. 1.7.2, right. Then he notes that, according to *El.* II.5,

$$u \cdot s + \text{sq. } q/2 = \text{sq. } p/2.$$

In view of the diagonal rule in *El.* I.47, this means that

$$\text{sq. } h = \text{sq. } p/2 - \text{sq. } q/2 = u \cdot s = A,$$

where h is the length of the upright side of the right triangle, and where A is the given area. Therefore, h is the side of a square with the given area.

Essentially, what Euclid does in his construction in *El.* II.14 is that he starts with *any* rectangle with the given area A, say one with the sides $u, s = A, 1$. He then constructs the diagram in Fig. 1.7.2, right, in the case when $p = u + s$. In this way, he manages to construct the side h of a square with the given area A, as the upright side of a right triangle. Therefore, proposition *El.* II.14 in the *inverted* form that Euclid chose to give to it (with u and s, hence also p and q, given from the beginning rather than A and p) may very well have replaced an original proposition *El.* II.14* in some earlier, now lost, version of the *Elements*, one which showed how to construct a solution u, s to *a rectangular-linear system of equations of type B1a*.

1.8. *El.* II.12-13, Constructive Counterparts to *El.* II.8

El. II.12

In obtuse-angled triangles the square on the side subtending the obtuse angle
is greater than the squares on the sides containing the obtuse angle
by twice the rectangle contained by one of the sides about the obtuse angle,
namely that on which the perpendicular falls, and the straight line cut off outside
by the perpendicular towards the obtuse angle.

El. II.13

In acute-angled triangles the square on the side subtending the acute angle
is less than the squares on the sides containing the acute angle
by twice the rectangle contained by one of the sides about the acute angle,
namely that on which the perpendicular falls, and the straight line cut off within
by the perpendicular towards the acute angle.

Just as the pair of propositions *El.* II. 9-10 were shown above to be concerned with pairs of right triangles *joined in two different ways along a common diagonal*, so the pair of propositions *El.* II.12, 13 are concerned with pairs of right triangles *joined in two different ways along a common upright side (perpendicular).* Thus, in Fig. 1.8.1, right (below), two right triangles are *added* to each other, joined along a common upright side, while in Fig. 1.8.1, left, one right triangle is *subtracted* from another right triangle, to which it is joined along a common upright side.

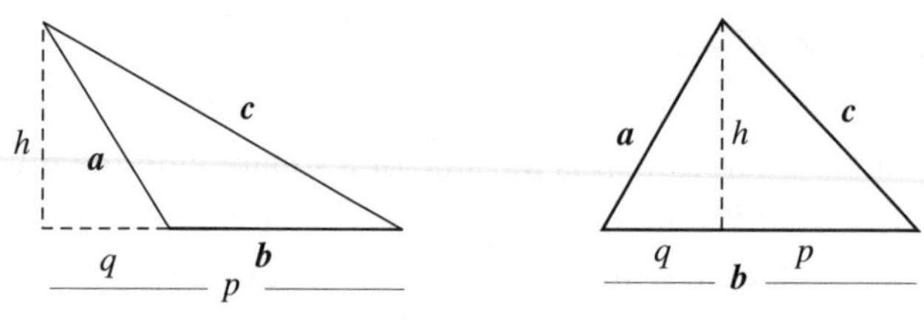

***El.* II.12 :**

sq. c = sq. a + sq. $b + 2\, b \cdot q$

B3b :

sq. p – sq. $q = \boldsymbol{D}$, $\quad p - q = \boldsymbol{b}$,

with $\boldsymbol{D}$ = **sq. *c* – sq. *a***

$\Rightarrow \quad 2\,\boldsymbol{b} \cdot q = \boldsymbol{D}$ – sq. $\boldsymbol{b}$, *etc.*

***El.* II.13 :**

sq. c = sq. a + sq. $b - 2\, b \cdot q$

B3a :

sq. p – sq. $q = \boldsymbol{D}$, $\quad p + q = \boldsymbol{b}$,

with $\boldsymbol{D}$ = **sq. *c* – sq. *a***

$\Rightarrow \quad 2\,\boldsymbol{b} \cdot q$ = sq. $\boldsymbol{b} - \boldsymbol{D}$, *etc.*

Fig. 1.8.1. Interpretations of the diagrams in *El.* II.12 and *El.* II.13.

With the notations introduced in Fig. 1.8.1, left, the proof of *El.* II.12 proceeds as follows:

$\text{sq.}\, p = \text{sq.}\, b + \text{sq.}\, q + 2\, b \cdot q$ *El.* II.4
$\text{sq.}\, p + \text{sq.}\, h = \text{sq.}\, b + \text{sq.}\, q + \text{sq.}\, h + 2\, b \cdot q$
$\text{sq.}\, c = \text{sq.}\, b + \text{sq.}\, a + 2\, b \cdot q$ *El.* I.47

Similarly in the case of *El.* II.13, with the notations in Fig. 1.8.1, right:

$\text{sq.}\, b + \text{sq.}\, q = 2\, b \cdot q + \text{sq.}\, p$ *El.* II.7
$\text{sq.}\, b + \text{sq.}\, q + \text{sq.}\, h = 2\, b \cdot q + \text{sq.}\, p + \text{sq.}\, h$
$\text{sq.}\, b + \text{sq.}\, a = 2\, b \cdot q + \text{sq.}\, c$ *El.* I.47
$\text{sq.}\, c = \text{sq.}\, b + \text{sq.}\, a - 2\, b \cdot q$

The purpose of ***El.* II.12** may have been to demonstrate that any *subtractive quadratic-linear system of equations of type B3b*:

$\text{sq.}\, p - \text{sq.}\, p = D,\ \ p - q = b,$ with D and b given,

can be solved as follows: Express D as a square-difference, for instance as

$D = D \cdot 1 = \text{sq.}\, c - \text{sq.}\, a$ with $c = (D + 1)/2,\ a = (D - 1)/2.$

(*Cf.* Fig. 1.1.2.) Then it follows from the result in *El.* II. 12 that

$2\, b \cdot q = D - \text{sq.}\, b.$

Therefore,

$q = (D - \text{sq.}\, b)/(2\, b),\ \ p = (p - q) + q = b + q = (D + \text{sq.}\, b)/(2\, b).$

In a similar way, the purpose of ***El.* II.13** may have been to demonstrate that any *subtractive quadratic-linear system of equations of type B3a*:

$\text{sq.}\, p - \text{sq.}\, p = D,\ \ p + q = b,$ with D and b given,

can be solved as follows: Express D as a square-difference,

$D = \text{sq.}\, c - \text{sq.}\, a.$

Then it follows from the result in *El.* II. 13 that

$2\, b \cdot q = \text{sq.}\, b - D,$

so that

$q = (\text{sq.}\, b - D)/(2\, b),\ \ p = (p + q) - q = b - q = (\text{sq.}\, b + D)/(2\, b).$

It may seem a bit strange that in *El.* II.12-13 the case of the obtuse-angled triangle precedes the case of the acute-angled triangle. The reason can be that, as pointed out above, the proof of *El.* II.12 makes use of *El.* II.4, while the proof of *El.* II.13 makes use of the *later* proposition *El.* II.7.

1.9. Summary. The Three Parts of *Elements* II

The discussion above aimed to demonstrate that *Elements* II can be divided into three distinct parts with obvious relations to the nine basic equations or systems of equations in OB metric algebra:

A. *El.* II.(1), 2, 3: related to the basic quadratic equations
B. *El.* II.4-8: related to the basic quadratic- or rectangular-linear systems of equations
C. *El.* II.9-14: related to the same quadratic- or rectangular-linear systems of equations

The question then naturally arises why work that was already done in part B of *Elements* II is repeated in a different way in part C. The answer to this question may be as follows:

It is possible that a lost Greek forerunner to *Elements* II, call it *Elements* II*, was written in imitation of a Babylonian theme text with the same subject. (See below, Sec. 1.12, for examples of OB theme texts.) Presumably, *Elements* II* contained only parts A and B, possibly with Babylonian style metric algebra diagrams rather than the lettered diagrams preferred by Euclid, and with solutions to concrete metric algebra problems instead of abstract geometric propositions. Then, somebody may have reacted to the circumstance that the solutions to the metric algebra problems in part B of *Elements* II* were *analytic and non-constructive*, in the sense that the diagrams associated with the forerunners to *El.* II.4-8 cannot be drawn accurately until *after* the solutions to the stated metric algebra problems have been found. Therefore, the non-constructive solutions in part B were complemented with alternative *synthetic and constructive* solutions in part C, consisting of forerunners to *El.* II.9-14.

Take, for instance, a renewed look at the pair ***El.* II.9-10**. Suppose that p is a *given length* and that $S = \text{sq. } d$ is the area of a square with *sides of given length d.* Then a solution to the metric algebra problem

B2a: $\text{sq. } u + \text{sq. } s = S = \text{sq. } d, \quad u + s = p$

can be constructed in the following way:

Draw a straight line of length p, as in Fig. 1.9.1, left. Bisect the straight line, and erect a perpendicular of length $p/2$ at its midpoint. Complete a half-square with the straight line of length p as its diagonal and base. Then draw a circle of radius d with its center at one of the endpoints of the given straight line. Draw a perpendicular to the given straight line from the point where the circle intersects that half-square. This perpendicular cuts the given straight line into two segments. Call the lengths of the segments u and s. Then s is also the length of the perpendicular from the point of intersection of the circle and the half-square. Therefore, it is clear that u, s is a solution to the stated metric algebra problem of type B2a.

As mentioned above (Fig. 1.6.1), this constructive *geometric* solution to the problem can be transformed into the following *metric* solution:

$u = p/2 +$ sqs. $(S/2 -$ sq. $p/2)$, $\quad s = p/2 -$ sqs. $(S/2 -$ sq. $p/2$.

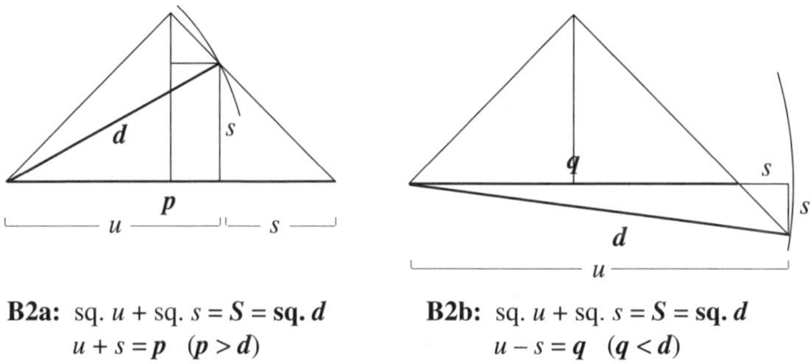

Fig. 1.9.1. Geometric constructions of solutions in possible forerunners to *El.* II.9, 10.

A similar constructive solution to the metric algebra problem of type B2b is illustrated in Fig. 1.9.1, right. It is a likely forerunner to *El.* II.10.

In a similar way, consider the following likely forerunners to the pair ***El.* II.11*** and ***El.* II.14*** (Fig. 1.7.2), the proposed forerunners to *El.* II.11 and II.14. First, suppose that q is a *given length* and that $A =$ sq. h is the *given area* of a square. Then a solution to the metric algebra problem

B1b: $\quad u \cdot s = A =$ sq. h, $\quad u - s = q$

can be constructed in the following way: Draw a rectangle with sides of length q and h, as in Fig. 1.9.2, left, and draw a semicircle with its center at the midpoint of one of the sides of length q, and passing through the two opposite vertices of the rectangle. Then the diameter of the circle is cut into three segments of which one is the side of the rectangle of length q. Let s be the common length of the remaining two segments, let $u = s + q$, and let $p = u + s$. Then $p/2$ is the length of the radius of the semicircle. Therefore, by the diagonal rule, sq. $p/2 -$ sq. $q/2 =$ sq. h. On the other hand,

sq. $p/2 -$ sq. $q/2 = u \cdot s$.

(See Fig. 1.1.2.) Consequently, $u \cdot s =$ sq. h, and it follows that u, s is a solution to the mentioned metric algebra problem of type B1b.

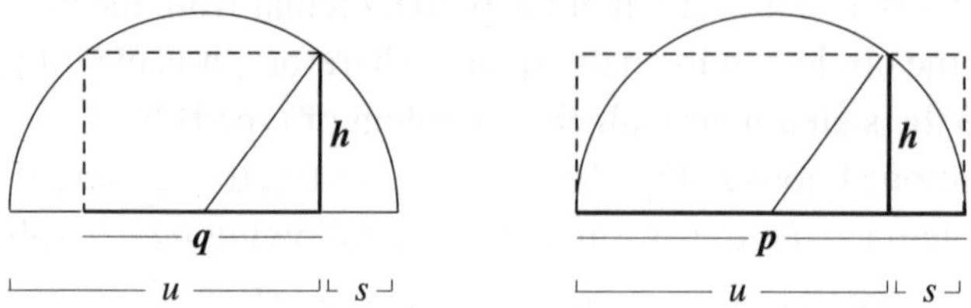

B1b: $u \cdot s = \boldsymbol{A} = \textbf{sq.}\, \boldsymbol{h},\ u - s = \boldsymbol{q}$ **B1a:** $u \cdot s = \boldsymbol{A} = \textbf{sq.}\, \boldsymbol{h},\ u + s = \boldsymbol{p}$

Fig. 1.9.2. Geometric constructions of solutions in possible forerunners to *El.* II.11*, 14*.

A similar constructive solution to the metric algebra problem of type B1a is illustrated in Fig. 1.9.2, right. It is a likely forerunner to *El.* II.14*.

Note: A shortcoming in the proposed constructive solutions to systems of equations of types B1a-b or B2a-b depicted in Figs. 1.9.1-2 is that they are based on the assumption that the square sides *d* and *h* of *S* and *A*, respectively, are known. Apparently, Euclid observed this shortcoming in the mentioned constructive solutions, and that is why he included a description of *the geometric construction of square sides* in his *El.* II. 14. Having inserted *El.* II.14 in *Elements* II, he did not bother to include also *El.* II.14* (Fig. 1.7.2, right), for which the diagram would be, essentially, the same.

Consider, finally, the following likely forerunners to the pair ***El.* II.12, *El.* II.13**. Suppose that *a, b, q are given lengths of the three sides of a triangle*. Then a geometric solution to the metric algebra problems

B3a-b: $\text{sq. } u - \text{sq. } s = D = \text{sq. } a - \text{sq. } b,\quad u - s = q\quad (\text{or}\quad u + s = p)$

can be constructed as in Fig. 1.9.3. The uncomplicated details of the argument are left to the readers.

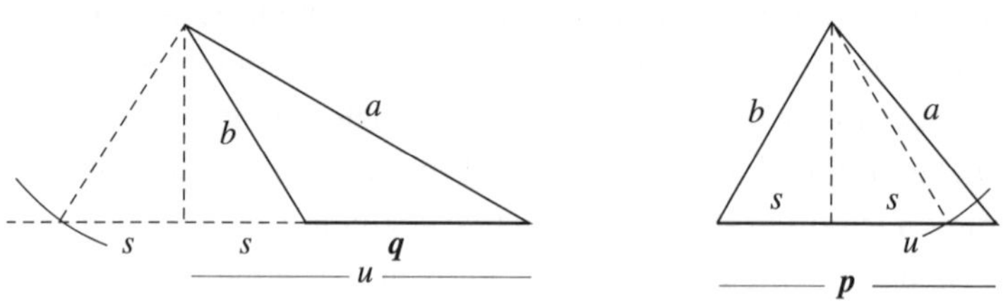

B3b: $\text{sq. } u - \text{sq. } s = \boldsymbol{D},\quad u - s = \boldsymbol{q}$
$\boldsymbol{D} = \text{sq. } a - \text{sq. } b$

B3a: $\text{sq. } u - \text{sq. } s = \boldsymbol{D},\quad u + s = \boldsymbol{p}$
$\boldsymbol{D} = \text{sq. } a - \text{sq. } b$

Fig. 1.9.3. Geometric constructions of solutions in possible forerunners to *El.* II.12, 13.

1.10. An Old Babylonian Catalog Text with Metric Algebra Problems

There does not exist any known OB mathematical text that is an exact parallel to *Elements* II, or to any one of the three parts of it. On the other hand, there do exist examples of *OB catalog texts or theme texts with metric algebra problems*, which therefore in a restricted sense can be called forerunners to *Elements* II. One such text is **BM 80209**, a small clay tablet from the OB city Sippar, now in the collections of the British Museum in London. The interpretation of that text given in Friberg, *JCS* 33 (1981) will be partly repeated here.

BM 80209 is a very special kind of theme text, namely a very brief but systematically arranged "catalog" of metric algebra problems of a number of different types, each represented by one or several numerical examples. There are no solution procedures and no answers to the stated problems.

Here is an abbreviated transliteration and translation of the text. In the transliteration, square brackets indicate destroyed parts of the text, Sumerian words are written in normal style, and Akkadian (that is, Babylonian) words in italics. (Sumerian terms were used in Babylonian mathematical texts in much the same way as words of Greek or Latin origin are used in modern mathematical texts.) Sexagesimal numbers are written as they are in the original text, without zeros and without any indication of where the fractional part of a number begins.

BM 80209

1. [···.ta *im*]*-ta-ḫar* a.šà *mi-nu-um*
2. [*šum-ma*] 20.ta *im-ta-ḫar* dal *mi-nu-um*
3. [*šum-m*]*a* 10.ta *im-ta-ḫar di-ik-šum mi-nu-um*
4. *šum-ma A* a.šà gúr *mi-nu-um*
5. *a-na* a.šà gúr *c* uš daḫ *P*
   i-na a.šà gúr *c* uš ba.zi *Q*
6. a.šà 2 gúr ul.gar *A* gúr ugu gúr 10 diri
7. a.šà gúr dal gúr *ù sí-ḫi-ir-ti* gúr ul.gar*-ma A*

1. ··· *each it is* equalsided. The field is what?
2. *If* 20 each it is equalsided, the transversal is what?
3. *If* 10 each it is equalsided, the expansion is what?

4. If A is the field, the arc is what?
5. To the field of the arc c (times) the length I added, P.
   From the field of the arc c (times) the length I tore off, Q.
6. The fields of 2 arcs I joined, S. Arc over arc 10 beyond.
7. The field of the arc, the transversal of the arc, the go-around of the arc I joined, S.

(In the translation, destroyed parts of the text are written in italics.)

For various reasons, it is advisable to use *literal translations* of Babylonian mathematical texts. In the translation above, the literal translations 'equalsided', 'field', and 'arc' correspond to the modern terms 'square', 'area', and 'circle'. The 'transversal' of a square is its diagonal, while the 'transversal' of a circle is its diameter. The circumference of a circle is called its 'arc', its 'length', or its 'go-around'. 'Tear off' means 'subtract', and 'arc over arc is 10 beyond' means that the circumference of one circle is 10 (length units) longer than the circumference of another circle.

The very convenient approximation π = appr. 3 is used in all Babylonian mathematical texts. More precisely, the area A and the diameter d of a circle are expressed as follows in terms of the arc (circumference) a:

$A = 5 \cdot a$, where '5' means $;05 = 5/60 = 1/12$ = appr. $1/4\pi$,
$d = 20 \cdot a$, where '20' means $;20 = 20/60 = 1/3$ = appr. $1/\pi$.

In addition to sexagesimal fractions, such as the circle constants '5' and '20', there are also two other kinds of fractions of numbers that appear in Babylonian mathematical texts. One kind is the "basic fractions"

3' (= 1/3)
2' (= 1/2)
3" (= 2/3)
6" (= 5/6)

for which there existed special signs in the cuneiform script. Another kind are the "reciprocals"

$1/n$, where $n = 4, 5, 6, \ldots$, often written in Sumerian in the form igi.*n*.gál.

In §§ 4-7 of the catalog text BM 80209, the coefficients A, P, Q, S, and c are allowed to take various values, so that there are several examples of each type of problem. In quasi-modern notations, the contents of BM 80209 can be described as follows. (The answers, which are not given in the text, are listed in the last column. A minor numerical error in the text is corrected here.)

BM 80209, table of contents (sexagesimal numbers with floating values)

1.	sq. $s = ?$	$s = [\cdots]$		
2.	sq. $s = A,\ \ d = ?$	$s = 20$		$d = 20 \cdot 1\ 24\ 51\ 10$
3.	expansion of $s = ?$	$s = 10$		(meaning unknown)
4.	5 sq. $a = A,\ \ a = ?$	$A =$ 8 20		$a = 10$
		$A =$ 2 13 20		$a = 40$
		$A =$ 3 28 20		$a = 50$
		$A =$ 5		$a = 1\ (\cdot\ 60)$
5.	5 sq. $a + c \cdot a = P,\ \ a = ?$	$c =$ 2'	$P = 8\ 25$	$a = 10$
	5 sq. $a - c \cdot a = Q,\ \ a = ?$		$Q = 8\ 15$	$a = 10$
		$c =$ 1	$P = 8\ 30$	$a = 10$
			$Q = 8\ 10$	$a = 10$
		$c =$ 1 3'	$P = 8\ 33\ 20$	$a = 10$
			$Q = 8\ 06\ 40$	$a = 10$
		$c =$ 1 2'	$P = 8\ 35$	$a = 10$
			$Q = 8\ 05$	$a = 10$
		$c =$ 1 3"	$P = 8\ 36\ 40$	$a = 10$
			$Q = 8\ 03\ 20$	$a = 10$
		$c =$ 1 1/4	$P = 8\ 332\ 30$	$a = 10$
			$Q = 8\ 07\ 30$	$a = 10$
		$c =$ 1 1/5	$P = 8\ 32$	$a = 10$
			$Q = 8\ 08$	$a = 10$
6.	5 sq. a + 5 sq. $b = S,\ \ a - b = 10,\ \ a, b = ?$		$S = 41\ 40$	$a = 20, b = 10$
			$S = 3\ 28\ 20$	$a = 40, b = 30$
			$S = 41\ 40$	$a = 50, b = 40$
			$S = 8\ 28\ 20$	$a = 1\ (\cdot\ 60), b = 50$
7.	5 sq. $a + 20\,a + a = B,\ \ a = ?$		$B = 8\ 33\ 20$	$a = 10$
			$B = 1$	$a = 20$
			$B = 1\ 55$	$a = 30$
			$B = 3\ 06\ 40$	$a = 40$

1.11. A Large Old Babylonian Catalog Text of a Similar Kind

Another similar, but much more extensive, OB catalog text with metric algebra problems without answers is ***TMS* 5**, from the ancient city Susa (Western Iran). Here is an abbreviated transliteration and translation, with several corrected readings of crucial but misunderstood words in the original edition of the text in Bruins and Rutten, *TMS* (1961):

***TMS* 5**

1a.	s nígin c uš-*ia mi-nu*
1b.	[c uš-*ia* b nígin *mi-nu*]
1c.	nígin *ù* c uš-*ia* gar.gar-*ma* e
1d.	nígin ugu c uš d diri
2a.	s nígin c a.šà *mi-nu*
2b.	c a.šà A nígin *mi-nu*
3a.	s nígin a.šà c uš *mi-nu*
3b	a.šà *ù* a.šà c uš gar.gar-*ma* S
3c.	a.šà ugu a.šà c uš D diri
4a.	*a-na* a.šà nígin-*ia* c uš daḫ-*ma* P
4b.	*i-na* a.šà nígin-*ia* c uš zi-*ma* Q
4c.	c nígin ugu a.šà D diri
4d.	c nígin *ki-ma* a.šà [···]-*ma*
5.	nígin.ba a.šà *ab-ni mi-nu* íb.si *ù* [···]-*ma*
6.	c a.šà *it-ba-al* íb.tag$_4$ a.šà D nígin *mi-nu*
7a.	p nígin *ki-di-tum* d *me-šé-tum* nígin *qer-bi-tum mi-nu*
7b.	q nígin *qer-bi-tum* d *me-šé-tum* nígin *ki-di-tum mi-nu*
7c.	p nígin *ki-di-tum* q nígin *qer-bi-tum* ul.gar a.šà 2 nigin *mi-nu*
7d.	a.šà 2 nigin ul.gar-*ma* S p nígin *ki-di-tum qer-bi-tum mi-nu*
7e.	a.šà 2 nigin ul.gar-*ma* S q nígin *qer-bi-tum ki-di-tum mi-nu*
7f.	a.šà 2 nigin ul.gar-*ma* S uš-*ši-na* gar.gar-*ma* b nigin *mi-nu*
[······································]	
8a.	[···] nígin *qer-bi-tim* [···] *qer-bi-tum* nigin *mi-nu*
8b.	[D] a.šà *dal-ba-ni* d *me-šé-tum* nígin *ki-di-tum ù qer-bi-tum mi-nu*
8c.	D a.šà *dal-ba-ni* c nígin *ki-di-tim* nígin *ki-di-tum qer-bi-tum mi-nu*
9a.	p nígin *ki-di-tum* q *múr* r nígin *qer-bi-tum* a.šà *dal-ba-an dal-ba-ni mi-nu*
9b.	a.šà *dal-ba-an dal-ba-ni* E uš-*ši-na* ul.gar-*ma* b nigin *mi-nu*
9c.	a.šà *dal-ba-an dal-ba-ni* E múr ugu nígin d nigin *mi-nu*
	4 22 mu.bi nigin.meš

1a. s is the equalside. c (times) my length is what?
1b. *c (times) my length is b. The equalside is what?*
1c. Equalside and c (times) my length I added, e.
1d. Equalside over c (times) the length is d beyond.
2a. s is the equalside. c (times) the field is what?
2b. c (times) the field is A. The equalside is what?
3a. s is the equalside. The field of c (times) the length is what?

3b. The field and the field of c (times) the length I heaped, S.
3c. The field over the field of c (times) the length is D beyond.

4a. To the field of my equalside c (times) the length I added, P.
4b. From the field of my equalside c (times) the length I tore off, Q.
4c. c (times) the equalside over the field is D beyond.
4d. c (times) the equalside is like the field [····]

5. (meaning not clear)

6. c (times) the field he took away. The remaining field is D. The equalside is what?

7a. p the outer equalside, d the distance. The inner equalside is what?
7b. q the inner equalside, d the distance. The outer equalside is what?
7c. p the outer equalside, q the inner equalside.
The join of the fields of the 2 equalsides is what?
7d. The fields of two equalsides I joined, S.
p is the outer equalside. The inner equalside is what?
7e. The fields of two equalsides I joined, S.
q is the inner equalside. The outer equalside is what?
7f. The fields of two equalsides I joined, S.
Their lengths I heaped, b. The equalsides are what?
..... (several problems missing)

8a. (badly preserved)
8b. D is the field between, d the distance.
The outer and inner equalsides are what?
8c. D is the field between.
c (times) the outer equalside is the inner equalside. The inner equalside is what?

9a. p is the outer equalside, q the middle, r the inner equalside.
The field between between is what?
9b. The field between between is E. Their lengths I joined, b.
The equalsides are what?
9c. The field between between is E. The middle over the <inner> equalside is d.
The equalsides are what?

The theme of *TMS* 5 is *problems for squares*. This is confirmed by the subscript which states that the text contains '4 22 (262) cases of squares'.

It is interesting to note that the cuneiform sign nígin, which in this text stands for 'equalside' has the form of a square. The related sign nigin, which stands for 'equalsides' has the form of two adjoining squares. Note also that it is difficult to establish the exact meaning of 'equalside'. Thus, for instance, 'the length of the equalside' means *the side of the square*, while 'the field of the equalside' means *the area of the square*.

In quasi-modern notations, the problems in *TMS* 5 can be explained as follows:

***TMS* 5, table of contents**

1 a.	s	given	$c \cdot s = ?$	20 values for c
1 b	$[c \cdot s$	given	$s = ?]$	20 values for c
1 c	$s + c \cdot s$	given	$s = ?$	20 values for c
1 d	$s - c \cdot s$	given	$s = ?$	17 values for c
2 a.	s	given	$c \cdot$ sq. $s = ?$	19 values for c
2 b	$c \cdot$ sq. s	given	$s = ?$	19 values for c
3 a.	s	given	sq. $(c \cdot s) = ?$	20 values for c
3 b	sq. s + sq. $(c \cdot s)$	given	$s = ?$	20 values for c
3 c	sq. s – sq. $(c \cdot s)$	given	$s = ?$	20 values for c
4 a.	sq. $s + c \cdot s$	given	$s = ?$	27 values for c
4 b	sq. $s - c \cdot s$	given	$s = ?$	27 values for c
4 c	$c \cdot s$ – sq. s	given	$s = ?$	3 values for c
4 d	$c \cdot s$ = sq. s		$s = ?$	1 value for c
5	....	(meaning not clear)	1 problem	
6	sq. $s - c \cdot$ sq. s	given	$s = ?$	5 values for c
7 a.	p and $(p - q)/2$	given	$q = ?$	1 problem
7 b	q and $(p - q)/2$	given	$p = ?$	1 problem
7 c	p and q	given	sq. p + sq. $q = ?$	1 problem
7 d	sq. p + sq. q and p	given	$q = ?$	1 problem
7 e	sq. p + sq. q and q	given	$p = ?$	1 problem
7 f	sq. p + sq. q and $p + q$	given	$p, q = ?$	1 problem
....			........(10 problems missing?)	
8 a.	p and $p - q$	given	$q = ?$	1 problem
8 b	sq. p – sq. q and $p - q$	given	$p, q = ?$	1 problem
8 c	sq. p – sq. q given, $q = c \cdot p$		$q = ?$	2 values for c
9 a.	p, m, q	given	sq. m – sq. $q = ?$	1 problem
9 b	sq. m – sq. q and $p + m + q$	given	$p, m, q = ?$	1 problem
9 c	sq. m – sq. q and $m - q$	given	$p, m, q = ?$	1 problem

Note: In 9b-c it is tacitly assumed that $p - q = q - r$.

In §§ 1-4 of *TMS* 5, the given values of the coefficient c are allowed to vary in the same way as the given values of the coefficient c in § 5 of BM 80209 (Sec. 1.10 above), but much more extensively. Here is a list of given values of c and the corresponding values of the solution s (the asked for length of the square side):

c	*s*	*c*	*s*	*c*	*s*
1	30	1	35	1	10 05
2		7		11 11	
3		2 7		2 11 11	
4		1 7		1	6 25
3"		1 2 7		11 7	
2'		7 1/7		2 11 7	
3'		7 2 1/7		1	12 50
4		1	4 05	3" 2' 3' 11 7	
3' 4		7 7		2 3" 2' 3' 11 7	
1 3"		2 7 7			
1 2'		1 7 7			
1 3'		1 2 7 7			
1 4		1	55		
1 3' 4		11			
2 2'		2 11			
3 3'					
4 4					

$35 = 5 \cdot 7$

$4\ 05 = 5 \cdot 7 \cdot 7$

$55 = 5 \cdot 11$

$10\ 05 = 5 \cdot 11 \cdot 11$

$6\ 25 = 5 \cdot 7 \cdot 11$

$12\ 50 = 2 \cdot 5 \cdot 7 \cdot 11$

Probably in order to save space, the values given for *c* in this text make use of some otherwise undocumented notations for fractions. Take, for instance the most complicated examples, those of the values 3" 2' 3' 11 7 and 2 3" 2' 3' 11 7. They appear in **§ 1 c**, in the two lines

nígin *ù* 3" 2' 3' 11 7 u*š-ia* gar.gar-*ma* 12 51 06 40
nígin *ù* 2 3" 2' 3' 11 7 u*š-ia* gar.gar-*ma* 12 52 13 20.

This means that

s + 3" 2' 3' 11 7 · s = 12 51 06 40, and s + 2 3" 2' 3' 11 7 · s = 12 52 13 20.

This and other examples together show that what is meant here is

3" 2' 3' 11 7 · $s = 2/3 \cdot 1/2 \cdot 1/3 \cdot 1/11 \cdot 1/7 \cdot s$

and

2 3" 2' 3' 11 7 · $s = 2 \cdot 2/3 \cdot 1/2 \cdot 1/3 \cdot 1/11 \cdot 1/7 \cdot s$.

It is likely that the student who got these equations as problems to solve was assumed to make use of the rule of false value, a frequently used method in Babylonian mathematics. He would then start with a tentative value for s, such as

$s^* = 7 \cdot 11 = 1\ 17\ (77)$.

Using this tentative value, and working with sexagesimal numbers in "relative place value notation" without zeros, he would then find that

$$2/3 \cdot 1/2 \cdot 1/3 \cdot 1/11 \cdot 1/7 \cdot 1\ 17 = 2/3 \cdot 1/2 \cdot 1/3 \cdot 1/11 \cdot 11 = 2/3 \cdot 1/2 \cdot 1/3 \cdot 1$$
$$= 2/3 \cdot 1/2 \cdot 20 = 2/3 \cdot 10 = 6\ 40.$$

Therefore, keeping track of the relative size of the computed fraction of 1 17, he could conclude that

$$s^* + 3''\ 2'\ 3'\ 11\ 7 \cdot s^* = 1\ 17 + 6\ 40 = 1\ 17\ 06\ 40,$$

where

$$1\ 17\ 06\ 40 = 1/10 \cdot 12\ 51\ 06\ 40.$$

This means that $s = 10 \cdot s^* = 10 \cdot 1\ 17 = 12\ 50$ is the correct solution to the first of the mentioned equations. It is left to the interested reader to show that it is also the solution to the second equation.

Note the following important connection between *TMS* 5 and the explanation of *Elements* II suggested in Secs. 1.2, 1.3, and 1.5 above: The problems in ***TMS* 5 § 4 a-c** are *basic quadratic equations* of types B4a, B4b, and B4c. Similarly, the problems in ***TMS* 5 § 7 f** (and probably the lost § 7 g) are *basic additive quadratic-linear system of equations* of types B2a (and B2b). Finally, the problem in ***TMS* 5 § 8 b** is a *subtractive quadratic-linear system of equations* of type B3b.

In ***TMS* 5 §§ 7-9** are also of interest in this connection, because they demonstrate quite clearly that OB mathematicians were familiar with the concepts of *concentric and parallel squares,* and s*quare bands*. (*Cf.* the discussion of Fig. 1.1.2 in Sec. 1.1 above.)

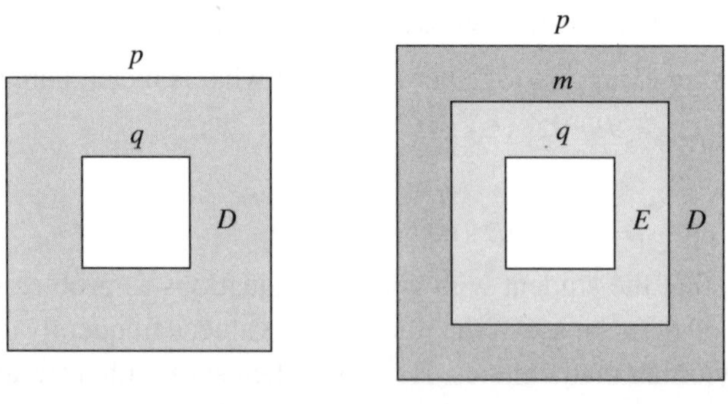

Fig. 1.11.1. The concentric squares and square bands in *TMS* 5 §§ 7, 8, and 9.

In *TMS* 5 § 7, and § 8 b, two squares have the sides 30 and 20, respectively. It is silently assumed that the two squares are concentric and parallel. The distance between the squares is 5.

In § 8 c, two cases are considered. In the first case, the area of the 'space between', that is of the square band, is 20 (· 60), and the length of the inner square is 1/7 of the length of the outer square, with 1/7 written simply as '7'. The solution procedure, which is not given in the text, is simple, since the length q of the inner square can be found as the solution to the equation

sq. $(7\,q)$ – sq. $q = 20$ (· 60), and sq. 7 – sq. 1 = 48,
so that 48 sq. $q = 20$ (· 60), hence sq. $q = 25$, and
$q = 5, p = 35$.

In the second case, the area of the square band is 16 40 (· 60), and the length of the inner square is 1/7 · 1/7 of the length of the outer square, with 1/7 · 1/7 written simply as '7 7'. In this case,

sq. $(49\,q)$ – sq. $q = 20$ (· 60), and sq. 49 – sq. 1 = 40 01 – 1 = 40 (· 60),
so that 40 sq. $q = 16\ 40$, hence sq. $q = 25$, and
$q = 5, p = 4\ 05$.

In TMS 5 § 9, there are three concentric squares and two square bands. It is silently assumed that the middle square is halfway between the other two. The sides of the three squares are simply 30, 20, and 10.

1.12. Old Babylonian Solutions to Metric Algebra Problems

1.12 a. Old Babylonian problems for rectangles and squares

The two OB catalog texts with metric algebra problems discussed in Secs. 1.10 and 1.11 above are well organized but lack both answers and explicit solution procedures to the stated problems.

BM 13901 (Neugebauer, *MKT 3* (1937); Høyrup, *LWS* (2002), 288) is of a different type, a *theme text* with metric algebra problems. It is a large text containing 23 exercises for squares, *each with a complete solution procedure*. The table of contents below, where the exercises are listed in their order of appearance in the text, reveals that BM 13901 is a mathematical "recombination text", by which is meant *a somewhat disorganized collection of more or less closely related mathematical exercises from a number of sources.*

BM 13901, table of contents		(sexagesimal numbers with floating values)
1 a.	**sq. $s + s = 45$**	**$s = 30$**
1 b	**sq. $s - s = 14\ 30$**	**$s = 30$**
1 c	sq. $s - 3'$ sq. $s + 3'\, s = 20$	$s = 30$
1 d	sq. $s - 3'$ sq. $s + s = 4\ 46\ 40$	$s = 20$
1 e	sq. $s + s + 3'\, s = 55$	$s = 30$
1 f	sq. $s + 3''\, s = 35$	$s = 30$
1 g	$7\, s + 11$ sq. $s = 6\ 15$	$s = 30$
2 a.	**sq. p + sq. $q = 21\ 40$, $p + q = 50$**	**$p = 30$, $q = 20$**
2 b	**sq. p + sq. $q = 21\ 40$, $p - q = 10$**	**$p = 30$, $q = 20$**
2 c	sq. p + sq. $q = 21\ 15$, $q = p - 1/7\, p$	$p = 3\ 30$, $q = 3$
2 d	sq. p + sq. $q = 28\ 15$, $p = q + 1/7\, q$	$p = 4$, $q = 3\ 3\ 0$
2 e	**sq. p + sq. $q = 21\ 40$, $p \cdot q = 10$**	**$p = 30$, $q = 20$**
2 f	sq. p + sq. $q = 28\ 20$, $q = 1/4\, p$	$p = 40$, $q = 10$
2 g	sq. p + sq. $q = 25\ 25$, $q = 3''\, p + 5$	$p = 30$, $q = 25$
4	sq. p + sq. q + sq. r + sq. $s = 27\ 05$, $q = 3''\, p$, $r = 2'\, q$, $s = 3'\, r$	$p = 30$, $q = 20$, $r = 15$, $s = 10$
1 h	sq. $s - 3'\, s = 5$	$s = 30$
3 a.	sq. p + sq. q + sq. $r = 10\ 12\ 45$, $q = 1/7\, p$, $r = 1/7\, q$	$p = 24\ 30$, $q = 3\ 30$, $r = 30$
3 b	sq. p + sq. q + sq. $r = 23\ 20$, $p - q = q - r = 10$	$p = 30$, $q = 20$, $r = 10$
2 h	sq. p + sq. q + sq. $(p - q) = 23\ 20$, $p + q = 50$	$p = 30$, $q = 20$
....	....	(3 exercises lost)
1 i	$4\, s$ + sq. $s = 41\ 40$	$s = 10$
3 c	sq. p + sq. q + sq. $r = 29\ 10$, $q = 3''\, p + 5$, $r = 2'\, q + 2\ 30$	$p = 30$, $q = 20$, $r = 10$

Four of the exercises in BM 13901 are closely associated with the theme of parts A and B of *Elements* II. (See Sec. 1.9 above.) These four exercises will be discussed separately below.

BM 13901 § 1 a, literal translation	explanation (relative values)
The field and my equalside I heaped, 45.	sq. $s + s = A = 45$
1, the going-out, you set.	Set $q = 1$
The halfpart of 1 you break.	$q/2 = 30$
30 and 30 you make eat each other.	sq. $q/2$ = sq. $30 = 15$
15 to 45 you add.	A + sq. $q/2 = 45 + 15 = 1$
1 makes 1 equalsided.	sqs. $1 = 1$
30 that you made eat itself,	$q/2 = 30$
inside 1 you tear out.	subtracted from $1 = 30$
30 is the equalside	$s = 30$.

See (Høyrup, *LWS* (2002), 50) for a transliteration of this text, and for a literal translation, differing in some details from the one proposed here. It is, by the way, not easy to find adequate literal translations of the terms in a Babylonian mathematical text, since there is often *no exact correspondence between Babylonian and modern mathematical terms*. Nevertheless, it is advisable to use literal translations, for the reason that OB mathematical terminology was not standardized. The fact that crucial elements of the terminology are different in texts from different sites and different periods ought to be apparent in the translations.

Besides, the use of non-literal translations can obscure important fine points of the text, such as the fact, first pointed out by Høyrup in *AoF* 17 (1990), that OB mathematicians used different terms for several different kinds of addition, several different kinds of multiplication, *etc.* Thus, for instance, in the text above, *when two lengths are multiplied with each other*, the term used is that the numbers for the two lengths "eat each other" (and become replaced by a number for an area).

Note in the explanation the use of the abbreviations sq. for the square of a length number and sqs. for the square side of an area number. (The use of modern notations for squares and square roots would be anachronistic.)

The term 'going-out' (Akk. *wāṣītum* 'that which goes out') in this text refers to the coefficient q in the quadratic equation sq. $s + q \cdot s = A$. It has to be understood as *a length number*, which explains why it is possible to add together the area number sq. s and the product $q \cdot s$. In the present case, when $q =$ '1', the phrase 'the field and my equalside' has to be understood as sq. $s + 1 \cdot s$, where both sq. s and $1 \cdot s$ are *area numbers*!

Note, by the way, that it is not absolutely clear what it means that in this text the going-out is equal to '1'. Høyrup is of the opinion that it means that $q = 1$ length unit, and is then forced to interpret the answer '30 is the equalside' as meaning that the computed side of the square is ;30 = 1/2 length unit. However, there is plenty of evidence that plane figures in OB mathematical texts were normally (but perhaps not always) thought of as *actual fields*, with the size of their sides in the range of *tens or sixties* of the length unit ninda (= about 6 meters). Since the situation is unclear, it may be a good idea to stay neutral on this issue and interpret '1' as *either* 1 *or* 1 00 = 1 · 60 and '30' as *either* 30 *or* ;30 = 30 · 1/60.

The quadratic equation in BM 13901 § 1a is of type **B4a**. The solution procedure can be interpreted as a combination of the ideas behind *El.* II.3 and II.6. (Fig. 1.2.3, left, and Fig. 1.4.2, right.) See Fig. 1.12.1 below:

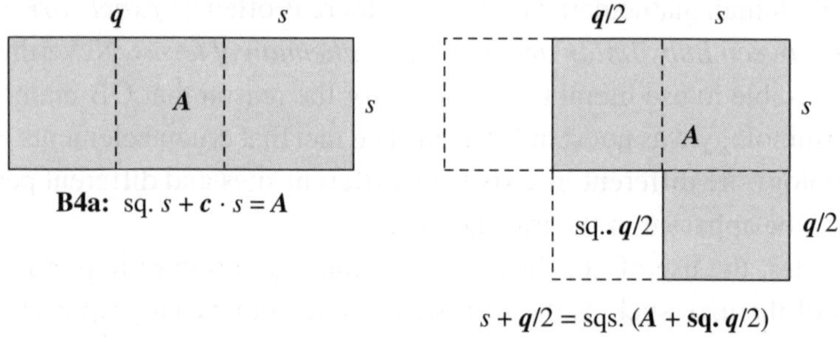

Fig. 1.12.1. A geometric explanation of the solution procedure in BM 13901 § 1 a.

Note that if it is assumed here that s = '30' = 30, then sq. s = '15' = 15 00 and consequently A = '45' = 45 00 and q = '1' = 1 00!

Now, consider instead **BM 13901 § 1 b** (Høyrup, *LWS* (2002), 52):

BM 13901 § 1 b, literal translation	explanation (floating values)
My equalside inside the field I tore out, 14 30.	sq. $s - s = A$ = 14 30
1, the going-out, you set.	set $q = 1$
The halfpart of 1 you break.	$q/2 = 30$
30 and 30 you make eat each other.	sq. $q/2$ = sq. 30 = 15
15 to 14 30 you add.	sq.q/2 + A = 15 + 14 30 = 14 30 15(!)
14 30 15 makes 29 30 equalsided.	sqs. 14 30 15 = 29 30
30 that you made eat itself,	Recall that $q/2 = 30$
to 29 30 you add. 30 is the equalside	30 added to 29 30 = 30, $s = 30$

The problem in BM 13901 § 1 b can be interpreted as a quadratic equation of type **B4b**, sq. $s - q \cdot s = A$, with q = '1'. The most likely interpretation of the solution procedure is that it is a combination of the ideas behind *El.* II.2 and II.6 (Figs. 1.2.2 and 1.4.2, right). See Fig. 1.12.2 below:

Note that in § 1b the computed value of u is again '30', but when s = 30, then in Fig. 1.12.2, left, the going-out q = '1' cannot possibly have the value 1 00, which is greater than 30. Indeed, in a geometric interpretation like the one in Fig. 1.12.2, the difference $s - q$ must be a (positive) length number. For this reason, the author of BM 13901 apparently chose to interpret 'the going-out is 1' in § 1 b as meaning that $q = 1$, not q = 1 00!

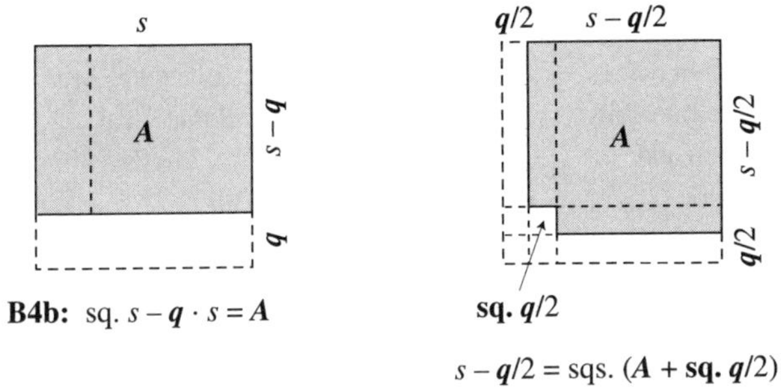

Fig. 1.12.2. A geometric explanation of the solution procedure in BM 13901 § 1 b.

Next, consider **BM 13901 § 2 a** (Høyrup, *op. cit.*, 66):

BM 13901 § 2 a, literal translation	explanation (floating values)
The fields of my two equalsides I heaped, 21 40,	sq. p + sq. $q = S = 21\ 40$
and my equalsides I heaped, 50.	$p + q = 2\ u = 50$
The halfpart of 21 40 you break.	$S/2 = 10\ 50$
10 50 you write down.	make a note of $S/2 = 10\ 50$
The halfpart of 50 you break.	$u = (p + q)/2 = 50/2 = 25$
25 and 25 you make eat each other.	sq. u = sq. $25 = 10\ 25$
10 25 inside 10 50 you tear out.	$S/2$ – sq. $u = 10\ 50 - 10\ 25 = 25$
25 makes 5 equalsided.	sqs. ($S/2$ – sq. u) = sqs. $25 = 5 = s$
5 to the first 25 you add,	$u + s = 25 + 5 = 30$
30 is the first equalside.	$p = 30$
5 inside the second 25 you tear out,	$u - s = 25 - 5 = 20$
20 is the second equalside.	$q = 20$

The problem in BM 13901 § 2 a can be interpreted as a quadratic-linear system of equations of type B2a, sq. p + sq. $q = S$, $p + q = 2\ u$. The solution procedure is based on the identity

sq. $s = S/2$ – sq. u when sq. p + sq. $q = S$, $p = u + s$ and $q = u - s$.

BM 13901 § 2 b (Høyrup, *op. cit.*, 68) is similar:

BM 13901 § 2 b, literal translation	explanation (floating values)
The fields of my two equalsides I heaped, 21 40.	sq. p + sq. $q = S = 21\ 40$
Equalside over equalside is 10 beyond.	$p - q = 10$
The halfpart of 21 40 you break.	$S/2 = 10\ 50$
10 50 you write down.	make a note of $S/2 = 10\ 50$
The halfpart of 10 you break.	$(p - q)/2 = s = 5$

5 and 5 you make eat each other.	sq. s = sq. 5 = 25
25 inside 10 50 you tear out.	$S/2$ – sq. s = 10 50 – 25 = 10 25
10 25 makes 25 equalsided.	sqs. ($S/2$ – sq. s) = 25 = u
25 you write down twice.	make two notes of u = 25
5 that you made eat itself	Recall that $s = 5$
to the first 25 you add,	$u + s = 25 + 5 = 30$
30 is the equalside.	$p = 30$
5 inside the second 25 you tear out,	$u - s = 25 - 5 = 20$
20 is the second equalside.	$q = 20$

The problem in BM 13901 § 2 b can be interpreted as a quadratic-linear system of equations of type B2b, sq. p + sq. $q = S$, $p - q = 2\,s$. The solution procedure is based on the identity

sq. $u = S/2$ – sq. s when sq. p + sq. $q = S$, $p = u + s$ and $q = u - s$.

In *LWS* (2002), 67-70, Figs. 10-12, Høyrup presents three different possible configurations in terms of squares and rectangles which the OB mathematicians may have used to prove identities like the ones mentioned above. There is, however, a fourth possible, and perhaps more plausible, configuration, which Høyrup did not consider in this connection (but which he did consider elsewhere, *op. cit.*, 259, Fig. 67).

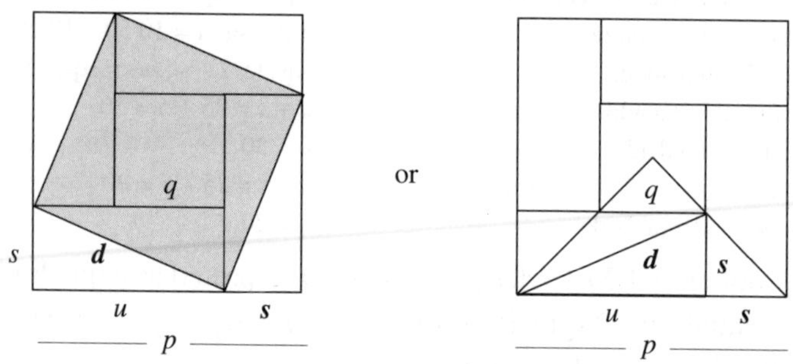

(sq. p + sq. q)/2 = $\mathbf{S}$/2 = sq. $\mathbf{d}$ = sq. u + sq. s
$p = u + s, \quad q = u - s$

Fig. 1.12.3. Left: A geometric explanation of BM 13901 § 2 a-b. Right: *El.* II.9.

Indeed, in Fig. 1.12.3 above, left,

sq. d (= the area of the oblique square) = (sq. p + sq. q)/2.

This is so because sq. d plus the areas of four right triangles = sq. p, while sq. d minus the areas of four right triangles = sq. q (see Fig. 2.3.2, right).

On the other hand, in view of the diagonal rule (Sec. 2.3), it is also true that

sq. d = sq. u + sq. s, where $u = (p + q)/2$, and $s = (p - q)/2$.

Therefore,

$S/2$ = sq. u + sq. s when sq. p + sq. $q = S$, $(p + q)/2 = u$ and $(p - q)/2 = s$.

The identity that can be derived in this way by use of the configuration in Fig. 1.12.3, left, can also derived by use of a birectangle as in the proof of *El.* II.9 (Fig. 1.6.1 above, left). That this is no coincidence is shown in Fig. 1.12.3 above, right.

MS 5112 is a large fragment of a mathematical recombination text with metric algebra problems, published in Friberg, *RC* (2007), Sec. 11.2 n. The text is late OB, maybe younger. It is inscribed on the obverse with a number of metric algebra problems for *squares*, and on the reverse with metric algebra problems for *rectangles*. There are explicit solution procedures for all the problems. One of the problems on the reverse is a rectangular-linear system of equations of type B1b:

MS 5112 § 11, literal translation	explanation
Length (and) front (I) made eat each other,	$u \cdot s = A$
1 èše the field.	= 1 èše = 10 00 square ninda
The length over the front is 10 beyond.	$u - s = q$ = 10 (ninda)
The length (and) the front are what?	u, s = ?
You with your doing:	Do it like this:
1/2 of 10 that the length over the front is beyond crush,	$q/2 = 10/2 = 5$
5 steps of 5 (make) eat (each other), 25.	sq. $q/2$ = sq. 5 = 25
To 10 the field add, 10 25.	A + sq. $q/2$ = 10 00 + 25 = 10 25
What is it equalsided?	sqs. (A + sq. $q/2$) = sqs. 10 25 = ?
25 each way equalsided.	sqs. (A + sq. $q/2$) = 25 = $p/2$
Twice write it down.	Write down 25 = $p/2$ twice.
5 that was eaten to the 1st 25 add, 30.	$p/2 + q/2 = 25 + 5 = 30$
30 ninda is the length.	$u = p/2 + q/2 = 30$
From the second 25 the 5 tear off, 20	.$p/2 - q/2 = 25 - 5 = 20$
20 ninda̱ is the front.	$s = p/2 - q/2 = 20$

The geometric model on which, apparently, both the question and the solution procedure in MS 5112 § 11 are based is obviously an OB forerunner to the construction in *El.* II.6 (see Figs. 1.4.1 and 1.4.2 above, right).

1.12 b. Old Babylonian problems for circles and chords

The examples discussed in Sec. 1.12 a above make it clear that parts A and B of *Elements* II (*El.* II.2-II.8) have many OB forerunners in the form of metric algebra problems for *squares and rectangles*. It is also easy to find examples of OB forerunners to part C of *Elements* II (*El.* II. 9-II.14), in the form of metric algebra problems for *right triangles and circles*. As suggested above, maybe the pair of exercises BM 13901 § 2 a-b is one such example. Further examples will be offered in the discussions of OB "igi-igi.bi problems" in Secs. 3.2-3 below, and in the discussion of an OB geometric algorithm in Appendix 1.

For some reason, there are few known metric algebra problems specifically for circles in the known corpus of OB mathematical texts.

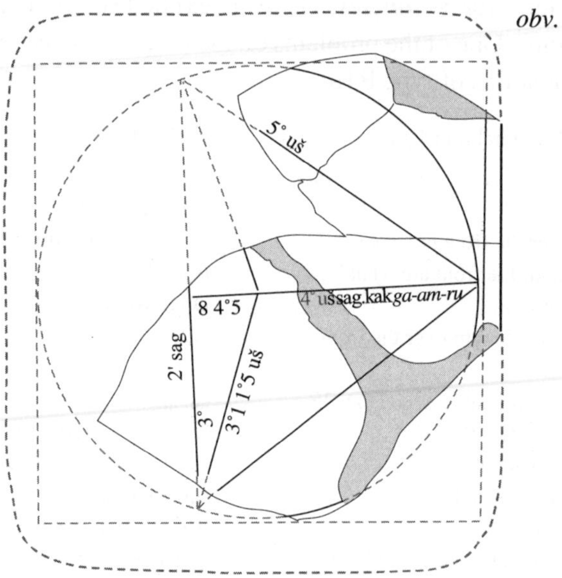

Fig. 1.12.4. *TMS* 1. A school boy's hand tablet with a diagram of a triangle in a circle.

An interesting first example is ***TMS* 1** (Bruins and Rutten, *TMS* (1961); Fig. 1.12.4 above). This is a relatively late OB "hand tablet" from the ancient city Susa, with a diagram of a "symmetric" (*isosceles*) triangle and its circumscribed circle. The triangle is constructed as two right triangles with the sides 50, 40, 30, glued together along a common side of length 40.

The object of the exercise was probably to compute the radius of the circumscribed circle. This can have been done, essentially, in the following way: Let $d/2$, $s/2$, $b/2$ be the sides of the small right triangle with $d/2$ = the radius and with $s/2 = 1/2$ the front of the symmetric triangle (in the diagram called 2' sag '1/2 of the front'). Then d and b can be found as the solutions to the following *subtractive quadratic-linear system of equations*:

sq. $d/2$ – sq. $b/2$ = sq. $s/2$ = sq. 30 = 15 00
$d/2 + b/2 = 40$ (the height of the symmetric triangle)

Apparently it was known that then

$d/2 - b/2$ = (sq. $d/2$ – sq. $b/2$)/($d/2 + b/2$) = 15 00 / 40 = 15 · 1;30 = 22;30, so that
$d/2$ = (40 + 22;30)/2 = 31;15, $b/2$ = (40 – 22;30)/2 = 8;45.

The correctly computed values are recorded in the diagram as '31 15 the length' and '8 45', respectively. It is, by the way, easy to check that the diagram is *amazingly accurate*. The person who made the diagram must have known quite well how to work with ruler and compass.

Presumably, he started by drawing, very carefully, a triangle with the sides proportional to 1 00, 50, 50, with the front 1 00 vertical and facing to the left, in agreement with an OB convention. Next, he found the midpoint on the front. (Euclid shows how to bisect a given straight line in *El.* I.10, with reference to the constructions in *El.*I.1 and I.9.) Then he drew a line from there to the opposite vertex of the triangle, a line which necessarily turned out to be horizontal. (*Cf.* the remark in Høyrup, *LWS* (2002), 265 that "the angle between the height and the base is as right as can be controlled on the photo".) The next step of the construction must have been to find the center of the circumscribed circle. How this was done is not known, of course, but it is likely that it was done by use of the method demonstrated by Euclid in *El.* IV.5, with reference to *El.* 1.11.)

The next example is taken from **MS 3049** (Friberg, *RC* (2007), Sec. 11.1), a small fragment of an OB mathematical recombination text, where only one exercise (**§ 1 a**) is preserved on the obverse:

MS 3049 § 1 a, literal translation	explanation
An arc *I curved,*	A circle
20 the transversal,	The diameter is $d = 20$
and 2 that which I went down.[11]	A chord is $p = 2$ below the top
The upper (= left) *transversal* (is) what?	The chord s = ?

You:	Do it like this:
20, the transversal, break, then 10 you see,	$d/2 = 10$
10, the descent that like a string is set.	10 = the "vertical" radius
Turn back, then solve(?).	Continue like this:
20, the transversal, break, 10 you see.	$d/2 = 10$
10, a copy, lay down,	Write down $d/2 = 10$ again
let (them) eat each other, then 1 40 you see.	sq. $d/2$ = 1 40
2, the upper descent, from 10, the descent	$d/2 - p =$
that like a string is set tear off, then 8 you see.	$10 - 2 = 8 = b/2$
8 let eat itself, then 1 04 you see.	sq. $b/2$ = 1 04
1 04 from 1 40 that you sa*w*	sq. $d/2$ – sq. $b/2$ = 1 40 – 1 04
tear off, then 36 you see	= 36 = sq. $s/2$
Its likeside let come up, then 6 you see.	sqs. 36 = 6 = $s/2$
To two repeat, then	$2 \cdot 6 =$
12, the upper *transversal*, you see.	s = the chord
Such is the do*ing*.	Done

The straightforward solution procedure is explained in Fig. 1.12.5, left Given are the diameter d of a circle and the distance p of a chord from the circumference of the circle along a radius orthogonal to the chord. The length of the chord is computed by use of the diagonal rule (see Ch. 2 below), applied to the triple $d/2$, $s/2$, $b/2$.

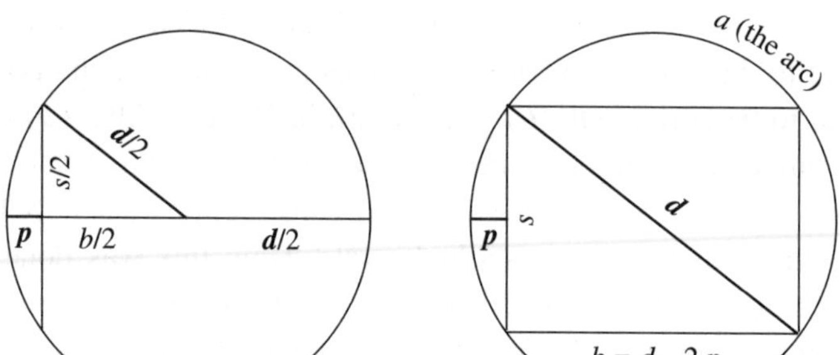

Fig. 1.12.5. Left: MS 3049 § 1 a. Right: BM 85194 # 21.

11. According to an OB convention, in cuneiform texts 'up' is to the left and 'down' to the right. The well known explanation is that the cuneiform script was originally written in vertical columns, but at some (unknown) point of time, the direction of writing seems to have changed so that texts became written in horizontal rows. After this rotation of the direction of writing, the meaning of 'up' and 'down' had changed correspondingly.

The strange way of calling half the diameter of the circle 'that (which) like a string is set' is not known from any other Babylonian mathematical text. It may refer to the fact that if a piece of string has one end point fixed, then upon rotation of the stretched string the other end point of the string describes a circle. Therefore a radius of a circle can be likened to a 'string'. (There is no competing word for the radius of a circle used in any other known mathematical cuneiform text.)

On the reverse of MS 3049, a subscript says that the text (originally) contained 6 problems for circles (and also 5 problems for squares, 1 for a triangle, etc.). Although only 1 of these 6 problems has happened to be preserved, it is a reasonable conjecture that the 6 circle problems resulted from the 6 possible ways of choosing 2 of the 4 parameters *d*, *p*, *s*/2, and *b*/2 as the *given pair of parameters* in the problem. In § 1 a, the given pair of parameters is *d* and *p*. The remaining possible choices of given pairs of parameters are *d* and *s*/2, *d* and *b*/2, *p* and *s*/2, *p* and *b*/2, *s*/2 and *b*/2. In *TMS* 1, by the way (Fig. 1.12.4), the given parameters are *s* and *d*/2 + *b*/2.

BM 85194, another OB mathematical recombination text contains two problems for circles, **## 21-22** (Høyrup, *LWS* (2002), 272):

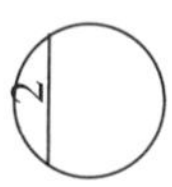

BM 85194 ## 21-22, literal translation	explanation
1 the arc,	The circumference is $a = 1\ 00$
2 that which I went down.	A chord is $p = 2$ below
The transversal (is) what?	The chord $s = ?$
You:	Do it like this:
2 square, 4 you see.	$2\,p = 2 \cdot 2 = 4$
4 from 20, the transversal, tear off,	$d = a/3 = 20,\ d - 2\,p = 20 - 4$
16 you see.	$= 16 = b$
20, the transversal, square, 6 40 you see.	sq. d = sq. $20 = 6\ 40$
16 square, 4 16 you see.	sq. b = sq. $16 = 2\ 24$
4 16 from 6 40 tear off, 2 24 you see.	sq. d – sq. $b = 6\ 40 - 2\ 24$
2 24 is what equalsided?	sqs. 2 24
12 equalsided, the transversal.	$= 12 = s$
Such is the doing.	Done

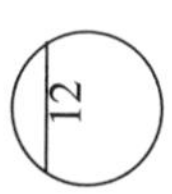

If an arc 1 I curved,	$a = 1\ 00$
12 the transversal.	$s = 12$
That which I went down?	$p = ?$
You:	Do it like this:
20, the transversal, square, 6 40 you see.	$d = a/3 = 20$, sq. d = sq. $20 = 6\ 40$

12 square, 2 24.	sq. *s* = sq. 12 = 2 24
From 6 40 tear off, 4 16 you see.	sq. *d* – sq. *s* = 4 16 = sq. *b*
16 is what equalsided? 4 equalsided.	sqs. 16 = 4 (error for sqs. 4 16 = 16)
(In) half 4 break, 2 you see,	1/2 · 4 = 2 (cheating)
2 that which you went down.	*p* = 2 (the correct answer)
The doing.	Done

The stated problem in BM 85194 # 21 is closely related to the problem in MS 3049 § 1. The only difference is that in BM 85194 # 21 the circumference $a = 3\ d$ is given (with the usual Babylonian approximation π = appr. 3), while in MS 3049 § 1 the diameter d is given directly. The straightforward solution procedure in # 21 is based on a geometric construction like the one in Fig. 1.12.5, right. It is an interesting variant of the solution procedure based on the construction in Fig. 1.12.5, left. Note that because the sides of the triangle in the circle to the right are twice as long as the sides in the right triangle in the circle to the left, it is "obvious" that in the figure to the right *the triangle with its diagonal along the diameter is a right triangle*. (*Cf.* a similar remark in Høyrup, *LWS* (2002), 274.)

In BM 85194 # 22, the stated problem is to find p when the circumference $a = 3\ d$ and the chord s are given. The solution is corrupt, but leads nevertheless to the correct answer (known in advance from # 21).

A dressed up problem, closely, although indirectly, related to the circle problems discussed above is problem **# 9** in **BM 85196**, like BM 85194 an OB mathematical recombination text from the ancient city Sippar. This is the well known "pole-against-a-wall problem", discussed before by several authors, for instance, Friberg, *HM* 8 (1981), Muroi, *KK* 30 (1991), Høyrup, *LWS* (2002), 275, Melville, *HM* 34 (2004), 151.

BM 85196 # 9, literal translation	explanation
A pole. 30, a reed, at a wall is placed equally.	*c* = 30 (;30 ninda = 1 reed)
Above, 6 it went down,	*s* = 6
below, *what did it move away*?	*b* = ?
You:	Do it like this:
30 square, 15 you see.	sq. *c* = 15
6 from 30 tear off, *24 you see.*	*c* – *s* = 30 – 6 = 24
24 square, 9 36 you see.	sq. (*c* – *s*) = sq. 24 = 9 36
9 36 from *15 tear off*, 5 24 you see.	sq. *c* – sq. (*c* – *s*) = 15 – 9 36 = 5 24
5 24 what is it equalsided?	sqs. 5 24
18 it is equalsided.	= 18

18 on the ground it moved away.	$= b$
If 18 on the ground,	Conversely, given that $b = 18$
above, what did it go down?	$s = ?$
18 square, 5 24 you see.	sq. b = sq. 18 = 5 24
5 24 from 15 tear off, 9 36 you see.	sq. c – sq. b = 15 – 5 24 = 9 36
9 36, what its it equalsided?	sqs. 9 36
24 it is equalsided.	$= 24 = a$
24 from 30 tear off,	$c - a = 30 - 24$
6 you see, (what) it went down.	$= 6 = s$
The doing.	Done

In this dressed up problem, the stated question is as follows:

> A wooden pole with the length 1 reed = 1/2 ninda (about 3 meters) at first stands upright against a wall of the same height. Then it starts sliding so that its upper end moves straight down ;06 ninda. How much does its lower end move along the ground?

The situation is illustrated in Fig. 1.12.6, left, where it is assumed that a pole of length c at first was standing upright along a wall of height c. Its top then slid down the distance s and its foot slid out a corresponding distance b. The set task is to find b if c and s are given. The connection between this dressed up problem and straightforward circle problems of the kinds discussed above is demonstrated in Fig. 1.12.6, right.

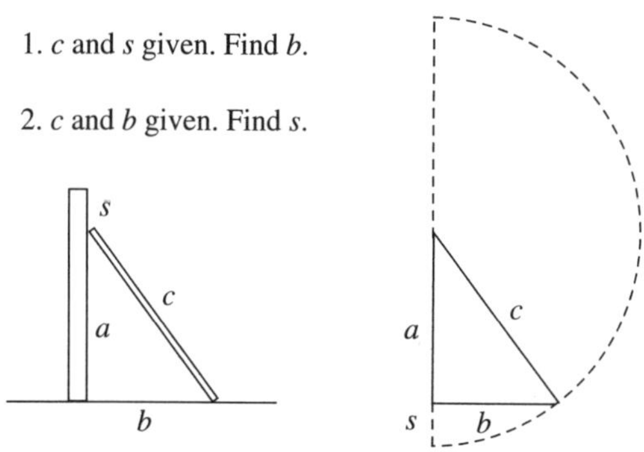

Fig. 1.12.6. BM 85196 # 9. A pole-against-a-wall problem.

The solution to the stated problem is obtained without effort by use of the diagonal rule. It is found that $b =$;18 ninda. This result is then checked by a reversal of the problem.

The pole-against-a-wall problem in the form given to it in BM 85169 # 9 is in itself quite trivial and uninteresting. Yet it is important for a couple of reasons. One reason is that dressed up problems like this one are quite rare in OB mathematics. The other reason is that the problem type reappears in a Seleucid mathematical recombination text and in an Egyptian mathematical recombination text, both from the latter third of the first millennium BCE (See Friberg, *UL* (2005), Sec. 3.1 b)

As mentioned above, the corpus of known OB metric algebra problems for circles and chords is small, compared to the related corpus of known metric algebra problems for squares and rectangles. Yet this fact may in part be due to unlucky circumstances. Thus, it is clear that all known OB problems for circles and chords are isolated exercises in mathematical recombination texts. It is likely that there once existed one or several extensive and well organized OB mathematical theme texts with relatively large numbers of such problems, from which exercises like MS 3049 § 1 a-[f], BM 85194 ## 21-22, and BM 85169 # 9 were borrowed. Be that as it may, there appears to be a close relation between on one hand such OB problems for circles and chords, and on the other hand *El.* II.11 and II.14, and their hypothetical forerunners *El.* II.11* and II.14* (Sec. 1.7 above).

1.12 c. Old Babylonian problems for non-symmetric trapezoids

The only known OB predecessors to *El.* II.12 and II.13 (see Fig. 1.8.1) can be found in **VAT 7351**, a mathematical cuneiform text from the ancient city Uruk. That text is extensively discussed in Friberg, *UL* (2005), Sec. 3.7 c.[12] Here is the text of the last one of the four exercises in that text:

VAT 7531 # 4, literal translation

2 43 30 the long length, 1 56 30 the short length,
1 37 30 the upper (= left) front, 1 30 30 the lower (= right) front.
Its area, how much it is, find out,
then to 5 brothers equally divide it, and (each) soldier show him his stake.

Properly speaking, VAT 7531 # 4 is an *assignment* rather than an exercise, since the question is not followed by a solution procedure and an answer. The object considered in the text is a quadrilateral field with the given lengths 2 43;30 and 1 56;30 (ninda), and the given fronts 1 37;30 and 1 30;30 (ninda). The field is to be divided equally among 5 brothers.

---

12. See now also the trapezoid diagonal problem in VAT 8393 in Appendix 1 below.

The form of the given field is not uniquely determined by the four sides. However, it must have been (silently) understood that the field should have the form of a trapezoid composed of a central rectangle and two flanking non-equal triangles (Fig. 1.12.7).

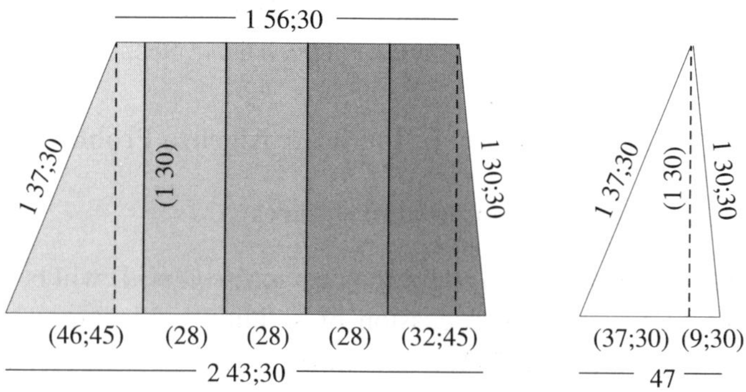

Fig. 1.12.7. VAT 7531 # 4. A trapezoidal field divided in five equal parts.

If the rectangle is removed, what remains is a non-symmetric (scalene) triangle with the sides 1 37;30, 1 30;30, and the base 47. Evidently, the OB author of this text was confident that his students knew how to compute the height of a non-symmetric (*scalene*) triangle! The way they would have done it was probably as follows: Let a, b, c be the sides of the triangle, and suppose that the height h against the side b divides b into the segments p and q, where p is greater than q. (See above, Fig. 1.8.1, right.) Then,

$p + q = b$ and
sq. c – sq. p = sq. a – sq. q (by the diagonal rule, since both are equal to sq. h).

This leads to a *quadratic-linear system of equations* for p and q:

$p + q = b$ and sq. p – sq. q = sq. c – sq. a.

This quadratic-linear system of equations can be solved by use of metric algebra. The solution can take several forms, for instance the following:

p = {sq. b + (sq. c – sq. a)}/(2 b), q = {sq. b – (sq. c – sq. a)}/(2 b).

With c, a, b = 1 37;30, 1 30;30, 47, these equations show that

p = 37;30, q = 9;30.

It is then easy to compute h = 1 30. The remaining part of the solution procedure for VAT 7531 # 4 is straightforward.

The result above shows that the triangle with the sides 1 37;30, 1 30;30, 47 is composed of two right triangles with the sides 1 37;30, 1 30, 37;30 = 7;30 · (13, 12, 5) and 1 30;30, 1 30, 9;30 = 30 · (3 01, 3 00, 19), glued together along a common side of length 1 30. This is clearly an OB predecessor to what is commonly known as "Heronic triangles". (*Cf.* the discussion of the pseudo-Heronic *Geometrica* 12 in Sec. 18.2 below.)

1.13. Late Babylonian Solutions to Metric Algebra Problems

1.13 a. Problems for rectangles and squares

The discussion above of *OB* forerunners to *Elements* II will be rounded off in this section with a discussion of solution procedures for metric algebra problems in **W 23291**, a *Late Babylonian* mathematical recombination text from Uruk, early in the second half of the first millennium BCE (Friberg, *BaM* 28 (1997)). W 23291 and the related text **W 23291-x** (Friberg, *et al.*, *BaM* 21 (1990)) are both concerned with the interesting topic of *a great variety of ways of measuring surface content.*

The first paragraph of W 23291 contains what looks like *the beginning of a well organized theme text with metric algebra problems.*

W 23291 § 1: Common seed measure and some basic problems in metric algebra

1 a The seed measure of a hundred-cubit-square. Metric squaring
1 b A rectangle of given front and seed measure. Metric division
1 c A square of given seed measure. Metric square side computation
1 d A rectangle of given side-sum and seed measure. Basic problem B1a
1 e A rectangle of given side-difference and seed measure. Basic problem B1b
1 f A square band of given width and seed measure. Basic problem B3b
1 g A circle of given seed measure divided into five circular bands of given width

In § 1 of W 23 291, the surface content of every square, rectangle, or other plane figure mentioned, is expressed in terms of "seed measure", by which is meant a capacity measure *proportional, in a certain ratio, to the area of the figure in question.* More precisely, the seed measure applied in § 1 is what may be called "common seed measure" (csm), the particular kind of seed measure characterized by the following igi.gub še.numun 'seed constant':

c_s = '20' = ;20 barig (= 1/3 barig) on each square of side 1 00 cubits (= 60 cubits). 13

The barig (Akk. *parsiktu*) was the "basic unit" of Late Babylonian capacity measure, in the sense that *sexagesimal multiples* of the barig were used in computations involving capacity measures and in references to metrological constants like the seed constant.

In the present text, just as in the related text W 23291-x, a dual *ninda-and-cubit format* is used in many of the solution procedures. What this means is that the solution of a given problem is presented twice, first in a "ninda section" where the ninda (= 6 m.) is the basic unit of length measure, then in a parallel "cubit section" where the cubit (= 1/2 m.) is the basic unit. In the cubit sections, the seed constant for common seed measure is '20' = 1/3 barig./sq. (60 c.), as explained above. In the ninda sections it is, equivalently,

c_s = '48' = 48 barig on each square of side 1 00 ninda (= 60 ninda).

The equivalence of the two alternative expressions is obvious, since

1 00 cubits = 5 ninda, so that sq. (60 cubits) = sq. (5 ninda) = 25 sq.ninda.

Therefore,

1/3 barig / sq. (60 cubits) = 1/3 barig / 25 sq.ninda
= 12 · 12 · 1/3 barig / sq. (60 ninda) = 48 barig / sq. (60 ninda).

The ninda section of a solution procedure is preceded by the phrase

šum-ma 5 *am-mat-ka* 'if 5 is your cubit'.

This phrase refers to the circumstance that when the ninda (= 12 cubits) is chosen as the basic unit of length measure, then 1 cubit is equal to 1/12 = 5/60 = ;05 of that basic unit. For a similar reason, the cubit sections are preceded by the phrase

šum-ma 1 *am-mat-ka* 'if 1 is your cubit'.

The seed measure of a hundred-cubit-square. Metric squaring

W 23291 § 1 a, literal translation explanation

---

13. Note that the use of zeros and separators in the transliteration of numbers in a mathematical cuneiform text tends to destroy the inherent simplicity of the definitions of various Babylonian mathematical and metrological "constants". So, for example, the seed constant for common seed measure was not understood as ;00 00 20 barig/sq.cubit. Nor was it understood as ;20 barig/sq.(60 cubits). Instead it was almost certainly understood as '20' times the area, with the silent understanding that *when the sides of a rectangle amount to a few sixties of cubits, then the seed measure of the rectangle amounts to a few barig*.

1 hundred cubits length, 1 hundred cubits front.	A square with the side $s = 100$ c.
What shall the seed be?	$C = ?$
If 5 is your cubit:	**If you count with ninda:**
8 20 is 1 hundred cubits.	$s = 100$ c. = 100/12 n. = 8;20 n.
8 20 steps of 8 20 go,1 09 26 40.	A = sq. 8;20 n. = 1 09;26 40 sq. n.
1 09 26 40 steps of 48 go,	$C = c_s \cdot A$ = '48' $\cdot A$
55 33 20, 5 bán 3 1/3 sìla of seed.	= ;55 33 20 barig = 5 bán 3 1/3 s.
If 1 is your cubit:	**If you count with cubits:**
1 40 is 100 cubits.	$s = 100$ c. = 1 40 c.
1 40 steps of 1 40 go, 2 46 40.	A = sq. 1 40 c. = 2 46 40 sq. c.
2 46 40 steps of 20 go, 55 33 20,	$C = c_s \cdot A$ = '20' $\cdot A$ = ;55 33 20 barig
5 bán 3 1/3 sìla of seed.	= 5 bán 3 1/3 sìla.

The problem stated and solved in § 1 a of W 23291 can be explained as follows: A square of side 100 cubits may be called a "hundred-cubit-square", or simply a "100-c.-square". As is shown by Late Babylonian metrological tables, notably *BE 20/1,* 30 (see Friberg, *GMS 3* (1993), 399), a "hundred-cubit" was occasionally used, in addition to the cubit and the ninda, as the *third* basic unit of Late Babylonian length measure. For this reason, it would be convenient to have at hand a *third* value of the seed constant for common seed measure, in addition to 48 barig./sq. (60 ninda) and ;20 barig./sq. (60 c), namely the common seed measure (csm) of a hundred-cubit-square. In W 23291 § 1 a this value is computed twice. In the ninda section, it is computed in the following way:

If, as in the present text, 1 n. = 12 c., then 1 c. = ;05 n. Therefore,

the *side* of the 100-cubit-square is $s = 100$ c. = 100 · ;05 n. = 8;20 n., so that

the *area* of the 100-cubit-square is A = sq. (100 c.) = sq. (8;20 n.) = 1 09;26 40 sq. n.

Note that all computations are carried out here in the traditional Babylonian way, that is by use of sexagesimal arithmetic. That is so in spite of the fact that in the statements of the problems in W 23291 § 1 linear measures are expressed as *decimal* multiples of the cubit!

Next, an application of the appropriate value of the seed constant proves that the *seed measure* of the 100.cubit-square is

C = 48 barig · 1 09;26 40 /sq. 1 00 = ;55 33 20 barig.

The final step of the computation is to convert this sexagesimal multiple of the barig into a *conventionally expressed capacity number*. This can be done, most conveniently, in the following way. (The computation is based on the fact that fractions of the barig when multiplied by a factor 6 yield multiples of the sub-unit bán, and that fractions of the bán when multiplied by another factor 6 yield multiples of the smaller sub-unit sìla.

C = ;55 33 20 barig = 6 · ;55 33 20 bán = 5;33 20 bán
= 5 bán + 6 · ;33 20 sìla = 5 bán 3;20 sìla = 5 bán 3 1/3 sìla.

In the *cubit section* of § 1 a, the computation of the common seed measure of a hundred-cubit-square proceeds in an entirely parallel way.

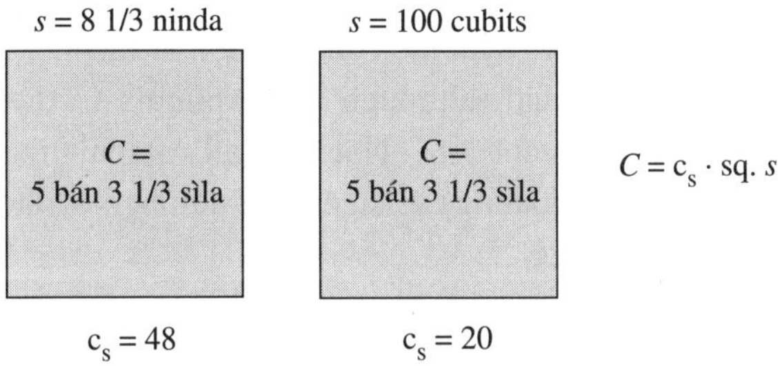

Fig. 1.13.1. W 23291 § 1 a. Metric squaring. The seed measure of a 100-cubit-square.

A rectangle of given front and seed measure. Metric division

W 23291 § 1 b, literal translation	explanation
1 hundred cubits front.	s = 100 cubits
The length, what shall it be long,	u = ?
so that there will be 1 gur of seed?	if, in addition, C = 1 gur = 5 barig
Since you do not know:	**Do it like this:**
The opposite of the front of the field raise,	Compute the reciprocal of the front
and steps of the opposite of	and multiply with the reciprocal
the seed constant you go,	of the seed constant
and the seed that was said to you go,	and multiply with the seed measure
the length you will see.	then you will se the length
If 5 is your cubit:	**If you count with ninda:**
8 20 is 1 hundred cubits.	s = 100 cubits = 8; 20 ninda
The opposite of 8 20, 7 12.	rec. s = rec. 8 20 = 7 12
7 12 steps of 1 15 go, 9.	rec. s · rec. c_s = 7 12 · 1 15 = 9
9 steps of 5 go, 45. 45 as much as	rec. s · rec. c_s · C = 9 · 5 = 45
the length of your field you will set.	u = 45
If 1 is your cubit:	**If you count with cubits:**
1 40 is 1 hundred cubits.	s = 1 40 cubits = 100 cubits
The opposite of 1 40, 36.	rec. s = rec. 1 40 = 36
36 steps of 3 go, 1 48.	rec. s · rec. c_s = 36 · 3 = 1 48

1 48 steps of 5 go, 9, that <for> 1 40 cubits	rec. s · rec. $c_s \cdot C$ = 1 48 · 5 = 9
as much as the length you will set.	= u, when s = 1 40 cubits

The statement of the problem in the first three lines of § 1 b is followed by a *general computation rule* headed by the phrase mu nu zu-*ú* 'since you do not know'. It is easily checked that the two parallel solution procedures in the ninda and cubit sections of the paragraph are two *different but equivalent numerical implementations* of this general computation rule.

The computation in each of the two cases is straightforward. The ninda section, for instance, begins with the computation of the reciprocals of the given front (100 cubits) and (although not explicitly in the text) of the reciprocal of the seed constant '48'. Note that all computations are carried out in terms of *relative* (floating) sexagesimal numbers without any indication of their absolute size.

The answer is given in relative sexagesimal numbers as

u = '45' in the ninda section and u = '9' in the cubit section.

Since the length is always greater than the front in Babylonian mathematical texts dealing with rectangles, the obvious interpretation of this result in relative numbers is that the length u is equal to 45 ninda = 9 00 cubits. It is easy to verify that, with this value for u,

the area A = 45 ninda · 8;20 ninda = 6 15 sq. ninda = ;06 15 · sq. (60 ninda).

Therefore, as required,

the seed measure C = 48 · ;06 15 barig = 5 barig.

The result of the dual computation is summarized in Fig. 1.13.2 below.

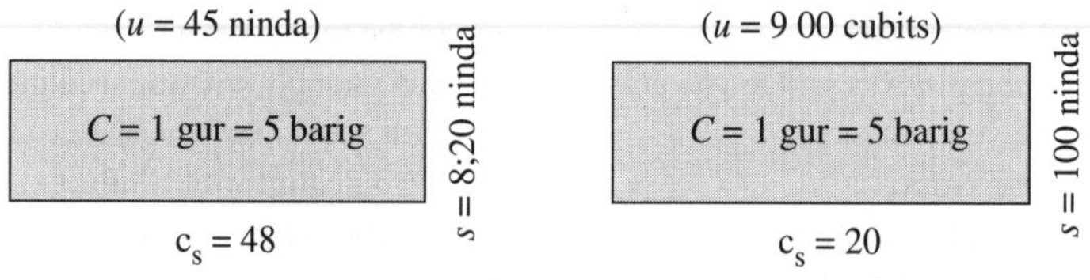

$$C = c_s \cdot u \cdot s \quad \Rightarrow \quad u = \text{rec. } c_s \cdot \text{rec. } s \cdot C$$

Fig. 1.13.2. W 23291 § 1 b. Metric division.

A square of given seed measure. Metric square side computation

W 23291 § 1 c, literal translation   explanation

The field, what each shall I make equalsided	Which are the equal sides (of a square)
so that 1 gur 2 bán will be the seed?	with seed measure 1 gur 2 bán?
Since you do not know:	**Do it like this:**
The seed that was said to you,	Take the mentioned seed measure
— what was said	multiply it with
steps of the seed constant,	the <reciprocal of> the seed constant
you go, the length.	compute the square side
If 5 is your cubit:	**If you count with ninda:**
5 20 steps of 1 15 go, 6 40,	$C \cdot$ res. c_s = 5 20 · 1 15 = 6 40
of which 20 each way take.	6 40 = sq. 20
20 ninda each way you make equalsided.	s = 20 <ninda> is the square side
If 1 is your cubit:	**If you count with cubits:**
5 20 steps of 3 go, 16,	$C \cdot$ res. c_s = 5 20 · 3 = 16
of which 4 each way take.	16 = sq. 4
2 hundred 40 cubits each	s = 4 00 = 240 <cubits>
you make equalsided.	is the square side

This exercise is quite straightforward. The given seed measure is

1 gur 2 bán = 5 1/3 barig = 5;20 <barig>,

and the computed square side is

20 ninda = 20 · 12 cubits = 240 cubits.

It is interesting that the cubits are counted decimally in the answer.

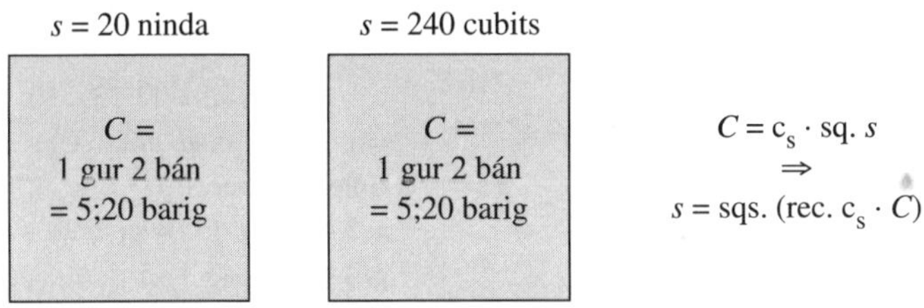

Fig. 1.13.3. W 23291 § 1 c. Metric square side computation.

A rectangle of given side-sum and seed measure. Basic problem B1a

W 23291 § 1 d, literal translation	explanation
A field of 1 bán seed.	C = 1 bán (= ;10 barig)
Length and front heap, it is 1 30 cubits.	$u + s$ = 1 30 cubits (= 7;30 ninda)
The length, what shall it be,	u = ?
and the front what shall it be?	s = ?
Since you do not know:	**Do it like this:**

1/2 to the heap, 1 30 cubits, raise,	1/2 · the sum 1 30
45 cubits equalsided,	= 45 cubits, squared
to the constant of seed [raise] it.	times the seed constant c_s
1 bán of seed out of it lift,	subtract C = 1 bán
the opposite of the constant you raise to it,	times rec. c_s
<and bring (out)>	<Compute the square side>
to 45 cubits add on, the length,	add 45 cubits, you get the length
from 45 cubits lift, the front.	Subtract 45 cubits, you get the front
If 5 is your cubit:	**If you count with ninda:**
Length and front, 7 30, 1/2 of it, 3 45, take.	$(u + s)/2 = p/2 = 7\ 30/2 = 3\ 45$
3 45 steps of 3 45 you go, 14 03 45	sq. $p/2$ = sq. 3 45 = 14 03 45
and steps of 48 go, 11 15,	c_s · sq. $p/2$ = 48 · 14 03 45 = 11 15
1 07 30 bán-measures.	= 1;07 30 bán
1 bán, 10, from 11 15 lift,	c_s · sq. $p/2 - C$ = 11 15 – 10
1 15 the remainder.	= 1 15 (= c_s· (sq. $p/2 - A$))
1 15 steps of 1 15,	times 1 15 (= rec. c_s)
1 33 45, of which 1 15 each take.	= 1 33 45 = sq. 1 15 (= sq. $q/2$)
1 15 to 3 45 join,	$p/2 + q/2$ = 3 45 + 1 15
5 ninda, the length,	= 5 = u
1 15 from 3 45 lift,2 30, the front.	$p/2 - q/2$ = 3 45 – 1 15 = 2 30 = s
If 1 is your cubit:	**If you count with cubits:**
1 30 cubits, 1/2 of it, 45, take,	$p/2$ = 1 30/2 = 45
45, steps of 45 go, 33 45,	sq. $p/2$ = sq. 45 = 33 45
33 45 steps of 20 go,	c_s · sq. $p/2$ = 20 · 33 45
the seed, 11 15.	= 11 15
1 bán, 10, from 11 15 lift,	c_s · sq. $p/2 - C$ = 11 15 – 10
1 15 the remainder.	= 1 15 (= c_s· (sq. $p/2 - A$))
1 15 steps of 3 go,	times 3 (= rec. c_s)
3 45, of which 15 each — take.	= 3 45 = sq. 15 (= sq. $q/2$)
15 to 45 join	$p/2 + q/2$ = 45 + 15
1-sixty cubits, the length,	= 1 · 60 cubits = u
from 45 lift, 30 cubits, the front.	$p/2 - q/2$ = 30 cubits = s

In the present exercise, § 1 d, just as in §§ 1 b and 1 c above, the question is followed by a *general computation rule* headed by the phrase mu nu zu-*ú* 'since you do not know'. It is interesting to note that in § 1 d the author of the text has not been quite successful in his formulation of a general computation rule, since he explicitly mentions the half-sum of the sides of the rectangle as '45 cubits' instead of just as '1/2 the heap'.

In the cuneiform text, there is no figure accompanying exercise § 1 d. Yet the wording of the solution procedure is such that there can be no doubt whatsoever that the author of the problem had in mind a *geometric* interpretation of the given problem and its solution. The most likely candidate for such an interpretation is based on a set-up like the one in Fig. 1.13.4 below, nearly identical with the set-up in Fig. 1.4.2 above, left, the suggested Babylonian style interpretation of the diagram in *El.* II.5. The only difference is the use of seed measure instead of area measure.

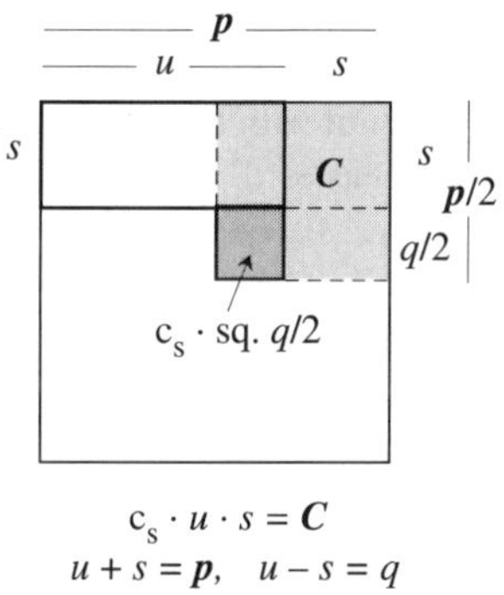

In the cubit section:

Given:
$p =$ 1 30 cubits, $C =$ 1 bán = ;10 barig
$c_s =$;20 barig/ sq. (1 00 cubits)

Computed:
$c_s \cdot$ sq. $q/2 = c_s \cdot$ sq. $\boldsymbol{p/2} - \boldsymbol{C} =$;01 15 barig
sq. $q/2 =$ 3 45 sq. c. = sq. (15 cubits)
$u = \boldsymbol{p/2} + q/2 =$ 45 c. + 15 c. = 1 00 c.
$s = \boldsymbol{p/2} - q/2 =$ 45 c. – 15 c. = 30 c.

Fig. 1.13.4. W 23291 § 1 d. A rectangle of given side-sum and seed measure.

A rectangle of given side-difference and seed measure. Type B1b

The problem stated in **W 23291 § 1 e** is to find the length u and front s of a rectangle, if the seed measure of the rectangle is 1 bán 4 sìla, and if the length exceeds the front by 10 cubits. This is a routine variation of the problem in § 1 d, and it can be solved by an obvious modification of the solution procedure in that paragraph. Thus, if the given side-difference is called $q =$ 10 cubits, then sq. $q/2 =$ 25 sq. cubits, and $c_s \cdot$ sq. $q/2 =$;08 20 barig, since $c_s =$;20 barig/sq. (60 cubits). On the other hand, the given seed measure of the rectangle is $C =$ 1 bán 4 sìla = ;16 40 barig, since (in this Late Babylonian text) 1 barig = 6 bán and 1 barig = 6 sìla.Therefore, $C + c_s \cdot$ sq. $q/2 =$;16 48 20 barig = $c_s \cdot$ sq. $p/2$. Hence, sq. $p/2 =$ 50 25 sq. cubits, and $p/2 =$ 55 cubits. Thus, finally, $u =$ (55 + 5) cubits = 1 00 cubits = 5 ninda, and $s =$ (55 – 5) cubits = 50 cubits = 4 ninda 2 cubits.

The problem in W 23291 § 1 e is of course, except for the use of seed measure instead of area measure, *a basic rectangular-linear system of equations of type B1b*. It is, therefore, related to *El.* II.6.

A square band of given width and seed measure. Type B3b

The statement of the problem in **W 23291 § 1 f** and the associated general computation rule are both completely lost. Nevertheless, the fortunate circumstance that almost the whole cubit section of the solution procedure is preserved allows a reconstruction of most of the problem.

W 23291 § 1 f, literal translation	explanation
....	....
If 1 is your cubit:	**If you count with cubits:**
10 is 10 cubits.	s = 10 cubits = 10
The opposite of 10 raise, 6.	rec. s = rec. 10 = 6
6 steps of 3 you go, 18.	rec. s · rec. $c_s = 6 \cdot 3 = 18$
18 steps of 10, 1 bán, raise, 3.	rec. s · rec. $c_s \cdot C = 18 \cdot 10 = 3$
3, its 4th raise, 45.	rec. s · rec. $c_s \cdot C/4 = 3/4 = 45 = u$
10 from 45 lift, 35 the remainder.	$u - s = 45 - 10 = 35$
35 cubits equalsided.	$q = 35$
[..]	[$q + 2\,s = 35 + 20 = 55 = p$]

The form of this solution procedure, and the position of W 23291 § 1 f in the text, between the better preserved § 1 e and § 1 g, makes it fairly certain that the problem stated in § 1 f was to find the sides of the squares bounding a square band, when the width (10 cubits) and the seed measure (1 bán) of the square band are given. The way in which a solution to this problem could be found is illustrated in Fig. 1.13.5 below.

The problem treated in W 23291 § 1f can be formulated as follows. Let the square band be interpreted as the difference of two parallel and concentric squares with the sides p and q, respectively. Then p and q can be computed as the solutions to the following *subtractive quadratic-linear system of equations of type B3b*:

$$c_s \cdot (\text{sq. } p - \text{sq. } q) = C = 1 \text{ bán}, \quad (p - q)/2 = s = 10 \text{ cubits}, \quad p, q = ?$$

In modern terminology, all that is required to solve a problem of this type is an application of the algebraic "conjugate rule"

$$\text{sq. } p - \text{sq. } q = (p + q) \cdot (p - q),$$

followed by a straightforward division. A *metric* counterpart of this algebraic conjugate rule can be based on the observation that a square band can be constructed in two ways, either as the space between two parallel (and, if so desired, concentric) squares with the sides p, q or as a ring of four rectangles with the sides *u*, *s*. In the present text, where surface content is measured in terms of seed measure, the resulting "metric conjugate rule" takes the following form:

$$C = c_S \cdot (\text{sq. } p - \text{sq. } q) = 4 \cdot c_S \cdot u \cdot s = 4 \cdot c_S \cdot (p+q)/2 \cdot (p-q)/2.$$

Accordingly, the recorded solution procedure in W 23291 § 1 f corresponds to the solution formula

$$u = (p+q)/2 = 1/c_S \cdot 1/s \cdot S/4, \quad p = u + s, \quad q = u - s.$$

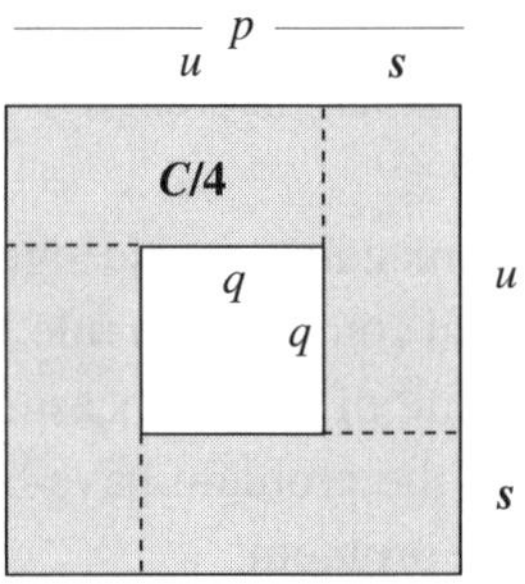

$c_s \cdot (\text{sq. } p - \text{sq. } q) = \boldsymbol{C}$

$c_s \cdot u \cdot s = \boldsymbol{C/4}$

In the cubit section.

Given:
$\boldsymbol{s} = (p-q)/2 = 10$ cubits, $\boldsymbol{C} = 1$ bán = ;10 barig
c_s = ;20 barig/sq. (1 00 cubits)

Computed:
$1/\boldsymbol{s} \cdot 1/c_s \cdot \boldsymbol{C/4} = 45$ cubits $= u = (p+q)/2$
$u - \boldsymbol{s} = 35$ cubits $= q$
$(q + 2\,\boldsymbol{s} = 55$ cubits $= p)$

Fig. 1.13.5. W 23291 § 1 f. A square band of given width and seed measure.

The metric algebra problem in W 23291 § 1 f is obviously closely related to *El.* II. 8, which can be seen if, for instance, Fig. 1.13.5 is compared with Fig. 1.5.2. Note however, that Euclid chose to operate with *non-concentric* parallel squares, and that, as a consequence of this choice, in *El* II.8 the difference between the two squares takes the form of a square corner (a *gnomon*) rather than that of a square band.

1.13 b. Problems for circles

A circle of given seed measure divided into five bands of equal width

W 23291 § 1 g, literal translation	explanation
A field of 1 barig seed I curved.	A circle of seed measure 1 barig
Steps 4, 1 each,	and four inner circles
the decrease came down.	with 1 ninda's distance
What each are the arcs I curved,	What are the arcs (circumferences)
from the outermost arc	of all the circles
to the innermost arc?	from the first to the last?

Since you do not know:	**Do it like this:**
1 steps of 6 go, 6.	$1 \cdot 6 = 6$
6 from [···] 30 lift, 24 the remainder,	$30 - 6 = 24$
the second arc.	the second arc.
6 from 24 lift, 18 the remainder,	$24 - 6 = 18$
18, the third arc.	the third arc
6 from 18 lift, 12 the remainder	$18 - 6 = 12$
12, the fourth arc.	the fourth arc
6 from 12 lift, 6 the remainder,	$12 - 6 = 6$
6, the fifth arc.	the fifth arc
6 is the innermost arc field	6 is arc of the innermost circle
he will take off.	since $6 - 6 = 0$

This exercise is only loosely related to the six preceding metric algebra problems. Note, in particular, that there is no general computation rule, and no separate ninda and cubit sections. (The basic unit of length measure is the ninda.) It is also likely that essential parts of the problem have been omitted both at the beginning and at the end of the problem.

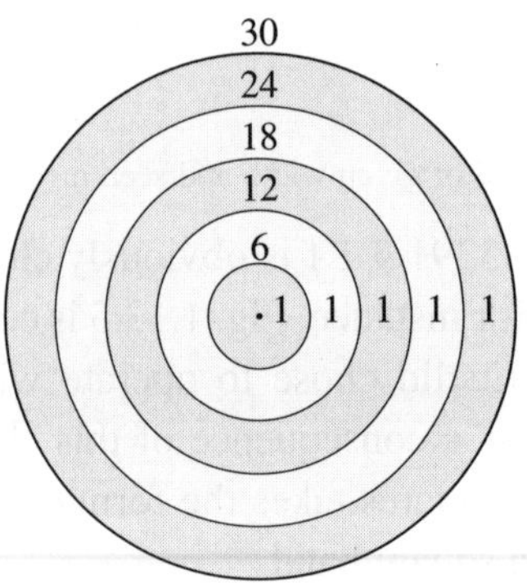

$C = c_s \cdot ;05 \cdot$ sq. a,
$c_s = 48$ barig/sq. (1 00 ninda)
$\Rightarrow$
$a =$ sqs. $(1/c_s \cdot 12 \cdot C)$
$d = ;20 \cdot a$

Example:
$C = 1$ barig
$\Rightarrow$
$a = 30$ ninda, $d = 10$ ninda

Fig. 1.13.6. W 23291 § 1 g. A circle of given seed measure. Five circular bands.

Thus, after it has been stated that the seed measure of the given circle is 1 barig, the arc a and the diameter d of the circle must have been computed, but this is not done explicitly in the text. To find the arc of a circle when the seed measure of the circle is given is a problem of the same type as the one in W 23291 § 1 c, to find the side of a square of given seed measure. The omitted computation should have had the following form:

$c_s \cdot ;05 \cdot$ sq. $a = C = 1$ barig, $c_s = 48$ barig/ sq. (60 ninda) $(1/4\pi =$ appr. $;05)$
$a =$ sqs. $(1/c_s \cdot 12\ C) =$ sqs. (15 00 sq. ninda) $= 30$ ninda,

$d = ;20 \cdot a = 10$ ninda ($1/\pi$ = appr. ;20)

The preserved part of the solution procedure can be explained as follows: If the width of each one of the circular bands is 1 ninda, then the diameter of each band is equal to the diameter of the preceding band minus 2 ninda, and the arc of each band is equal to the arc of the preceding band minus 6 ninda (counting with π = appr. 3). This gives the arcs listed in the text, 30, 24, 18, 12, and 6 ninda.

The computation of the seed measure of each one of the five circular bands has, for some reason, been omitted from the text of § 1g.

In the closely related Late Babylonian mathematical recombination text **W 23291-x** (Friberg, *et al.*, *BaM* 21 (1990)) there are, among other things, parallels to four of the seven exercises in W 23291 § 1. It is interesting to compare the parallel exercises with each other, for the reason that the exercises in W 23291-x resemble OB mathematical exercises more than what the corresponding exercises in W 23291 do.

Here is, first, the text of the parallel in W 23291-x to W 23291 § 1 g:

A circle of given circumference divided into five bands of equal width

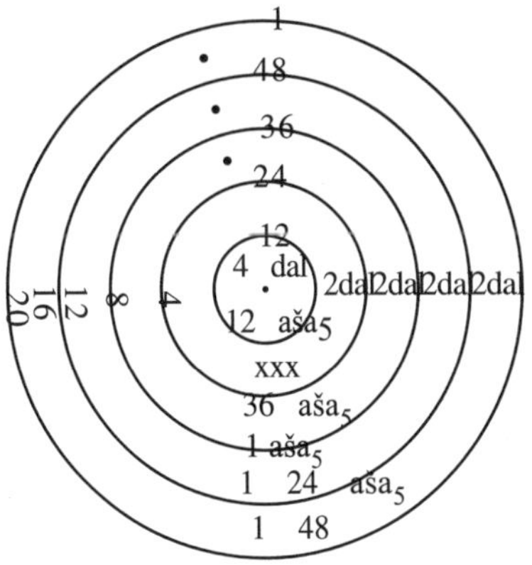

a_1 = 100 ninda	d_1 = 20 ninda
a_2 = 48 ninda	d_2 = 16 ninda
a_3 = 36 ninda	d_3 = 12 ninda
a_4 = 24 ninda	d_4 = 8 ninda
a_5 = 12 ninda	d_5 = 4 ninda

A_1 = 1 48 šar
A_2 = 1 24 šar
A_3 = 1 00 šar
A_4 = 36 šar
A_5 = 12 šar

Fig. 1.13.7. The diagram associated with W 23291-x § 2.

W 23291-x § 2, literal translation	explanation
1 (= the first) arc-field 1(60) ninda I curved.	A circle of arc 1 00 ninda
Steps, 4, 2 ninda each	4 inner circles with a distance of

as decrease I made come up.	2 ninda between each pair
What each are the fields?	The areas between the circles = ?
(Fig.)	(Fig.)
54 steps of 2, 1 48,	54 · 2 = 1 48
1(iku) 8 šar is the outermost decrease.	= 1 iku 8 šar (the area of band 1)
42 steps of 2, 1 24,	42 · 2 = 1 24
1/2(iku) 34 šar is the 2nd decrease.	= 1/2 iku 34 šar (the area of band 2)
30 steps of 2, 1,	30 · 2 = 1 00
1/2(iku) 10 šar is the 3rd decrease.	= 1/2 iku 10 šar (the area of band 3)
18 steps of 2, 36,	18 · 2 = 36
36 šar is the 4th decrease.	= 36 šar (the area of band 4)
12 steps of 12, 2 24,	12 · 12 = 2 24
2 24 steps of 5 go, 12,	2 24 · 5 = 12
12 šar is the 5th and innermost decrease.	= 12 šar (area of the innermost circle)
Heap them, all of them are 3(iku).	Check: 3 iku = the total area

In this exercise, four circular bands, all of width 2 ninda, are broken off from a circle of given arc length 1 00 n. The exercise is illustrated by a diagram, exhibiting the arcs of the five circles bounding the circular bands, the diameters of those circles, and the *area measures* of the four circular bands and the innermost circular core. In the solution procedure, only the computation of the area measures of the circular bands is expressly indicated. The use of traditional area measure as well as the absence of a general computation rule and of separate ninda and cubit sections are some conspicuous features of the first three exercises on W 23291-x, including this one. It is likely that these initial exercises were copied with only superficial changes from some OB mathematical text. The parallel text W 23 291 § 1 g, on the other hand, may be viewed as a Late Babylonian *revised edition* of a text like W 23291-x § 2, with the OB area measure replaced by the Late Babylonian seed measure.

In this connection it may be noted that it is likely that the purpose of computations with the ninda as the basic length unit in the ninda sections of Late Babylonian mathematical exercises was to make students familiar with the *Old Babylonian* way of counting, so that they would be able to understand Old Babylonian mathematical texts. In Late Babylonian non-mathematical texts, the *cubit* is always the basic length measure.

The other parallels to W 23291 § 1 in W 23291-x are the exercises in § 4 of the latter text. They are reproduced here, in literal translation.

W 23291-x § 4 a-d

§ 4 a. Rules for the computation of areas of rectangles and square sides

Reeds, such that	Reed measure (surface content) when
1-ninda-reed length, 1-ninda-reed front	length and front both = 1 ninda
is 1 šar.	makes 1 šar.
If 5 is your cubit:	**If you count with ninda:**
The line steps of ditto and steps of 1 go.	The area of a square = sq. $s \cdot 1$
Steps of 1, each take.	The side of a square is sqs. $(1 \cdot A)$
If 1 is your cubit:	**If you count with cubits:**
The line steps of ditto and steps of 25 go.	The area of a square = sq. $s \cdot 25$
Steps of 2 24, each take.	The side of a square is sqs. $(2\ 24 \cdot A)$

§ 4 b. Example of metric squaring

1 *ṣuppān* the length,	Length and front both equal to
and 1 *ṣuppān* the front.	1 *ṣuppān*
What are the šar?	What is that in šar?
If 5 is your cubit:	**If you count with ninda:**
5 is the *ṣuppān*.	5 (ninda) = 1 *ṣuppān*
5 steps of 5 go, 25, 25 šar.	$5 \cdot 5 \cdot 1 = 25$ (sq. ninda) = 25 šar
If 1 is your cubit:	**If you count with cubits:**
1 is the *ṣuppān*.	1 (· 60 cubits) = 1 *ṣuppān*
1 steps of 1 go, 1,	$1 \cdot 1 = 1$ (· sq. (60 cubits))
1 steps of 25 go, 25, 25 šar.	$1 \cdot 1 \cdot 25 = 25 = 25$ šar

§ 4 c. Example of metric square side computation

[···] of 25 šar.	A *square* of 25 šar
The equalside shall be what?	What is the square side?
If 5 is your cubit:	**If you count with ninda:**
Each of 25 take.	The square side of 25 (sq. ninda)
<a *ṣuppān* is the equalside>.	= 5 (ninda) <= 1 *ṣuppān*>
If 1 is your cubit:	**If you count with cubits:**
25 steps of 2 24 go,	$25 \cdot 2\ 24$
1, of which each take,	= 1 (· sq. (60 cubits)), the square side
a *ṣuppān* is the equalside.	= 1 (· 60 cubits) = 1 *ṣuppān*

§ 4 d. Example of metric division

The front is 4 (ninda).	$s = 4$ (ninda).
The length, what shall it be long,	$u = ?$
so that it is 20 šar?	if, in addition, $A = u \cdot s = 20$ šar
If 5 is your cubit:	**If you count with ninda:**
The 4th-part, 15,	$1/s = 1/4 = ;15$

15 steps of 20 go, 5,	$1/s \cdot A = 15 \cdot 20 = 5$ (ninda)
a *ṣuppān*, it is long.	$u = 5$ (ninda) = 1 *ṣuppān*
If 1 is your cubit:	**If you count with cubits:**
The 48th-part, 1 15,	($s = 48$ cubits), $1/s = 1/48 = {;}01\ 15$
1 15 steps of 2 24 go, 3.	$1\ 15 \cdot 2\ 24 = 3$
3 steps of 20 go, 1.	$3 \cdot 20 = 1$ ($\cdot$ 60 cubits)
< a *ṣuppān*, it is long.>	<$u = 1$ ($\cdot$ 60 cubits) = 1 *ṣuppān*>

It is obvious that W 23291-x § 4 is another example of (the beginning of) a theme text with metric algebra problems, just like W 23291 § 1. The brief and idiomatic style of the text makes the literal translation quite hard to read, so that the explanation in the right column is indispensable.

Anyway, this is what is going on here: In § 4 a, a rule is first formulated for the computation of areas of *rectangles* in terms of the unit šar = 1 square-ninda. When lengths are expressed in terms of ninda, the rule is simply that $A = 1 \cdot u \cdot s$. However, when lengths are expressed in terms of cubits, the rule takes the form $A = {;}00\ 25 \cdot u \cdot s$, for the reason that

1 sq. cubit = 1 sq. (;05 ninda) = ; 00 25 sq. ninda = ;00 25 šar.

In § 4 a, a rule is formulated also for the computation of the "square side" of a given area. When lengths are expressed in terms of ninda, the rule is simply that the length of the square side is s = sqs. $(1 \cdot A)$, a length number such that sq. $s = A$. However, when lengths are counted in cubits, the rule is that the square side is s = sqs. $(2\ 24 \cdot A)$, for the reason that

1 šar = 1 sq. ninda = 1 sq. (12 cubits) = 2 24 sq. cubits.

In § 4 b-d, examples of the most basic metric algebra problems are worked through. The computations are quite simple although they are somewhat complicated by repeated references to the OB length unit

1 *ṣuppān* = 5 ninda = 1 00 cubits.

A Seleucid pole-against-a-wall problem

The OB pole-against-a-wall problem in BM 85196 # 9 (see above, Fig. 1.12.6) has a counterpart in **BM 34568 # 12** (Høyrup, *LWS* (2002), 391 ff), an isolated exercise in a large mathematical recombination text from the Seleucid period in Mesopotamia (the last third of the 1st millennium BCE).

The question in this exercise can be rephrased as:

A reed of unknown length at first stands upright against a wall of the same height.
Then it starts sliding so that its upper end moves straight down 3 cubits.
At the same time, its lower end moves away from the wall 9 cubits.
What is the length of the reed, how far up the wall does the reed reach?

With the notations in Fig. 1.12.6 above, the question takes the form

$s = 3$ cubits, $b = 9$ cubits. $c = ?$, $a = ?$

The obvious way of solving this problem would be to proceed as follows:

sq. c – sq. a = sq. b = sq. 9, $c - a = s = 3$.

This is *a subtractive quadratic-linear system of equations of type B3b*. In BM 34568 # 12, the solution to this problem is given in the form

c = (sq. b + sq. s)/2 · 1/s = (sq. 9 + sq. 3)/2 · 1/3 = 45 · 1/3 = 15,
sq. a = sq. c – sq. b = sq. 15 – sq. 9 = 2 24, a = sqs. 2 24 = 12.

One way in which the solution can have been obtained in this form is illustrated in Fig. 1.13.8 below. The problem is then interpreted as a problem for a "semichord" in a semicircle. If the semichord, of length b, divides the diameter of the semicircle in two parts of lengths u and s, then

$(u + s)/2 = c$ (the radius), $(u - s)/2 = a$, and consequently
$u \cdot s$ = sq. $\{(u + s)/2)$ – sq. $\{(u - s)/2)$ = sq. b.

(*Cf.* the proof of *El.* II.14 and Fig. 1.7.2, right.) It follows that

$c \cdot s = (u + s)/2 \cdot s = (s \cdot u$ + sq. $s)/2$ = (sq. b + sq. s)/2, so that c = (sq. b + sq. s)/2 · 1/s.

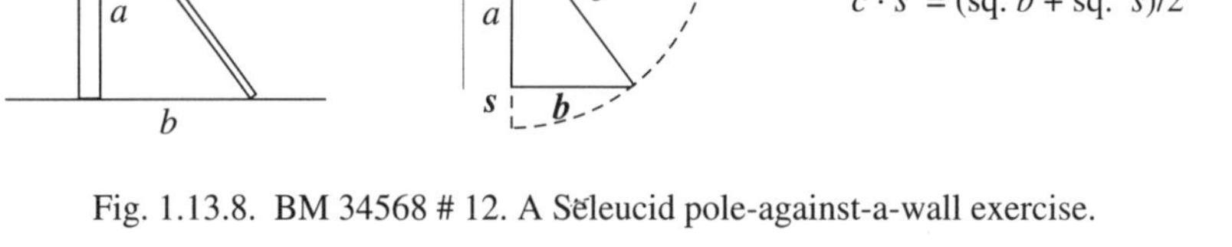

Fig. 1.13.8. BM 34568 # 12. A Seleucid pole-against-a-wall exercise.

Note that the same geometric configuration, with various permutations of the given parameters, is behind the three pole-against-a-wall problems in BM 34568 # 12 (b and s given) and BM 85196 # 9 (c and s, or c and b given), as well as behind the proposed forerunners *El.* II.11* and *El.* II 14* to *El.* II.11 and *El.* II 14 (b and a, or c and b given; see Fig. 1.7.2).

It is interesting that further examples of the pole-against-a-wall problem appear in § 8 g-h of *P.Cairo J. E. 89127-30, 89137-43*, an Egyptian demotic mathematical text from the third century BCE (Parker, *DMP* (1972) ## 30-31; Friberg, *UL* (2005), Sec. 3.1 b). The solution method is the same in the demotic text as in BM 34568 # 12.

The problem type reappears in § 1of *Liber Mahameleth*, a Latin manuscript based on Islamic sources, compiled in Spain in the 12th century by a Christian traveller (Sesiano, *Cent.* 30 (1987)). Here is the first part of the text of the third exercise in § 1 (my translation):

> Another example. If a ladder, I don't know how long, standing against a wall of the same height and moved 6 cubits from the foot of the wall descends from the top of the wall two cubits, then how much is the length?
> You do it like this: Multiply 6 with itself, and 2 with itself, and subtract the smaller product from the larger, and 32 will remain. Of which the half, which is 16, divide by 2 cubits, and 8 will come out. To which add 2 cubits, and it makes 10, and so much is the height of the ladder or the wall.

In this exercise, $b = 6$ cubits, $s = 2$ cubits, and the solution is given as

$$c = (\text{sq. } b - \text{sq. } s)/\ 2s + s = (\text{sq. } 6 - \text{sq. } 2)/\ 2 \cdot 2 + 2 = 8 + 2 = 10.$$

This is clearly *not* the same solution method as the one in BM 34568 # 12. (*Cf.* also Tropfke, *GE 4* (1940), Sec. 4.2.3.1.1.)

Seleucid parallels to *El.* II.14* (systems of equations of type B1a)

AO 6484 is another large Seleucid mathematical recombination text of mixed content. In that text, § 7 is a series of four "igi-igi.bi problems" (Friberg, *RC* (2007), Appendix 7). The most interesting of those problems is § 7 a, because of the extreme values of the given data in that exercise.

AO 6484 § 7 a, literal translation	explanation
igi and igi.bi 2 00 00 33 20.	igi + igi.bi = *p* = 2 00 00 33 20
igi and igi.bi *how much …*	igi and igi.bi = ?
· 30 go, then 1 00 00 16 40.	*p*/2 = 1 00 00 16 40
1 00 00 16 40 · *1 00 00 16 40 go,*	sq. *p*/2 = sq. 1 00 00 16 40
then 1 00 00 33 20 04 57 46 40.	= 1 00 00 33 20 04 57 46 40
1 from inside (it) remove,	sq. *p*/2 – 1
then remains 33 <20> 04 37 46 40.	= 33 <20> 04 37 46 40
What · *what may I go,*	sqs. 33 <20> 04 37 46 40

then 33 <20> 04 37 46 40?	= ?
44 43 20 · 44 43 20 go,	sq. 44 43 20
then 33 <20> 04 37 46 *40.*	= 33 <20> 04 37 46 40
44 43 20 to 1 00 00 16 40 repeat,	1 00 00 16 40 + 44 43 20
then1 00 45, the igi.	= 1 00 45 = igi
44 04 43 20 from 1 00 00 16 40 rem*ove,*	1 00 00 16 40 – 44 43 20
then 59 15 33 20, the igi.bi.	= 59 15 33 20 = igi.bi

In this exercise, the terms igi and igi.bi denote a "reciprocal pair" of sexagesimal numbers, by which is meant any pair of (positive) sexagesimal numbers such that their product is equal to '1' (any power of 60). Therefore, the question in the exercise can be interpreted as *a rectangular-linear system of equations of type B1a* of the following *special form*:

igi · igi.bi = 1,   igi + igi.bi = 2 00 00 33 20.

Presumably the Seleucid mathematicians used some kind of geometric model to help them find a solution procedure, just like their OB predecessors had done. Two candidates for such a model are shown in Fig. 1.13.9 below. The one to the left is the "square-difference model" related to *El.* II.5 (Fig. 1.4.2, left). The one to the right is the "semi-chord model", related to *El.* II.14* (Fig. 1.7.2, right).

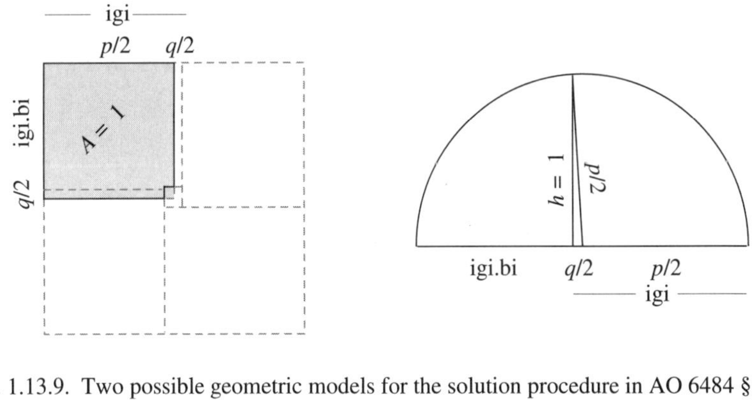

Fig. 1.13.9. Two possible geometric models for the solution procedure in AO 6484 § 7 a.

Assume that the given number 2 00 00 33 20 in AO 6484 § 7 a (written with a special sign for internal zeros) can be interpreted as, for instance, 2;00 00 33 20 (2 plus a very small fractional part). Then the successive steps of the solution procedure in the text can be explained as follows:

$p/2$ = 2;00 33 20 / 2 = 1;00 16 40

sq. $p/2$ = 1;00 00 33 20 04 57 46 40
sq. $q/2$ = sq. $p/2 - 1$ = ;00 00 33 20 04 37 46 40
$q/2$ = ;00 44 43 20
igi = $p/2 + q/2$ = 1;00 16 40 + ;00 44 43 20 = 1;00 45 (= 81/80)
igi.bi = $p/2 - q/2$ = 1;00 16 40 – ;00 44 43 20 =;00 59 15 33 20 (= 80/81)

The curious choice of data is best explained by the semi-chord model. Apparently, the purpose of the exercise was to show that an *extremely thin right triangle* can be constructed by letting the sides of the triangle be

$c, b, a = p/2, 1, q/2$ = (igi + igi.bi)/2, 1, (igi – igi.bi)/2,

where igi and igi.bi are the sides of a *nearly square rectangle* with the area 1. (*Cf.* Friberg, *RC* (2007), Appendix 8, Fig. A8.5.))

1.14. Old Akkadian Square Expansion and Square Contraction Rules

It is known (see Friberg, *CDLJ* (2005:2), Figs. 8 and 10) that already mathematicians in the Old Akkadian period in Mesopotamia (ca. 2340-2200 BCE) may have been familiar with the "square expansion rule"

sq. $(u + s)$ = sq. u + sq. $s + 2\, u \cdot s$,

and with the closely related "square contraction rule"

sq. $(u - s)$ = sq. u + sq. $s - 2\, u \cdot s$.

These rules are clearly the *Old Akkadian forerunners to El. II.4 and II.7*. (Compare Fig. 1.14.1 below with Fig. 1.3.2 above.)

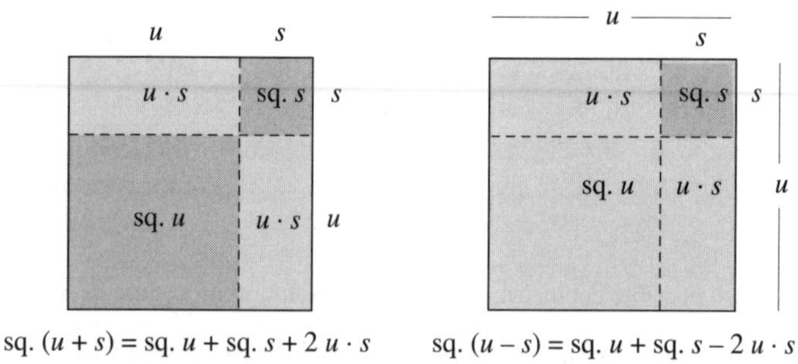

Fig. 1.14.1. The Old Akkadian square expansion and square contraction rules.

Thus, for instance, in the Old Akkadian mathematical exercise ***DPA* 36** (Friberg, *op. cit.*, Fig. 7), the area is given of a square with the side

11 ninda 1 1/2 1/4 seed-cubit = 10 ninda + 1/8 · 10 ninda + 1/4 seed-cubit.

(1 n. = 6 seed-cubits.) The area was probably computed by use of a repeated application of the square expansion rule, as follows:

1. sq. (10 n. + 1/8 · 10 n.) = sq. 10 n. + 2 · 1/8 · sq. 10 n. + sq. 1/8 · sq. 10 n.
2. sq. (10 n. + 1/8 · 10 n. + 1/4 s.c.)
   = sq. (10 n. + 1/8 · 10 n.) + 2 · (10 n. + 1/8 · 10 n.) · 1/4 s.c. + sq. 1/4 s.c.

For the details of the computation, which is quite complicated because of the involvement of various Old Akkadian units for length and area measure, the reader is referred to Friberg, *op. cit.*

Similarly, in the Old Akkadian mathematical exercise ***DPA* 37**, for instance (Fig. 5.3.2 below), the area is given of a square with the side

1 šár ninda 5 géš ninda – 1 seed-cubit (1 šár = 60 · 60, 1 géš = 60).

The area was probably computed by use of an application of the square expansion rule, followed by an application of the square contraction rule. For the complicated details of the computation, see Friberg, *op. cit.*

It is known through a number of examples that the mentioned rules were applied in various situations also by OB mathematicians.

1.15. The Long History of Metric Algebra in Mesopotamia

The oldest known examples of metric algebra are applications of a "field expansion procedure" in proto-cuneiform texts from the end of the 4th millennium BCE (Friberg, *AfO* 44/45 (1997/98); *UL* (2005), Fig. 2.1.15.) The aim of the field expansion procedure seems to have been to find rectangles of given area with the lengths of the sides of the rectangle in a given ratio.

Next in time, in the small corpus of known mathematical texts from the Old Akkadian (Sargonic) period, c. 2340-2200 BCE, there are several known, quite elaborate examples of *metric squaring* (such as the ones mentioned in Sec. 1.14 above) and *metric division*, possibly also an even more elaborate example of the *metric computation of a side of a square with given area.* Moreover, although the known examples of Old Akkadian metric squaring and metric division problems are written only one or two at a time on small clay tablets, they appear to have been excerpted from systematically arranged theme texts. (*Cf.* Friberg, *CDLJ* (2005:2).)

In the large corpus of OB mathematical texts, metric algebra is, as is well known, one of the most popular subjects. The extensive discussion in

Secs. 1.10-1.12 above shows that there are several known examples of *well organized OB theme texts with metric algebra problems,* in particular metric algebra problems for one, two, or several squares.

In Sec. 1.13 it was shown that examples exist also of *well organized Late Babylonian/Seleucid theme texts with metric algebra problems,* resembling such OB theme texts. Several features suggest that those Late Babylonian/Seleucid texts were written in direct imitation of OB models.

Thus, for instance, the problem for concentric circles in W 23291-x § 2 is indistinguishable, at least in translation, from an OB mathematical text. It measures length in ninda and surface content in square ninda (šar), although in Late Babylonian cuneiform texts lengths are normally measured in cubits or reeds (= 7 cubits) and surface content in either seed measure or "reed measure" (the length of a rectangle with the given surface content and with one side equal to precisely 1 reed).

Also the fragment of a theme text in W 23291 § 4 measures surface content in šar, expressly defined as 1 square ninda. It shows its dependence on an OB archetype by having a separate ninda section, and in the cubit section the cubit is 1/12 of a ninda, which implies that the cubit is 1/6 of a reed, as in OB texts, not 1/7.

The problem for concentric circles in W 23291 § 4 g is more removed from its OB archetype by measuring surface content in terms of seed measure, but it still measures lengths in ninda. The metric algebra problems in W 23291 § 1 b-f also measure surface content in terms of seed measure and have separate ninda and cubit sections, with the cubit equal to 1/12 ninda in the cubit sections.

Summing up, it is now possible to conclude that metric algebra problems were studied systematically in Mesopotamian scribe schools during a time span of at least 2000 years, from the Old Akkadian to the Late Babylonian period. The investigation has also shown that, at least in some respects, *Late Babylonian mathematics was directly influenced by OB mathematics, actually in the same way that OB mathematics must have been inspired by Old Akkadian mathematics.* This is not an unexpected conclusion, and it is supported but other facts not mentioned here. Still, it is remarkable, since the terminology used in Late Babylonian mathematical texts is in many ways different from the terminology used in corresponding OB mathematical texts.

Thus, when *Elements* II, or more likely a lost Greek forerunner to *Elements* II was written in imitation of some oriental archetype, it was only the last link in an extremely long chain of theme texts with metric algebra problems. The heated debate over the question whether some of the propositions in *Elements* II were Greek *geometric reformulations* of Babylonian *algebra* can now be laid to rest. In reality, *Elements* II appears instead to have been a direct *translation* into non-metric and non-numerical "geometric algebra" of key results from Babylonian *metric algebra.* It is noteworthy that, in spite of this translation, Greek geometric algebra still relied on *the same geometric models* as Babylonian metric algebra.

Chapter 2

El. I.47 and the Old Babylonian Diagonal Rule

2.1. Euclid's Proof of *El.* I.47

The proposition ***El.* I.47** begins with the following statement:

In right-angled triangles the square on the side subtending the right angle is equal to the squares on the sides containing the right angle.

The well known diagram accompanying the proposition (see, for instance, Heath, *ETBE 1* (1956)) is reproduced below:

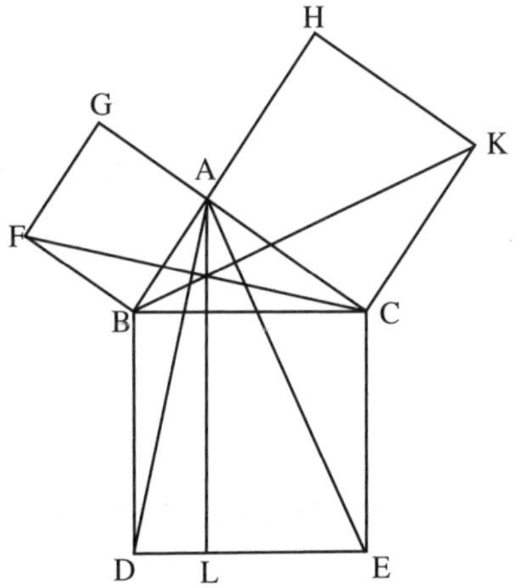

Fig. 2.1.1. The diagram accompanying *El.* I.47.

In spite of the complicated diagram, Euclid's proof is relatively simple. It begins with a careful construction of the diagram. The rest of the proof can, essentially, be divided into the following steps:

1. The triangle ABD is equal to the triangle FBC (*El.* 1.4)
2. The rectangle BL is equal to twice the triangle ABD (*El.* I.41)
3. The square GB is equal to twice the triangle FBC (*El.* I.41)

4. Therefore, the rectangle BL is equal to the square GB
5. Similarly, the rectangle CL is equal to the square HC
6. Therefore, the whole square BDEC is equal to the two squares GB, HC

Steps 1-3 of Euclid's proof are separately illustrated below, in a set of unlettered diagrams

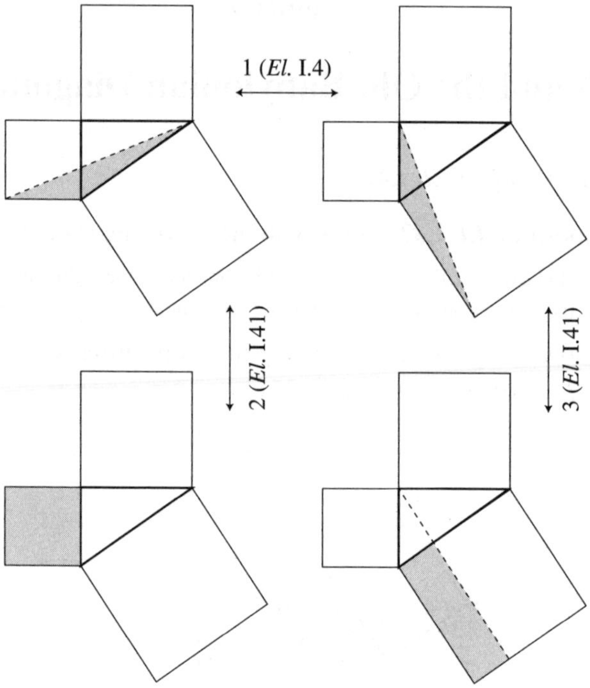

Fig. 2.1.2. Steps 1-3 of Euclid's proof of *El.* I.47

2.2. Pappus' Proof of a Generalization of *El.* I.47

An alternative, but closely related proof of *El.* I.47 was given by Pappus in ***Collections*** **IV.1** (see, for instance, Heath, *ETBE 1* (1956), 366). Pappus showed that his proof could be used also for the proof of a generalized proposition, where the right-angled triangle is replaced by an arbitrary triangle, and where the given squares on two of the sides of the triangle are replaced by arbitrary given parallelograms on the two sides. Undoubtedly, the new proof is more interesting than the generalization of *El.* I.47.

Pappus' new proof is illustrated by the following diagram:

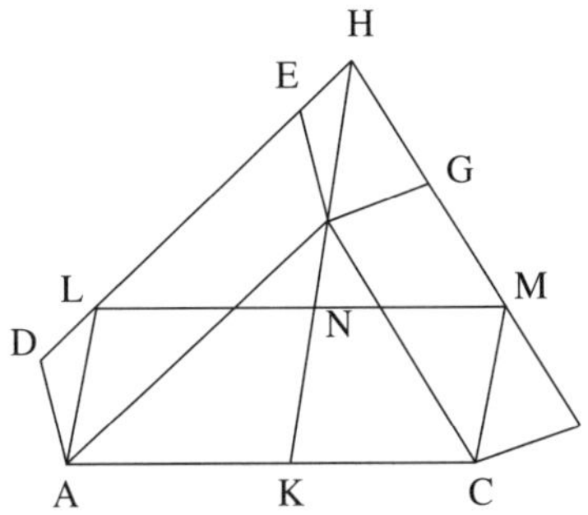

Fig. 2.2.1. The diagram illustrating Pappus' proof of his generalization of *El.* I.47.

In Fig. 2.2.2 below, the basic idea in Pappus' proof is illustrated by a sequence of diagrams in the most interesting case, that of given squares on the length and the front of a right triangle. The right triangle is interpreted here as one half of a rectangle.

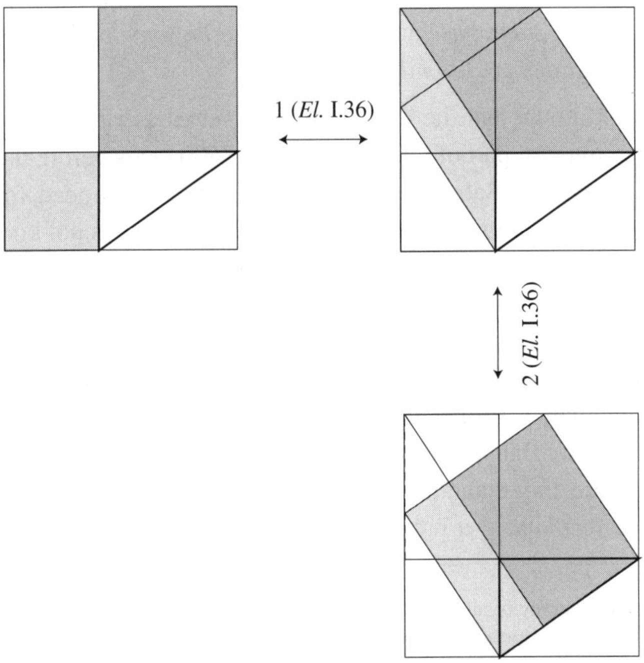

Fig. 2.2.2. An illustration of the basic idea in Pappus's proof of *El.* I.47.

2.3. The Original Discovery of the OB Diagonal Rule for Rectangles

Euclid's own proof of *El.* I.47 is based on two propositions from *Elements* I, namely *El.* I.4 and *El.* I.41.

***El.* I.4** is a *congruence theorem* stating that if *two sides and the angle contained by those two sides* in one triangle are equal to two sides and the angle contained by those two sides in another triangle, then also the *remaining side* and the *remaining angles* in the first triangle are equal to the remaining side and the remaining two angles in the other triangle, and the two triangles are '*equal*', presumably in the sense that they have the same area. Euclid's proof of *El.* I.4 is far from a wonder of lucidity.

***El.* I.41** is a *transformation theorem* stating that if a parallelogram and a triangle have *the same base* and stay *between the same parallels*, then the parallelogram is '*equal' to twice the triangle*.

Pappus' proof of *El.* I.47, which is simpler than Euclid's proof, is based on only one proposition from *Elements* I, namely *El.* I.36.

***El.* I.36** is another *transformation theorem*, stating that if two parallelograms have *the same base* and stay *between the same parallels*, then the two parallelograms are '*equal*'.

It is well known that the "diagonal rule" stated as a proposition in *El.* I.47 was an integral part of OB mathematics 1500 years before the time of Euclid.[14] However, Babylonian mathematicians felt no need for formal statements of theorems and formal proofs. Therefore, it is not known how they would have formulated a proof of their diagonal rule.[15] Nevertheless, it is clear that they could never have contemplated proofs like the ones formulated by Euclid and Pappus, illustrated in Figs. 2.1-2 and 2.2.1-2 above, since concepts such as *angles* and *parallelograms* were unknown in Babylonian mathematics. Propositions such as *El.* I.4, *El.* I.36, and *El.* I.41 cannot have had any Babylonian counterparts.

More interesting than the question of how Babylonian mathematicians could have formulated a proof of the diagonal rule, if they had had the

---

14. A attempted survey of all known Old or Late Babylonian applications of the diagonal rule can be found in Friberg, *RC* (2007), Appendix 8, Sec. A8 f.

15. In an attempted search for the original proof of the rule there would be no shortage of candidates. Thus, in Loomis, *The Pythagorean Proposition* (1968), there are listed 109 "algebraic" proofs of the proposition, 225 "geometric" proofs, *etc.*

inclination to do so, is the question how they actually can have *discovered* the rule in the first place. A step towards a possible answer to this question is the observation that Pappus' proofs of *El.* I.47, in the slightly modified version of it shown in Fig. 2.2.2, is closely related to the proof "by inspection" shown in Fig. 2.3.1 below:

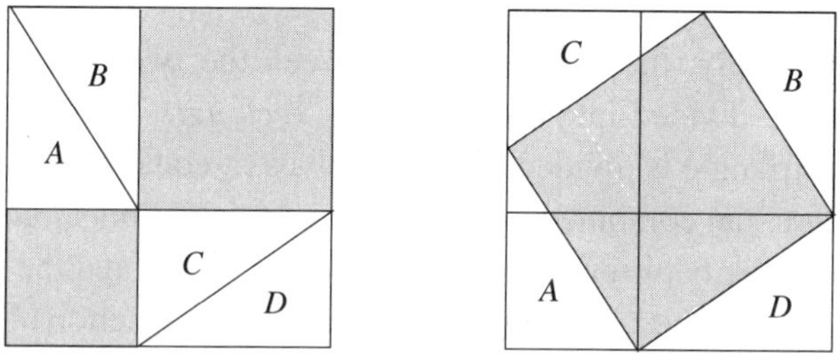

Fig. 2.3.1. A proof by inspection closely related to Pappus' proof.

The diagram in Fig. 2.3.1, right, may, in its turn, be closely related to the actual *accidental* discovery of the diagonal rule, which may have taken place in the way described below:

OB mathematicians were familiar with the idea of concentric (and parallel) squares, and called a square band bounded by two concentric squares a 'field between'. Various mathematical exercises for two or three concentric squares are listed in the OB catalog text Bruins and Rutten (1961) *TMS* 5 (Sec. 1.11 above). In ***TMS* 5 § 9 b-c** it is silently assumed that *one square is halfway between two other squares* in the sense that *the distance from the outer square to the middle square is equal to the distance from the middle square to the inner square*.

There is, however, another way in which a middle square can be said to be halfway between two given concentric (and parallel) squares. That is when *the area between the outer square and the middle square is equal to the area between the middle square and the inner square*. Suppose that an OB mathematician tried to figure out how to construct a square halfway, in this sense, between two given squares. How could he do it? There are two answers to this question:

1. The *square band* between the two given concentric squares can be divided by the diagonals of the outer square into *a ring of four trapezoids*. (see Fig. 2.3.2 below, left), The problem is therefore reduced to the prob-

lem of finding a transversal parallel to the base of a given trapezoid and dividing the trapezoid into two parts of equal area. This "trapezoid bisection problem" and its solution were both well known in OB mathematics. It appears to have been known even by Old Akkadian mathematicians, five hundred years before the time of the OB mathematicians. (See Sec. 11.3 a below.)

2. Alternatively, the square band between the two given concentric squares can be divided into *a ring of four rectangles* (Fig. 2.3.2 below, right). Each triangle is divided into two halves of equal area by its diagonal. Therefore, the combined area of *the inner square plus four half rectangles* is halfway between the area of the outer square and the area of the inner square. Moreover, it is naively obvious from inspection of Fig. 2.3.2, right, that the combination of the inner square plus the four half rectangles is an *obliquely placed square* touching the outer square at four points. Actually, it is *the square on the diagonal* of any one of the four rectangles.

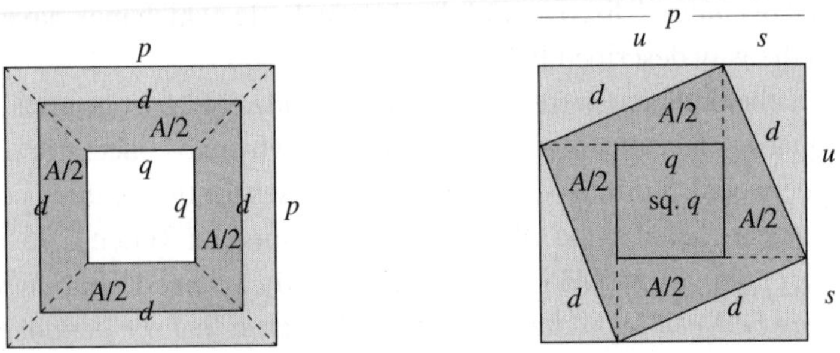

$A = (\text{sq. } p - \text{sq. } q)/4$
$\text{sq. } d = (\text{sq. } p + \text{sq. } q)/2$

$\text{sq. } d = (\text{sq. } p + \text{sq. } q)/2$
$= \{\text{sq. } (u + s) + \text{sq. } (u - s)\}/2$
$= \text{sq. } u + \text{sq. } s$

Fig. 2.3.2. A square halfway in area between two given concentric squares.

Consequently, the area of the square on the diagonal d, expressed *in terms of the pair of square sides p and q* is

$$\text{sq. } d = (\text{sq. } p + \text{sq. } q)/2.$$

On the other hand, expressed *in terms of the "dual pair" of rectangle sides u and s*, the area of the square on the diagonal is

$$\text{sq. } d = \text{sq. } q + 4 \cdot A/2 = \{\text{sq. } (u + s) + \text{sq. } (u - s)\}/2 = \text{sq. } u + \text{sq. } s.$$

Therefore, the Babylonian diagonal rule may very well have been accidentally discovered by someone who was actually more interested in finding a square halfway in area between two given concentric squares! If this was really the way in which the diagonal rule was discovered, it is also *the first proof* the diagonal rule. Note that this putative first proof of the diagonal rule is in addition an obvious candidate to being *a forerunner of the proofs ascribed to Pappus and Euclid* (Figs. 2.1.2 and 2.2.2)!

2.4. Chains of Triangles, Trapezoids, or Rectangles

There is no direct evidence that OB mathematicians found and proved (to their own satisfaction) the diagonal rule in the way suggested above. There is, however, plenty of indirect evidence. Take, for instance the following entry in the OB mathematical "table of constants" Bruins and Rutten, ***TMS* 3** (= **BR**):

57 36 igi.gub *šà* šár	57 36, the constant of the šár	BR 30

As first shown by Vaiman, *VDI* 15 (1961), the entry may refer to the area of a geometric figure in the form of a ring of right triangles, vaguely resembling the cuneiform number sign šár = 60 · 60 (Fig. 2.4.1, right). In what probably was a standard example, such a figure could be composed of four right triangles with the sides 1 00, 48, 36 = 12 · (5, 4, 3). It is shown in Fig. 2.4.1, left, how the area of the šár-figure in this standard example can be computed either as the sum of the areas of the four right triangles, or as the difference of the areas of an outer square of side 1 00 and an inner square of side 12. In either case, the area is found to be 57 36.

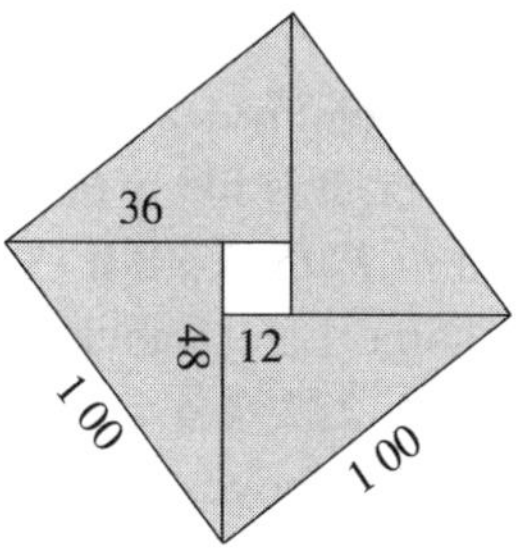

4 · 1/2 · 36 · 48
= 4 · 14 24
= 57 36 (sq. ninda)
or
sq. 1 00 – sq. 12
= 1 00 00 – 2 24
= 57 36

Fig. 2.4.1. Left: The šár-figure. Right: The šár-sign.

Parenthetically it may be remarked here that the OB mathematicians, who often used the powerful method of *a systematic variation of a basic idea*, had found an interesting variation also of the idea of a ring of four right triangles as in Fig. 2.4.1, right, or of four rectangles, as in Fig. 2.3.2, right. The OB round clay tablet **MS 2192** (Friberg, *RC* (2007), Sec. 8.2 a), contains a diagram of a triangular band bounded by two equilateral triangles, and divided into a ring of three trapezoids. Apparently, this is an assignment, with the student's task being to compute the area of the triangular band. A first step towards the solution of the problem has been taken with the notation '35' near the edge of the clay tablet. It can be interpreted as an indication that the area of the triangular band is 35 times as large as the area of the inner triangle. Indeed, since the side of the outer triangle is 6 times larger than the side of the inner triangle, it follows that the area of the triangular band is 35 = sq. 6 – 1 times larger than the area of the inner triangle.

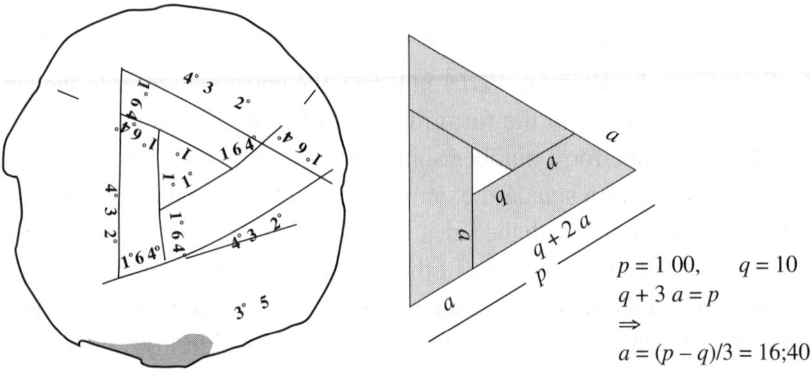

Fig. 2.4.2. MS 2192. An equilateral triangular band divided into three trapezoids.

There are no explicit solution procedures for the problems dealing with concentric squares stated in §§ 7-9 of *TMS* 5. However, it is likely that an OB mathematician would have solved the problem in ***TMS* 5 § 8 b** (Sec. 1.11 above), for instance, in the same way that some Late Babylonian mathematician solved a corresponding problem in **W 23291 § 1 f** (Sec. 1.13 a). Compare Fig. 2.4.3 below, left, with Fig. 1.13.5 above.

The problem stated without solution in ***TMS* 5 § 7 f** (Sec. 1.11) is solved explicitly in **BM 13901 § 2 a** (Sec. 1.12 above), apparently by use of a method illustrated by the diagram in Fig. 2.4.3, right. The first step of that method is based on the observation that the inner square plus four half rectangles is equal to the square on the diagonal of one of the rectangles, and that the square on the diagonal plus four more half rectangles is equal to the outer square. The second step of the solution procedure in BM 13901 § 2 a makes use of the diagonal rule.

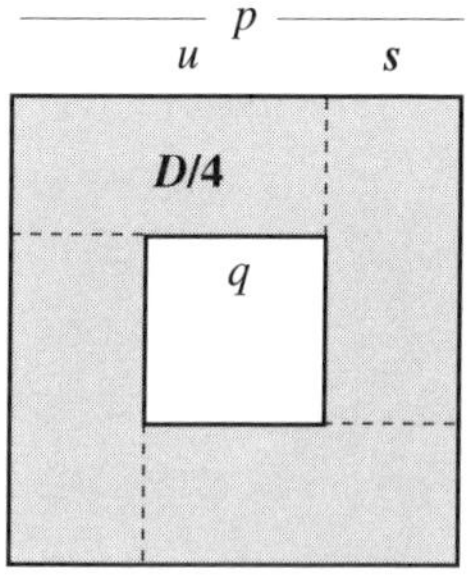

sq. p – sq. q = $\boldsymbol{D}$
$(p - q)/2 = \boldsymbol{s}$

TMS 5 § 8 b & W 23291 §1 f

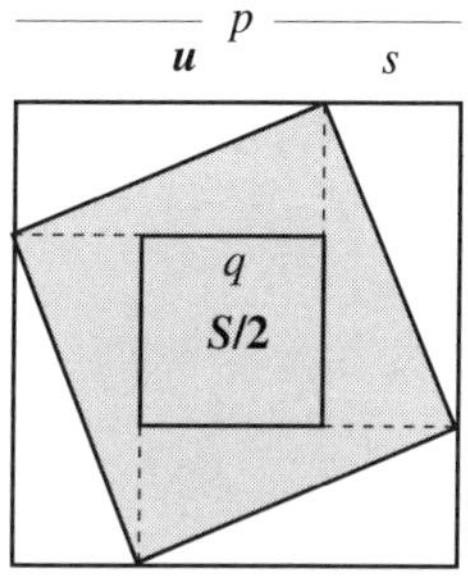

(sq. p + sq. q)/2 = sq. $\boldsymbol{d}$ = sq. u + sq. s
$(p + q)/2 = \boldsymbol{u}$

TMS 5 § 7 f & BM 13901 § 2 a

Fig. 2.4.3. Quadratic-linear systems of equations and square bands as rings of rectangles.

Thus, the square band divided into a ring of four rectangles, which was assumed above to be a geometric model behind the discovery and first proof of the diagonal rule, appears to have played a role also in other connections in OB metric algebra.

Chapter 3

Lemma *El.* X.28/29 1a, Plimpton 322, and Babylonian igi-igi.bi Problems

3.1. Greek Generating Rules for Diagonal Triples of Numbers

Euclid's Generating Rule in Lemma *El.* X.28/29 1a

Lemma *El.* X.28/29 1a poses the following construction problem:

To find two square numbers such that their sum is also a square.

Although what is asked for here is two "numbers" (positive integers), the ensuing construction is accompanied by a *geometric* diagram:

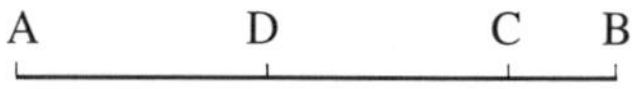

Here AB and BC represent two numbers, assumed to be *both odd* or *both even*. In addition, it is assumed that AB and BC are either two *square numbers*, or, more generally, *similar plane numbers*. The latter assumption means, essentially, that here is some "plane number" $t = h \cdot k$ and some numbers m and n such that

$$\text{AB} = m\,h \cdot m\,k, \quad \text{BC} = n\,h \cdot n\,k.$$

Consequently,

$$\text{AB} = t \cdot \text{sq.}\ m, \quad \text{BC} = t \cdot \text{sq.}\ n.$$

(It is not clear to me why Euclid prefers to talk about similar plane numbers instead of numbers proportional to a pair of square numbers.)

Euclid lets AC be bisected at D. In view of the mentioned assumptions, CD is then a *number*, and the product of AB, BC is a *square number*. The latter fact is proved in *El.* IX.1. Moreover, in view of *El.* II.6,

The product of AB, BC together with the square on CD is equal to the square on BD.

Thus, the product of AB and BC, and the square on CD, are, as requested, two square numbers such that their sum is also a square.

In a rather implicit way, lemma *El.* X.28/29 1a is a general generating rule for an infinite number of "diagonal triples" of numbers (positive integers), corresponding to the three sides of equally many right(-angled) triangles, or to the two sides and the diagonal of equally many rectangles.

The Generating Rules Attributed to Pythagoras and Plato

In *HGM II* (1981 (1921)), 79-82, Heath gives a brief account of two related generating rules, one attributed to Pythagoras, the other to Plato, both described in Proclus' commentary to Euclid's *Elements* I (Friedlein/ Proclus, *In Primum Euclidis Elementorum Commentarii* (1873)).

According to Heath, *the generating rule ascribed to Pythagoras* "amounts to the statement that"

$$m^2 + (1/2\,(m^2 - 1))^2 = (1/2\,(m^2 + 1))^2, \quad \text{where } m \text{ is any odd number.}$$

Heath suggests that Pythagoras found this construction rule by observing that $2\,a + 1$ is the "*gnomon* of dots" put around a^2 to make $(a + 1)^2$. Therefore, if also $2\,a + 1$ is a square number, say $2\,a + 1 = m^2$, it follows that

$$a = 1/2\,(m^2 - 1), \quad \text{and} \quad a + 1 = 1/2\,(m^2 + 1),$$

which gives the generating rule in this case.

Similarly, according to Heath, *the generating rule ascribed to Plato* amounts to the statement that

$$(2\,m)^2 + (m^2 - 1)^2 = (m^2 + 1)^2, \quad \text{where } m \text{ is an arbitrary number.}$$

Heath suggests that Plato found this alternative construction by observing that $4\,a$ is the *gnomon* of dots put around $(a - 1)^2$ to make $(a + 1)^2$. Therefore, if also $4\,a$ is a square number, say $4\,a = (2\,m)^2$, it follows that

$$a = m^2, \quad \text{so that } a + 1 = m^2 + 1, \quad \text{and} \quad a - 1 = m^2 - 1,$$

which gives the generating rule in this second case.

Heath further observes that both these generating rules are special cases of the generating rule given by Euclid in lemma *El.* X.28/29 1a,which, essentially, amounts to the statement that

$$(t\,m\,n)^2 + ((t\,m^2 - t\,n^2)/2)^2 = ((t\,m^2 + t\,n^2)/2)^2,$$
when $t\,m^2$, $t\,n^2$ are both odd or both even.

Metric Algebra Derivations of the Greek Generating Rules

It is instructive to investigate how the generating rules mentioned above can be derived by use of *geometric diagrams* like the ones in *El.* II rather than by use of squares and *gnomons* of dots or pebbles.

Since *El.* II.6 was used by Euclid for his construction in lemma *El.* X.28/29 1a, the obvious choice of such a diagram is the one in Fig. 1.4.1, right, or its metric algebra counterpart in Fig. 1.4.2, right.

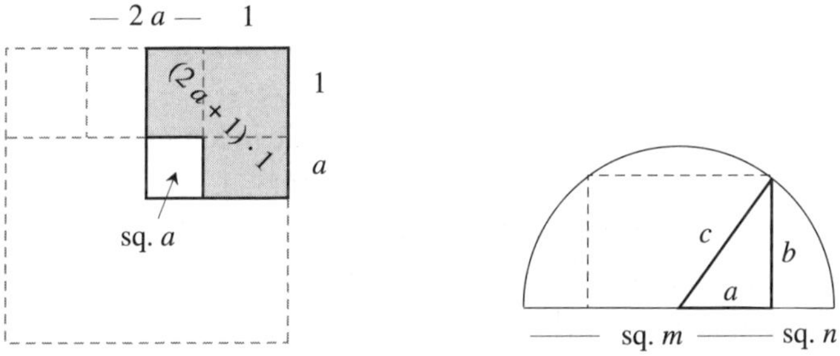

sq. $(a + 1)$ – sq. $a = (2\,a + 1) \cdot 1$,
$(2\,a + 1) \cdot 1 =$ sq. m
$\Rightarrow$
$c, b, a =$
(sq. $m + 1)/2$, m, (sq. $m - 1)/2$

$c + a =$ sq. m, $c - a =$ sq. n,
sq. $b =$ sq. $m \cdot$ sq. n
$\Rightarrow$
$c, b, a =$
(sq. m + sq. $n)/2$, $m \cdot n$, (sq. m – sq. $n)/2$

Fig. 3.1.1. Geometric derivations of the generating rules of Pythagoras and Euclid

The diagram in Fig. 3.1.1, left, is related to the diagram used for the proof of *El.* II.6. It shows that the square difference sq. $(a + 1)$ – sq. a is equal to a square corner (a *gnomon*) of area $(2\,a + 1) \cdot 1$. Just as in Heath's proposed explanation of the generating rule ascribed to Pythagoras, if the area of the square corner is equal to the area of a square, say sq. m, then

$2\,a + 1 =$ sq. m, so that $a =$ (sq. $m - 1)/2$ and $a + 1 =$ (sq. $m + 1)/2$.

Note that this is an *analytic* argument: The diagram can be drawn only if it is assumed that the numbers a and $a + 1$ are already known.

The alternative diagram in Fig. 3.1.1, right, is related to the diagram used for the constructions in *El.* II.14 (Fig. 1.7.1, right) and *El.* II.14* (Fig. 1.7.2, right). The diagram shows that the sides c, b, a of a right triangle can be explicitly constructed as follows, by use of a completely *synthetic* argument: Choose m and n as arbitrary numbers, *both odd or both even*, and draw a semicircle with the diameter sq. m + sq. n. From the point where the diameter is divided in two parts of lengths sq. m and sq. n, erect a perpendicular. The radius of the semicircle ending at the point where the perpendicular cuts the circle is then the diagonal (the *hypotenuse*) of a right triangle with the sides

$$c, b, a = (\text{sq. } m + \text{sq. } n)/2,\ m \cdot n,\ (\text{sq. } m - \text{sq. } n)/2.$$

Indeed, c is half the diameter, $a = c$ – sq. n, and (*cf.* the proof of *El.* II.14)

$$\text{sq. } b = \text{sq. } m \cdot \text{sq. } n = \text{sq. } m \cdot n, \quad \text{so that} \quad b = m \cdot n.$$

In the special case when $n = 1$, the construction in Fig. 3.1.1, right, is an alternative to the construction in Fig. 3.1.1, left.

Euclid's slightly more general generating rule in lemma *El.* X.28/29 1a, corresponds to the case when in Fig. 3.1.1, right, sq. m and sq. n are multiplied by a number t. With $t = 2$, this generalization takes care of the case when one of m and n is odd, the other even.

It is interesting that Euclid never makes use of the generating rule in lemma *El.* X.28/29 1a. Actually, the reason for the insertion of the lemma after *El.* X.28 seems to be the brief remark at the end of the proof of that lemma, to the effect that if AB and BC (corresponding to the segments $c + a$ and $c - a$ of the diameter of the semicircle in Fig. 3.1.1, right) are *not* similar plane numbers, then the difference of the squares on BD and DC (corresponding to $(c + a) \cdot (c - a)$ = sq. c – sq. a in Fig. 3.1.1, right) is *not* a square number. It is this *negative* version of the lemma that is used in *El.* X.29-30.

3.2. Old Babylonian igi-igi.bi Problems

MS 3971 is an interesting OB mathematical recombination text from the ancient city Uruk (Friberg, *RC* (2007), Sec. 10.1). MS 3971 § 2 was discussed in Sec. 2.4 above. **MS 3971 § 3** is a series of five "igi-igi.bi problems", clearly related to *an OB generating rule for diagonal triples.* An interesting difference between the Greek and the OB generating rules is that Greek mathematicians were interested only in producing triples of "numbers" (integers), while OB mathematicians customarily worked with sexagesimal numbers in relative (floating) place value notation.

Here is the text of one of the problems in MS 3971 § 3:

MS 3971 § 3 e, literal translation	explanation
The 5th.	The 5th example.
1 12 the igi, 50 the igi.bi.	igi = 1 12, igi.bi = 50.
1 12 and 50 *heap*, 2 02.	igi + igi.bi = 2 02.
1/2 of 2 02 break, 1 01.	(igi + igi.bi)/2 = 1 01.
1 01 (make) butt (itself), 1 02 01.	sq. (igi + igi.bi)/2 = 1 02 01.
1 from 1 02 01 tear off, 2 01 it gives.	sq. (igi + igi.bi)/2 – 1 = 2 01
2 01 makes 11 equalsided.	= sq. 11.
11, the 5th front.	The 5th front (of a right triangle) = 11.

The igi-igi.bi problems in the OB text MS 3971 § 3 are clearly related to the igi-igi.bi problems in the Seleucid text AO 6484 § 7 (Sec. 1.13 above, Fig. 1.13.9). However, in MS 3971 § 3 the values of igi and igi.bi are given and the half-difference (igi – igi.bi)/2 is computed, while in AO 6484 § 7 the sum igi + igi.bi is given, and the values of igi and igi.bi are computed. In spite of the different goals for the computations in the two cases, the two possible geometric models remain the same (Fig. 3.2.1).

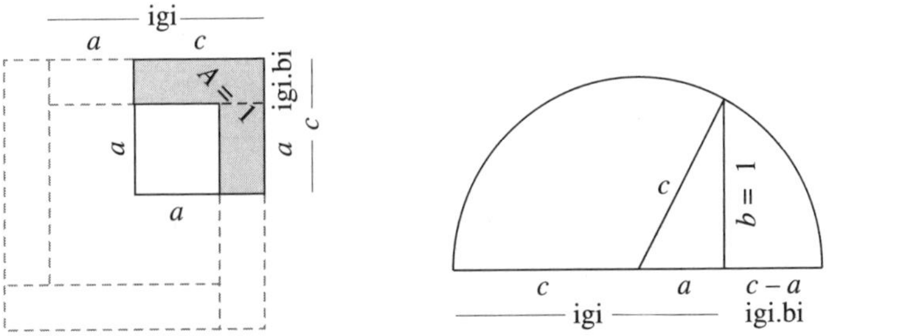

c, b, a = (igi + igi.bi)/2, 1, (igi – igi.bi)/2

$c + a$ = igi, $c - a$ = igi.bi, sq. c – sq. a = igi · igi.bi = 1

Fig. 3.2.1. Two possible geometric models for the solution procedure in MS 3971 § 3.

The five examples in MS 3971 § 3 demonstrate how every given pair igi, igi.bi of reciprocal sexagesimal numbers can be used for the construction of a right triangle, corresponding to the generating rule

c, b, a = (igi + igi.bi)/2, 1, (igi – igi.bi)/2.

Note that in these examples, the value of a is not computed directly by use of this generating rule, but via an application of the diagonal rule. However, in this way the need for a verification of the result is avoided.

The diagonal triples constructed by use of the OB generating rule displayed above are "normalized" in the sense that the middle term is 1.

3.3. Plimpton 322: A Table of Parameters for igi-igi.bi Problems

Plimpton 322 is the name of a famous OB mathematical table text from the ancient Mesopotamian city Larsa. It is a large fragment, probably the right half or two-thirds of a clay tablet of a very unusual format, much more broad than it is tall. There are four columns of numbers on the preserved part of the clay tablet, each with its own heading.

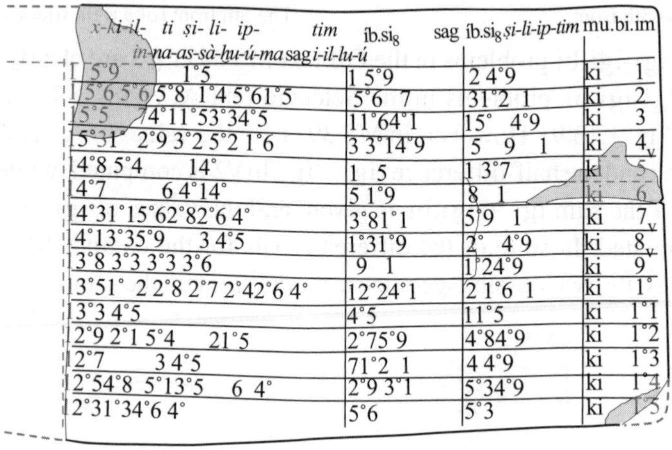

Fig. 3.3.1. Plimpton 322. A large fragment of an Old Babylonian mathematical table text.

The meaning of the headings is far from obvious. Nevertheless they can, at least tentatively, be translated as follows:

*The square of the hol*der for the diagonal (from) which 1 is subtracted,
then <the square of the holder for > the front comes up.
The square side of <the square of the holder for > the front.
The square side of <the square of the holder for > the diagonal.
Its line number.

A detailed analysis and explanation of the text can be found in Friberg, *RC* (2007), Appendix 7. There it is shown that the whole table text is a systematically arranged list of numerical parameters for fifteen igi-igi.bi problems of the same kind as either the five igi-igi.bi problems in MS 3971 § 3 or the four related problems in AO 6484 § 7 (Sec. 1.13 above).

As will be explained below, the mysterious term "holder" can be interpreted as a name for an intermediate result in the solution of a rectangular-linear system of equations. Thus, what the heading above the first preserved column tries to say, in a rather awkward way, is that the numbers in that column are the values of sq. (igi + igi.bi)/2 = sq. *c*, where *c* is the diagonal, and that if 1 is subtracted from sq. (igi + igi.bi)/2, then the result is sq. (igi – igi.bi)/2 = sq. *a*. (*Cf.* the geometric model in Fig. 3.2.1.)

The preserved columns on Plimpton 322 are reproduced below in transliteration, with the errors in the text corrected.

[a.šà *takī*]*lti ṣiliptim* *ša* 1 *inassaḫuma* sag *illû*	íb.si$_8$ sag	íb.si$_8$ *ṣiliptim*	mu. bi.im
1 59 00 15	1 59	2 49	ki.1
1 56 56 58 14 50 06 15	56 07	3 13	ki.2
1 55 07 41 15 33 45	1 16 41	1 50 49	ki.3
1 53 10 29 32 52 16	3 31 49	5 09 01	ki.4
1 48 54 01 40	1 05	1 37	ki.5
1 47 06 41 40	5 19	8 01	ki.6
1 43 11 56 28 26 40	38 11	59 01	ki.7
1 41 33 45 14 03 45	13 19	20 49	ki.8
1 38 33 36 36	8 01	12 49	ki.9
1 35 10 02 28 27 24 26 40	1 22 41	2 16 01	ki.10
1 33 45	45	1 15	ki.11
1 29 21 54 02 15	27 59	48 49	ki.12
1 27 00 03 45	2 41	4 49	ki.13
1 25 48 51 35 06 40	29 31	53 59	ki.14
1 23 13 46 40	56	53	ki.15

Here follows, in addition, a tentative reconstruction of the columns from the missing left half or third of the clay tablet.

igi	igi.bi	*takīlti ṣiliptim*	*takīlti sag*
2 24	25	1 24 30	59 30
2 22 13 20	25 18 45	1 23 46 02 30	58 27 17 30
2 20 37 30	25 36	1 23 06 45	57 30 45
2 18 53 20	25 55 12	1 22 24 16	56 29 04
2 15	26 40	1 20 50	54 10
2 13 20	27	1 20 10	53 10
2 09 36	27 46 40	1 18 41 20	50 54 40
2 08	28 07 30	1 18 03 45	49 56 15
2 05	28 48	1 16 54	48 06
2 01 30	29 37 46 40	1 15 33 53 20	45 56 06 40
2	30	1 15	45
1 55 12	31 15	1 13 13 30	41 58 30
1 52 30	32	1 12 15	40 15
1 51 06 40	32 24	1 11 45 20	39 21 30
1 48	33 20	1 10 40	37 20

Presumably, each line of the table text contains all the numerical parameters for an igi-igi.bi problem. In the example of line 1, for example, the listed parameters can be used to set up the following exercise:

Chosen parameters:	igi = 2 24, igi.bi = 25
Equations:	igi + igi.bi = 2 49, igi · igi.bi = 1
Solution procedure:	c = (igi + igi.bi)/2 = 1 24;30
	sq. c = sq. 1 24;30 = 1 59 00;15
	sq. a = sq. c – 1 = 59 00;15
	a = sqs. 59 00;15 = 59;30
Check:	igi = c + a = 1 24;30 + 59;30 = 2 24
	igi.bi = c – a = 1 24;30 – 59;30 = 25

Note that the value 1 24;30 for the diagonal c = (igi + igi.bi)/2 appears repeatedly in this solution procedure. First it is squared, then it is committed to memory to be used again in the final pair of operations. It is probably this being committed to memory that gave it its name *takīlti ṣiliptim* 'the *holder* for the diagonal', since a Babylonian phrase for 'commit it to memory' was, literally, 'let it *hold* your head'.

The values 2 24, 25, 1 24;30, and 59;30 appearing in the solution procedure above correspond to the numbers 2 24, 25, 1 24 30, and 59 30 in line 1 of the four lost columns on Plimpton 322. The value 1 50 00;15 corresponds to the number 1 50 00 15 in line 1 of the first preserved column, and the value 50 00;15 is obtained by subtraction of 1, as instructed in the heading over that column.

Thus, it remains to explain only what the purpose was of the numbers in line 1 of columns 2 and 3 of the preserved part of the clay tablet. An interesting answer to this question is that they probably played an important role in the computation of the square sides of 50 00;15 and 1 50 00;15. Indeed, it is known (see the explicit examples in Friberg, *RC* (2007), Appendix 8, Sec. A8 a) that OB mathematicians had invented a clever *factorization method* as a convenient shortcut in computations of square sides. To find the square side of 50 00;15, they could operate as follows:

Since 59 00 15 = 15 · 4 · 59 00 15 = 15 · 3 54 01,
it follows that sqs. 59 00 15 = sqs. 15 · sqs. 3 54 01.

Here, sqs. 15 (00) = 30, and sqs. 3 54 01 could be computed as follows (see Sec. 16.7 below):

sqs. 3 54 01 = appr. 2 (00) – 4 / 2 · 2 = 2 (00) – 1 = 1 59, and sq. 1 59 = 3 54 01.

Putting the two results together, one finds that

sqs. 59 00 15 = sqs. 15 · sqs. 3 54 01 = 30 · **1 59** = 59 30.

Similarly,

Since 1 59 00 15 = 15 · 4 · 1 59 00 15 = 15 · 7 54 01,
it follows that sqs. 1 59 00 15 = sqs. 15 · sqs. 7 54 01,
where sqs. 7 54 01 = appr. 3 (00) – 1 06 / 2 · 3 = 3 (00) – 11 = 2 49, sq. 2 49 = 7 54 01.
Therefore, sqs. 1 59 00 15 = 30 · **2 49** = 1 24 30.

The trick was apparently to first remove obvious square factors from the given number and then compute the square side of the remaining "factor-reduced" number. That square side can be called the "factor-reduced core" of the square side of the given number. Thus, in line 1 of Plimpton 322, the numbers 1 59 and 2 49 in columns 2 and 3 of the preserved part of the clay tablet can be interpreted as the factor-reduced cores of the square sides of 59 00 15 and 1 59 00 15, respectively.

It is clear that knowing in advance such factor-reduced cores would greatly simplify the computations necessary in each case for the solution of the igi-igi.bi problems with the data given in, for instance, (the lost) column 3 on Plimpton 322. This is certainly true in the case of the igi-igi.bi problem associated with line 10 of Plimpton 322, where the need arises to compute the square sides of the "many-place" sexagesimal numbers 35 10 02 28 27 24 26 40 and 1 35 10 02 28 27 24 26 40.

Another interesting question is how the 15 examples of parameters for igi-igi.bi problems listed on Plimpton 322 were chosen. More precisely, which is the origin of the 15 pairs of reciprocal sexagesimal numbers in (the lost but reconstructed) columns 1-2 on the clay tablet? There is an astonishingly simple answer to this question. (*Cf.* Friberg, *HMath* 8 (1981), *RC* (2007), Appendix 7.) Take, for instance, 2 24, the first igi value in Plimpton 322 (the first number in the reconstructed column 1). Since

2 24 · 5 = 10 (00) + 2 (00) = 12 (00).

2 24 can be written (in relative sexagesimal place value notation) as

2 24 = 12 · igi 5 (= 12 / 5).

Similarly, in the case of 2 22 13 20, the second igi value in column 1,

2 22 13 20 · 3 = 7 06 40, 7 06 40 · 3 = 21 20, 21 20 · 3 = 1 04.

Therefore,

2 22 13 20 = 1 04 · igi 27 (= 1 04 / 27).

And so on. Summing up the results, one finds that

2 24 = 12 · igi 5	2 22 13 20 = 1 04 · igi 27	2 20 37 30 = 1 15 · igi 32
2 18 53 20 = 2 05 · igi 54	2 15 = 9 · igi 4	2 13 20 = 20 · igi 9
2 09 36 = 54 · igi 25	2 08 = 32 · igi 15	2 05 = 25 · igi 12
2 01 30 = 1 21 · igi 40	2 = 2 · igi 1	1 55 12 = 48 · igi 25
1 52 30 = 15 · igi 8	1 51 06 40 = 50 · igi 27	1 48 = 9 · igi 5

Therefore, all the igi values appearing in (the reconstructed) column 1 of Plimpton 322 can be written in the form igi = m · igi n (= m/n), where n varies between 1 and 54, while m varies between 2 and 2 05. The obvious conclusion is that the author of Plimpton 322 decided to use only those values igi n from the OB table of reciprocals for which $1 \le n < 1\ 00$ (= 60). (As is well known, if igi n appears in the OB table of reciprocals, then n must be a *regular* sexagesimal number, that is a sexagesimal integer with no other factors than powers of 2, 3, and 5.)

A continued analysis shows (*cf.* Friberg, *HMath* 8 (1981); *RC* (2007), Appendix 8) that the list of igi values in (the reconstructed) column 1 of Plimpton 322 consists of *all* numbers of the form igi = m · igi n (= m / n), where

n and m are regular sexagesimal numbers, with
$1 \le n < 1\ 00$, and $1\ 48 \le m \cdot$ igi $n \le$ sqs.2 + 1 = appr. 2 24.

The condition that m · igi $n \le 2\ 24$ ensures that in Fig. 3.2.1, right, the side a, the *front* of the right triangle, will always be *shorter* than the side $b = 1$, the *length* of the right triangle. This is in agreement with a well known convention in all mathematical cuneiform texts. The fact that $1\ 48 \le m$ · igi n is simply a consequence of the circumstance that the table on Plimpton 322 ends at the lower edge of the clay tablet, just when the descending series of igi values in column 1 has happened to reach the value 1 48.

The brief discussion above has demonstrated that the table of parameters for 15 igi-igi.bi problems on Plimpton 322 was based on a cleverly and systematically arranged series of applications of *the Old Babylonian generating rule*

c, b, a = (igi + igi.bi)/2, 1, (igi – igi.bi)/2.

This generating rule, by the way, fails to be completely general only because, in agreement with Babylonian conventions, the front a is supposed to be shorter than the length b, and the values of c and a are supposed to be *regular* sexagesimal numbers.

In a similar way, of course, Euclid's generating rule

$c, b, a = t \cdot \{(\text{sq. } m + \text{sq. } n)/2, m \cdot n, (\text{sq. } m - \text{sq. } n)/2\}$, where $t = 1$ or 2.

fails to be general only because, in agreement with Greek conventions, the values of c, b, and a are supposed to be (positive) *integers*.

There is a simple connection between the OB and the Greek generating rules. Indeed, since in the Babylonian generating rule igi is supposed to be of the form $m \cdot$ igi n, it follows that (in modern notations)

(igi + igi.bi)/2, 1, (igi – igi.bi)/2 =
$(m/n + n/m)/2$, 1, $(m/n - n/m)/2$ =
$t \cdot \{(\text{sq. } m + \text{sq. } n)/2, m \cdot n, (\text{sq. } m - \text{sq. } n)/2\}$,
where $t = 1/(m \cdot n)$.

In this connection, it is important to point out that OB mathematicians were well aware of the fact that if c, b, a are the diagonal and the sides of a rectangle or a right triangle, a diagonal triple, then also all (positive) multiples $t \cdot (c, b, a)$ are diagonal triples. This fact is demonstrated by the OB exercise **MS 3971 § 4** (Friberg, *RC* (2007), Sec. 10.1 d), which is a scaling problem for diagonal triples.

MS 3971 § 4, literal translation	explanation
7, the diagonal (*ṣiliptum*).	The diagonal c = 7 (00)
The length and the front are what?	The length b and the front a = ?
5, 4, 3, the square sides (íb.si$_8$).	Let, for instance, c^*, b^*, a^* = 5, 4, 3
5 release, 12 it gives.	igi 5 = 12
12 to 4 lift, 48 it gives.	igi 5 · (5, 4, 3)
12 to 3 lift, 36 it gives.	= 1 (00), 48, 36
48 to 7 lift, 5 36, the length.	7 · 1 (00), 48, 36
36 to 7 lift, 4 12, the front.	= 7 (00), 5 36, 4 12

In this exercise, the briefly stated problem is, apparently, to find a rectangle with the diagonal '7'. The solution is given in three steps. In the first step, an arbitrary diagonal triple c^*, b^*, a^* is chosen, namely the well known triple 5, 4, 3. It is interesting to note that the three numbers 5, 4, 3 are referred to as íb.si$_8$ 'square sides'. It is tempting to try to explain this surprising designation as a reference to a construction like the one in Fig. 2.1.1, where the diagonal and the sides of a right triangle are all adorned with squares!

In the second step of the solution procedure, the diagonal triple 5, 4, 3 is scaled up by the factor igi 5 = 12. The result is the triple 1 00, 48, 36.

Finally, in the last step of the solution procedure, this diagonal triple, in its turn, is scaled up by the factor 7. The result is, of course, the diagonal triple 7 00, 5 36, 4 12 (since $7 \cdot 48 = 336 = 5\ 36$ and $7 \cdot 36 = 252 = 4\ 12$).

Chapter 4

The Lemma *El.* X.32/33 and an OB Geometric Progression

4.1. Division of a right triangle into a pair of right sub-triangles

The lemma *El.* X.32/33 begins with the following statement:

Let ABC be a right-angled triangle having the angle A right,
and let the perpendicular AD be drawn. Then
the rectangle CB, BD is equal to the square on BA,
the rectangle BC, CD is equal to the square on CA,
the rectangle BD, DC is equal to the square on AD,
the rectangle BC, AD is equal to the rectangle BA, AC.

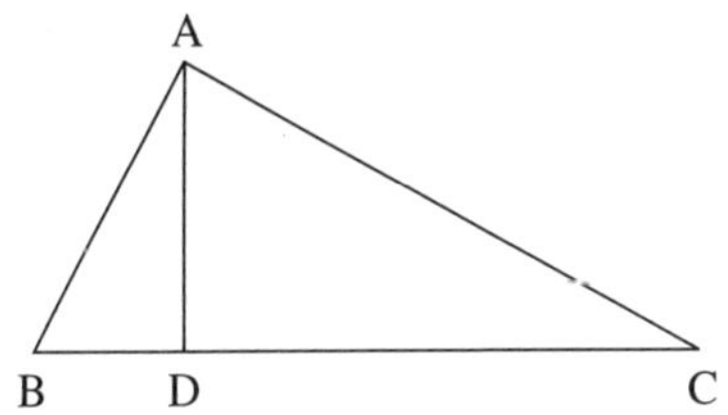

1) CB · BD = sq. BA
2) BC · CD = sq. CA
3) BD · DC = sq. AD
4) BC · AD = BA · AC

Fig. 4.1.1. The diagram illustrating the lemma *El.* X. 32/33.

The proof proceeds, essentially, as follows:

1) The triangle ABD is similar to the triangle ABC	*El.* VI.8
CB : BA = BA : BD	*El.* VI.4
CB · BD = sq. BA	*El.* VI.17
2) The triangle ADC is similar to the triangle ABC	*El.* VI.8
BC : CA = CA : CD	*El.* VI.4
BC · CD = sq. CA	*El.* VI.17
3) AD is the mean proportional between BD and DC	*El.* VI.8, Por.
BD : DA = AD : DC	
BD · DC = sq. AD	*El.* VI.17

4) The triangle ABD is similar to the triangle ABC — *El.* VI.8
BC : CA = BA : AD — *El.* VI.4
BC · AD = BA · AC — *El.* VI.16

4.2. A Metric Algebra Proof of the Lemma *El.* X.32/33

As indicated above, Euclid's proof of the lemma *El.* X.32/33 is based on Propositions *El.* VI.4, 8, 16, and 17, which in their turn are based on the theory of proportions in *Elements* V. In particular, the first step of the proof is the observation that "the triangles ABD, ADC are similar both to the whole ABC and to each other" (*El.* VI.8).

An Old Babylonian mathematician can easily have found results closely related to the four equations in Fig. 4.1.1 above by a simple use of metric algebra arguments, for instance as follows. He would naturally interpret a given right triangle with the sides c, b, a as one half of a rectangle with the sides b, a and the diagonal c, as in Fig. 4.2.1 below.

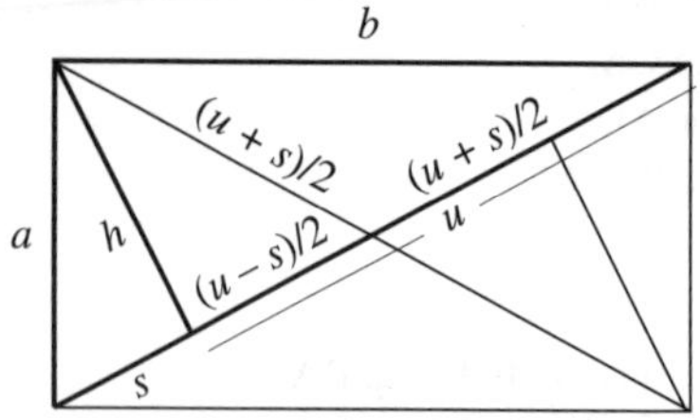

1) sq. h = sq. $(u+s)/2$ – sq. $(u-s)/2 = u \cdot s$
2) sq. b = sq. u + sq. h = sq. $u + u \cdot s = u \cdot c$
3) sq. a = sq. h + sq. $s = u \cdot s$ + sq. $s = s \cdot c$
4) $b \cdot a = c \cdot h$ (= half the area)

Fig. 4.2.1. A metric algebra proof of the lemma *El.* X.32/33 .

Suppose that he wanted to compute the height h against the diagonal, as well as the segments u, s into which the diagonal is cut by the height h. Since h is then the upright in a right triangle with the base $(u-s)/2$ and the diagonal $(u+s)/2$, he could proceed in the following way, using nothing but the Old Babylonian diagonal rule:

1) sq. h = sq. $(u+s)/2$ – sq. $(u-s)/2 = u \cdot s$ (as in Fig. 1.1.2 or *El. II*.14)
2) sq. b = sq. u + sq. h = sq. $u + u \cdot s = u \cdot c$, hence u = sq. b / c
3) sq. a = sq. h + sq. $s = u \cdot s$ + sq. $s = s \cdot c$, hence s = sq. a / c

These results corresponds to the first three of the four equations in the lemma *El.* X.32/33. The fourth equation can be proved quite simply by observing that if A is the area of the triangle, then

4) $b \cdot a = 2\,A = h \cdot c$

Instead of using the prior knowledge that *the height against the diagonal divides the given right triangles into two right triangles similar to each other and to the whole* (*El.* VI.8), in order to prove the four equations in the lemma *El.* X.32/33, as Euclid did, an Old Babylonian mathematician can have proceeded in the opposite direction, using equations 2)-4) above to prove the mentioned similarity relations. Indeed,

2) $\Rightarrow$ $b/c \cdot b = \text{sq. } b \,/\, c = u$

3) $\Rightarrow$ $a/c \cdot a = \text{sq. } a \,/\, c = s$

4) $\Rightarrow$ $a/c \cdot b = (a \cdot b) \,/\, c = (h \cdot c) \,/c = h$ and $b/c \cdot a = (b \cdot a) \,/\, c = (h \cdot c) \,/\, c = h$

Therefore,

$a, h, s = a/c \cdot (c, b, a)$ and $b, u, h = b/c \cdot (c, b, a)$

This means that *a given right triangle is divided by the height against the diagonal into two right sub-triangles similar to itself but scaled down with the scale factors b/c and a/c, respectively.*

This "height-against-the-diagonal rule" was known in Old Babylonian mathematics, as shown by the discussion of IM 55357 in Sec. 4.3 below. The rule can have been found by use of metric algebra in the way suggested above, but that is, for the moment, only a reasonable conjecture.

4.3. An Old Babylonian Chain of Right Sub-Triangles

IM 55357 (Baqir, *Sumer* 6 (1950), Høyrup, *LWS* (2002), 231) is a mathematical single problem text from the site Tell Harmal, near Baghdad. It is one of the oldest known OB mathematical cuneiform texts.

IM 55357, literal translation	explanation

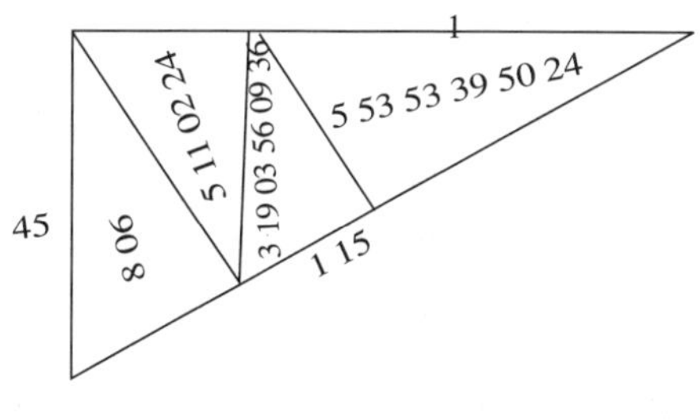

A peg-head.	A (right) triangle with the sides c, b, a.
1 the length, 1 15 the long length,	$b = 1$ (00), $c = 1$ 15
45 the upper front, 22 30 the complete field,	$a = 45, A = a \cdot b\,/2 = 22$ 30

From 22 30 the complete field,	Divide A into
8 06 the upper field,	$B = 8\ 06$ (see Fig. 4.3.1 below)
5 11 02 24 the next field,	$C = 5\ 11{:}02\ 24$
3 19 03 56 09 36 the 3rd field,	$D = 3\ 19;03\ 56\ 09\ 36$
5 53 53 39 50 24 the lower field.	$E = 5\ 53;53\ 39\ 50\ 24$
The upper length, the middle length,	What are then
the lower length, and the descendant are what?	the sides of the sub-triangles?
You, to know the doing,	Do it like this:
the opposite of 1, the length, resolve,	$1/b = 1/1(00)$
to 45 raise it, 45 you see.	$a/b = 45/1\ (00) = ;45$
45 to 2 raise, 1 30 you see.	$2\ a/b = 1;30$
To 8 06, the upper field, raise it, 12 09 you see.	$2\ a/b \cdot B = 1;30 \cdot 8\ 06 = 12\ 09$
12 09, what is it equalsided?	sqs. 12 09 = ?
27 is the equalside. **27 is the front.**	sqs. 12 09 = $\mathbf{27 = s_1}$
27 break, 13 30 you see.	$s_1/2 = 27/2 = 13;30$
The opposite of 13 30 resolve,	$B\ /\ (s_1/2)$
to 8 06, the upper field, raise it, **36 you see,**	$= 8\ 06\ /\ 13;30$
the length next to the length 45, the front.	$\mathbf{= 36 = h_1}$
Turn around.	The next part of the computation:
The length 27 of the upper peg-head	$c - s_1 =$
from 1 15 tear out, 48 it leaves.	$1\ 15 - 27 = 48$
The opposite of 48 resolve, 1 15 you see.	$h_1\ /\ (c - s_1) = h_1\ /\ u_1$
1 15 to 36 raise, 45 you see.	$= 36/48 = ;45$
To 2 raise, 1 30 you see.	$2\ h_1\ /\ u_1 = 1;30$
1 30 to 5 11 02 24 raise,	$2\ h_1\ /\ u_1 \cdot C =$
7 46 33 36 you see.	$1;30 \cdot 5\ 11;02\ 24 = 7\ 46;33\ 36$
7 46 33 36, what is it equalsided?	sqs. 7 46;33 36 = ?
21 36 it is equalsided.	sqs. 7 46;33 36
21 36 is the front of the 2nd peg-head.	$\mathbf{= 21\ 36 = s_2}$
The halfpart of 21 36 break, 10 48 you see.	$s_2/2 = 21\ 36\ /\ 2 = 10\ 48$
the opposite of 10 48 resolve, to ·· ··· ··	$C\ /\ (s_2/2) =\ (28;48)$

As so often in OB mathematical problem texts, the question in this exercise is very vaguely stated. Luckily, the situation is clearly described by a diagram accompanying the text. The diagram is explained in Fig. 4.3.1 below. Given are the diagonal and sides of a right triangle, c, b, a = 1 15, 1 00, 45, and the areas of four right sub-triangles, B, C, D, E = 8 06, 5 11;02 24, 3 19;03 56 09 36, 5 53;53 39 50 24. Apparently, the goal of the text was the computation of the heights h_1, h_2, h_3, and of the segments s_1, s_2, s_3 and u_1, u_2.

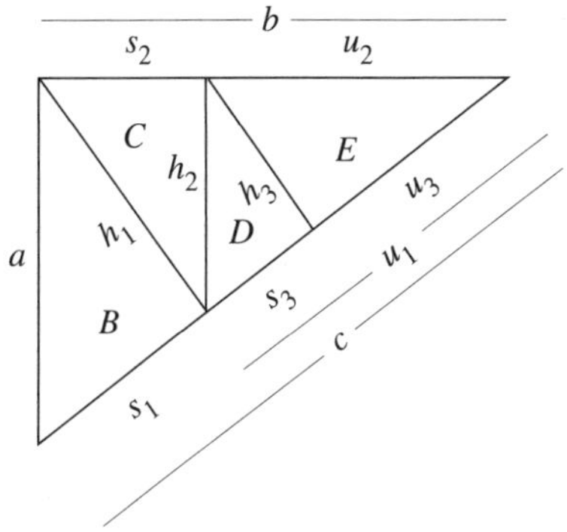

b = the 'length'
c = the 'long length'
a = the 'upper front'
$B + C + D + E$ = the 'complete field'
B = the 'upper field', the 'upper peghead'
C = the 'next field', the '2nd peghead'
D = the '3rd field'
E = the 'lower field'
h_1 = the 'upper length'
h_2 = the 'middle length'
h_3 = the 'lower length'
u_2 = the 'descendant'

Fig. 4.3.1. Explanation of the solution procedure in IM 55357.

It is many times a good idea to *try to find out how the author of an OB mathematical text conceivably constructed the data appearing in the question of the text.* In the case of IM 55357, the diagonal triple c, b, a = 1 15, 1 00, 45 = 15 · (5, 4, 3) is the OB standard example of such a triple, but where do the complicated values of B, C, D, E come from? The obvious answer to that question is that OB mathematicians knew that in Fig. 4.2.1 above, according to the mentioned *height-against-the-diagonal rule*,

$a, h, s = a/c \cdot (c, b, a)$ and $b, u, h = b/c \cdot (c, b, a)$.

Indeed, consider again Fig. 4.3.1, and let A be the area of the whole right triangle. According to the height-against-the-diagonal rule

$a, h_1, s_1 = a/c \cdot (c, b, a),.$

Therefore, the area B of the first right sub-triangle is

B = sq. $a/c \cdot A$ = sq.(45 / 1 15) · 45 00 / 2 = sq. ;36 · 22 30 = ;36 · 13 30 = 8 06.

The area of the remaining part of the whole triangle is

$A - B$ = sq. $b/c \cdot A$ = sq. ;48 · 22 30 = ;48 · 18 00 = 14 24.

A new application of the height-against-the diagonal rule shows that

C = sq. $a/c \cdot (A - B)$ = sq. ;36 · 14 24 = ;36 · 8 33;24 = 5 11;02 24.

Consequently,

$A - B - C$ = sq. $b/c \cdot (A - B)$ = sq. ;48 · 14 24 = ;48 · 11 31;12 = 9 12;57 36.

In the third step of the computation,

D = sq. $a/c \cdot (A - B - C)$ = sq. ;36 · 9 12;57 36 = 3 19;03 56 09 36.

Then, finally,

$A - B - C - D = \text{sq. } b/c \cdot (A - B - C) = \text{sq. } ;48 \cdot 9\ 12;57\ 36 = 5\ 53;53\ 39\ 50\ 24.$

These numbers were computed correctly. The algorithm could have been continued indefinitely, but the procedure was halted when there was no more space in the diagram to write additional sexagesimal numbers.

Note that *B*, *C*, *D* are the three first terms of a geometric progression of sexagesimal numbers with the common ration sq. $b/c \cdot$ sq. a/c. Actually, the three right sub-triangles with the areas *B*, *C*, *D* can be interpreted as the three first terms of a geometric progression *in the literal sense*!

Now to the actually recorded solution procedure for the stated problem in IM 55357. It begins with the computation of s_1 and h_1, evidently as the solutions to the following simple rectangular-linear system of equations:

$h_1 \cdot s_1 / 2 = B$ (an area equation)

$s_1 = a/b \cdot h_1$ (a similarity equation).

The solution follows immediately:

sq. $s_1 = 2\ a/b \cdot B$, so that

$s_1 = \text{sqs. } (2\ a/b \cdot B) = 27$, and $h_1 = B / (s_1 / 2) = 36$.

The next step of the algorithm proceeds similarly. Therefore,

sq. $s_2 = 2\ h_1 / u_1 \cdot B$, so that

$s_2 = \text{sqs. } (2\ h_1 / u_1 \cdot C) = 21;36$, and $h_2 = C / (s_2 / 2) = 28;48$.

There was no more space on the clay tablet for further computations.

Geometrically, the first step of the solution procedure can be explained as in Fig.4.3.2 below.

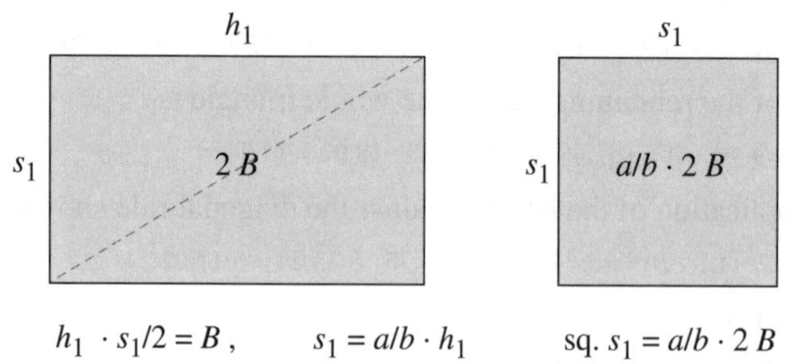

Fig. 4.3.2. IM 55357. Geometric interpretation of the solution procedure.

Chapter 5

Elements X and Babylonian Metric Algebra

5.1. The Pivotal Propositions and Lemmas in *Elements* X

Book X of Euclid's *Elements* is notoriously difficult, partly because the presentation of the results is purely synthetic, with no preceding analysis. That is why in Taisbak's *Coloured Quadrangles* (1982), for instance, a relatively reader-friendly version of *El.* X is achieved only through a complete reorganization of the order of appearance of the various definitions and propositions. Another very valuable introduction to *El.* X, with a similar reorganization of the material, is Knorr's brief paper *BAMS* 9 (1983).

Below, only those aspects of El. X will be dwelt upon which are in some way related to metric algebra of the Babylonian type. As it turns out, all the pivotal propositions and lemmas in El. X are of that kind. Two of the lemmas, by the way, have already been discussed above, lemma *El.* X.28/29 1a in Chapter 3, and lemma *El.* X.32/33 in Chapter 4.

In order to facilitate for the readers the understanding of the discussion below, a concise introductory outline of the contents of *El. X* is first presented here. In this outline the embarrassing repetitiveness of *El. X* in its original form is deliberately suppressed by grouping together propositions for "binomials" (sums of pairs of expressible straight lines, commensurable in square only) with parallel propositions for "apotomes" (differences of such pairs of straight lines). In addition, in several places in this outline, separate but parallel propositions for each one of six distinct classes of inexpressible sums or differences of straight lines are grouped together.

The following convenient abbreviations will be used here:

a* com *b and ***a* inc *b*** mean *a* is commensurable (or incommensurable) with *b*.
a* · *b means, as usual, (the area of) the rectangle with the sides *a* and *b*, and
sq. *a* means (the area of) the square on *a* and **sqs. *A*** means the square side of *A*.

A Concise Outline of the Contents of *Elements* X

Def. X.I.1	Commensurable magnitudes.
Def. X.I.2	Straight lines commensurable in square (only).
Defs. X.I.3-4	Expressible rectangles and straight lines relative to an assigned straight line.
X.1	The exhaustion principle.
X.2-13	About commensurable or incommensurable magnitudes.
X.13/14	Geometric construction of the square difference sq. u – sq. v.
X.15-16	About sums of commensurable or incommensurable magnitudes.
X.16/17	If $a + b = c$, $a \cdot b = A$, then $c \cdot b - sq.\ b = A$.
X.17-18	If $a + b = u$, $a \cdot b =$ sq. $v/2$, $a > b$, then a com b if and only if u com w, where $w =$ sqs. (sq. u– sq. v).
X.19-20	Rectangles with sides that are expressible and commensurable.
X.21	Rectangles with sides that are expressible and commensurable in square only (medial rectangles), and their square sides (medial straight lines).
X.21/22	$a : b =$ sq. $a : a \cdot b$. (*Cf. El.* II.3, Fig. 1.2.1, right.)
X.22-28	About medial rectangles and medial straight lines.
X.28/29 1a	Construction of diagonal triples of "numbers". (*Cf.* Fig. 3.1.1, right.)
X.28/29 1b	Construction of pairs of numbers such that the difference of their squares is (or is not) a square number. (Used in X.29.)
X.28/29 2	Construction of pairs of numbers such that the sum of their squares is not a square number. (Used in X.30.)
X.29	Construction of expressible straight lines a, b commensurable in square, $a > b$, such that $c =$ sqs. (sq. a – sq. b) com a. (Used in X.31-32.)
X.30	Construction of expressible straight lines a, b commensurable in square, $a > b$, such that $c =$ sqs. (sq. a – sq. b) inc a. (Used in X.33.)
X.31	Construction of medial straight lines c, d commensurable in square, $c > d$, $c \cdot d$ expressible, such that $v =$ sqs. (sq. c – sq. d) com c. (Used in X.34.)
X.32	Construction of medial straight lines c, d commensurable in square, $c > d$, $c \cdot d$ medial, such that $v =$ sqs. (sq. c – sq. d) com c. (Used in X.35.)
X.32/33	If the height h against the diagonal of a right triangle with the sides c, b, a, with $b > a$, cuts c into the segments u and s, with $u > s$, then $c \cdot s =$ sq. a, $c \cdot u =$ sq. b, $u \cdot s =$ sq. h, and $c \cdot h = a \cdot b$. (Used in X.33.)

X.33-35	Construction of pairs of straight lines (used in X.39-41).
X.36-41	Sums of class 1-6 (not defined until here) are inexpressible.
X. 73-78	Differences of class 1-6 (not defined until here) are inexpressible.
X.41/42	If $u + s = u' + s'$, $u > s$, $u' > s'$, $u > u'$, then sq. u + sq. s > sq. u'+ sq. s'.
X.42-47	Sums of class 1-6 are uniquely split into their terms.
X. 79-84	Differences of class 1-6 are uniquely split into their terms.
Defs. X.II.1-6	1st to 6th binomials.
Defs. X.III.1-6	1st to 6th apotomes.
X.48-53	Construction of examples of 1st to 6th binomials.
X. 85-90	Construction of examples of 1st to 6th apotomes.
X.53/54	sq. u + sq. s + 2 $u \cdot s$ = sq. $(u + s)$, and $u \cdot s$ is the mean proportional to sq. u and sq. s. (Used in X.33.)
X.54-59	Sqs. {(1st to 6th binomial) · expressible line} = sum of class 1-6.
X.91-96	Sqs. {(1st to 6th apotome) · expressible line} = difference of class 1-6.
X.60-65	{Sq. (sum of class 1-6)} / expressible line = 1st to 6th binomial.
X.97-102	{Sq. (difference of class 1-6)} / expressible line = 1st to 6th apotome.
X.66-70	Sum commensurable with sum of class 1-6 is of the same class.
X.103-107	Difference commens. with difference of class 1-6 is of the same class.
X.71	Sqs. (expressible area + medial area) = sum of class 1-2 or 4-5.
X.108	Sqs. (expressible area – medial area) = difference of class 1 or 4.
X.72	Sqs. (sum of two incommensurable medial areas) = sum of class 3 or 6.
X.109	Sqs. (medial area – expressible area) = difference of class 2 or 5.
X. 110	Sqs. (medial area – medial area) = difference of class 3 or 6.
X.72b	Binomials and apotomes are distinct kinds of inexpressible straight lines.
X.111a,b	Similarly, medial straight lines, sums of class 1-6, and differences of class 1-6 are 13 distinct kinds of inexpressible straight lines.
X.112-113	An expressible area / a 1st to 6th binomial = a cognate 1st to 6th apotome. An expressible area / a 1st to 6th apotome = a cognate 1st to 6th binomial.
X.114	A binomial · a cognate apotome = an expressible rectangle.
X.115	Generalizations: Medials of medials and so on.

5.2. Binomials and Apotomes, Majors and Minors

Of the thirteen kinds of inexpressible straight lines considered in *Elements* X, only four kinds actually appear as straight lines in known geometric figures, namely binomials and apotomes, "majors" and "minors". In this section, a condensed version of *Elements* X, rephrased in terms of metric algebra, will be concerned only with propositions and lemmas having to do with *binomials and majors*. The closely parallel propositions related to *apotomes and minors* will not be mentioned.

The most convenient starting point for a discussion of *Elements* X is *El.* X.17-18. The following definitions and propositions (slightly rephrased here) are assumed to be known in that proposition and its proof:

El. X.Def. I 1. Two straight lines or two areas are called *commensurable* if they both are (integral) multiples of some straight line or area.

El. X.Def. I 2. Two straight lines are called *commensurable in square only* if they are incommensurable but their squares are commensurable.

El. X.Defs. I 3-4. Let an arbitrarily chosen straight line e^* be called *expressible*. Then an area is called an *expressible area* if it is commensurable with sq. e^*, and a line segment is called an *expressible straight line* if its square is an expressible area.

El. X.5-8. Two magnitudes are commensurable if and only if they have to each other the ratio that a number (a positive integer) has to a number.

***El.* X.17-18** (rephrased in terms of metric algebra)

If $a + b = u$, $a \cdot b =$ (sq. v)/4, with $a > b$, $u > v$, then a com b if and only if u com w, where $w =$ sqs. (sq. $u -$ sq. v).

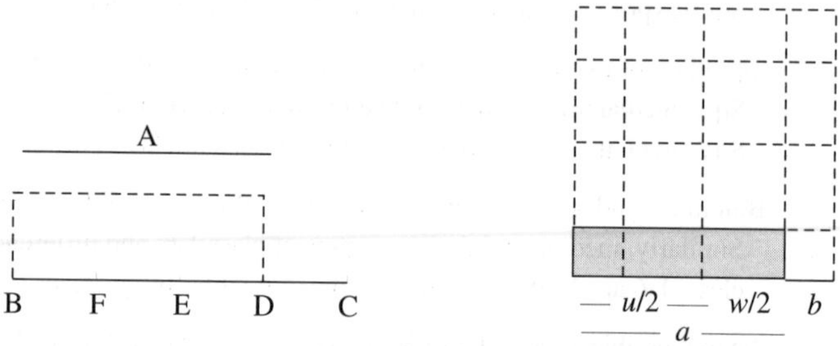

Fig. 5.2.1. Left: The figure in *El.* X.17. Right: An explanation, in terms of metric algebra.

In a metric algebra interpretation of the proof of *El.* X.17, let the given straight lines A and BC in Fig. 5.2.1, left, be called v and u, respectively. Also, let the sides BD and DC of the rectangle with the area equal to "the fourth part of the square on the less" (sq. $v/2$) be called a and b. Then

$a + b = u, \quad a \cdot b =$ sq. $v/2$.

The proof proceeds in analogy with the proof of *El.* II.5 (see Fig. 1.4.2, left). The result of the procedure, in terms of metric algebra, is that

$a = u/2 + w/2$, $b = u/2 - w/2$, where sq. $w/2$ = sq. $u/2$ – sq. $v/2$
and, consequently, sq. w = 4 sq. $u/2$ – 4 sq. $v/2$ = sq. u – sq. v.

See Fig. 5.2.1, right. It is clear that w can be identified with FD in Euclid's diagram. In view of this result, it follows that

a com b if and only if $u/2 + w/2$ com $u/2 - w/2$, that is, if and only if u com w.

After *El.* X.17-18 follows a series of simple propositions:

El. X.19. If p, q is a pair of expressible straight lines, with p com q, then the rectangle $p \cdot q$ is expressible.

The proof is based on the observation that $p : q$ = sq. $p : p \cdot q$.

El. X. 20. Conversely, if an expressible area is "applied to" an expressible straight line, then the resulting width is expressible and commensurable with the first straight line.

In terms of metric algebra: If the area $p \cdot q$ and the length p *of* a rectangle are expressible, then also the width q is expressible, and p com q.

The proof is, again, based on the observation that $p : q$ = sq. $p : p \cdot q$.

The next few propositions consider the case when p and q are incommensurable.

El. X.21. If p, q is a pair of expressible straight lines, with p inc q, then the rectangle $p \cdot q$ is inexpressible. Therefore, also the side s of a square equal (in area) to the rectangle $p \cdot q$ is inexpressible. Let $p \cdot q$ and s be called a *medial area* and a *medial straight line*, respectively.

The proof is similar to the proof of *El.* X. 19.

El. X. 22. Conversely, if a medial area is applied to an expressible straight line, then the resulting width is expressible and incommensurable with the first straight line.

The construction in *El.* X. 30 is an auxiliary result preparing for the explicit construction, in *El.* X. 33 below, of the terms of a *major* straight line:

***El.* X.30** (rephrased in terms of metric algebra)

How to find two expressible straight lines u and v, with $u > v$, such that u inc v and sq. u – sq. v = sq. w, where w inc u.

In the proof, it is assumed that sq. m and sq. n are two square numbers (integers) such that their sum is not a square number. Such square numbers are constructed in the lemma *El.* X.28/29 2. However, see the critical remark in Knorr, *BAMS* 9 (1983), 58. Evidently it is enough to observe that, for instance, 4 + 16 = 20, where 4 and 16, but not 20, are square numbers.

Now, let u and v be two straight lines with

u expressible, $u > v$, and sq. u : sq. v = (sq. m + sq. n) : sq. m.

Then u and v are both expressible, but commensurable in square only, since the squares of commensurable straight lines are to each other as two square numbers (*El.* X.9).

A third straight line w such that sq. w = sq. u – sq. v is then constructed as the third side of a right triangle inscribed in a semicircle with diameter u (lemma *El.* X.13/14). It is observed that then

sq. u : sq. w = (sq. m + sq. n) : sq. n.

Therefore, u and w are incommensurable.

The next step, in *El.* X.33, of the explicit construction of a major straight line is preceded by the lemma *El.* X.32/33 (already discussed in Chapter 4 above).

***El.* X.33** (rephrased in terms of metric algebra)

How to find two straight lines p, q incommensurable in square and such that the sum sq. p + sq. p. is expressible but the rectangle $p \cdot q$ medial.

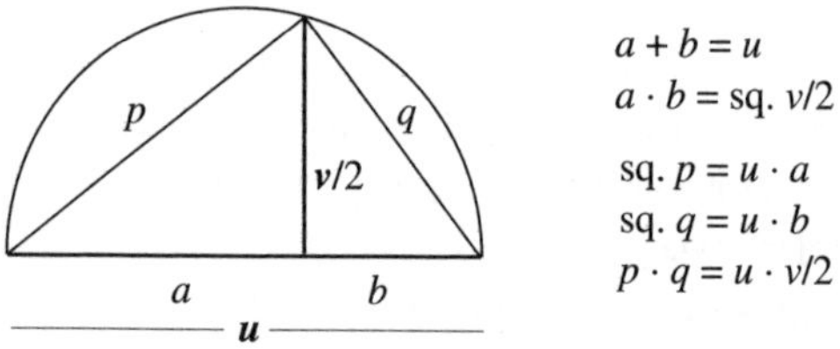

Fig. 5.2.2. The diagram in *El.* X.33, explained in terms of metric algebra.

The proof starts with two expressible straight lines u and v, with $u > v$, such that u inc v and sq. u – sq. v = sq. w, where w inc u (*El.* X.30).

Next, a related pair of straight lines a, b is constructed as the solutions to the rectangular-linear system of equations

$a + b = u$, $a \cdot b$ = sq. $v/2$ $(a > b)$.

See Fig. 5.2.2 above, and compare with *El.* II.14* (Fig. 1.7.2, right.) Note that here, according to *El.* X.17-18, a inc b, since by assumption w inc u.

Then follows the crucial step of implying the lemma *El.* X.32/33 (see Fig. 4.1.1 above), according to which

$u \cdot a$ = sq. p, $u \cdot b$ = sq. q, and $u \cdot v/2 = p \cdot q$, with p and q as in Fig. 5.2.2.

The pair of straight lines p, q constructed in this way enjoys the following series of properties:

sq. p : sq. $q = u \cdot a : u \cdot b = a : b$, so that sq. p inc sq. q, since a inc b,
sq. p + sq. q = sq. u is expressible, since u is assumed to be expressible,
$p \cdot q = u \cdot v/2$ is a medial rectangle, since u, v are expressible, u inc v.

Therefore, the desired construction is accomplished.

Now, at last, the main actors in this play are formally introduced:

El. X.36. The sum of two expressible straight lines p, q commensurable in square only is inexpressible. Let it be called a *binomial*.

El. X.39. The sum of two straight lines p, q incommensurable in square, and with sq. p + sq. q expressible but $p \cdot q$ medial, is inexpressible. Let it be called a *major*.

(Correspondingly,

El. X.73.The difference of two expressible straight lines p, q commensurable in square only is inexpressible.. Let it be called an *apotome*.

El. X.76. The difference of two straight lines p, q incommensurable in square, and with sq. p + sq. q expressible but $p \cdot q$ medial, is inexpressible. Let it be called a *minor*.)

The following lemma is interesting.

Lemma *El.* X.41/42 (rephrased in terms of metric algebra)

Let a straight line be divided into two parts in two different ways, so that it is equal to either $p+q$ or $p'+q'$, and suppose that $p>p'$. Then also sq. p + sq. q > sq. p' + sq. q'.

The diagram accompanying the proof of this lemma is unhelpful. The proof can be explained more easily by reference to the more informative diagram in Fig. 5.2.3 below.

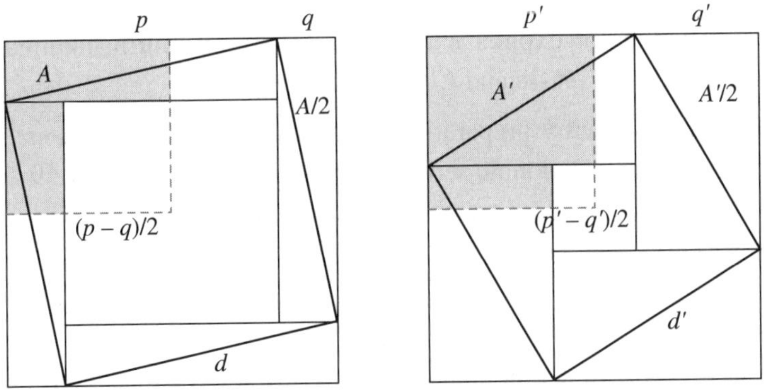

$p+q=p'+q'$, $p>p' \Rightarrow (p-q)/2 > (p'-q')/2 \Rightarrow A < A' \Rightarrow$ sq. d > sq. d'

Fig. 5.2.3. Explanation of the proof of *El.* X.41/41 .

Expressed in terms of metric algebra, the proof proceeds as follows:

If $p + q = p' + q'$, and $p > p'$, then $q < q'$.
Consequently, $(p - q)/2 > (p' - q')/2$, and sq. $(p - q)/2 >$ sq. $(p' - q')/2$.
Then also $A = p \cdot q < A' = p' \cdot q'$, and it follows that
sq. p + sq. q = sq. d = sq. $(p - q) - 2A >$ sq. $(p' - q') - 2A'$ = sq. d' = sq. p' + sq. q'.

This lemma is used in the proof of

El. X.42. The terms p, q of a binomial straight line are uniquely determined.

Namely, if the binomial can be expressed in two different ways, as $p + q = p' + q'$, then also sq. $(p + q)$ = sq. $(p' + q')$, and it follows that

(sq. p + sq. q) – (sq. p' + sq. q') = $2\,p' \cdot q' - 2\,p \cdot q$.

In this equation, the left side is an expressible area, while the right side is the difference between two medial areas. This is impossible (*El.* X.26), unless both sides of the equation are equal to zero. However, that cannot happen in view of the lemma.

A corresponding uniqueness theorem for major straight lines, with a similar proof, is

El. X.45. The terms p, q of a major straight line are uniquely determined.

Now it is revealed that there are, actually, six mutually exclusive types of binomials. Of those, only two are of interest here:

El. X.Def. II 1. Given an expressible straight line e, a binomial $u + v$, $u > v$, is called a *first binomial* (with respect to e) if u com e, and if sq. u – sq. v = sq. w, where w com e.

El. X.Def. II 4. Given an expressible straight line e, a binomial $u + v$, $u > v$, is called a *fourth binomial* (with respect to e) if u com e, and if sq. u – sq. v = sq. w, where w inc e.

Constructions of explicit examples of first and fourth binomials are demonstrated in *El.* X.48 and *El.* X.51.

After these lengthy preparations, the main results of *Elements* X are finally presented in a double series of propositions, in *El.* X.54-59 and *El.* X. 60-65. Of interest here are only *El.* X.54, 57 and *El.* X.60, 63.

***El.* X.54** (rephrased in terms of metric algebra)

A rectangle formed as the product of a given expressible straight line e and a *first binomial* (with respect to e) is equal (in area) to the square of a *binomial.*

If $u + v$ is a first binomial, the proposition says that there exists a binomial $p + q$ which is the “square side” of $(u + v) \cdot e$ in the sense that

$(u + v) \cdot e$ = sq. $(p + q)$.

The proof starts by recalling that when $u + v$, $u > v$, is a *first binomial* (with respect to e), then

u com v in square only, u com e, and sq. u – sq. v = sq. w, where w com u.

The proof continues by letting a, b be solutions to the following *rectangular-linear system of equations of type B1a*:

$a \cdot b$ = sq. $(v/2)$, $\quad a + b = u$.

Then it can be shown, as in Sec. 1.4 (*cf.* also Fig. 5.2.1 above), that

$(a - b)/2$ = sqs. (sq. $u/2$ – sq. $v/2$) = sqs. (sq. u – sq. v) $/2 = w/2$.

Since by assumption w com u, it follows that also $(a - b)$ com $u = a + b$. Therefore, obviously, as in *El.* X.17,

a com b.

Now, as in *El.* II.14, it is possible to find straight lines p, q such that

$a \cdot e$ = sq. p, and $b \cdot e$ = sq. q.

On the other hand, since $a \cdot b$ = sq. $(v/2)$ it is clear that $v/2$ is a mean proportional between a and b ($a : v/2 = v/2 : b$). Then also $v/2 \cdot e$ is a mean proportional between $a \cdot e$ and $b \cdot e$ ($a \cdot e : v/2 \cdot e = v/2 \cdot e : b \cdot e$). Since $a \cdot e$ = sq. p, and $b \cdot e$ = sq. q, this means that $v/2 \cdot e$ is a mean proportional between sq. p and sq. q. Consequently (as shown in lemma *El.* X.53/54)

$v/2 \cdot e = p \cdot q$.

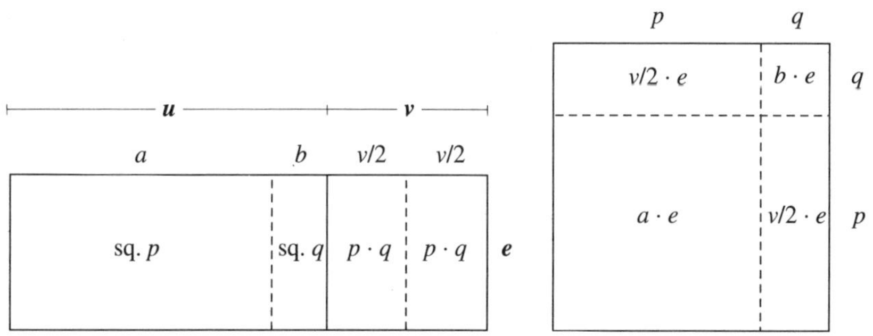

Fig. 5.2.4. The diagrams in *El.* X.54, in the style of metric algebra. u, v, e are given.

Therefore, as in Fig. 5.2.4 above, the square with the side $p + q$ can be divided into two unequal squares and two equal rectangles with the areas $a \cdot e$, $b \cdot e$, and $v/2 \cdot e$, respectively. Consequently, as desired,

sq. $(p + q) = (a + b + 2 \cdot v/2) \cdot e = (u + v) \cdot e$.

It remains to be shown that the constructed square side $p + q$ is a *binomial*, namely that

***p* and *q* are expressible straight lines, commensurable in square only.**

This is done in the following way:

1. a com b and $u = a + b$ com e $\Rightarrow$ a and b com u com e $\Rightarrow$ a and b expr.

2. a com b, both expr. $\Rightarrow$ $a \cdot e$ com $b \cdot e$, both expr $\Rightarrow$ sq. p com sq. q, both expr.

3. u com e, sq. u com sq. v $\Rightarrow$ v expr., but
a com u and u inc v $\Rightarrow$ a inc $v/2$ $\Rightarrow$ $a \cdot e$ inc $v/2 \cdot e$ $\Rightarrow$ sq. p inc $p \cdot q$ $\Rightarrow$ p inc q.

***El.* X.57** (rephrased in terms of metric algebra)

A rectangle formed as the product of a given expressible straight line e and a fourth binomial (with respect to e) is equal (in area) to the square of a major.

This time, Euclid starts by recalling that when $u + v$, $u > v$, is a fourth binomial (with respect to e), then

***u* com *v* in square only, *u* com *e*, and sq. *u* – sq. *v* = sq. *w*, where *w* inc *u*.**

He continues as in *El.* X.54, finding a solution p, q to the equation

sq. $(p + q) = (a + b + 2 \cdot v/2) \cdot e = (u + v) \cdot e$.

Also, since by assumption w inc u, it follows that $(a - b)$ com $u = a + b$. Therefore, as in *El.* X.18, a inc b.

It remains to be shown that the constructed square side $p + q$ is a major, namely that p and q are expressible straight lines, incommensurable in square, with sq. p + sq. q expressible but $p \cdot q$ medial. This is done in the following way:

1. a inc b $\Rightarrow$ $a \cdot e$ inc $b \cdot e$ $\Rightarrow$ sq. p inc sq. q $\Rightarrow$ p, q incommensurable in square.
2. $u = a + b$ com e $\Rightarrow$ $(a + b) \cdot e$ expressible. $\Rightarrow$ sq. p + sq. q expressible.
3. v inc u $\Rightarrow$ v inc e $\Rightarrow$ $v/2$ inc e $\Rightarrow$ $v/2 \cdot e$ medial $\Rightarrow$ $p \cdot q$ medial.

Note that now, in the light of *El.* X.57, the seemingly unmotivated construction in *El.* X.33 can be understood as the construction of a major as the square side of the product of an expressible straight line e and a fourth binomial (with respect to e) $u + v$ in the special case when $e = u$.

El. X.60, *El.* X.63 are the converses to *El.* X.54, *El.* X.57:

***El.* X.60** (rephrased in terms of metric algebra)

The square on a *binomial* $p + q$
applied to an arbitrarily given expressible straight line e
is equal to a *first binomial* $u + v$ (with respect to e).

If $p + q$ is a binomial, with $p > q$ as usual, then p, q are expressible straight lines commensurable in square only. Now, for an arbitrarily chosen expressible straight line e, it is possible to construct a rectangle with e as one side and with the area $a \cdot e =$ sq. p. In the language of *El.* I.44, a rectangle with given area sq. p can be applied to the given straight line e. Similarly, sq. q and $p \cdot q$ (twice) can be applied to straight lines parallel with e, so that a diagram such as the one in Fig. 5.2.5 below is formed.

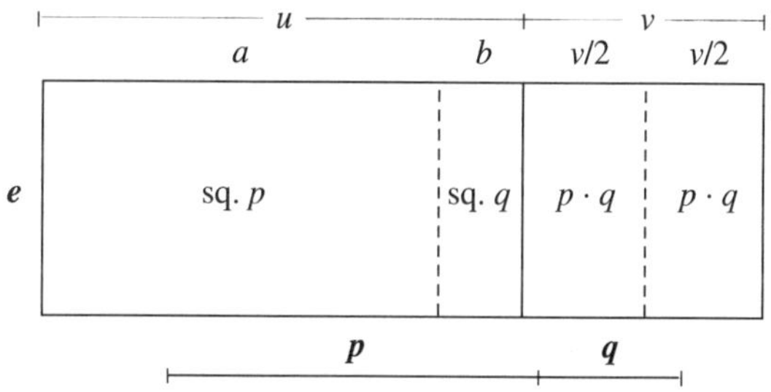

Fig. 5.2.5. The figure in *El.* X.60, presented in the style of metric algebra. p, q, e are given.

From here on, the proof is straightforward. First, since p, q are expressible, sq. p and sq. q are expressible and commensurable, so that the sum sq. p + sq. q is expressible. Since also e is expressible, it follows that the side $u = a + b$ is expressible and commensurable with e (*El.* X.20). Next, since p, q are commensurable in square only, the rectangles $p \cdot q$ and $2\,p \cdot q$ are medial (*El.* X.21). Therefore, v is expressible and incommensurable with e (*El.* X.22). Consequently, u, v are expressible and commensurable in square only. It follows that $u + v$ is a binomial.

It remains to be shown that $u + v$ is a *first* binomial. Now, it has already been shown that u com e. In addition,

$$\text{sq.}\,p : p \cdot q = p \cdot q : \text{sq.}\,q \quad (\text{Lemma } \textit{El.}\ \text{X.53/54}) \Rightarrow \quad a \cdot e : v/2 \cdot e = v/2 \cdot e : b \cdot e$$
$$\Rightarrow \quad a : v/2 = v/2 : b \;\Rightarrow\; a \cdot b = \text{sq.}\,v/2.$$

Also,

$$\text{sq.}\,p \text{ com sq.}\,q \;\Rightarrow\; a \cdot e \text{ com } b \cdot e \;\Rightarrow\; a \text{ com } b,$$

and

$$\text{sq.}\,p + \text{sq.}\,q > 2\,p \cdot q \;\Rightarrow\; u > v.$$

All that now is needed in order to complete the proof of *El.* X.60 is an application of *El.* X.17, which shows that if $w =$ sqs. (sq. u – sq. v), then

w com u, where u com e, so that also w com e.

***El.* X.63** (rephrased in terms of metric algebra)
The square on a *major* $p + q$
applied to an arbitrarily given expressible straight line e
is equal to a *fourth binomial* $u + v$.

There is no need to give here the details of the proof of *El.* X.63, which is closely parallel to the proof of *El.* X.60.

The logical end of the investigation comes in *El.* X.72b/*El.* X.111, with the observation that an apotome cannot be equal to a binomial, that a binomial cannot be equal to a major, *etc*. Therefore, the various classes of inexpressible sums and differences discussed in *Elements* X are non-overlapping. The proofs are simple and straightforward.

Two further propositions deserve to be mentioned here:

***El.* X.112** (rephrased in terms of metric algebra)
If a rectangle with expressible area is applied to a *binomial*,
then the other side is a cognate *apotome* with
a com c, b com d, and $a : b = c : d$.

***El.* X.114** (rephrased in terms of metric algebra)
A rectangle formed as the product of a binomial $a + b$ and cognate apotome $c - d$
with a com c, b com d, and $a : b = c : d$ has an expressible area.

In Knorr, *BAMS* 9 (1983), 55, the proofs of these two propositions are deservedly called "monstrously complicated". (Actually, only the proof of *El.* X.112 is quite complicated, while the proof of *El.* X.114 is based on the result in *El.* X.112.) Knorr shows that it is easy to find much simpler proofs. In the case of *El.* X.114, for instance, it is clear that the condition that $a : b = c : d$ implies that $a \cdot d = b \cdot c$. Therefore,

$$(a + b) \cdot (c - d) = a \cdot c + b \cdot c - a \cdot d - b \cdot d = a \cdot c - b \cdot d.$$

Since a, b, c, d are expressible and a com c, b com d, the products $a \cdot c$ and $b \cdot d$ are expressible areas. Therefore, also the product $(a + b) \cdot (c - d)$ is expressible, and the proof of *El.* X.114 is complete.

Knorr (*op. cit.*) further makes the remark that *El.* X.112-114 are not formulated generally for all classes of inexpressible sums and differences of straight lines discussed in *Elements* X. This is because there is no clear cut generalization of this kind. Thus, for instance, the area of a rectangle formed as the product of a major and a minor is medial, not expressible.

5.3. Euclid's Application of Areas and Babylonian Metric Division

Three elementary geometric operations that play important roles in *Elements* X are 1) the "binomial rule" which says that a square can be split into two unequal squares and two equal rectangles (as in *El.* II.4 and *El.* Lemma *El.* X.53/54 2) the construction of a square "equal" (in area) to a given rectilineal figure, and 3) the "application" of a figure (of given area) to a given straight line.

Thus, for instance, in *El. El.* X.54 it is said "let the square SN be constructed equal to the parallelogram AH, and the square NQ equal to GK". In the language of metric algebra this corresponds to constructing two squares sq. p and sq. q equal in area to the rectangles $a \cdot e$ and $b \cdot e$, as illustrated in Fig. 5.2.4. That an operation of this kind is possible is guaranteed by *El.* II.14. See Sec. 1.7 above.

An example of the third kind of elementary geometric operation is *El.* X.60, which begins with the statement that "The square on a binomial straight line applied to an expressible straight line produces as width a first binomial". See Fig. 5.2.5 above. Another example is *El.* X.20, which begins with the statement: "If an expressible area is applied to a given expressible straight line, it produces as width an expressible straight line commensurable with the given straight line". That an operation of this second kind is possible is guaranteed by the following pair of propositions:

***El.* I.43**

In any parallelogram the complements of the parallelograms about the diagonal are equal to one another.

***El.* I.44**

To a given straight line to apply, in a given rectilineal angle, a parallelogram equal to a given triangle.

The two proposition are formulated in meaningless generality, mentioning parallelograms, instead of simply rectangles. At the same time I.44 is unnecessarily restricted, mentioning (the area of) a triangle, instead of an arbitrary area. Essentially, what is meant is

I.43. In any rectangle the complements of two rectangles about the diagonal are equal.

I.44. To apply a rectangle of given area to a straight line of given length.

I.43 is illustrated by a diagram which, in metric algebra notations, corresponds to the rectangle with a diagonal in Fig. 5.3.1, left. The 'rectangles about the diagonal' are $A' + B'$ and $A'' + B''$, and the 'complements' of those rectangles are A and B. The proof of the proposition is simple: Since $A' = B'$, $A'' = B''$, and $A + A' + A'' = B + B' + B''$, it follows that also $A = B$.

The requested construction in I.44 is accomplished, essentially, in the following way: In Fig. 5.3.1, right, $A = a \cdot b$ is the given rectangle, and a' the given straight line. The first step of the construction is to complete the small rectangle with the sides a and a'. Then the diagonal of this small rectangle is extended until it intersects an extension of the right side of the given rectangle A. It is then easy to complete the rectangle with the sides b and b', and after that the rectangle B with the sides a' and b'. In view of I.43, B has the same area as A. Therefore, the construction is finished.

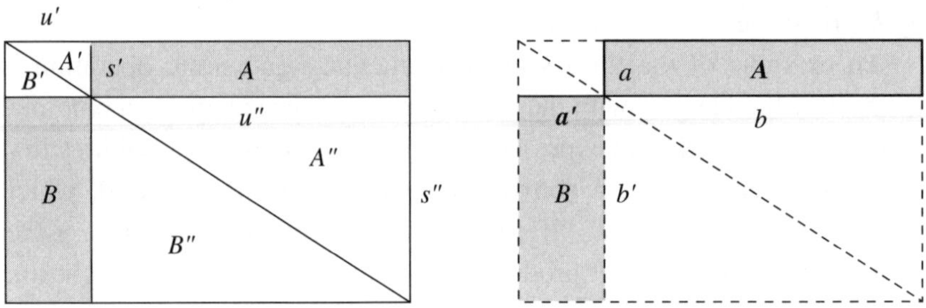

Fig. 5.3.1. The diagrams in *El.* I.43 and I.44, presented in the style of metric algebra.

The Babylonian counterpart to *El.* I.43 is the frequently used "OB similarity rule", which says that (in Fig. 5.3.1, left)

$s'' = f \cdot u''$, where $f = s'/u'$ (f is called the 'feed' of the triangle A').

The Babylonian counterpart to the construction in *El.* II.14 of "a square equal to a given figure" is the *computation of the square side* of a square of given area, also appearing frequently in OB mathematical texts.

Finally, the Babylonian counterpart to the "application of area" in I.44 is Babylonian *metric division*, the computation of the length of the second side of a rectangle when the area and the length of one side are given.

It is interesting in this connection that among the oldest known Mesopotamian texts with mathematical exercises is a group of small clay tablets from the Sargonic or Old Akkadian period in Mesopotamia (2340-2200 BCE). See the discussions in Friberg, *CDLJ* 2005:2 and *RC* (2007),

Appendix 6. The topics of this group of exercises are 1) metric division (*Friberg, CDLJ* 2005:2, §§ 2-3, Figs. 1-4), 2) computations, by use of the binomial rule, of the areas of squares with given sides (*op. cit.* § 4.3-4, Figs. 7-12), and 3) computation of the side of a square with given area (*op. cit.* § 4.7, Fig. 13). In several cases, the computations are complicated, due to an intentionally nasty choice of numerical data.

Evidently, the purpose of the exercises was to train the computational skills of the students and to increase their ability to deal with the complicated Mesopotamian systems of measures for lengths and areas. Two examples of such Old Akkadian mathematical exercises are shown below.

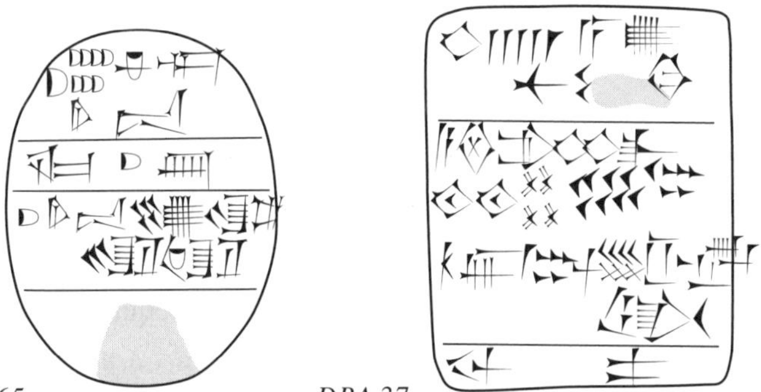

TMH 5, 65 *DPA* 37

Fig. 5.3.2. Two Old Akkadian mathematical exercises.

In ***TMH* 5, 65**, Fig. 5.3.2, left, a rectangle has the given length 1 · 60 + 7 1/2 ninda, and the given area 1 · 60 + 40 (= 100) square ninda. The width of the rectangle has been computed (metric division). The result, which is quite complicated, is recorded in the last two lines of the text. In ***DPA* 37**, Fig. 5.3.2, right, a square has the given side length 1 · 60 · 60 + 5 · 60 – 1/6 ninda. The area of the square has been computed. The result, again quite complicated, is recorded in three lines of the text.

The examples highlight an important and all pervasive difference between Sumerian/Old Akkadian/Babylonian mathematics and the kind of mathematics that one meets in Euclid's *Elements*: *Mesopotamian mathematics abounds with examples of meticulous computations with complicated numbers, while no numbers other than small integers are allowed in the deliberately non-numerical argumentation in the Elements.*

5.4. Quadratic-Rectangular Systems of Equations of Type B5

The key result in *Elements* X is the proof in *El.* X.54 and *El.* X.57 of the two related statements that if e is a given expressible straight line, and if $u + v$ is a first or fourth binomial (with respect to e), then there exists a binomial or major $p + q$, respectively, which is the "square side" of $(u + v) \cdot e$ in the sense that

$$(u + v) \cdot e = \text{sq.}\ (p + q).$$

The central part of the proof is the solution of the following "quadratic-rectangular" system of equations:

sq. p **+ sq.** $q = u \cdot e$**,** $2\,p \cdot q = v \cdot e$**, where** u**,** v**, and** e **are given straight lines.**

The solution method is based on the observation that if one sets

$\text{sq.}\ p = a \cdot e$, and $\text{sq.}\ q = b \cdot e$,

as in Fig. 5.2.4, right, then the pair a, b must be the solution to the following rectangular-linear system of equations of type B1a:

$a \cdot b = \text{sq.}\ (v/2)$, $a + b = u$, where u and v are given straight lines.

Therefore, the solution to the original system of equations is

$p = \text{sqs.}\ (a \cdot e)$, $q = \text{sqs.}\ (b \cdot e)$, where $a, b = u/2 \pm \text{sqs.}\ (\text{sq.}\ u - \text{sq.}\ v)\ /2$.

In *El.* X.33, an example of a major $p + q$ is constructed. The essential part of the construction is the solution of the *quadratic-rectangular system*

sq. p **+ sq.** q **= sq.** u**,** $2\,p \cdot q = v \cdot u$**, where** u**, and** v **are given straight lines.**

This is a special case of the quadratic-rectangular system in *El.* X.54 and *El.* X.57. The solution method, however, is quite different. See Fig. 5.2.2.

Quadratic-rectangular systems of equations of the same kind as in *El.* X.54, *etc.*, appear also in four OB mathematical exercises. They will be discussed individually below.

BM 13901(Neugebauer, *MKT 3* (1935), 1 *ff.*) is an important mathematical recombination text with 24 problems for one, two, or several squares. The problem in **BM 13901 # 12** (Høyrup, *LWS* (2002),71) is a quadratic-rectangular system of equations for two unknowns, solved in a surprising way by use of metric algebra.

BM 13901 # 12, literal translation	explanation
The fields of my two equalsides	Two squares with the sides p and q

I heaped, 21 40.	sq. p + sq. $q = S$ = 21 40
My equalsides I made eat each other, 10.	$p \cdot q = P$ = 10 (00)
The halfpart of 21 40 you break.	$S/2$ = 21 40 / 2 = 10 50
10 50 and 10 50 you make eat each other,	sq. $S/2$ = sq. 10 50
1 57 46 40. (error!)	= 1 57 21$^{!}$ 40
10 and 10 you make eat each other, 1 40.	sq. P = sq. 10 (00) = 1 40 (00 00)
Inside 1 57 46 40 you tear (it) out.	sq. $S/2$ – sq. P = 17 21$^{!}$ 40
17 46 40 makes 4 10 equalsided.	sqs. 17 21$^{!}$ 40 = 4 10
4 10 to one 10 50 you add.	10 50 + 4 10 = 15 (00)
15 makes 30 equalsided.	sqs. 15 (00)
30 is the first equalside.	= 30 = p
4 10 inside the second 10 50 you tear out.	10 50 – 4 10 = 6 40
6 40 makes 20 equalsided.	sqs. 6 40
20 is the second equalside.	= 20 = q

The quadratic-rectangular system of equations here is of the type

B5: sq. p + sq. $q = S$, $p \cdot q = P$.

It can be understood as an additional OB *basic metric algebra problem*, beyond the ones discussed in Sec. 1.1 above.

Apparently, the first step of the solution procedure is to (silently) set

sq. $p = a$, sq. $q = b$.

Then the original quadratic-rectangular system of equations for p and q is replaced by the following *basic rectangular-linear* system for a and b:

$a \cdot b$ = sq. P, $a + b = S$.

This is a system of equations of type B1a. Therefore, in the usual way,

sq. $(a - b)/2$ = sq. $(a + b)/2 - a \cdot b$ = sq. $S/2$ – sq. P, and

$a, b = (a + b)/2 \pm (a - b)/2 = S/2 \pm$ sqs. (sq. $S/2$ – sq. P).

Since sq. $p = a$ and sq. $q = b$, the result of the solution procedure in BM 13901 # 12 can be expressed in quasi-modern notations as follows:

p, q = sqs. a, sqs. b = sqs. {$S/2 \pm$ sqs. (sq. $S/2$ – sq. P)}.

With S = 21 40 and P = 10 00, the corresponding numerical answer is

p, q = sqs. {10 50 ± sqs. (sq. 10 50 – sq. 10 00)} = sqs. (10 50 ± 4 10) = 30, 20.

That $p = 30$, $q = 20$ is, actually, the known *standard answer* for most OB metric algebra problems for two squares. Apparently that is why the author of this exercise managed to find the correct answer, in spite of having calculated sq. 10 50 incorrectly as 1 57 46 40 instead of 1 57 21 40!

The same problem appears as **MS 5112 § 2**, an exercise in a large OB mathematical recombination text with metric algebra problems for squares and rectangles (Friberg, *RC* (2007), Sec. 11.2 e). Interestingly, the solution method is not the same as in BM 13901 # 12:

MS 5112 § 2 c, literal translation	explanation
The fields of 2 samesides (I) heaped, 21 40.	sq. p + sq. $q = S = 21\ 40$
Sameside with sameside (I made) eat, 10.	$p \cdot q = P = 10$ (00)
The samesides are what?	$p, q = ?$
You with your doing:	Do it like this:
10 that sameside with sameside	$P = 10$
(were made) eat to 2 repeat, 20.	$2\ P = 20$ (00)
From 2140 the fields of the samesides	$S - 2\ P =$
tear off, 1 40 is the remainder.	1 40
The equalside of 1 40 resolve, 10.	sqs. $(S - 2\ P)$ = sqs. $1\ 40 = 10 = p - q$
Its 1/2 crush, 5.	$(p - q)/2 = 5$
Steps of 5 (make) eat, 25.	sq. $(p - q)/2 = 25$
To 10, that sameside with sameside	$P = 10$
were made eat, add, 10 25.	P + sq. $(p - q)/2 = 10\ 25$
What is it equalsided?	sqs. $\{P$ + sq. $(p - q)/2\}$ =
25 each way it is equalsided.	$25 = (p + q)/2$
25 to 2 inscribe.	Write it down twice
To the 1st 25, 5 that (was made) eat add, 30.	$(p + q)/2 + (p - q)/2 = 30$
30 ninda each way, the 1st.	$p = 30$ ninda
From the 2nd 25, 5 tear off, 20.	$(p + q)/2 - (p - q)/2 = 20$
20 ninda each way, the 2nd.	$q = 20$ ninda

This is again a quadratic-rectangular system of equations of the type

B5: sq. p + sq. $q = S, \quad u \cdot s = P$.

A possible geometric interpretation of the solution procedure is illustrated in the last but one diagram in Fig. 5.4.1 below. In a square with the side $p + q$, two squares with the sides p and q are inscribed, in opposite corners. The combined area of two rectangles, both with the sides p, q is subtracted. What remains is then a square with the side $p - q$ and the area

sq. $(p - q)$ = sq. p + sq. $q - 2\ p \cdot q = S - 2\ P$.

This square is then quartered, and a square corner with the area $P = p \cdot q$ is added to the quartered square. The result is another quartered square, with the side $(p + q)/2$. Since now both the half-sum and the half-difference of p and q are known, p and q can be computed in the usual way.

The result of the solution procedure in MS 5112 § 2 can be expressed as follows, in quasi-modern notations:

$$\boldsymbol{p, q = \text{sqs.}\,\{(S + 2\,P)/4\} \pm \text{sqs.}\,\{(S - 2\,P)/4\}.}$$

With S = 21 40 and P = 10 00, the corresponding numerical answer is

$$p,\ q = \text{sqs. } 10\ 25 \pm \text{sqs. } 25 = 25 \pm 5 = 30,\ 20.$$

In *El.* X.54, *El.* X.57, BM 13901 # 12 (see Fig. 5.4.1 B and C below), and MS 5112 § 2 c (Fig. 5.4.2 D), a quadratic-rectangular system of equations of type B5 is connected with various *problems for two squares*. The same kind of quadratic-rectangular system of equations is connected with various problems for *the diagonal and the area of a rectangle or a right triangle*. Thus, in *El.* X.33 (Fig. 5.4.1 A), the diagonal u and area $A = v/2 \cdot u/2$ of a right triangle are known, with $v/2$ denoting the height of the right triangle against the diagonal. The situation is described by the equations

$$\text{sq. } p + \text{sq. } q = \text{sq. } u, \quad p \cdot q = 2\,A = v/2 \cdot u.$$

Two OB clay tablets with closely related quadratic-rectangular systems of equations of type B5 for rectangles or right triangles are known, namely **IM 67118**, also known as Db_2-146, and **MS 3971 § 2** (Friberg, *RC* (2007), Sec. 10.1 b). The text of MS 3971 § 2 is reproduced below. (See also the illustrating diagram in Fig. 5.4.1 E.)

MS 3971 § 2, literal translation	explanation
1 15 the cross-over,	d, the diagonal (of a rectangle), = 1 15
45 the field.	A, the area, = 45 (00)
The length and the front are what?	The length u and the front s = ?
1 15 (make) butt (itself) 1 33 45 it gives.	sq. d = sq. 1 15 = 1 33 45
45, the field, to 2 you repeat, 1 30.	$2\,A$ = 1 30 (00)
1 30 to a 33 45 *join, 3* 03 45.	sq. $d + 2\,A$ = 3 03 45
3 03 45 makes 1 45 *equ*alsided.	sqs. (sq. $d + 2\,A$) = 1 45 = p
1/2 of 1 45 break, *52 30 it gives.*	$p/2$ = 52;30
52 30 (make) butt (itself), 45 *56* 15 it gives.	sq. $p/2$ = 45 56;15
45, the field, from 45 *56 15* tear off,	sq. $p/2 - A$ = 45 56;15 – 45 (00)
56 15 it gives.	= 56;15 = sq. $q/2$
56 15 <makes> 7 30 equalsided.	sqs. 56;15 = 7;30 = $q/2$
7 30 to 52 30 join, 1, the length, it gives.	$p/2 + q/2$ = 52;30 + 7; 30 = 1 (00) = u
from 52 30 tear off, 45, the front, it gives.	$p/2 - q/2$ = 52;30 – 7; 30 = 45 = s

A. *El.* X.33:

A quadratic-rectangular system of equations:

sq. p + sq. q = sq. $\mathbf{u}$, $p \cdot q = \mathbf{v}/2 \cdot \mathbf{u}$

Solution:

sq. $p = a \cdot \mathbf{u}$, sq. $q = b \cdot \mathbf{u} \Rightarrow$

$a + b = \mathbf{u}$, $a \cdot b$ = sq. $\mathbf{v}/2 \Rightarrow$

p, q = sqs. ($\{\mathbf{u}/2 \pm$ sqs. (sq. $\mathbf{u}/2$ – sq. $\mathbf{v}/2)\} \cdot \mathbf{u}$)

= sqs. ($\{\mathbf{u}/2 \pm w/2\} \cdot \mathbf{u}$)

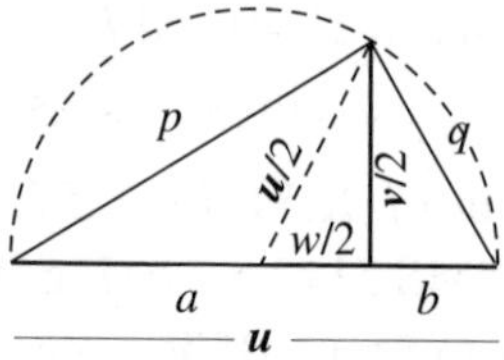

B. *El.* X.54, X.57:

A quadratic-rectangular system of equations:

sq. p + sq. $q = \mathbf{u} \cdot \mathbf{e}$, $p \cdot q = \mathbf{v}/2 \cdot \mathbf{e}$

Solution:

sq. $p = a \cdot \mathbf{e}$, sq. $q = b \cdot \mathbf{e} \Rightarrow$

$a + b = \mathbf{u}$, $a \cdot b$ = sq. $\mathbf{v}/2 \Rightarrow$

p, q = sqs. ($\{\mathbf{u}/2 \pm$ sqs. (sq. $\mathbf{u}/2$ – sq. $\mathbf{v}/2)\} \cdot \mathbf{e}$)

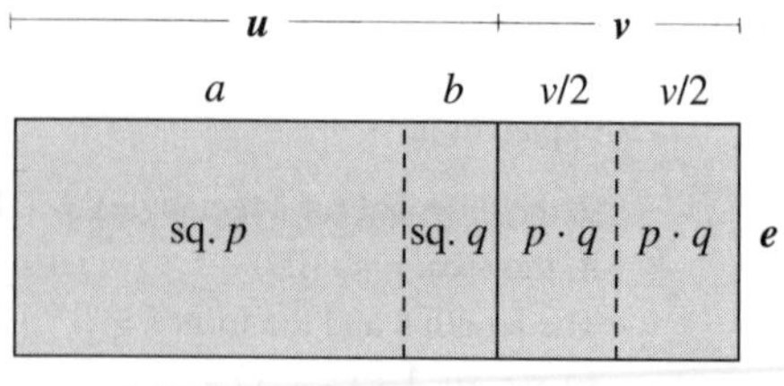

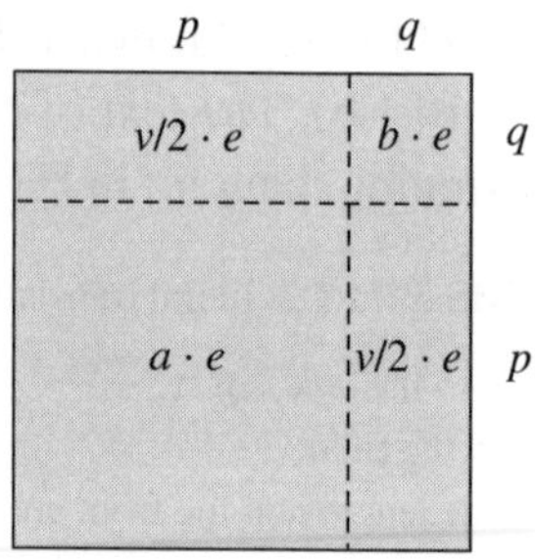

C. BM 13901 # 12:

A quadratic-rectangular system of equations:

sq. p + sq. $q = \mathbf{S}$, $p \cdot q = \mathbf{P}$

Solution:

sq. $p = a$, sq. $q = b \Rightarrow$

$a + b = \mathbf{S}$, $a \cdot b$ = sq. $\mathbf{P} \Rightarrow$

p, q = sqs. $\{\mathbf{S}/2 \pm$ sqs. (sq. $\mathbf{S}/2$ – sq. $\mathbf{P})\}$

Geometric interpretation:
As in **B**, but with $\mathbf{e} = 1$?

Fig. 5.4.1. Three ways of solving a quadratic-rectangular system of equations of type B5.

D. MS 5112 § 2 c:

A quadratic-rectangular system of equations:

sq. p + sq. $q = \mathbf{S}$, $p \cdot q = \mathbf{P}$

Solution:

sq. $(p - q) = \mathbf{S} - 2\,\mathbf{P}$, sq. $\{(p + q)/2\} = \mathbf{P}$ + sq. $(p - q)/2 \Rightarrow$

p, q = sqs. $\{(\mathbf{S}/2 + \mathbf{P})/2\}$ ± sqs. $\{(\mathbf{S}/2 - \mathbf{P})/2\}$

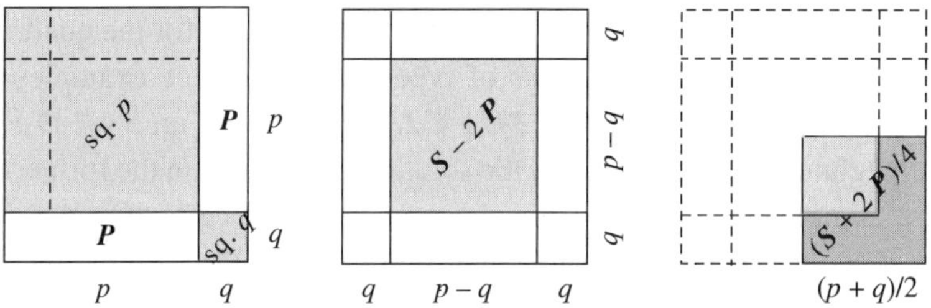

E. IM 67118 & MS 3971 § 2:

A quadratic-rectangular system of equations:

sq. u + sq. s = sq. $\mathbf{d}$, $u \cdot s = \mathbf{A}$

Solution:

sq. $(u - s)$ = sq. $\mathbf{d} - 2\,\mathbf{A}$, sq. $\{(u + s)/2\}$ = sq. $(u - s)/2 + \mathbf{A} \Rightarrow$

u, s = sqs. $\{($sq. $\mathbf{d}/2 + \mathbf{A})/2\}$ ± sqs. $\{($sq. $\mathbf{d}/2 - \mathbf{A})/2\}$

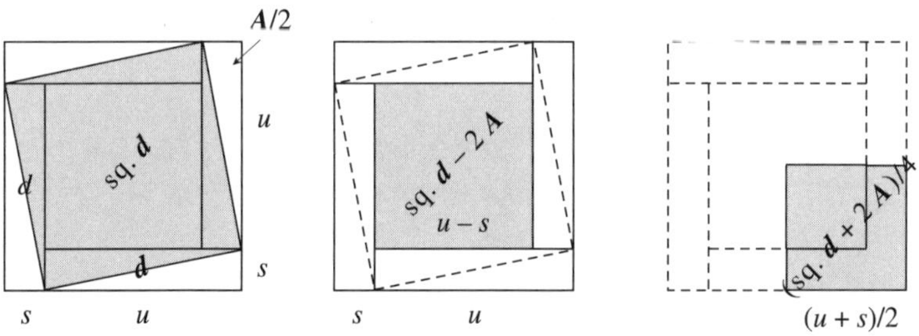

Fig. 5.4.2. Two more ways of solving a quadratic-rectangular system of equations.

(The diagram in Fig. 5.4.2 E illustrates the text IM 67118, which differs from MS 3971 § 2 only in that the first step in IM 67118 is to compute sq. $d - 2\,A$, rather than sq. $d + 2\,A$, as in MS 3971 § 2.)

The form of the solution in MS 3971 § 2 is of the same form as the solution in MS 5112 § 2 c (see above):

$$\boldsymbol{u, s = \{\text{sqs. (sq. } d + 2\,A)\}/2 \pm \{\text{sqs. (sq. } d - 2\,A)\}/2.}$$

With $d = 1\ 15$ and $A = 45\ 00$, the corresponding numerical answer is

$$u,\ s = \text{sqs. } (3\ 03\ 45)\ /2 \pm \text{sqs. } (3\ 45)\ 2 = 52;30 \pm 7;30 = 1\ 00,\ 45.$$

It is clear that the solution procedures for the quadratic-rectangular systems of equations of type B5 in the four examples *El.* X.33, *El.* X.54, *El.* X.57, and BM 13901 # 12, exhibited in Fig. 5.4.1 A-C, are closely related to each other. Similarly, the solution procedures for the quadratic-rectangular systems of equations of type B5 in the three examples MS 5112 § 2 c, IM 67118, and MS 3971 § 2, exhibited in Fig. 5.4.2 D-E are closely related to each other. Yet the solution procedures in the former cases are completely different from those in the latter cases. On the other hand, when the data are the same, the solutions ought to be the same, too, in all the considered cases. Now, the system of equations

$$\text{sq. } p + \text{sq. } q = S, \quad p \cdot q = P$$

has the solution

$$p,\, q = \text{sqs. } \{S/2 \pm \text{sqs. (sq. } S/2 - \text{sq. } P)\}$$

in the text BM 13901 # 12, but the solution

$$p,\, q = \text{sqs. } \{(S/2 + P)/2\} \pm \text{sqs. } \{(S/2 - P)/2\}$$

in the text MS 5112 § 2 c. Therefore, the obvious conclusion is that

$$\boldsymbol{\text{sqs. } \{S/2 \pm \text{sqs. (sq. } S/2 - \text{sq. } P)\} = \text{sqs. } \{(S/2 + P)/2\} \pm \text{sqs. } \{(S/2 - P)/2\}.}$$

Chapter 6

Elements IV and Old Babylonian Figures Within Figures

6.1. *Elements* IV, a Well Organized Geometric Theme Text

Book IV of Euclid's *Elements* is quite brief, with only 16 propositions, all concerned with *figures within figures*:

An Outline of the Contents of *Elements* IV

1 To inscribe a given *straight line*, not greater than the diameter, in a given circle.

2 To inscribe a *triangle* of given form in a given circle.
3 To circumscribe a triangle of given form around a given circle.
4 To inscribe a circle in a given triangle.
5 To circumscribe a circle around a given triangle.

6 To inscribe a *square* in a given circle.
7 To circumscribe a square around a given circle.
8 To inscribe a circle in a given square.
9 To circumscribe a circle around a given square.

10 To construct a triangle with each of the angles at the base double the remaining angle.
11 To inscribe a regular *pentagon* in a given circle.
12 To circumscribe a regular pentagon around a given circle.
13 To inscribe a circle in a given regular pentagon.
14 To circumscribe a circle around a given regular pentagon.

15 To inscribe a regular *hexagon* in a given circle.

16 To inscribe a regular *15-gon* in a given circle.

Most of these propositions are simple constructions. Only IV.10-11 are more interesting, in particular IV.10, which is a quite ingenious construction of a special triangle needed for the subsequent construction in IV.11.

In terms of metric algebra, the construction in ***El.* IV.10** begins with a straight line d cut in two parts in the way described in *El.* II.11 (Sec. 1.7 above), so that the rectangle contained by the whole and the smaller segment is equal to the square on the greater segment. If the two segments are called s and d' (for reasons that will become clear later), with $s > d'$, then s and d' satisfy the equation

$$(s + d') \cdot d' = \text{sq. } s.$$

A circle is drawn with the divided straight line d as a radius, and so that the larger segment s emanates from the center of the circle (Fig. 6.1.1, left). Next, an isosceles triangle is constructed, with the divided straight line as one of its legs, and with a chord of the circle, of length s, as its base. From the endpoint of the chord, a straight line is drawn to the point where the original straight line is cut in two parts. A circle is circumscribed around a second triangle formed by this straight line, the segment s, and the second straight line of length d [IV.5]. It is observed that, in view of III.37, the base s of the first triangle is a tangent to the circumscribed circle. And so on, with reference to, among other things, I.32 and III.32. The result of the complicated construction is the desired triangle with the two angles at its base each twice as large as the remaining angle

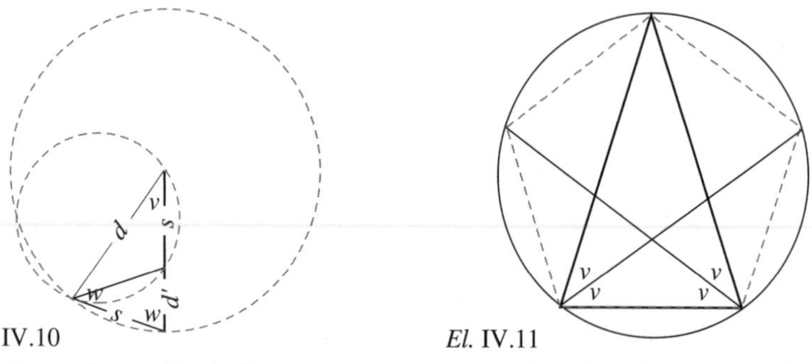

El. IV.10

$d = s + d', \quad s > d', \quad d \cdot d' = \text{sq. } s$

$\Rightarrow \quad w = 2\,v$

El. IV.11

To inscribe a regular pentagon in a given circle

Fig. 6.1.1. Euclid's method for the inscription of a regular pentagon in a circle.

In ***El.* IV.11**, a regular pentagon inscribed in a given circle is constructed in the following way. First, a triangle of the same form as the triangle constructed in IV. 10 is inscribed in the circle [IV.2]. Then. two chords in the circle are drawn, bisecting the two equal angles at the base of the triangle (Fig. 6.1.1, right). The three vertices of the triangle and the endpoints of the two chords then determine the positions of the five vertices of the regular pentagon. End of the construction

As mentioned, figures within figures play a dominant role in *Elements* IV. In IV.2-5, for instance, it is required that a triangle of given form shall be *inscribed in or circumscribed around a given circle*, alternatively that *a circle shall be inscribed in or circumscribed around* a given triangle. In IV.6-9, the same four types of constructions are repeated, with the triangle replaced by a square. In IV.10-14, the square is replaced by a regular pentagon. In IV.15-16, finally a regular hexagon and a regular 15-gon are inscribed in circles. From the point of view of Babylonian mathematics, *Elements* IV is a typical "theme text". There are several known OB theme texts of various kinds, some of them discussed in Secs. 1.10-1.12 above. There are reasons to believe that well organized theme texts were the "original" OB mathematical texts, while other types of OB mathematical texts contain more or less extensive excerpts from such theme texts.

Note that an important difference between *Elements* IV and an OB theme text is that all the propositions in *Elements* IV are *non-numerical construction problems*, while all exercises in OB geometric texts are *computational with emphasis on metric relationships*.

Figures within figures play a dominant role also in *Elements* XIII. (See Chapter 7 below.) However, *Elements* XIII is a less clear cut case than *Elements* IV. The propositions in *Elements* XIII have a double purpose. On one hand, *Elements* XIII.13-17 is a continuation of *Elements* IV, with elaborate *constructions of the five regular polyhedrons inscribed in a given sphere*. On the other hand, just as important is that *Elements* XIII contains a number of results showing how to *express* the length of *the side of an inscribed regular polygon* (a pentagon or equilateral triangle) or *the edge of an inscribed regular polyhedron* in terms of the radius or diameter of the given circle or sphere.

6.2. Figures Within Figures in Mesopotamian Mathematics

Geometric objects (other than triangles, squares, rectangles, and trapezoids) occur relatively infrequently in Babylonian mathematics. Moreover, in many cases, clay tablets with drawings of geometric figures appear to have been hastily written on roughly shaped "hand tablets" either by young and inexperienced students, or by students listening inattentively to their teachers' explanations and making careless notes of what they saw and heard. For these reasons, it is difficult to get a clear picture of how much Babylonian mathematicians really knew about geometry.

Nevertheless, it is clear that figures within figures was one of the favorite geometric themes in Babylonian mathematics. This will be shown below by means of a number of examples.

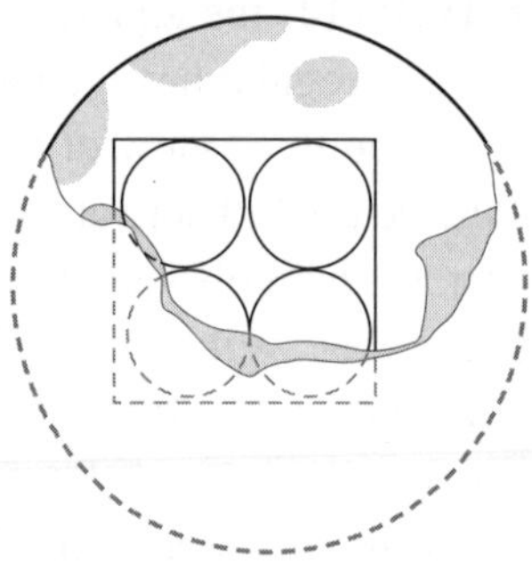

Fig. 6.2.1. A diagram on a Mesopotamian clay tablet from the Early Dynastic IIIa period.

The oldest known example of a geometric diagram on a Mesopotamian clay tablet is ***TSS* 77** (Jestin 1937). It is a fragment of a round hand tablet, from the ancient site Shuruppak, dateable to the Early Dynastic IIIa period (c. 2600-2500 BCE). It shows four circles inscribed in a square.

The same figure reappears in the exercise **BM 15285 # 36** (Neugebauer, *MKT 1* (1935), 137-142; Robson, *MMTC* (1999), Appendix 2), two large fragments of a famous OB geometric theme text with originally 41 briefly formulated problems, all illustrated by diagrams of squares divided into smaller pieces. The purpose of the problems is, in each case, to compute the areas of all the pieces.

In the outlines of obverse and reverse of BM 15285 in Figs. 6.2.2-3 below, the text belonging to each exercise is shown only as a gray rectangle, because the copy of the clay tablet is scaled down so much that the text would be unreadable, anyway.

Instead, the texts belonging to the exercises are given below in literal translation, in the cases when they are sufficiently well preserved.

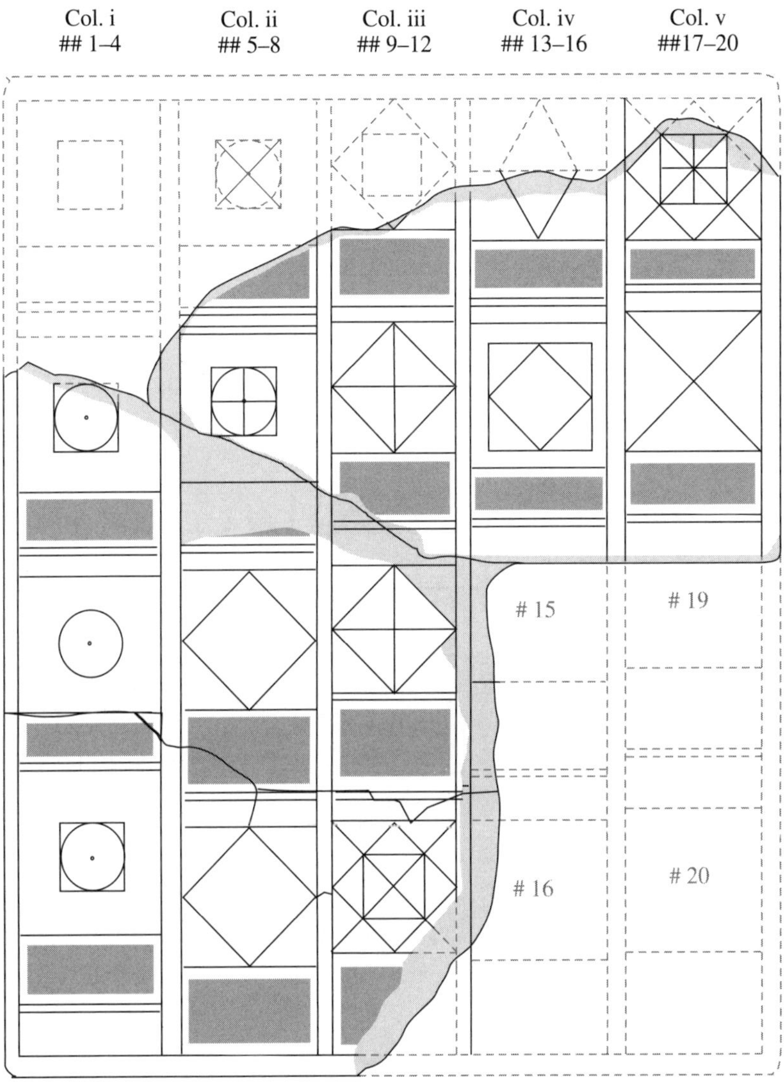

Fig. 6.2.2. The obverse of BM 15285, an Old Babylonian geometric theme text.

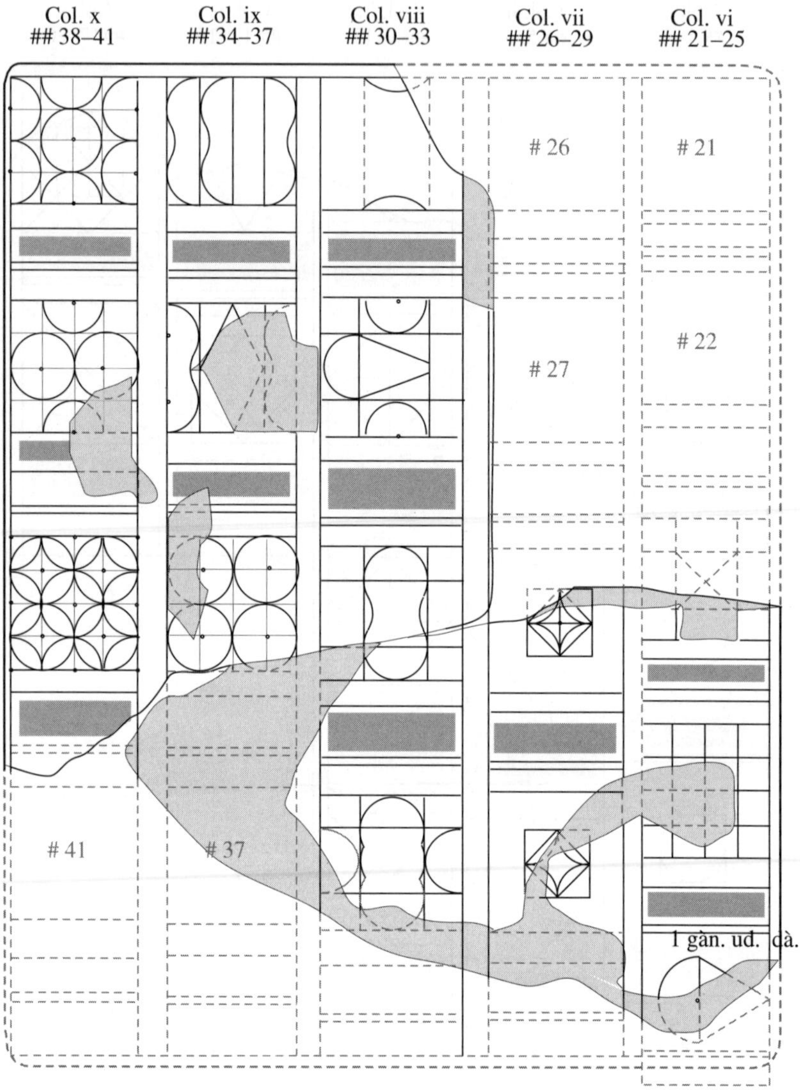

Fig. 6.2.3. The reverse of BM 15285. (The order of the columns is reversed here.)

	BM 15285, literal translation	explanation
...	....	....
2.	1 uš the equalside.	Given a square with the side 1 00 (ninda).
	A bit(?) I pushed inwards, then	A certain distance(?) inwards (from the sides
	a second equalside I drew.	of the square) a second square is drawn.
	Inside the equalside, an arc I drew.	A circle is drawn inside the (second) square.
	Their areas are what?	Compute the areas of the pieces.
3.	1 [uš the equalside.]	*Given a square with the side* 1 00 *(*ninda*).*
	A bit I pushed inwards, then	A certain distance inwards (from the sides
	an arc I drew.	of the square) a circle is drawn.
	Their areas are what?	Compute the areas of the pieces.
4.	1 uš the equalside.	Given a square with the side 1 00 (ninda).
	Inside the equalside	Inside it
	an arc I drew.	a circle is drawn.
	The arc that I drew touches	The circle is inscribed in
	an equalside.	a (second) square.
	Their areas are what?	Compute the areas of the pieces.
5.	[1 uš the equalside.]	*Given a square with the side 1 00 (*ninda*).*
	[Inside it] a second [equalside.]	A second *square is drawn inside it.*
	[Inside the second equalside]	*Inside the second square*
	4 peg-heads, 1 arc I drew.	are drawn 4 triangles and 1 circle.
	Their areas are what?	Compute the areas of the pieces.
6.	[1 uš the equalside.]	*Given a square with the side 1 00 (*ninda*).*
	[Inside it a second equalside.]	*A second square (is drawn) inside it).*
	[Inside it the second equalside]	*Inside the second square*
	[4 equalsides (and) 1 arc I drew.]	*are drawn 4 triangles and 1 circle.*
	Their areas are what?	Compute the areas of the pieces.
7.	1 uš the equalside.	Given a square with the side 1 00 (ninda).
	Inside it a second equalside I drew.	A second square is drawn inside it.
	The equalside that I drew	The second square
	touches the outer equalside.	is inscribed (obliquely) in the given square.
	Their areas are what?	Compute the areas of the pieces.
8.	1 uš the equalside.	Given a square with the side 1 00 (ninda).
	Inside it 4 peg-heads, 1 equalside.	Inside it 4 triangles and 1 square (are drawn).
	The equalside that I drew	The (second) square
	touches the second equalside.	is inscribed (obliquely) in the given square.
	Their areas are what?	Compute the areas of the pieces.
9.	1 uš the equalside.	Given a square with the side 1 00 (ninda).
	Inside it an equalside I drew.	Inside it a square (is drawn).
	[The equalside] that I drew	*The (second) square*

touches the equalside.	is inscribed (obliquely) in the given square.
Inside the second equalside	Inside the second square
a third equalside I drew.	is drawn a third square.
(The equalside) that I drew	The (third) square
touches the equalside.	is inscribed (obliquely) in the (second) square.
Their areas are what?	Compute the areas of the pieces.
10. 1 uš the equalside.	Given a square with the side 1 00 (ninda).
Inside it 8 peg-heads I drew.	Inside it 8 triangles (are drawn).
Their areas are what?	Compute the areas of the pieces.
11. 1 uš the equalside.	Given a square with the side 1 00 (ninda).
Inside it an equalside I drew.	Inside it a second square is drawn.
The equalside that I drew	The (second) square
touches the equalside.	is inscribed (obliquely) in the (given) square.
Inside the equalside	Inside the (second) square
4 peg-heads I drew.	are drawn 4 triangles.
12. 1 uš the equalside.	Given a square with the side 1 00 (ninda).
Inside it 16 [peg-heads I drew].	Inside it 16 *triangles are drawn.*
Their areas are what?	Compute the areas of the pieces.
13. 1 uš the equalside. Inside it	Given a square with the side 1 00 (ninda).
4 ox-heads, 2 peg-heads I drew.	Inside it 4 trapezoids, 2 triangles are drawn.
Their areas are what?	Compute the areas of the pieces.
14. 1 uš the equalside.	Given a square with the side 1 00 (ninda).
Half I pushed inwards,	Halfway in (?) (from the sides of the given
then an equalside I drew.	square) another square is drawn.
Inside the second equalside	Inside the second square
a third equalside I drew.	a third square is drawn (obliquely).
Their areas are what?	Compute the areas of the pieces.
... .	....
17. 1 uš the equalside.	Given a square with the side 1 00 (ninda).
12 peg-heads 4 equalsides I drew.	(Inside it) 12 triangles, 4 squares are drawn.
Their areas are what?	Compute the areas of the pieces.
18. 1 uš the equalside.	Given a square with the side 1 00 (ninda).
Inside 4 peg-heads I drew.	Inside it 4 triangles are drawn.
Their areas are what?	Compute the areas of the pieces.
... .	....
23. 1 uš the equalside.	Given a square with the side 1 00 (ninda).
Inside it 4 equalsides,	Inside it 4 squares,
4 diagonals, 4 peg-heads.	4 rectangles and 4 triangles are drawn.
Their areas are what?	Compute the areas of the pieces.

24. [1 uš] the equalside. Inside it 16 equalsides I drew. Their areas are what?	Given a square with the side *1 00* (ninda). Inside it 16 squares are drawn. Compute the areas of the pieces.
25. A crescent.	A semicircle. ....
... .	....
28. 1 [uš the equalside.] A bit I pushed inwards, then an equalside I drew. Inside the equalside that I drew 1 lyre-window. Their areas are what?	*Given a square with the side* 1 00 (ninda). A certain distance inwards (from the given square) a square is drawn. Inside that square there is 1 concave square. Compute the areas of the pieces.
... .	....
30. 1 uš the equalside. A bit I pushed inwards, then a lyre field(?) I drew. Their areas are what?	Given a square with the side 1 00 (ninda). A certain distance inwards (from the given square) a concave rectangle(?) is drawn. Compute the areas of the pieces.
31. 1 [uš] the equalside. [Inside it] 2 crescents, 1 peg-head, 1 peg-crescent(?), 1 diagonal, 4 equalsides. Their areas are what?	Given a square with the side 1 *00* (ninda). *Inside it* there are 2 semicircles, 1 triangle, 1 circle segment(?), 1 rectangle, and 4 squares. Compute the areas of the pieces.
32. [1] uš the equalside. Inside it 2 diagonals, 1 ··· field, 4 equalsides. Their areas are what?	Given a square with the side *1 00* (ninda). Inside it there are 2 rectangles, 1 ····-figure, and 4 squares. Compute the areas of the pieces.
... .	....
34. 1 uš the equalside. Inside it 3 bow fields, 1 diagonal. Their areas are what?	Given a square with the side 1 00 (ninda). Inside it there are 3 bow figures. and1 rectangle. Compute the areas of the pieces.
35. 1 uš the equalside. Inside it 2 bow fields, 1 ····-field, 4ox-heads ··· . Their areas are what?	Given a square with the side 1 00 (ninda). Inside it there are 2 bow figures, 1 ····-figure, 4 trapezoids (sic!) ··· . Compute the areas of the pieces.
... .	....
38. 1 uš the equalside. Inside it 1 arc, 6 crescents. Their areas are what?	Given a square with the side 1 00 (ninda). Inside it there are 1 circle and 6 semicircles. Compute the areas of the pieces.

39. 1 uš the equalside. 2 circles, 2 crescents, 4 [equalsides]. Their areas [are what]?	Given a square with the side 1 00 (ninda). (Inside it) 2 circles, 2 semicircles, 4 squares. Compute the areas of the pieces.
40. [1 uš the equalside.] Inside it 4 peg-heads, [16] boat fields, [5] lyre-windows. Their areas are what?	*Given a square with the side 1 00* (ninda*).* Inside it there are 4 (concave) triangles, *16* boat-figures, and *5* concave squares. Compute the areas of the pieces.

As a geometric theme text, BM 15285 is not particularly well organized, and the exercises are in general exceedingly simple. Yet the text is interesting for several reasons, not least because it quite explicitly gives the names (mostly Sumerian) of a number of geometric figures. A few of those figures will be discussed below (Fig. 6.2.6 and Chapter 12).

Some of the diagrams illustrating the exercises occur more than once. Thus the diagrams for problems ## 2 and 4 are identical, and so are the diagrams for ## 7-8, and for ## 10-11. The reason is probably that the author of the text wanted to teach his students that the same diagram can be interpreted in more than one way. Note by the way, how much the situation in ## 7-8, 10-11, and 18, resembles the famous geometric passage in Plato's ***Meno*, 82 B - 85 B** (see Heath, *HGM 1* (1981), 297), where Socrates tries to get a slave boy to figure out on his own how a square can be constructed that is twice as large as a given square.

Particularly interesting are **BM 15285 ## 36 and 40**. The text under the diagram for # 36 is lost, but presumably it was of the following form:

> Given a square with the side 1 00 (ninda). Inside it there are 8 (concave) triangles, 4 circles, and 1 concave square. Compute the areas of the pieces.

As a help for the drawing of the diagram, and also for the computation of the areas of the pieces, there are weakly drawn guide lines in the diagrams for ## 36 and 40, dividing the given squares into 16 smaller squares.

An OB school boy could find the solution to problem # 36 in the following way, for instance: The guide lines show that each one of the four circles is contained in a square with the side 30. Therefore, the diameter of each circle is also 30, and the arc (the circumference) is, approximately, 3 · 30 = 1 30. Hence, the area of each circle is, approximately, 1/12 · sq. 1 30 = ;05 · 2 15 (00) = 11 15. On the other hand, the area of one of the circumscribed squares is sq. 30 = 15 (00), which is 3 45 more than the area of the circle inside it. This means that the area of each one of the four "concave triangles" in the corners of one of the small squares, outside the circle, is 3 45 / 4 = 56;15. Thus, the area 1 (00 00) of the given square with the side 1 (00) is the sum of the following sub-areas:

the total area of four small circles = appr. 4 · 11 15 = 45 (00),
the area of the central "concave square" (lyre-window) = appr. 4 · 56;15 = 3 45,
the total area of four "double concave triangles" = appr. 4 · 2 · 56;15 = 7 30,
the total area of four "single concave triangles" = appr. 4 · 56;15 = 3 45.

A similar computation in the case of problem # 40 would yield the result that the area of the given square is the sum of the following sub-areas:

the total area of five concave squares = appr. 5 · 3 45 = 18 45,
the total area of four double concave triangles = appr. 4 · 2 · 56;15 = 7 30,
the total area of four single concave triangles = appr. 4 · 56;15 = 3 45,
the total area of 16 boat-figures = appr. 16 · 1 52;30 = 30.

Note that the area of any one of the boat-figures can be computed as 1/4 of the difference between the area of a circle and the area of a concave square inscribed in the circle, that is, as appr. (11 15 – 3 45)/4 = 7 30/4 = 1 52;30. Note also that, for some reason, the four single concave squares are not mentioned in the text of problem 40.

Indirectly related to the geometric theme text BM 15285 is a quite well known sequence of entries in the OB mathematical "table of constants" **BR** = Bruins and Rutten, ***TMS* 3** (1961):

5 igi.gub *šà* gúr	5, constant of the arc	BR 2
20 dal *šà* gúr	20, transversal of the arc	BR 3
10 *pi-ir-ku šà* gúr	10, crossline of the arc	BR 4
15 igi.gub *šà ús-ka*$_4$*-ri*	15, constant of the crescent	BR 7
40 dal *šà ús-ka*$_4$*-ri*	40, transversal of the crescent	BR 8
20 *pi-ir-ku šà ús-ka*$_4$*-ri*	20, crossline of the crescent	BR 9
6 33 45 igi.gub *šà pa-na-ak-ki*	6 33 45, constant of the bow	BR 10
52 30 dal *šà pa-na-ak-ki*	52 30, transversal of the bow	BR 11
15 *pi-ir-ku šà pa-na-ak-ki*	15, crossline of the bow	BR 12
13 07 30 igi.gub *šà* gán giš.má	13 07 30, constant of the boat field	BR 13
52 30 dal *šà* gán giš.má	52 30, transversal of the boat field	BR 14
30 *pi-ir-ku šà* gán giš.má	30, crossline of the boat field	BR 15
13 20 igi.gub *šà* a.šà še	13 20, constant of the barleycorn field	BR 16
56 40 dal *šà* a.šà še	56 40, transversal of the barleycorn field	BR 17
23 20 *pi-ir-ku šà* a.šà še	23 20, crossline of the barleycorn field	BR 18
16 52 30 igi.gub *šà* igi.gu$_4$	16 52 30, constant of the ox-eye	BR 19

52 30 dal *šà* igi.gu$_4$	52 30, transversal of the ox-eye	BR 20
30 *pi-ir-ku šà* igi.gu$_4$	30, crossline of the ox-eye	BR 21
26 40 igi.gub *šà a-pu-sà-am-mi-ki*	26 40, constant of the lyre-window	BR 22
1 20 dal *šà a-pu-sà-mi-ki*	1 20, transversal of the lyre-window	BR 23
33 20 *pi-ir-ku šà a-pu-sà-mi-ki*	33 20, crossline of the lyre-window	BR 24
15 *a-pu-sà-mi-ik-ki šà* 3	15, the lyre-window of 3	BR 25

In this systematically organized sequence of entries, three numerical parameters are given for each one of seven geometric figures, namely, in this order, the 'arc' (circle), the 'crescent' (semicircle), the 'bow', the 'bow field', the 'barleycorn field', the 'ox-eye', and the 'lyre-window' (concave square).

The first three entries, those for the circle, can be explained as follows: For a *circle* with given arc *a*, the remaining three parameters, namely the area *A*, the diameter *d*, and the radius *r,* can be computed as follows:

$A = (1/4\pi \cdot \text{sq. } a =)$ appr. $1/12 \cdot \text{sq. } a = ;05 \cdot \text{sq. } a$ BR 2
$d = (1/\pi \cdot a =)$ appr. $1/3 \cdot a = ;20 \cdot a$ BR 3
$r = (1/2\pi \cdot a =)$ appr. $1/6 \cdot a = ;10 \cdot a$ BR 4

For a *semicircle* with the arc *a*, the area *A*, the diameter *d*, and the radius *r,* the four parameters are connected through the following equations:

$A = 1/4 \cdot a \cdot d = ;15 \cdot a \cdot d$ BR 7
$d = (2/\pi \cdot a =)$ appr. $2/3 \cdot a = ;40 \cdot a$ BR 8
$r = (1/\pi \cdot a =)$ appr. $1/3 \cdot a = ;20 \cdot a$ BR 9

And so on. See Robson, *MMTC* (1999), Chapter 3, for details. Note that the 'bow' in BR 10-12 is not identical with the 'bow field' mentioned in BM 15285 ## 34-35!

An OB mathematical problem mentioning the 'lyre-window' (BR 22-24) will be discussed below (Fig. 6.2.6). The 'barleycorn field' and the 'ox-eye' (BR 16-18, 19-21) will play prominent roles in Chapter 12 below.

Now, consider again the OB geometric theme text BM 15285. Its connection with the theme of *Elements* IV is particularly obvious in the two exercises # 2 and # 4, illustrated by identical diagrams. The text in # 2 says 'Inside the equalside, an arc I drew', while the text in # 4 says 'The arc that I drew touches an equalside'. In other words, the difference between # 2 and # 4 is that in the former exercise a circle is *inscribed in* a square, while in the latter exercise a square is *circumscribed around* a circle!

MS 3050 and **MS 3051** are two OB hand tablets published in Friberg, *RC* (2007), Figs. 8.2.2 and 8.1.1. Both are inscribed with geometric diagrams and scribbled numbers (see Figs. 6.2.4-5 below).

obv.

Fig. 6.2.4. MS 3050. A square with diagonals inscribed in a circle (Old Babylonian).

obv.

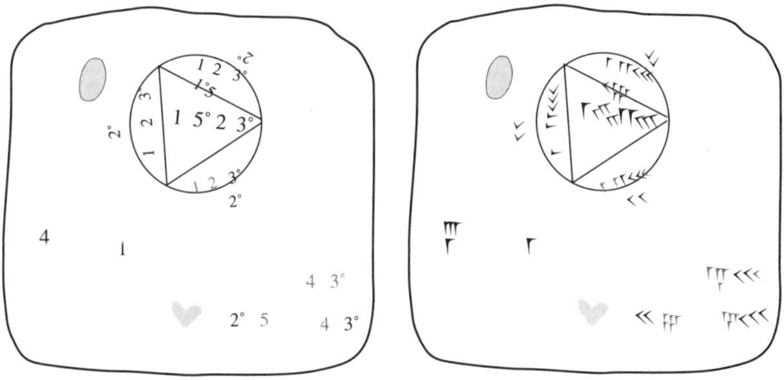

Fig. 6.2.5. MS 3051. An equilateral triangle inscribed in a circle (Old Babylonian).

The diagram in MS 3050 shows a square with diagonals, inscribed in a circle, while the diagram in MS 3051 shows an equilateral triangle inscribed in a circle. In agreement with an OB convention, the 'fronts' of the inscribed square and the inscribed equilateral triangle both face to the left. (In a *Late* Babylonian diagram showing an equilateral triangle, the triangle would have been shown standing on its base.)

It is not easy to make sense of the all scribbled numbers on MS 3050, even if some of them seem to suggest that the diameter of the circle (= the diagonal of the square) was assumed to have the length 1 (00).

The diagram on MS 3051 is amazingly exact. The sides of the triangle are nearly equal. The circle passes through two of the three vertices of the triangle and passes close by the third vertex. It is clear that a compass must have been used in the construction of the figure, although there are no remaining traces of the point of the compass. It is also clear that the accurate construction of a figure of this kind would be difficult without a good understanding of basic geometric principles (*cf. El.* IV.2 and IV.5).

The equilateral triangle divides the circumference of the circle in three equal parts. In the diagram on MS 3051, they are all marked with the number '20'. That means that the whole circumference of the circle is 1 (00).

Presumably, the students who drew the diagrams on MS 3050 and MS 3051 were supposed to compute the areas of the inscribed square and the equilateral triangle, as well as the areas of the four circle segments outside the square, and of the three circle segments outside the triangle. Unfortunately, no Babylonian mathematical text are known that contain the details of such computations.

On the other hand, in the Egyptian demotic text *P. Cairo* (the 3rd c. BCE), which is strongly influenced by Babylonian mathematics (see Friberg, *UL* (2005), Sec. 3.1), there are two exercises with precisely that kind of computations. Thus, in ***P.Cairo*** **§ 12** (*op. cit.*, Sec. 3.1 k), a square is inscribed in a circle with given diameter $d = 30$ cubits and given area $A = 675$ square cubits (= 3/4 · sq. 30). In the exercise, very good approximations of the areas of the inscribed square and of the four circle segments are computed, and it is shown that the sum of these sub-areas is almost precisely equal to the area of the circle. In the first step of the computation, for instance, the area of the inscribed square is computed as half the square on the diagonal of the square, equal to the diameter of the circle.

In ***P.Cairo*** **§ 11** (*op. cit.*, Sec. 3.1 j), an equilateral triangle of side $s = 12$ (divine) cubits is inscribed in a circle. In a number of steps, the following numerical parameters are computed: 1) the height of the equilateral triangle, 2) the area of the equilateral triangle, 3) the height of a circle segment, 4) the area of a circle segment, 5) the sum of the sub-areas, 6) the diameter of the circle, 7) the circumference of the circle, 8) the area of the circle, 9) the (small) difference between the results in 5) and 8).

It is likely that an Old (or Late) Babylonian school boy would have performed the computations in the same way, provided that he had reached a sufficiently advanced stage of his education in mathematics.

Another category of OB mathematical problems for figures within figures are related to diagrams like ## 1-6 and 28-29 in the geometric theme text BM 15285 (see Figs. 6.2.2-3 above) In this category of problems, a figure is inscribed *in the middle of another figure*, a given distance away from the sides of that figure.

The best example of a problem of this kind is ***TMS* 21** (Bruins and Rutten (1961); Friberg, *RC* (2007), Sec. 8.2), a difficult problem text only recently explained by Muroi in *Sciamvs* 1 (2000). In *TMS* 21 a, a concave square is inscribed in the middle of a square, at a distance of 5 ninda from all the four sides of the square. The a.šà dal.ba.na, the 'field between', bounded on one side by the square and on the other side by the concave square, is given as 35 (00 sq. ninda). What is then the side of the square?

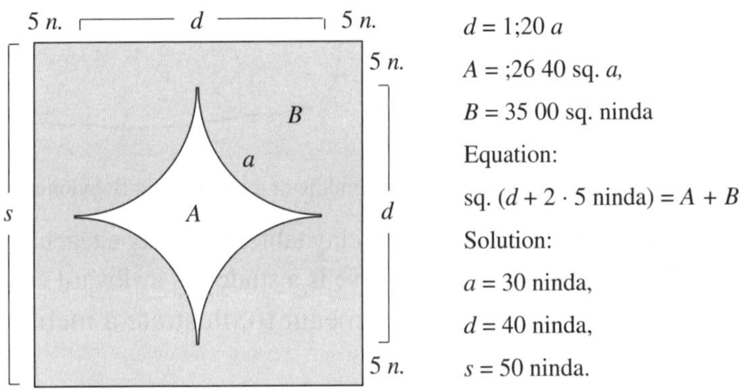

Fig. 6.2.6. *TMS* 21a. A concave square in the middle of a square (Old Babylonian).

The problem is solved in the following way: According to ## 22-23 in the table of constants BR (see above), the diagonal and the area of a concave square are '1 20' and '26 40' times a certain length, actually the length a of one of the circular arcs bounding the concave square. Therefore, the diagonal and the area are $d = 1;20 \cdot a$ and $A = ;26\ 40 \cdot$ sq. a. At the same time, the side of the square is $s = d + 2 \cdot 5$ (ninda). Consequently,

sq. $(1;20 \cdot a + 2 \cdot 5$ n.$) = ;26\ 40 \cdot$ sq. $a + 35\ (00)$ sq. n.

This is a quadratic equation for the unknown a, which is shown in the text to have the solution $a = 30$ ninda. Hence, the diagonal $d = 40$ ninda, and the side of the square is $s = 40$ ninda $+ 2 \cdot 5$ ninda $= 50$ ninda.

In the problem text *TMS* 21, there is no diagram illustrating the exercise. At the other end of the scale, there are examples of exercises of the same kind (figures inscribed in the middle of figures) in the form of hand tablets with diagrams but no text other than some numbers. One such text is **YBC 7359** (Friberg, *RC* (2007), Fig. 2.8.9).

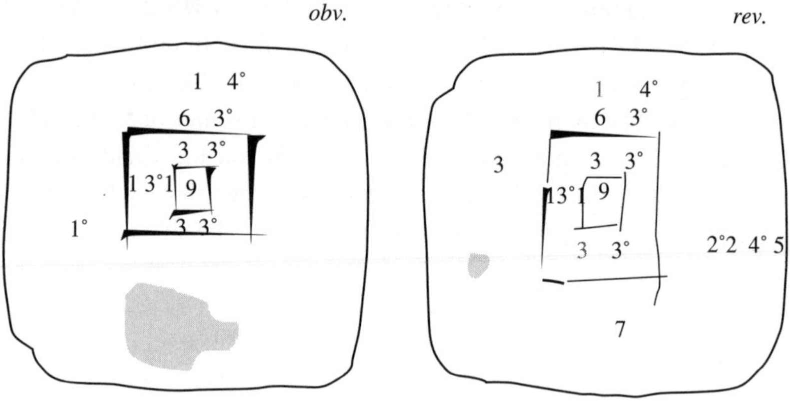

Fig. 6.2.7. YBC 7359. A square in the middle of a square (Old Babylonian).

The diagram on the obverse of this clay tablet is clearly a teacher's neat model, while the diagram on the reverse is a student's awkward copy.

Apparently, the diagrams were meant to illustrate a metric algebra problem of the following form:

The area B between two (concentric and parallel) squares is 1 31.
The distance d between the squares is 3;30. Find the sides p and q of the squares.

It is easy to find the solution, $p = 10$, $q = 3$.

Another text of the same kind is **MS 2985** (Friberg, *op. cit.*, Fig. 8.1.1) shown in Fig. 6.2.8 below, both in a cuneiform hand copy and in a "conform" transliteration. In that text, a circle is inscribed in the middle of a square. Some scribbled numbers appear to indicate that the circle is inscribed a distance $b = 15$ ninda from all the sides of the square. The value B of the area between the circle and the square is not indicated. Anyway, if the diagram on MS 2985 illustrates a problem of the same kind as the problem in *TMS* 21, then that problem can be reduced to a quadratic equation of the form

$$\text{sq.}\,(;20 \cdot a + 2 \cdot b) - ;05 \cdot \text{sq.}\,a = B, \quad \text{where } a \text{ is the circumference of the circle.}$$

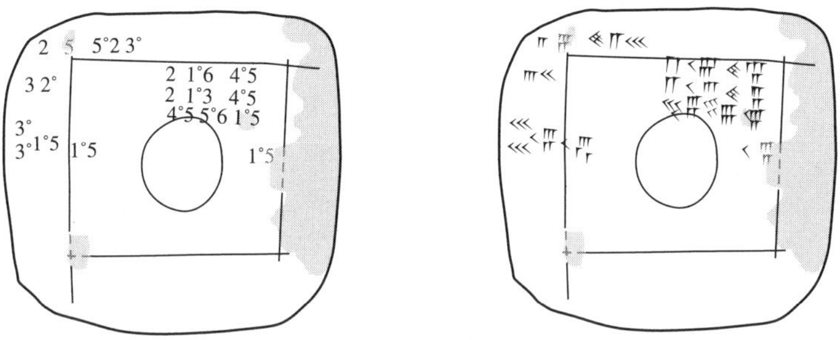

Fig. 6.2.8. MS 2985. A circle in the middle of a square (Old Babylonian).

Yet another example of a problem of the same kind may be given by **MS 1938/2** (Friberg, *op. cit.*, Fig. 8.2.14), a fragment of a clay tablet with a diagram showing what appears to be a circle inscribed in the middle of a regular hexagon.

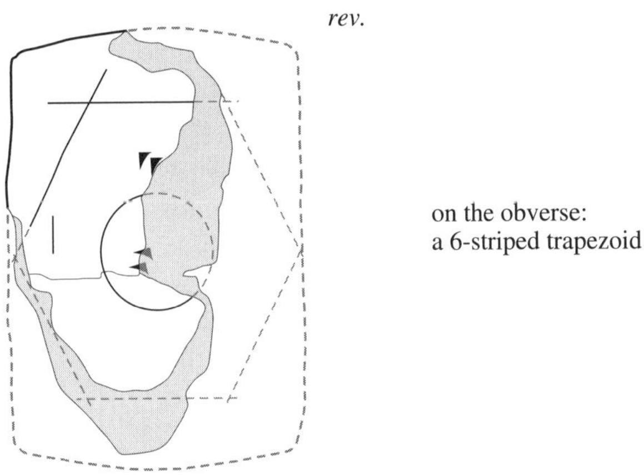

Fig. 6.2.9. MS 1938/2. A circle in the middle of a regular hexagon (Old Babylonian).

The examples mentioned above suggest that there once may have existed two OB *geometric theme texts*, one with *problems for figures inscribed in figures*, and another with *problems for figures in the middle of figures*.

If such theme texts really existed, they were Old Babylonian precursors to the theme text *Elements* IV.

Chapter 7

El. VI.30, XIII.1-12, and Regular Polygons in Babylonian Mathematics

7.1. *El.* VI.30: Cutting a Straight Line in Extreme and Mean Ratio

The special division of a given straight line which appeared first in *El.* II.11 (Fig. 1.7.1), and then again in *El.*IV.10 (Fig. 6.1.1), is belatedly given its rather peculiar name at the beginning of *Elements* VI:

El. VI Def. 3A given straight line is *cut in extreme and mean ratio* when the whole straight line is to the greater part as the greater part is to the smaller.

In all of *Elements* VI, a straight line cut in this way appears only in

El. VI.30 To cut a given straight line in extreme and mean ratio.

The construction in ***El.* VI.30** is quite indirect. It can be explained as follows, in terms of metric algebra:

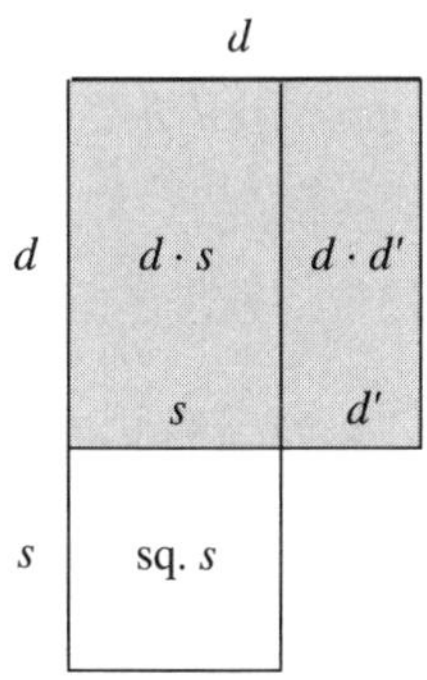

sq. $s + d \cdot s =$ sq. $d \quad \Rightarrow$
sq. $s =$ sq. $d - d \cdot s = d \cdot d'$
$\Rightarrow \quad d : s = s : d'$

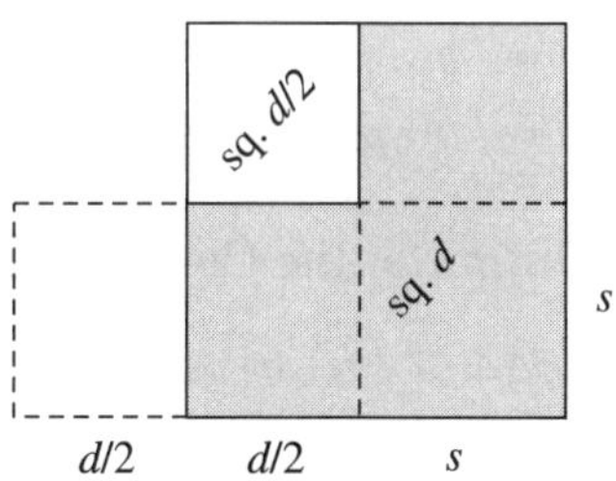

sq. $s + d \cdot s =$ sq. $d \quad \Rightarrow$
sq. $(s + d/2) =$ sq. $d +$ sq. $d/2$
$= 5$ sq. $d/2$
(another proof is given in XIII.1)

Fig. 7.1.1. Explanation of the argumentation in *El.* VI.30, in terms of metric algebra.

Given is a straight line d, and it is required to cut it in two parts in extreme and mean ratio. If the two parts are called s and d', then according to the procedure in *El.* VI.30, s can be found "through the application to a straight line of length d of a rectangle equal to sq. d and exceeding by a square". This is a way of describing a *quadratic equation,* which was introduced in the preceding proposition, *El.* VI.29. What it means is that

$$\text{sq. } s + d \cdot s = \text{sq. } d.$$

It is shown geometrically (see Fig. 7.1.1 above, left) that if s satisfies this equation, then also sq. $s = d \cdot d'$, so that $d : s = s : d'$, as required.

The solution to the mentioned quadratic equation is not explicitly given in *El.* VI.30, so the construction of a straight line divided in extreme and mean ratio remains incomplete. However, a procedure for the geometric solution of a quadratic equation of type B4a: sq. $s + d \cdot s = P$ is demonstrated in *El.* VI.29. (See Sec. 10.3.) When $P =$ sq. d, as in *El.* VI.30, the geometric solution as in *El.* VI.29 (Fig. 7.1.2, right) can be interpreted as follows:

$$\text{sq. } s + d \cdot s = \text{sq. } d \;\Rightarrow\; \text{sq. } (s + d/2) = \text{sq. } d + \text{sq. } d/2 \;\Rightarrow\; \text{sq. } (s + d/2) = 5 \cdot \text{sq. } d/2.$$

(*Cf.El.* XIII.1.) More concisely, with a modern standard notation,

$$s = \phi \cdot d, \quad \text{where,} \quad \phi = (\sqrt{5} - 1)/2.$$

7.2. Regular Pentagons and Equilateral Triangles in *Elements* XIII

Book XIII of Euclid's *Elements* contains 18 propositions, concerned with *straight lines cut in extreme and mean ratio* (XIII.1-6), with *regular polygons inscribed in circles* (XIII.7-12), and with *regular polyhedrons inscribed in spheres* (XIII.13-18).

An Outline of the Contents of *El.* XIII.1-12

1 If $d = s + d'$ is *cut in extreme and mean ratio*, $s > d'$, then sq. $(s + d/2) = 5$ sq. $d/2$.

2 Conversely, if sq. $(s + d/2) = 5$ sq. $d/2$, and if d is cut in extreme and mean ratio, then s is the greater of the parts into which d is cut.

3 If $d = s + d'$ is cut in extreme and mean ratio, $s > d'$, then.sq. $(d' + s/2) = 5$ sq. $s/2$.

4 If $d = s + d'$ is cut in extreme and mean ratio, $s > d'$, then sq. d + sq. $d' = 3$ sq. s.

5 If $d = s + d'$ is cut in extreme and mean ratio, $s > d'$, then $d + s$ is also cut in extreme and mean ratio, with d being the greater of the two parts of $d + s$.

6 If $d = s + d'$ is cut in extreme and mean ratio, and if d is *expressible*, then the two parts s and d' into which d is cut are *apotomes*.

7 If three angles of an equilateral *pentagon* are equal, then all the five angles are equal.
8 Consecutive diagonals in a (regular) pentagon cut each other in extreme and mean ratio. The greater of the two parts of each is equal to the side of the pentagon.
9 The sum of the sides of a hexagon and a decagon inscribed in the same circle is a straight line cut in extreme and mean ratio. The greater part is the side of the hexagon.
10 If a pentagon, a hexagon, and a decagon are inscribed in the same circle, the square on the side of the pentagon is equal to the sum of the squares on the sides of the hexagon and the decagon.
11 The side of a *pentagon* inscribed in a circle with *expressible diameter* is a *minor*.
12 If an *equilateral triangle* is inscribed in a circle, the square on the side of the triangle is three times the square on the radius of the circle.

The geometric proofs of ***El.* XIII.1-5** are given in terms of rectangles and squares. More conveniently, *El.* XIII.1, for instance, can be proved by use of the diagonal rule applied as in *El.* II.11 (see Fig. 1.7.1 above). *El.* XIII.1 is also an immediate consequence of *El.* VI.30 (see Fig. 7.1.1).

In ***El.* XIII.6** it is stated that if $d = s + d'$ is an expressible straight line cut in extreme and mean ratio, then the two parts s and d' are apotomes in the sense of *Elements* X. For the proof, it is noted that sq. $(s + d/2) = 5$ sq. $d/2$ [*El.* XIII.1]. Therefore, $d/2$ and $s + d/2$ are expressible straight lines, commensurable in square only. Since $s = (s + d/2) - d/2$, it follows that s is an *apotome*. Moreover, sq. $s = d \cdot d'$, where d is expressible and s an apotome. Therefore, d' is a *first apotome* (with respect to d) [*El.* X.97].

The important connection between the regular pentagon and straight lines cut in extreme and mean ratio is demonstrated in ***El.* XIII.8**. See the diagram in Fig. 7.2.1 top, left. There two diagonals, obviously both of the same length d, cut each other. Let the two parts of each diagonal be called a and b, with $a > b$. It is claimed in the proposition that d is cut in extreme and mean ratio, and that $a = s$. In the proof, it is observed that the angle w' is twice the angle v [*El.* I.32]. At the same time, the angle w is twice the angle v [*El.* III.28, ***El.* VI.33**]. Hence, $w = w'$, and the triangle a, s, b is isosceles, so that $a = s$. Furthermore, the triangle with the sides b, b, s is similar to the triangle with the sides $a + b, s, s$. Therefore, $(a + b) : s = s : b$. This means that $d = a + b$ is cut in extreme and mean ratio, so that the proof is complete. Since it was shown that $a = s$, b may now be called d'. Then the result can be stated in the form

$$d = s + d', \quad s > d', \text{ and } \quad d : s = s : d'.$$

Note the reappearance in the proof of *El.* XIII.8 of the special triangle which played a crucial role in *El.* IV.10-11 (Fig. 6.1.1). The same special triangle appears again in the proof of ***El.* XIII.9** (Fig. 7.2.1), which says that the sum $r + s_{10}$, where r is the radius of a circle, s_{10} the side of the inscribed regular decagon, is divided in extreme and mean ratio, with r as the greater of the two parts. The result is used in the proof of *El.* XIII.16.

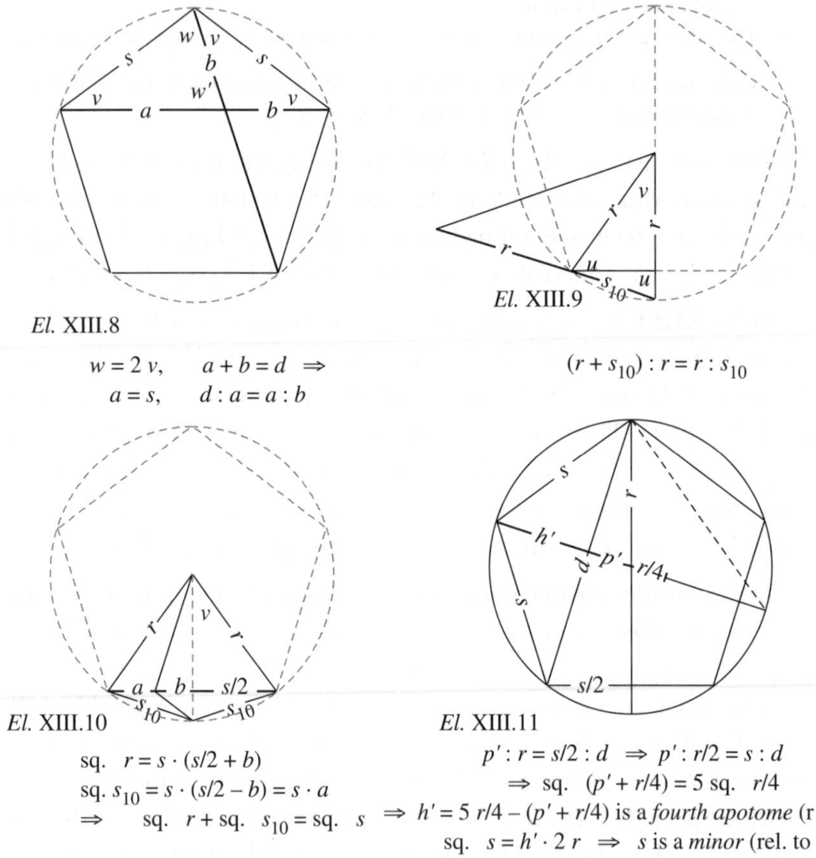

Fig. 7.2.1. Metric algebra versions of Euclid's diagrams in *El.* XIII.8-11.

In ***El.* XIII.10** (see again Fig. 7.2.1), it is proved that the sum of the square on the radius r of a circle and the square on the side s_{10} of the inscribed decagon is equal to the square of the side of the inscribed pentagon. The complicated proof again makes extensive use of angles in various triangles. The result of *El.* XIII.10 is used repeatedly in *El.* XIII.16, the construction of an icosahedron inscribed in a given sphere.

The last of the propositions concerned with regular pentagons is ***El.* XIII.11**. The proof of *El.* XIII.11 starts with the observation that if h is the height of the pentagon, then the triangle with the sides d, h, $s/2$ is similar to the triangle with he sides r, $d/2$, p' (see the last diagram in Fig. 7.2.1). Therefore, $p' : r = s/2 : d$, so that $p' : r/2 = s : d$. Hence, in view of *El.* XIII.8 and *El.* XIII.1, $p' + r/2$ is cut in extreme and mean ratio, and

$$\text{sq. } (p' + r/4) = 5 \text{ sq. } r/4.$$

Now, let the diameter $2\,r$ of the circle be an assigned expressible straight line in the sense of *Elements* X. Then also $r + r/4 = 5\,r/4$ is an expressible straight line. On the other hand,

$$\text{sq. } (5\,r/4) = 25 \text{ sq. } r/4 = 5 \text{ sq. } (p' + r/4).$$

Therefore, $5\,r/4$ and $p' + r/4$ are *both expressible*, but *commensurable in square only*. Now, consider the height h' against the base of the isosceles triangle with the sides d, s, s (Fig. 7.2.1 bottom, right). It is clear that

$$h' = r - p = 5\,r/4 - (p' + r/4).$$

Consequently, h' is an *apotome* in the sense of *Elements* X, with respect to the radius r. More precisely, h' is a *fourth apotome* in the sense of the following definition

El. X.Def. III 4. Given an expressible straight line e, an apotome $u - v$, $u > v$, is called a *fourth apotome* (with respect to e) if u com e, and if sq. u – sq. v = sq. w, where w inc e.

Indeed, with $e = 2\,r$, $u = 5\,r/4$, $v = p' + r/4$, it is clear that u com e, and that

$$\text{sq. } u - \text{sq. } v = \text{sq. } (5\,r/4) - \text{sq. } (p' + r/4) = (25 - 5) \cdot \text{sq. } r/4 = 20 \cdot \text{sq. } r/4.$$

Clearly, sq. $(5\,r/4)$: sq. $(p' + r/4) = 25 : 20 = 5 : 4$ is not the ratio of a square number to a square number. Therefore, w = sqs. (sq. u – sq. v) and $5\,r/4$ are incommensurable, as required [*El.* X.9].

The last step of the proof of *El.* XIII.11 makes use of the observation that the height against the diagonal in the right triangle with the diagonal $2\,r$ and the short side s cuts off a right triangle with the sides s, $d/2$, and h'. Therefore, it follows from the lemma *El.* X.32/33 (see Chapter 4 above) that

$$\text{sq. } s = h' \cdot 2\,r, \quad \text{where } 2\,r \text{ is expressible and } h' \text{ is a fourth apotome.}$$

Hence, in view of *El.* X.94, the side s of a regular pentagon inscribed in a given circle is a *minor*, with respect to r, the radius of the circle.

In ***El.* XIII.12** it is shown that the square of the side of an *equilateral triangle* inscribed in a circle is 3 times the square on the radius of the circle. Therefore, *if the diameter of the circle is expressible, then also the side of the inscribed equilateral triangle is expressible*. This consequence of the result in *El.* XIII.12 is, apparently, so obvious that it is not explicitly stated in the text. The simple proof of XIII.12 is demonstrated in Fig. 7.2.2, left.

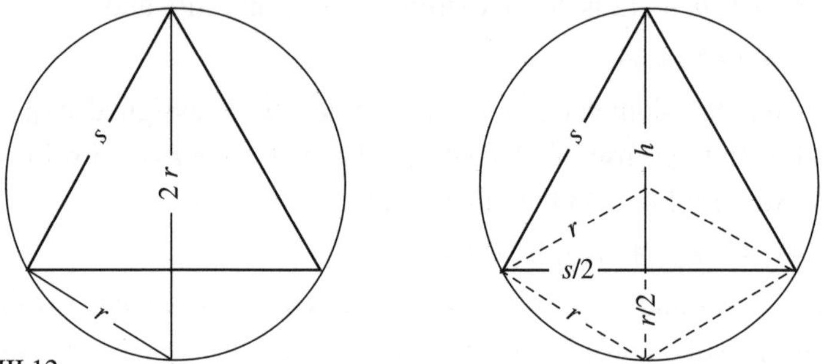

El. XIII.12

sq. s = sq. 2 r – sq. r = 3 sq. r

sq. h = sq. s – sq. s/2 = 3/4 sq. s
h = sqs. 3 · s/2, r = 2/3 · h = sqs. 3 · s/3
A = (sqs. 3)/4 · sq. s

Fig. 7.2.2. The equilateral triangle in *El.* XIII.12 (left) and in a Babylonian exercise (right).

The way in which equilateral triangles were treated in Babylonian mathematics (Fig. 7.2.2, right) will be discussed in Sec. 7.7 below.

7.3. An Extension of the Result in *El.* XIII.11

A quite big apparatus appears to be needed for what may be called the "metric analysis" (in disguise) of the regular pentagon in *El.* XIII.8-11 (Fig. 7.2.1). The appearance is deceptive, for the truth is that *El.* XIII.8-10 are needed only for the construction and metric analysis in *El.* XIII.16 of an icosahedron inscribed in a given sphere. For the study of the pentagon alone is needed only *El.* XIII.11, with its relatively uncomplicated proof.

Also *El.* XIII.11 seems to be included in *El.* XIII principally because it will be needed in *El.* XIII.16. It is probably for this reason that the metric analysis of the pentagon in *El.* XIII.11 is only half-finished. Indeed, the detailed computation of an expression for the *side* of a regular pentagon inscribed in a circle of given radius is not followed by a similar computation of the *diagonal*, which would have been easy to provide, using a rather obvious variation of the method in *El.* XIII.11. (*Cf.* Knorr, *BAMS* 9 (1983), 48.)

The diagram in Fig. 7.3.1 below is a further development of the last diagram in Fig. 7.2.1 above, the metric algebra counterpart to Euclid's own diagram in *El.* XIII.11. The new notations that are introduced in Fig. 7.3.1, in addition to the notations used in Fig. 7.2.1, are h for the height of the pentagon, and p for $h - r$. The observation in *El.* XIII.11 that the two triangles with the sides r, $d/2$, p' and d, h, $s/2$ are similar is echoed here by the new observation that the triangles with the sides r, $s/2$, p and s, h', $d/2$ are similar.

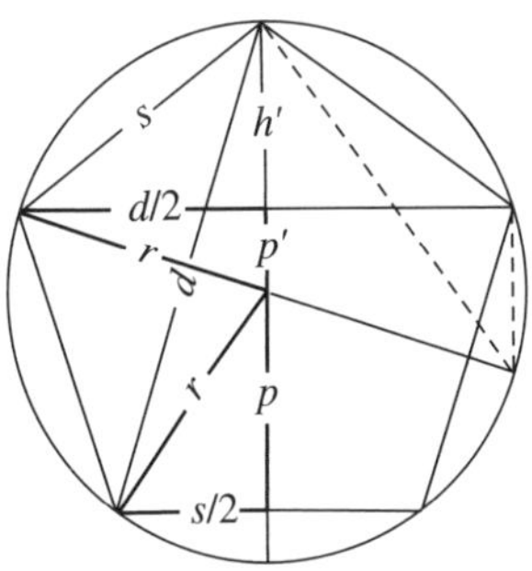

$p' : r = s/2 : d \quad \Rightarrow p' : r/2 = s : d$
$p : r = d/2 : s \ \Rightarrow (p - r/2) : r/2 = (d - s) : s = s : d$
$\Rightarrow p - r/2 = p'$
$p = r/2 + p'$ is cut in *extreme and mean ratio*,
$\Rightarrow$ sq. $(p' + r/4) =$ sq. $(p - r/4) = 5$ sq. $r/4$

$h' = r - p' = 5\ r/4 - (p' + r/4)$ is a *fourth apotome*
$h = r + p = 5\ r/4 + (p - r/4)$ is a *fourth binomial*

sq. $s = h' \cdot 2\ r$, $s =$ sqs. $(5 -$ sqs. $5)/2 \cdot r$ is a *minor*
sq. $d = h \cdot 2\ r$, $d =$ sqs. $(5 +$sqs. $5)/2 \cdot r$ is a *major*

Fig. 7.3.1. Metric analysis of the pentagon, in terms of the *radius*.

From the similarity of the two pairs of triangles it follows that

$p' : r = s/2 : d$, and $p : r = d/2 : s$, respectively.

In view of XIII.8, $d = s + d'$ is cut in extreme and mean ratio with s as the greater part. Therefore, $d \cdot d' =$ sq. s, where $d' = d - s$. Hence,

$p' : r/2 = s : d = \phi$ and $(p - r/2) : r/2 = (d - s) : s = s : d = \phi$.

Consequently, in view of XIII.1,

sq. $(p' + r/4) = 5$ sq. $r/4$, and sq. $(p - r/4) =$ sq. $\{(p - r/2) + r/4\} = 5$ sq. $r/4$.

On the other hand,

$h' = r - p' = 5\ r/4 - (p' + r/4)$, and $h = r + p = 5\ r/4 + (p - r/4)$.

Therefore, as shown in XIII.11, h' is a *fourth apotome* (with respect to r). The same kind of arguments show that h is a *fourth binomial* (with respect to r). This qualitative result can be replaced by the *explicit* result that

$h' = (5 -$ sqs. $5) \cdot r/4$, and $h = (5 +$ sqs. $5) \cdot r/4$.

Now, in the diagram in Fig. 7.3.1, s is the short side in a right triangle with the diagonal $2\ r$ and the height $d/2$ against the diagonal, and d is the long side in a right triangle with the diagonal $2\ r$ and the height $s/2$ against the diagonal. Therefore, in view of lemma *El.* X.32/33 (Chapter 4 above),

$$\text{sq. } s = h' \cdot 2\ r = (5\ r/4 - (p' + r/4)) \cdot 2\ r, \quad \text{where} \quad \text{sq. } (p' + r/4) = 5 \text{ sq. } r/4,$$

and

$$\text{sq. } d = h \cdot 2\ r = (5\ r/4 + (p - r/4)) \cdot 2\ r, \quad \text{where} \quad \text{sq. } (p - r/4) = 5 \text{ sq. } r/4.$$

Consequently, s is a *minor* (with respect to r), as stated in *El.* XIII.11, and d is a *major* (with respect to r). This qualitative result, too, can be replaced by a corresponding *explicit* result. Indeed, if for the sake of increased clarity the side and diagonal of the pentagon are called s_5 and d_5, then

$$\mathbf{s_5 = \text{sqs. } \{(5 - \text{sqs. } 5)/2 \cdot r\}} \quad \text{and} \quad \mathbf{d_5 = \text{sqs. } \{(5 + \text{sqs. } 5)/2 \cdot r\}.}$$

Incidentally, the diagram in Fig. 7.3.1 and the sub-diagrams in Fig. 7.3.2 below show that if s_{10} and d_{10} are the side and the "third" diagonal of the regular *decagon* inscribed in the same circle as the pentagon, then

$$\text{sq. } s_{10} = (r - p) \cdot 2\ r, \quad \text{and} \quad \text{sq. } d_{10} = (r + p') \cdot 2\ r.$$

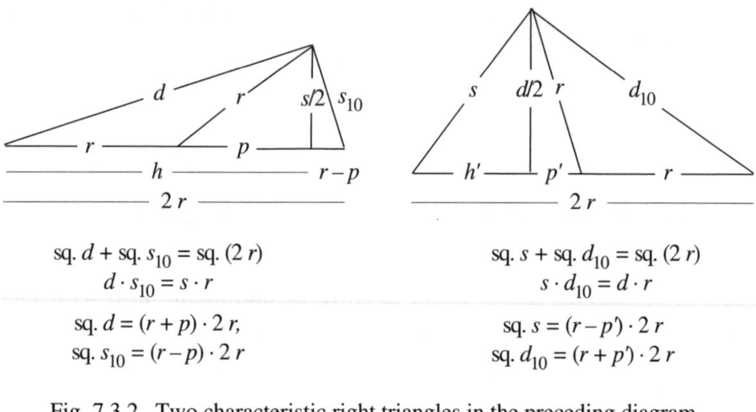

$$\text{sq. } d + \text{sq. } s_{10} = \text{sq. } (2\ r) \qquad \text{sq. } s + \text{sq. } d_{10} = \text{sq. } (2\ r)$$
$$d \cdot s_{10} = s \cdot r \qquad s \cdot d_{10} = d \cdot r$$

$$\text{sq. } d = (r + p) \cdot 2\ r, \qquad \text{sq. } s = (r - p') \cdot 2\ r$$
$$\text{sq. } s_{10} = (r - p) \cdot 2\ r \qquad \text{sq. } d_{10} = (r + p') \cdot 2\ r$$

Fig. 7.3.2. Two characteristic right triangles in the preceding diagram.

Here, $r - p$ is a *first apotome*, in the sense of the following definition

El. X.Def. III 1. Given an expressible straight line e, an apotome $u - v$, $u > v$, is called a *first apotome* (with respect to e) if u com e, and if sq. u – sq. v = sq. w, where w com e.

Indeed,

$$r - p = 3\ r/4 - (p - r/4), \quad \text{where}$$
$$\text{sq. } (p - r/4) = 5 \text{ sq. } r/4, \quad \text{sq. } 3 > 5, \quad \text{and} \quad \text{sq. } 3\ r/4 - \text{sq. } (p - r/4) = \text{sq. } r/2.$$

In view of *El.* X.91 (a parallel to *El.* X.54; see Fig. 5.2.4), s_{10} is then an *apotome*, say $s_{10} = u - v$, with

$$\text{sq.}\ (u - v) = \text{sq.}\ s_{10} = \{3\ r/4 - (p - r/4)\} \cdot 2\ r,$$

and consequently

$$\text{sq.}\ u = a \cdot r \quad \textit{and} \quad \text{sq.}\ v = b \cdot r,$$

where

$$a + b = 3\ r/2 \quad \text{and} \quad a \cdot b = \text{sq.}\ (p - r/4) = 5\ \text{sq.}\ r/4.$$

These equations for a and b are easy to solve, and the result is that

$$a = 5 \cdot r/4,\ b = r/4, \quad \text{so that} \quad \text{sq.}\ u = 5/4\ \text{sq.}\ r,\ v = 1/4\ \text{sq.}\ r.$$

Consequently, the explicit form of the *apotome* s_{10} is

$$\boldsymbol{s_{10} = (\text{sqs. } 5 - 1)/2 \cdot r.}$$

Similarly, the explicit form of the *binomial* d_{10} is

$$\boldsymbol{d_{10} = (\text{sqs. } 5 + 1)/2 \cdot r.}$$

7.4. An Alternative Proof of the Crucial Proposition *El.* XIII.8

It is an interesting question whether any of the properties of a regular pentagon mentioned in *Elements* XIII can have been known by (Old) Babylonian mathematicians. In this connection, it is important to observe that the proof of the crucial proposition *El.* XIII.8, which says that *two intersecting diagonals in a pentagon cut each other in extreme and mean ratio, with the greater part equal to the side of the pentagon*, makes essential use of ***El.* VI.33**, a proposition stating that "In equal circles angles have the same ratio as the circumferences on which thy stand". This proposition cannot have been known to Babylonian mathematicians, who apparently were totally ignorant of the concept of angles based on circular arcs. Actually, by the way, *El.* VI.33 is strangely isolated from the rest of *Elements* VI. It is much closer associated with *Elements* III, which in its entirety is outside the scope of Babylonian mathematics.

On the other hand, an OB mould shows the image of *a pentagram* of entangled wild men (see Fig. 7.9.7 below), and an entry in an OB table of constants mentions an approximation to the area of a normalized *regular pentagon* (Sec. 7.8). These two facts together make it clear that OB mathematicians knew at least about the *existence* of pentagrams and regular pentagons. In view of this circumstance, anyone familiar with the general character of OB mathematics, in particular its extreme readiness to consider *all* imaginable aspects of a given mathematical situation, is forced to draw the conclusion that OB mathematicians must have tried to compute *the lengths of the diagonals in a regular pentagon, and the lengths of their segments,* using methods available to them. Although nothing is known about how they did that, the discussion below aims to show which methods they conceivably *may* have used.

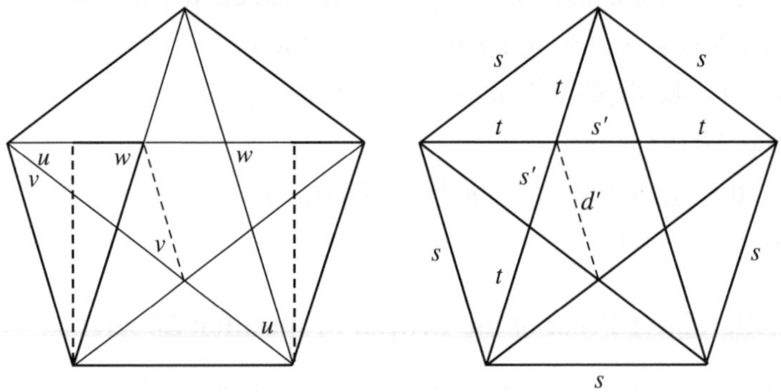

$w = u + v, \quad v = u \quad \Rightarrow \quad u = v = w/2$

$t = d'$ and $t + s' = s \Rightarrow s = d' + s', \quad d = s + d'$ and $d/s = s/d'$

Fig. 7.4.1. A hypothetical Babylonian alternative to the procedure in *El.* XIII.8.

Consider a regular pentagon with its diagonals, as in Fig. 7.4.1, left. The diagonals form a pentagram with a *central body* in the form of a smaller regular pentagon, and with five *arms* in the form of symmetric (isosceles) triangles. Let u be the angle between two diagonals meeting at a vertex of the pentagon, and let v be the angle between a diagonal and a side. (Although the Babylonians were not familiar with the general concept of angles, they were in a certain way familiar with angles in right triangles and with angles in symmetric triangles, probably understood as double right triangles.) Also, let w be the smaller of the two angles between intersecting diagonals. Then it is clear that $w = u + v$, because the right triangle indicated in the right half of the pentagon with one angle equal to w is similar to the right triangle indicated in the left half of the pentagon with one angle equal to $u + v$. In addition, $u = v$, because the angle v between a side and a diagonal in the inner pentagon is clearly equal to the angle u between two diagonals meeting at a vertex of the outer polygon. Indeed, v and u are angles at the bases of two similar symmetric triangles. Since at the same time $u = v$ and $w = u + v$, it follows that $u = v = w/2$.

Now, let the sides and diagonals of the larger pentagon be called s and d, and the sides and diagonals of the smaller pentagon s' and d', as in Fig. 7.4.1, right. Also, let t denote the side of an arm of the pentagram which is formed by the diagonals. Then it is clear that

$$t = d', \quad \text{because} \quad u = v, \quad \text{and that} \quad t + s' = s, \quad \text{because} \quad u + v = w.$$

Therefore also

$$s = d' + s' \quad \text{and} \quad d = 2\,d' + s' = s + d'.$$

In addition, the three symmetric triangles with the sides (d, d, s), (s, s, d'), and (d', d', s') are similar because $w = u + v$. Therefore,

$$d : s = s : d' = d' : s'.$$

This means that both $d = s + d'$ and $s = d' + s'$ are cut in what Euclid calls "extreme and mean ratio".

The observation that the diagonal d of a pentagon can be cut in three pieces as $d = d' + s' + d'$, with two *extreme* pieces of length d' and one *middle* piece of length s', at the same time as $s = d' + s'$ is cut in "extreme and mean ratio", provides a previously lacking explanation of this curious expression. Indeed, the observation shows that a more appropriate translation of the obscure Greek phrase (ἄκροσ καὶ μέσοσ λόγοσ) may be "extreme and middle ratio"!

It must be understood that one never meets terms like "angles" or "ratios" in Babylonian mathematical texts. Instead, the Babylonians preferred to think in terms of the "feed" of a right or symmetric triangle, meaning the "front" divided by the "length" (alternatively, the height). Thus, for instance, the fact that the diagonal of a regular pentagon is cut in extreme and mean ratio with the side as the greater part would have been expressed by a Babylonian mathematician, essentially, in the following way:

$$d = s + d', \quad d' = f \cdot s, \quad \text{where the feed } f = s/d.$$

(In modern terminology, this particular ratio is usually called ϕ.)

7.5. Metric Analysis of the Regular Pentagon in Terms of its Side

In the OB mathematical table of constants BR mentioned in Sec. 7.8 below, an approximation for the area of the pentagon is given, in the case when the side of the pentagon is equal to 1 (00). This is in agreement with the Babylonian convention that *constants for geometric figures should be given for the special case when a prominent side of the figure is normalized to the value '1'*. With this convention in mind, it is natural to assume that a Babylonian mathematician wanting to compute the lengths of various segments in a regular pentagon would do that in the special case when the *side* s = '1'. In the terminology of *Elements* X, for a Babylonian mathematician it would be natural to choose *the side of the pentagon*, not the radius of the circumscribed circle, as the *assigned expressible straight line*.

For this reason, it may be worthwhile to investigate what the result will be of a metric analysis of the regular pentagon *with respect to the side s*.

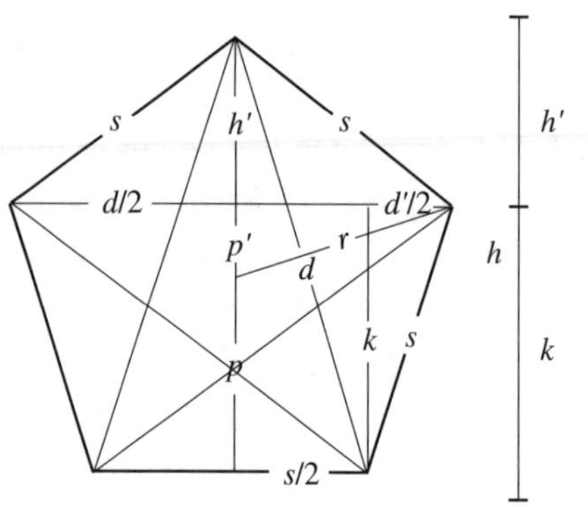

$s = f \cdot d,\quad d' = f \cdot s,\quad f = (\text{sqs. } 5 - 1)/2$

$\text{sq. } h = \text{sq. } d - \text{sq. } s/2$
$\text{sq. } k = \text{sq. } s - \text{sq. } d'/2$
$\text{sq. } h' = \text{sq. } s - \text{sq. } d/2$

$h = \text{sqs. } (5 + 2 \text{ sqs. } 5) \cdot s/2$
$k = \text{sqs. } \{(5 + \text{sqs. } 5)/2\} \cdot s/2$
$h' = \text{sqs. } \{(5 - \text{sqs. } 5)/2\} \cdot s/2$

$A = h \cdot s/2 + 2\, h' \cdot d/2$
$r = \text{sq. } d \,/\, 2\, h$

$A = \{\text{sqs. } (5 + 2 \text{ sqs. } 5) + \text{sqs. } (10 + 2 \text{ sqs. } 5)\} \cdot \text{sq. } s/2$
$r = \text{sqs. } \{(5 + \text{sqs. } 5)/10\} \cdot s$

Fig. 7.5.1. Metric analysis of the pentagon, in terms of the *side* of the pentagon.

As shown in Fig. 7.5.1, it follows from three straightforward applications of the diagonal rule that, with the notations in that figure,

$\text{sq. } h = \text{sq. } d - \text{sq. } s/2,$
$\text{sq. } k = \text{sq. } s - \text{sq. } d'/2,$
$\text{sq. } h' = \text{sq. } s - \text{sq. } d/2.$

The way to proceed from here in the style of *El.* XIII.11 (Sec. 7.2 above) will be discussed below. However, in terms of quasi-modern notations,

$\text{sq. } h = \{\text{sq. } (\text{sqs. } 5 + 1) - 1\} \cdot \text{sq. } s/2 = (5 + 2 \text{ sqs. } 5) \cdot \text{sq. } s/2,$
$\text{sq. } k = \{4 - \text{sq. } (\text{sqs. } 5 - 1)/2\} \cdot \text{sq. } s/2 = (5 + \text{sqs. } 5)/2 \cdot \text{sq. } s/2,$
$\text{sq. } h' = \{4 - \text{sq. } (\text{sqs. } 5 + 1)/2\} \cdot \text{sq. } s/2 = (5 - \text{sqs. } 5)/2 \cdot \text{sq. } s/2.$

Here, it is easy to check that both (5 + 2 sqs. 5) and (5 + sqs. 5)/2 are *fourth binomials*, and that (5 – sqs. 5)/2 · *s* is a *fourth apotome*. Therefore,

h = sqs. {(5 + 2 sqs. 5) · s/2} is a *major* (with respect to s),
k = sqs. {(5 + sqs. 5)/2 · s/2} is a *major* (with respect to s),
h' = sqs. {(5 – sqs. 5)/2 · s/2} is a *minor* (with respect to s).

It is interesting that it follows directly from the geometric situation that

$h = k + h'$.

Therefore,

sqs. (5 + 2 sqs. 5) · s/2 = sqs. (5 + sqs. 5)/2 · s/ + sqs. (5 – sqs. 5)/2 · s/2.

This is a surprising geometric demonstration of the way in which the major h = sqs. (5 + 2 sqs. 5) · s/2 can be split into a sum of two inexpressible parts, one a major, the other a minor. Note, that thereby $h = k + h'$ is cut in extreme and mean ratio, just like the diagonal d. Note also that *the major k and the minor h' (with respect to s) are the greater and smaller parts, respectively, of h*. This may be a previously unobserved explanation of the curious terms "major" and "minor"! (Compare with Knorr's explanation in *BAMS* 9 (1983), 49, that the origin of the term is that $d + s$ is cut in extreme and mean ratio with the major d and the minor s (with respect to r) as the greater and smaller parts, respectively.)

It is now easy to compute also the *area* of the regular pentagon:

$A = h \cdot s/2 + 2\,h' \cdot d/2 = h \cdot s/2 + 2\,k \cdot s/2$
= {sqs. (5 + 2 sqs. 5) + sqs. (5 + sqs. 5)/2} · sq. s/2.

Finally, the *radius* r can be computed, by use of the equation

r = sq. d / 2 h = {(3 + sqs. 5)/2 · sq. s} / {sqs. (5 + 2 sqs. 5)} · s}.

The situation described by this equation is not covered by *El.* X.112, a proposition dealing only with the case of an *expressible* area applied to a *binomial* straight line. Yet, it is clear that, in modern notations,

$$\text{sq. } s / \sqrt{(5 + 2\sqrt{5})} \cdot s = \{\sqrt{(5 - 2\sqrt{5})} / \sqrt{(25 - 20)}\} \cdot s = \sqrt{(5 - 2\sqrt{5})} / \sqrt{5} \cdot s.$$

Therefore,

$$r = \{(3 + \sqrt{5})/2 \cdot \sqrt{(5 - 2\sqrt{5})} / \sqrt{5}\} \cdot s = \sqrt{\{(7 + 3\sqrt{5})/2 \cdot (5 - 2\sqrt{5})/5\}} \cdot s.$$

It follows that

r = sqs. {(5 + sqs. 5)/10· s}.

(*Cf.* the previous result in Fig. 7.3.1 that s = sqs. {(5 – sqs. 5)/2} · r.)

An alternative, and simpler way to compute r in terms of s is to start by showing, by use of the diagram in Fig. 7.5.2, left, that

$$\text{sq. } 2\,r - \text{sq. } s_{10} = \text{sq. d}, \quad \text{with} \quad s_{10} = f \cdot r, \ \ f = s/d.$$

The details of the computation are left to the reader.

The computations above of d, h, k, h', and r in terms of s make use only of metric algebra operations familiar to OB mathematicians (a quadratic equation to compute $f = (\text{sqs. } 5 - 1)/2$, the diagonal rule to compute h, k, h', and a metric division to compute r). Therefore, all the mentioned straight lines in the regular pentagon, as well as the area of the pentagon, could be (and maybe were) computed by Babylonian mathematicians, although probably only with suitable approximations for the square sides.

All the results obtained above can be found equally well by use of methods more close to the methods used in the proof of *El.* XIII.11 The key observation is that *when s is the assigned expressible straight line*, then the straight line

$$d_m = d - s/2 = d' + s/2$$

plays the same role as the one played by the straight line

$$p_m = p - r/4 = p' + r/4$$

in the case when r is the assigned expressible straight line (Fig. 7.3.1). Indeed, since $d = s + d'$ is cut in extreme and mean ratio, it follows that (as in *El.* XIII.1)

$$\text{sq. } d_m = \text{sq. } (d - s/2) = \text{sq. } (d' + s/2) = 5 \text{ sq. } s/2.$$

Therefore,

$$\text{sq. } d = \text{sq. } (d_m + s/2) = 6 \text{ sq. } s/2 + d_m \cdot s = (6\,s/2 + 2\,d_m) \cdot s/2, \quad \text{and}$$
$$\text{sq. } d' = \text{sq. } (d_m - s/2) = 6 \text{ sq. } s/2 - d_m \cdot s = (6\,s/2 - 2\,d_m) \cdot s/2.$$

Consequently (*cf.* Fig. 7.5.1),

$$\text{sq. } h = \text{sq. } d - \text{sq. } s/2 = (5\,s/2 + 2\,d_m) \cdot s/2,$$
$$\text{sq. } k = \text{sq. } s - \text{sq. } d'/2 = (5\,s/2 + d_m)/2 \cdot s/2,$$
$$\text{sq. } h' = \text{sq. } s - \text{sq. } d/2 = (5\,s/2 - d_m)/2 \cdot s/2.$$

Here it is easy to check that $5\,s/2 + 2\,d_m$ and $5\,s/2 + d_m$ are *fourth binomials* while $5\,s/2 - d_m$ is a *fourth apotome*. Therefore, it follows, again, that h and k are *majors*, while h' is a *minor*, with respect to s.

Furthermore, since $p = r/2 + p'$ is cut in extreme and mean ratio (see Fig.

7.5.1), and since $k = p + p'$ (see again Fig. 7.5.1), it follows that

$$k = 2\,p - r/2 = 2\,(p - r/4) = 2\,p_{\rm m}.$$

Moreover,

$$\text{sq. } p_{\rm m} = \text{sq. } (p - r/4) = 5 \text{ sq. } r/2 \quad \text{(Fig. 7.3.1).}$$

Therefore,

$$\text{sq. } k = 4 \text{ sq. } p_{\rm m} = 5 \text{ sq. } r/2.$$

Now, since both $h = k + h'$ and $p = r/2 + p'$ are cut in extreme and mean ratio, the triples h, k, h' and $p, r/2, p'$ are proportional. Therefore,

$$\text{sq. } p = 1/5 \text{ sq. } h = (5\,s/2 + 2\,d_{\rm m})/5 \cdot s/2,$$
$$\text{sq. } r/2 = 1/5 \text{ sq. } k = (5\,s/2 + d_{\rm m})/10 \cdot s/2,$$
$$\text{sq. } p' = 1/5 \text{ sq. } h' = (5\,s/2 - d_{\rm m})/10 \cdot s/2.$$

Consequently, p and r are *majors*, and p' a *minor*, with respect to s.

Easy alternative proofs of *El.* XIII.9-10 follow from the results obtained above. Indeed, as shown by the triangle in Fig. 7.3.2, left,

$$s_{10} : 2\,r = s/2 : d, \quad \text{so that} \quad s_{10} : r = s : d.$$

Therefore, as in *El.* XIII.9, the sum $r + s_{10}$ is cut in extreme and mean ratio, obviously with r as the greater part. Moreover,

$$\text{sq. } s_{10} + \text{sq. } d = \text{sq. } 2\,r.$$

Therefore, as in *El.* XIII.10,

$$\text{sq. } r + \text{sq. } s_{10} = 5 \text{ sq. } r - \text{sq. } d = (10\,s/2 + 2\,d_{\rm m}) \cdot s/2 - (6\,s/2 + 2\,d_{\rm m}) \cdot s/2 = \text{sq. } s.$$

7.6. Metric Analysis of the Regular Octagon

It is demonstrated by the existence of a clay tablet with a drawing of an *octagram with its diagonals* (Fig. 7.8.2 below) that the Babylonians were familiar with octagrams, and therefore probably also with octagons. Hence, it may be of interest to make *a metric analysis of the regular octagon, in imitation of the metric analysis above of the regular pentagon.*[16]

In the diagram in Fig. 7.6.1 below, s is the side of the octagon, e the

16. Note that also Vitrac, with a completely different approach to *Elements* X in his *EA 3* (1998), suggests (*op. cit.*, 73-86) that a study of straight lines in the regular octagon may have played an important role in the prehistory of the Greek classification of inexpressible straight lines.

"first diagonal", d the "second diagonal", a the side of the arm of the inscribed octagram formed by all second diagonals, and r the radius of the circumscribed circle. There is also an "inner octagon" circumscribed by all the second diagonals in the given octagon, which together form an inscribed regular octagram. The straight lines in the inner octagon corresponding to s, e, d, a, r are called s', e', d', a', r'.

It is easy to see in Fig. 7.6.1 that the side a of a arm of the octagram is also the side of a (half) square with the diagonal s. Therefore,

sq. s = 2 sq. a, sq. $(2\,a)$ = 2 sq. s.

The second diagonal d can be expressed in terms of the sides s and a:

$\boldsymbol{d = 2\,a + s}$ so that d = (sqs. 2 + 1) $\cdot\, s$.

It is easy to see in Fig. 7.6.1 also that $s = a + a'$, and that $a = s' + a'$. Therefore, the pair s', a' depends linearly on the pair s, a:

$\boldsymbol{s' = 2\,a - s}$, $\boldsymbol{a' = s - a}$ so that s' = (sqs. 2 – 1) $\cdot\, s$, a' = (2 – sqs. 2)/2 $\cdot\, s$.

Hence d is a *binomial*, and $\boldsymbol{s'}$ and a' *apotomes*, with respect to s.

Conversely, the pair s, a depends linearly on the pair s', a':

$\boldsymbol{s = 2\,a' + s'}$, $\boldsymbol{a = s' + a'}$ so that s = (sqs. 2 + 1) $\cdot\, s'$, a = (2 + sqs. 2)/2 $\cdot\, s'$.

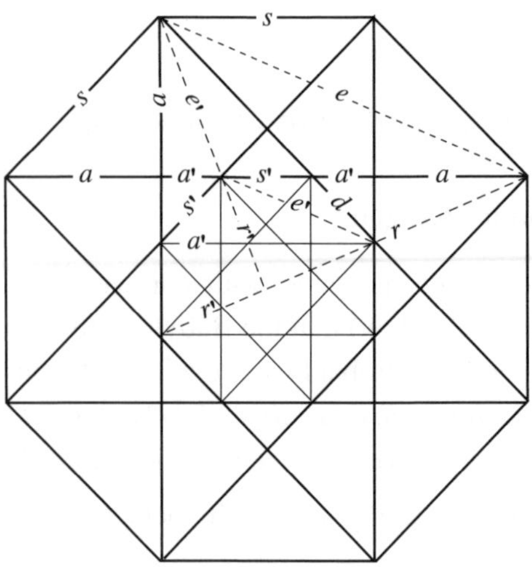

Straight lines in the given octagon.
s = the side of the octagon
e = the first diagonal
d = the second diagonal
a = the side of the arm of the octagram
r = the radius

Straight lines in the inner octagon.
s' = the side of the octagon
e' = the first diagonal
a' = the side of the arm of the octagram
r' = the radius

Fig. 7.6.1. Straight lines in the regular octagon.

Evidently, all straight lines in the octagon parallel to one of the *sides* of the octagon (in particular, all segments of the *second diagonals*) can be expressed as linear combinations of s and a. This means that they can be computed as sums or differences of (integral) multiples of s and a.

In contrast to this, segments of the *diameters* or *first diagonals* of the regular octagon depend in a more complicated way on s and a. Consider, for instance, the radius r of the given octagon, and the radius r' of the inner octagon. The characteristic right triangle in Fig. 7.6.2. left has these segments as orthogonal sides and the segments $2\,a$ and $s/2$ as the diagonal and the altitude against the diagonal, respectively. (Cf. Fig. 7.3.2.)

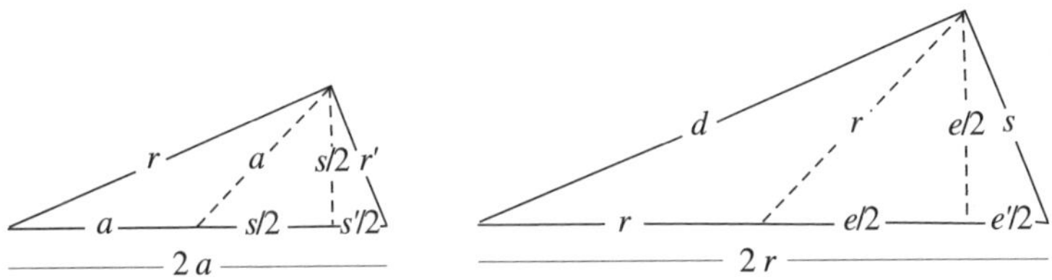

$\text{sq.}\ r + \text{sq.}\ r' = \text{sq.}\ (2\,a),\quad r \cdot r' = a \cdot s$

$\text{sq.}\ r = (a + s/2) \cdot 2\,a = (s + a) \cdot s$
$\text{sq.}\ r' = (a - s/2) \cdot 2\,a = (s - a) \cdot s$

$\text{sq.}\ d + \text{sq.}\ s = \text{sq.}\ (2\,r),\quad d \cdot s = e \cdot r$

$\text{sq.}\ d = (r + e/2) \cdot 2\,r$
$\text{sq.}\ s = (r - e/2) \cdot 2\,r$

Fig. 7.6.2. A quadratic-rectangular system of equations for r and r' in terms of s and a.

From this observation it follows at once that

$\text{sq.}\ r + \text{sq.}\ r' = \text{sq.}\ (2\,a) = 2\ \text{sq.}\ s$ (by the diagonal rule),

$r \cdot r' = a \cdot s$ (both rectangles equal to twice the area of the triangle).

Therefore, r and r' are solutions to a "quadratic–rectangular" system of equations of type B5, with data depending on s and a. *Cf.* Sec. 5.4 above, and in particular, the first diagram in Fig. 5.4.1, which is the metric algebra equivalent of the diagram in *El.* X.33. The solutions can be obtained directly from the diagram, in view of the lemma *El.* X.32/33:

$\text{sq.}\ r = (a + s/2) \cdot 2\,a = (s + a) \cdot s$, where $s + a$ is a *fourth binomial* w. respect to s,
$\text{sq.}\ r' = (a - s/2) \cdot 2\,a = (s - a) \cdot s$, where $s - a$ is a *fourth apotome* w. respect to s.

In other words,

$r = \text{sqs.}\ \{(s + a) \cdot s\} = \text{sqs.}\ (2 + \text{sqs.}\ 2)/2 \cdot s,$
$r' = \text{sqs.}\ \{(s - a) \cdot s\} = \text{sqs.}\ (2 - \text{sqs.}\ 2)/2 \cdot s,$

Obviously, then, r is a *major*, and r' a *minor*, with respect to s.

Now, consider instead the *first diagonals* e and e' in Fig. 7.6.1. Clearly,

$\text{sq.}\ e = 2 \cdot \text{sq.}\ r,$ and $\text{sq.}\ e' = 2 \cdot \text{sq.}\ r'.$

Therefore,

$$e = \text{sqs. } \{(2\,s + 2\,a) \cdot s\} = \text{sqs. } (2 + \text{sqs. } 2) \cdot s,$$
$$e' = \text{sqs. } \{(2\,s - 2\,a) \cdot s\} = \text{sqs. } (2 - \text{sqs. } 2)/2 \cdot s.$$

On the other hand, $e = r + r'$ and $e' = r - r'$. Therefore,

$$\text{sqs. } \{(2\,s + 2\,a) \cdot s\} = e = r + r' = \text{sqs. } \{(s + a) \cdot s\} + \text{sqs. } \{(s - a) \cdot s\},$$
$$\text{sqs. } \{(2\,s - 2\,a) \cdot s\} = e' = r - r' = \text{sqs. } \{(s + a) \cdot s\} - \text{sqs. } \{(s - a) \cdot s\}.$$

Explicitly,

$$\text{sqs. } (2 + \text{sqs. } 2) \cdot s = \text{sqs. } (2 + \text{sqs. } 2)/2 \cdot s + \text{sqs. } (2 - \text{sqs. } 2)/2 \cdot s,$$
$$\text{sqs. } (2 - \text{sqs. } 2) \cdot s = \text{sqs. } (2 + \text{sqs. } 2)/2 \cdot s - \text{sqs. } (2 - \text{sqs. } 2)/2 \cdot s.$$

This means that the major e = sqs. (2 + sqs. 2) · s with respect to s can be split into a sum of two inexpressible parts $r + r'$, one a major, the other a minor, in the same way as the height h in the pentagon is a major with respect to s which can be split into a sum of inexpressible parts $k + h'$, one a major, the other a minor (see Sec. 7.4 above).

Consider now the case when the radius r, rather than the side s, is the "assigned" straight line in the octagon. Evidently, all straight lines in the octagon parallel to one of the *diameters* or one of the *first diagonals* can be expressed as linear combinations of e and r. This means that they can be computed as sums or differences of (integral) multiples of e and r.

In contrast to this, straight lines in the octagon parallel to the *sides* of the octagon depend in a more complicated way on e and r. Consider, in particular, the second diagonal d and the side s of the octagon. The characteristic triangle for d and s (Fig. 7.6.2, right) has these segments as orthogonal sides and the segments 2 r and $e/2$ as the diagonal and the altitude against the diagonal, respectively. From this observation it follows at once that

$$\text{sq. } d + \text{sq. } s = \text{sq. } (2\,r), \quad d \cdot s = e \cdot r$$

The situation in Fig. 7.6.2, right, is clearly a perfect parallel to the situation in Fig. 7.6.2, left. Therefore, the same arguments as above, with obvious modifications, can be used to show that, for instance,

$$d = \text{sqs.}\{(2\,r + e) \cdot r\}, \quad s = \text{sqs. } \{(2\,r - e) \cdot s\}, \quad \text{and so on.}$$

Evidently, *the metric analysis above of the octagon is to a large part parallel to the corresponding metric analysis of the pentagon, only consid-*

erably simpler. It is also clear that a metric analysis of this kind, possibly for the pentagon, but more obviously for the octagon, would have been *well within the competence of OB mathematicians*!

7.7. Equilateral Triangles in Babylonian Mathematics

In *El.* XIII.12, the side of an equilateral triangle is expressed *in terms of the radius* of the circumscribed circle as follows (see Fig. 7.2.2, left):

sq. s = sq. 2 r – sq. r = 3 sq. r.

In Babylonian mathematics, on the other hand, the side s was the main parameter of an equilateral triangle. Other parameters were expressed *in terms of the side*. Thus, Babylonian mathematicians computed the height h and the area A of an equilateral triangle as follows (see Fig. 7.2.2, right):

sq. h = sq. s – sq. $s/2$ = 3 sq. $s/2$, h = sqs. 3 · $s/2$,
$A = h \cdot s/2$ = sqs. 3 · sq. $s/2$ = (sqs. 3)/4 · sq. s.

No Babylonian mathematical text is known where the radius r of the circumscribed circle is computed in terms of the side, but clearly

$r = 2/3 \cdot h$ = (sqs. 3)/3 · s.

A curious name for an equilateral triangle appears in the OB table of constants **G = IM 52916** (Goetze, *Sumer* 7 (1951)):

sag.kak-*kum*	A peghead (triangle),	
ša sa-am-na-[*tu na*]*-ás-ḫa*	the one with an eigh*th tor*n out,	
26 15 [*i-gi-gu-bu-šu*]	26 15 *its constant.*	G *rev.* 7'

The curious name refers to the fact that the height h of an equilateral triangle with the side s can be given in the form

h = sqs. 3 · $s/2$ = (appr.) 7/4 · $s/2$ = 1;45 · $s/2$ = ;52 30 · s = (1 – ;07 30) · s = s – 1/8 · s.

Correspondingly, the area A of an equilateral triangle with the side s is

$A = h \cdot s/2$ = (appr.) ;52 30 · $s/2$ = ;26 15 (7/16) · sq. s.

It is interesting that the same way of computing the height of an equilateral triangle is employed also in the Kassite (post-Old-Babylonian) mathematical text MS 3876 (Sec. 8.3 below). Even more interesting is that the method was still known in the Late Babylonian period, somewhat after the middle of the 1st millennium BCE. This is shown by a passage in the Late Babylonian mathematical text **W 23291** (Friberg, *BaM* 28 (1997)):

W 23291 § 4 b

1 gán.sag.kak ur.a	1 peghead-field, equilateral,
ša 8-*šú na-as-ḫu*	the one with an 8th torn out.
mi-ḫi-il-tú a.rá ki.2 *ù*	Stroke steps of ditto, and
a.rá 2[6 1]5 du	steps of *26 15* go.

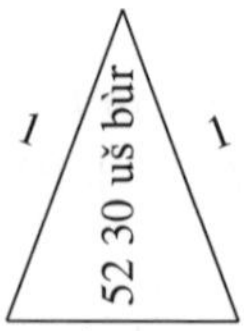

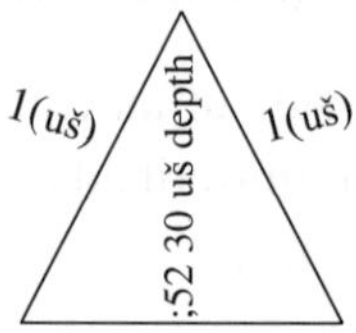

1 uš a.an ur.a	1 length each way, equilateral,
ḫé en aša$_5$.ki.ḫá	What shall the field be?
1 a.[rá 1 1]	1 step*s of 1 (is) 1.*
[1] a.[rá] [2]6 15 du-*ma* 26 15	1 steps *of* 26 15 go, then 26 15
2$_{èše}$ 3$_{iku}$ aša$_5$ 2'$_{iku}$ aša$_5$ 25 šar	2 èše 3 1/2 iku 25 šar.

This text begins with a *general method* for the computation of the area of an equilateral triangle. (Such explicit statements of a general method are exceedingly rare in mathematical texts from the OB period.) Then follows an *explicit example*, preceded by an *illustrating diagram*. Essentially, the method is identical with the one given in the OB table of constants G, *rev.* 7´. Note, however, the introduction of the new term 'stroke', possibly meaning 'straight line'! Thus, 'stroke steps of ditto' means 'the straight line times itself', which here apparently stands for sq. *S*. (In the preceding exercise § 4 a, 'where the area of a symmetric triangle is computed, the similar phrase 'stroke steps of stroke' stands for the height times the front', that is for $h \cdot s$.) There is also here a new term for the height, which is called ušbùr, literally 'the length of the depth'.

In the explicit example, the side of the equilateral triangle is '1 length' = 1 00 ninda. The area is then computed as

1 (00) · 1 (00) ·;26 15 = 26 15 (sq. ninda) = 2 èše 3 1/2 iku 25 šar.

Note that here the numerical result, 26 15 sq. ninda, is converted to the otherwise abandoned *Old Babylonian area measure,* with 1 èše = 6 iku, 1 iku = 100 šar, and 1 šar = 1 square ninda. This is one of several indications that *Late Babylonian mathematicians attempted to continue the traditions of Old Babylonian mathematics*!

7.8. Regular Polygons in Babylonian Mathematics

The following three entries in the OB table of constants ***TMS* 3** = **BR**, Bruins and Rutten (1961), demonstrate that OB mathematicians were familiar with regular polygons and possessed methods for the (approximate) computation of the area of such polygons:

1 40	igi.gub	šà	sag.5	1 40	the constant	of a	5-front	BR 26
2 37 30	igi.gub	šà	sag.6	2 37 30	the constant	of a	6-front	BR 27
3 41	igi.gub	šà	sag.7	3 41	the constant	of a	7-front	BR 28

The area of, for instance, a *regular hexagon* (a "6-front"), *normalized* so that the length of each side is equal to '1' (= 60), is here computed as the sum of the areas of six equilateral triangles with the side '1'. Explicitly,

A_6 = appr. 6 · ;26 15 · sq. 1 00 = 6 · 26 15 = 2 37 30,

where the following relatively good *approximation* is used:

sqs. 3 /4 = appr. {2 – 1/(2 · 2)} · 1/4 = 7/4 · 1/4 = 1;45 ·1/4 = ;26 15.

The value in this OB table of constants for the area of a *normalized regular pentagon* was, apparently, obtained as follows. If the side of the pentagon is '1', then the circumference of the circumscribed circle is *approximately* equal to 5 · 1 00 = 5 00. Consequently, the diameter of the circle is *approximately* equal to ;20 · 5 00 = 1 40. The area of the regular pentagon can therefore be computed as the sum of the areas of five symmetric triangles with the base 100 and the side *approximately* equal to 1 40 · 1/2 = 50 (the radius of the circumscribed circle). The height in each triangle is easily computed by use of the diagonal rule and is equal to 40. Hence, the area of a normalized regular pentagon is:

A_5 = appr. 5 · 30 · 40 = 5 · 20 00 = 1 40 00.

This result is recorded in the entry BR 26 as '1 40'.

Similarly, in the case of a *normalized regular heptagon*, the circumference of the circumscribed circle is *approximately* equal to 7 00. The diameter is then approximately equal to ;20 · 7 00 = 2 20, so that the radius will be approximately equal to 1 10. The height can then be computed as

h_7 = sqs. (sq. 1 10 – sq. 30) = sqs. 1 06 40 = appr. 1 00 + 6 40/2 00 = 1 03;20.

Hence, the area of a normalized heptagon is:

A_7 = appr. 7 · 30 · 1 03;20 = 7 · 31;40 = 3 41;40 = appr. 3 41.

***TMS* 2** (Bruins and Rutten (1961); Fig. 7.4.1 below) is a square clay tablet with diagrams showing a 6-front on the obverse and a 7-front on the reverse. Circumscribed circles appear to have been drawn by use of a compass as an aid for the construction, then erased when they were no longer needed. Only vague traces of the circles are now remaining.

Apparently, when the area of a *normalized* geometric figure was given as an entry in an OB table of constants, it was silently understood that *the areas of similar geometric figures are proportional to the squares of their basic lengths.* In particular, in the case of an n-front with $n = 3$ (an equilateral triangle), 4 (a square) 5, 6, or 7, the basic length clearly was the left-most, or "upper", side of the n-front. Now, on the obverse of *TMS* 2, it is indicated that the upper side of the figure is 30. The side of a *normalized* 6-front is 1 00, twice as much. Therefore, the area of the 6-front in the diagram is only one fourth of the area of a normalized 6-front. Also, the area of the "upper" equilateral triangle in the 6-front is one fourth of the area of a normalized equilateral triangle, so that

A(triangle) = appr. 1/4 · 26 15 = 6 33;45.

This value is recorded inside the upper equilateral triangle. A third recorded number in the same diagram is 30 for the length of a radius. It is possible that the area of the equilateral triangle was recorded on the broken off piece of the clay tablet.

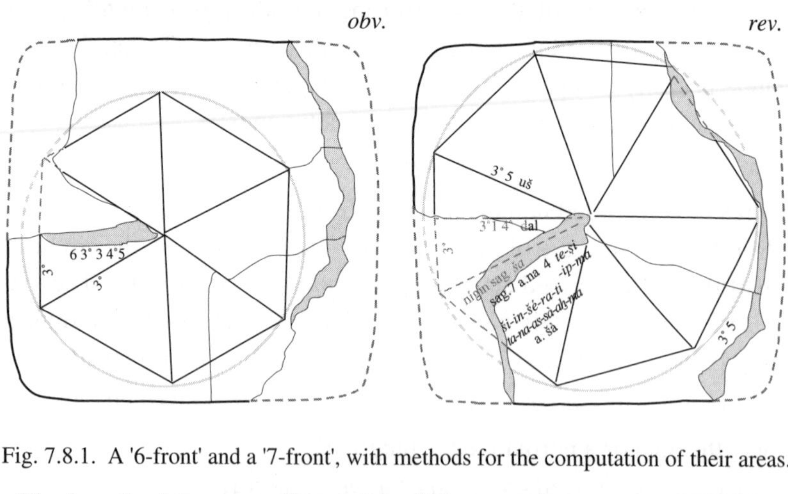

Fig. 7.8.1. A '6-front' and a '7-front', with methods for the computation of their areas.

The length of the upper front of the 7-front on the reverse of *TMS* 2 was also set equal to 30, although the number, probably inscribed close to the leftmost side, is now lost. As a consequence, the length of the circumference of the circumscribed circle was equal to *approximately* 7 · 30 = 3 30, so that the diameter was *approximately* 1 10, and the radius 35. The area of the 7-front could then be computed as the sum of the areas of 7 symmetric triangles with the front 30 and the "length" 35. The notation 35 uš '35, the length' is still readable close to one side of the upper triangle.

The next step was to compute the approximate height of the upper triangle as h_7 = appr. ;30 · 1 02;30 = 31;40. (Why is explained below.) A notation under the height of the upper triangle in the diagram,

[31 40 d]al	'31;40, the transversal',

is almost completely lost. Only the last half of the sign dal is preserved.

One would now expect to find the total area of the upper triangle or of the whole 7-front recorded in the diagram. That is not the case. Instead one finds a somewhat cryptic inscription, interpreted as follows by Robson in *MMTC* (1999), 49:

[nígin sag *šà*]	*The square of the front* (the side) of
sag.7 a.na *4 te-ṣi-ip-ma*	the 7-front by 4 you repeat, then
ši-in-šé-ra-ti	the twelfth
ta-na-as-sà-aḫ-ma	you tear out, then
a.šà	the field (the area).

What this means is that the area of a 7-front (heptagon) can be computed as

A_7 = appr. 4 · sq. s – 1/12 of 4 · sq. s = 4 · sq. s – ;20 · sq. s = 3;40 ·sq. s.

In other word, you get the area of the 7-front if you first multiply the square of the front by 4, then reduce the result by a twelfth of its value. This computation rule is a handy variant of the more formal computation rule

A_7 = sq. s · 3;40.

(Compare with the entry '3 41 the constant of a 7-front' in BR 28).

Another indication that OB mathematicians were interested in regular polygons is a drawing on the roughly shaped hand tablet **IM 51979** (Fig. 7.8.2 below; published here by courtesy of F. Al-Rawi).

The drawing shows an *octagram*, formed by all the "second diagonals" of a regular octagon (*cf.* Fig. 7.6.1 above). Included in the drawing are also all the diameters of the octagram, which are also diameters of the regular octagon. Note that the whole octagram can be drawn with a continuous movement of the stylus, just like a pentagram (Fig. 7.4.1 above).[17]

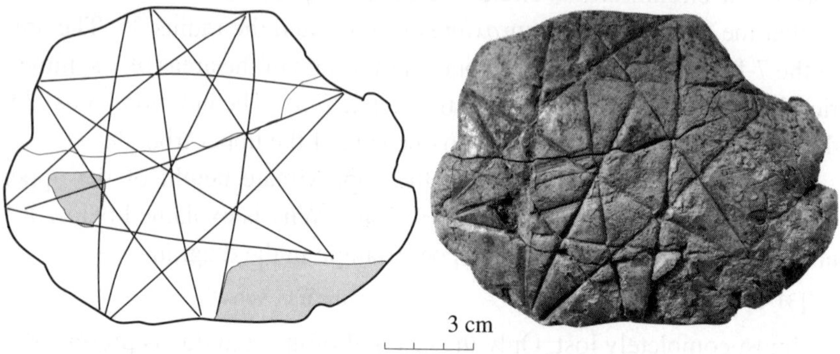

Fig. 7.8.2. IM 51979. An Old Babylonian drawing of an octagram with its diagonals.

7.9. Geometric Constructions in Mesopotamian Decorative Art

Interest in complicated geometric configurations arose early. It is, for instance, well known that there are many examples of appealing geometric patterns in decorative art from various periods in the history of Mesopotamia. A few particularly intriguing examples will be shown below.

The seven drawings in Figs.7.3.1-2 below, all reproduced from Legrain, *UE 3* (1936), are copies of seal imprints on various objects of clay from the ancient city Ur, excavated from layers *below* the famous royal cemetery at Ur, and dated by Legrain to the proto-Sumerian Jemdet Nasr period in Mesopotamia, around the beginning of the 3rd millennium BCE.

The first of the drawings, ***UE 3*, 78**, is a picture of *conjugate pairs of hyperbolas, complete with asymptotes and diameters*. No attempt will be made here to explain the presence of this design in an archaic seal imprint.

Less surprising are the examples of drawings of *pentagrams* in ***UE 3*, 105, 227, 398**. The *eight-petalled rosette* or *eight-pointed star*, as in *UE 3*, 286 and *UE 3*, 393, can be found as one of the details in many of the seal imprints copied in *UE 3*.

17. An unpublished OB hand tablet from Haddad is inscribed with an octagram formed by all the "*first* diagonals" of a regular octagon, and its four diameters (Farouk Al-Rawi, personal communication). The octagram is depicted in Appendix 2 below, Fig. 1 g,6.

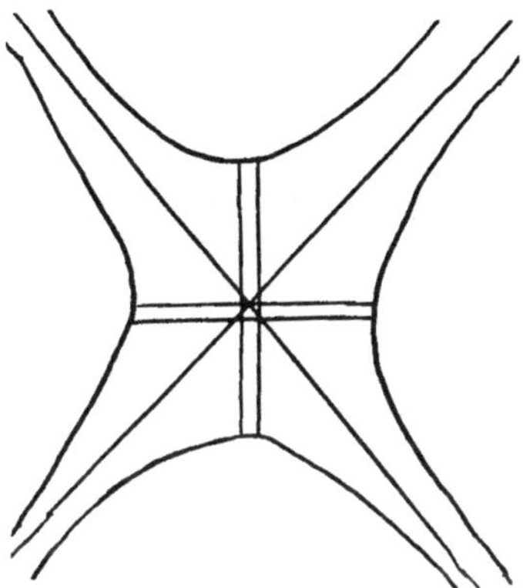

***UE 3*, 78**

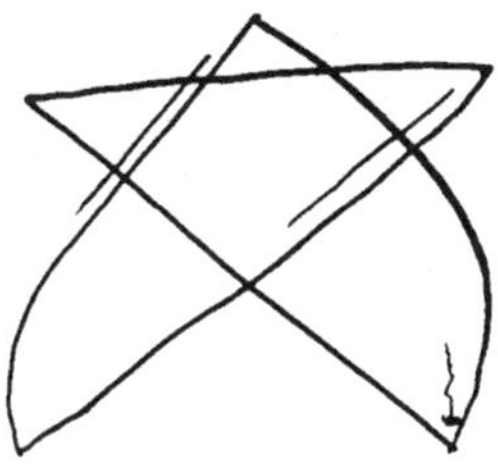

***UE 3*, 105**

***UE 3*, 284**

***UE 3*, 227**

***UE 3*, 274**

Fig. 7.9.1. Copies of seal imprints from layers below the royal cemetery at Ur.

UE 3, 286

UE 3, 393

UE 3, 398

Fig. 7.9.2. Additional copies of seal imprints from layers below the royal cemetery at Ur.

Snakes in various configurations was another common motive. In ***UE 3*, 284**, the geometric design behind the snake motive may have been a *rectangle with its diagonals*.

Similarly, the geometric design behind the figure of two entangled acrobats in ***UE 3*, 274 and *UE 3*, 286** may have been a *concave square* such as the one depicted in Fig. 6.2.6 above. See the attempted explanation in Fig. 7.9.3 below.

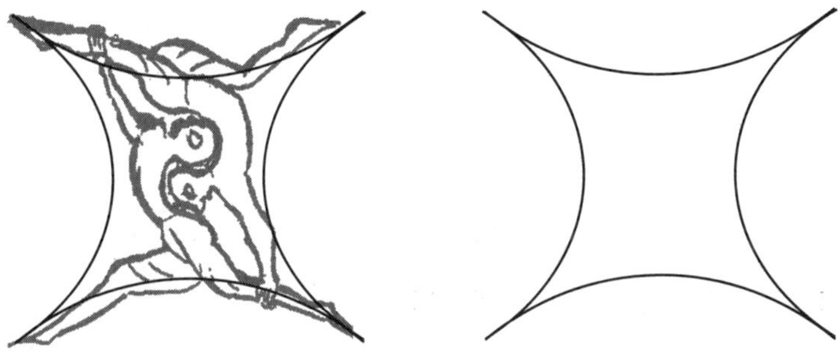

Fig. 7.9.3. *UE 3*, 274, 286. Attempted explanation of the figure of two entangled acrobats.

The central motive in *UE 3*, 393 is four entangled acrobats. The geometric design behind this motive may have been a *ring of four right triangles*, like the one depicted in Fig. 2.4.1, left, above. See Fig. 7.9.4 below.

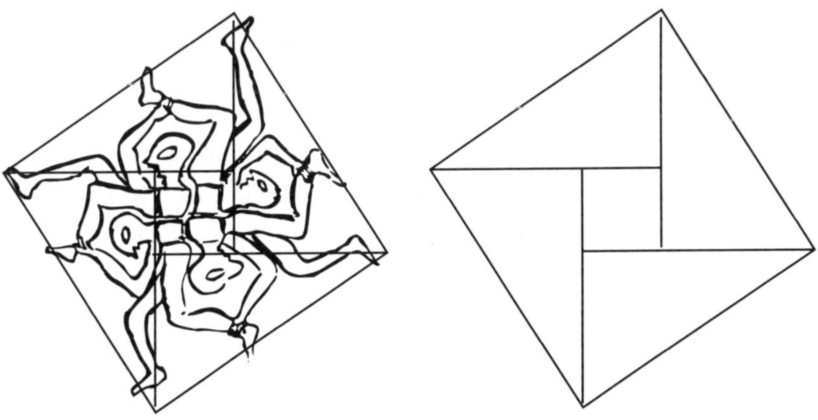

Fig. 7.9.4. *UE 3*,393. Attempted explanation of the figure of four entangled acrobats.

UE 3, 398 (Fig. 7.9.2, bottom) is a seal imprint with a very complex design, maybe an imprint of a royal seal. The design is a mix of several proto-cuneiform signs (among them a pentagram), probably spelling the names of several important cities, and three "wind-mills" of the same type as the four entangled acrobats in *UE 3*, 393 (Fig. 7.9.4 above). One of the wind-mills is composed of two human heads and two ox heads. The second wind-mill seems to be composed of four human legs, and the last wind-mill (which also appears in several other seal imprints published in *UE 3*) seems to be composed of two tools of some kind and two animal legs.

The seal imprint ***UE 3*, 518** (Fig. 7.9.5 below), contains a brief cuneiform inscription, mentioning the name of Mesannepada, a king of the First Dynasty of Ur. (This seal imprint is from a layer immediately *above* the royal cemetery at Ur (c. 2600 BCE).)

Fig. 7.9.5. *UE 3*, 518. An imprint of the royal seal of Mesannepada.

Below the cuneiform inscription on *UE 3*, 518, there is a wind-mill design in the form of 4 entangled men armed with knives. The underlying geometric design may be *a concave square with its diagonals and its circumscribed square,* as shown in Fig. 7.9.6 below.

Another example, in Fig. 7.9.7 below, is not a seal imprint but a a small marble plaque (apparently a mold) from Old Babylonian Babylon adorned with five intricately entangled bearded men (**VA 5953**; Andrae, *BPK* 58 (1937)). The underlying mathematical design is clearly a *pentagram enclosing a central regular pentagon.*

The twelve-pointed star shown in Fig. 7.9.8, finally, is a copy of a drawing in the Seleucid astrological text **O 176** (Thureau-Dangin, *TCL 6* (1922), text 13; see the commentary in Rochberg-Halton, *ZA* 77 (1987)).

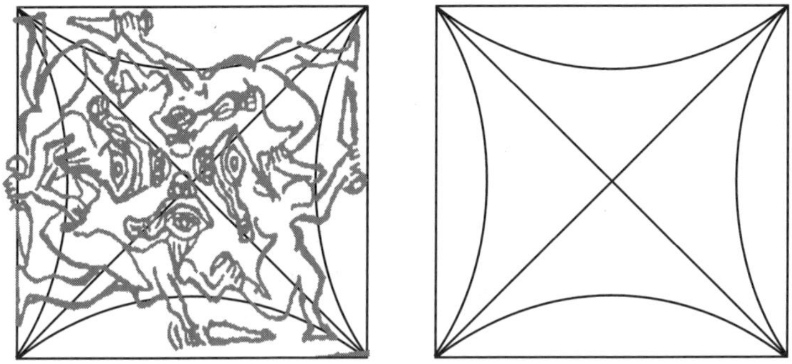

Fig. 7.9.6. Attempted explanation of the figure of four entangled armed men.

Fig. 7.9.7. VA 5953. An Old Babylonian mold showing a pentagram of bearded men.

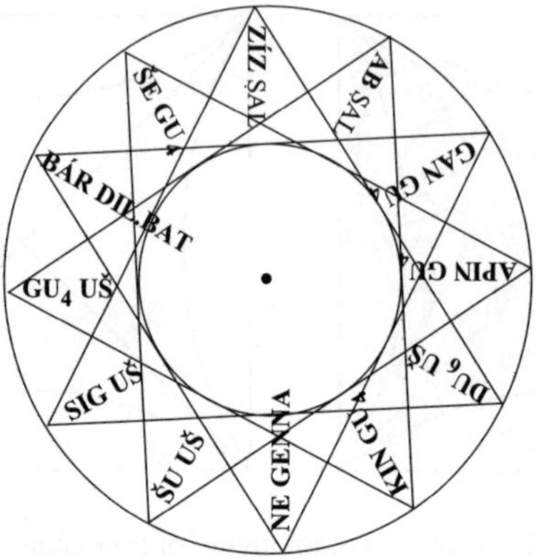

(Abbreviated) month names:

BÁR	(I)	ŠU	(IV)	DU_6	(VII)	AB	(X)
GU_4	(II)	NE	(V)	APIN	(VIII)	ZÍZ	(XI)
SIG	(III)	KIN	(VI)	GAN	(IX)	ŠE	(XII)

(Abbreviated) planet/god names:

DIL.BAT (Venus/Ishtar)	UŠ (Saturn/Ninurta)	GENNA(?) (also Saturn?)
GU_4 (Mercury/Nabû)	ṢAL (Mars/Nergal)	

Fig. 7.9.8. O 176. A twelve-pointed star with inscribed and circumscribed circles, month names and planet (or god) names. The astrological meaning of this diagram is unknown.

Chapter 8

El. XIII.13-18 and Regular Polyhedrons in Babylonian Mathematics

8.1. Regular Polyhedrons in *Elements* XIII

Four of the five regular polyhedrons are defined in ***El.*** **XI.Defs. 25-28**:

25 A *cube* is a solid figure bounded by *six equal squares.*

26 An *octahedron* is a solid figure bounded by *eight equal and equilateral triangles.*

27 An *icosahedron* is a solid figure bounded by *twenty equal and equilateral triangles.*

28 A *dodecahedron* is a solid figure bounded by *twelve equal, equilateral, and equiangular pentagons.*

To these definitions should be added the omitted definition

... A *tetrahedron* is a solid figure bounded by *four equal and equilateral triangles.*

There is no further mention of polyhedrons (other than the cube) in *El.* XI. Book XIII of Euclid's *Elements*, however, contains 6 propositions concerned with *regular polyhedrons inscribed in spheres* (XIII.13-18).

An Outline of the Contents of *El.* XIII.13-18

13 To construct a *tetrahedron* (a 'pyramid') inscribed in a given sphere, and to prove that the square on the diameter of the sphere is *one and a half times* the square on the edge of the tetrahedron.

14 To construct an *octahedron* inscribed in a given sphere, and to prove that the square on the diameter of the sphere is *two times* the square on the edge of the octahedron.

15 To construct a *cube* inscribed in a given sphere, and to prove that the square on the diameter of the sphere is *three times* the square on the edge of the cube.

16 To construct an *icosahedron* inscribed in a given sphere, and to prove that the edge of the icosahedron is a *minor*.

17 To construct a *dodecahedron* inscribed in a given sphere, and to prove that the edge of the dodecahedron is an *apotome.*

18 To set out the edges of *the five regular polyhedrons*, and to compare them with each other.

There is also a **postscript to *El.*XIII.18** saying that

No other figure, besides the mentioned five figures, can be constructed which is bounded by equilateral and equiangular figures equal to one another.

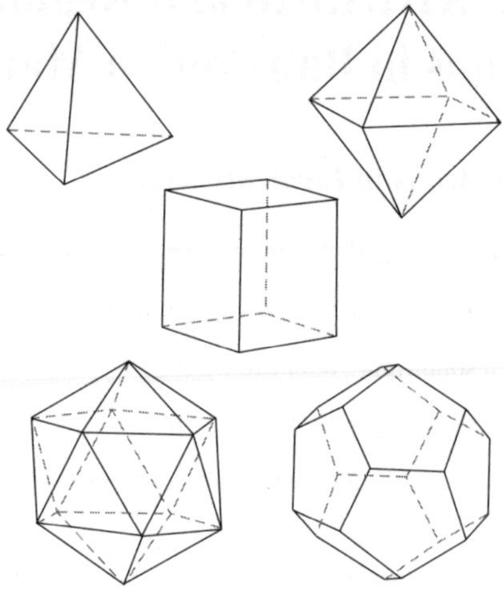

Fig. 8.1.1. The five regular polyhedrons.

Note: In *El.*XIII.13-15, the diameters of the spheres circumscribed around a regular tetrahedron, octahedron, or cube (hexahedron) are shown to be expressible *in terms of the edges* of the figures. In *El.*XIII.16 -17, on the other hand, the edges of a regular icosahedron and dodecahedron are shown to be inexpressible *in terms of the diameters* of the circumscribed spheres. A possible explanation for this lack of consistency may be that in a now lost precursor to Euclid's *Elements* the diameters of the circumscribed spheres for *all* the five regular polyhedrons had been expressed in terms of the edges of the figures, *following the Babylonian tradition*!

In ***El.* XIII.12** it is shown (see Fig. 7.2.2 above) that the square of the side of an *equilateral triangle* inscribed in a circle is 3 times the square on the radius of the circle. Therefore, if the diameter of the circle is expressible, then also the side of the inscribed equilateral triangle is expressible.

This result is used in ***El.* XIII.13** to show a result which (silently) implies that if the diameter of a given sphere is expressible, then also the edge of an inscribed 'pyramid', meaning a regular *tetrahedron,* is expressible.

The proof of *El.*XIII.13 is purely *synthetic*, as are also the proofs of *El.*XIII.14-17. The proof begins with the 'setting out' of the diameter AB of the given sphere (Fig. 8.1.2, left), cut at the point C so that *AC = 2 BC*. In the circle with AB as diameter, the perpendicular CD is erected, and the straight line DA is drawn.

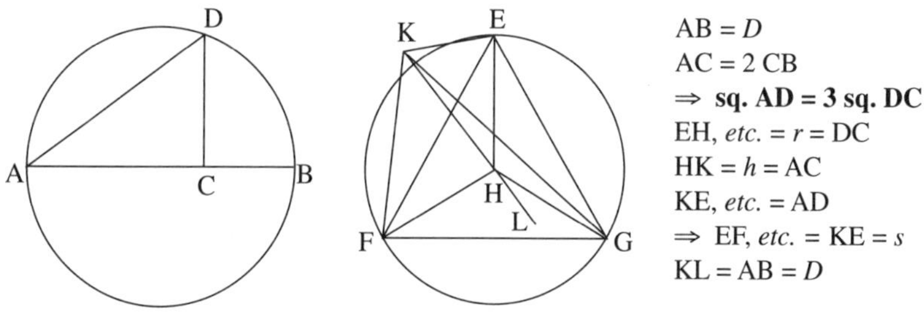

Fig. 8.1.2. *El.* XIII.13. Construction of a regular tetrahedron inscribed in a given sphere.

Next a circle with the radius CD is drawn, and an equilateral triangle EFG is inscribed in the circle (Fig. 8.1.2, right). From the centre H of the circle, a perpendicular HK equal to AC is erected, and the straight lines KE, KF, KG are drawn, *clearly all equal to DA*.

Now, since AC = 2 BC, it follows that AB = 3 BC, and therefore

sq. AD : sq. DC = AB · AC : AC · BC = AB : BC = 3 : 1, so that sq. AD = 3 · sq. DC.

This follows easily from, for instance, Lemma *El.*X.32/33 (Sec. 4.1 above). Since also sq. FE = 3 · sq. EH, according to XIII.12 (Fig. 7.2.2), and EH = DC, it follows that EF = AD, and then also FG = AD, and GE = AD. Hence, a regular tetrahedron has been constructed with the side AD.

The next step of the proof is to continue KH with a straight line HL equal to CB. It is then easy to show that a semicircle with the diameter KL passes through E. When this semicircle is rotated around its diameter, it generates a sphere passing through all the vertices of the tetrahedron.

The diameter KL of the sphere is equal to AB, and BA : AC = 3 : 2.

Therefore,

sq. BA : sq. AD = BA : AC = 3 : 2.

Since BA is the diameter of the sphere and AD is the edge of the inscribed tetrahedron, it follows that *the diameter of the given sphere is one and a half times the edge of the inscribed regular tetrahedron.*

The straightforward *analysis* which must have preceded the synthetic construction of a regular tetrahedron in *El.*XIII.13 is presented in Fig. 8.1.3 below, in metric algebra notations.

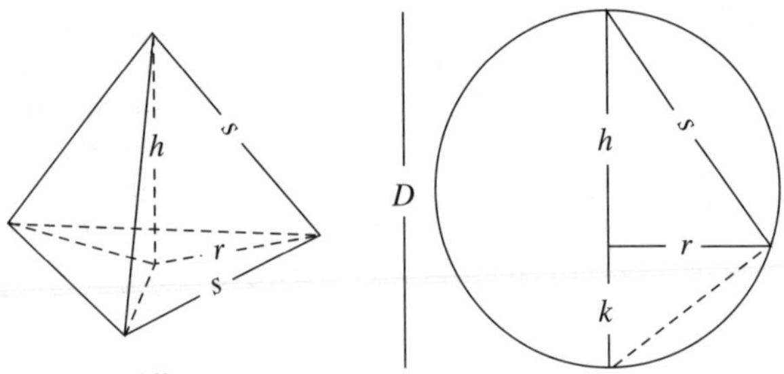

sq. r = 1/3 sq. s
sq. h = sq. s – 1/3 sq. s = 2/3 sq. s
D : h = $h \cdot D$: sq. h = sq. s : sq. h = 3/2
$h = 2/3 \cdot D$, $k = 1/3 \cdot D$, sq. D = sq. $(3/2 \cdot h) = 3/2 \cdot$ sq.

Fig. 8.1.3. The missing analysis in *El.* XIII.13.

The synthetic proof of ***El.* XIII.14** begins with the 'setting out' of the diameter AB of the given sphere (Fig. 8.1.4, left), cut at the point C so that $AC = BC$. In the circle with AB as diameter, the perpendicular CD is erected, and the straight line DB is drawn.

Next, a square EFGH is drawn with the side s = DB. The center K of the square is constructed as the point common to the diagonals of the square, and two perpendiculars KM, KL, both equal to EK are drawn. Joining M and L to the four vertices of the square EFGH completes the construction of the octahedron.

Clearly, LM = AB = D. Hence, LM is equal to the diameter of the given sphere, the constructed octahedron is inscribed in the given sphere, and sq. D = 2 sq. s. Therefore, if the diameter of a sphere is expressible, then also the edge of an inscribed regular *octahedron* is expressible.

The analysis that must have preceded this synthetic construction is simple and obvious, and will not be repeated here.

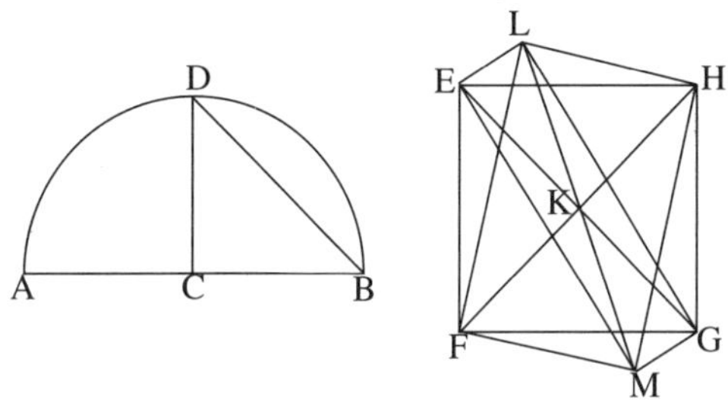

AB = D
AC = CB
⇒ **sq. AB = 2 sq. BD**
EF, *etc.* = s = BD
KM, KL = EK
⇒ LE, *etc.* = EF = s
LM = AB = D

Fig. 8.1.4. *El.* XIII.14. Construction of a regular octahedron inscribed in a given sphere.

The synthetic proof of ***El.* XIII.15** begins with the 'setting out' of the diameter AB of the given sphere (Fig. 8.1.5, left), cut at the point C so that $AC = 2\ BC$. In the circle with AB as diameter, the perpendicular CD is erected, and the straight line DB is drawn.

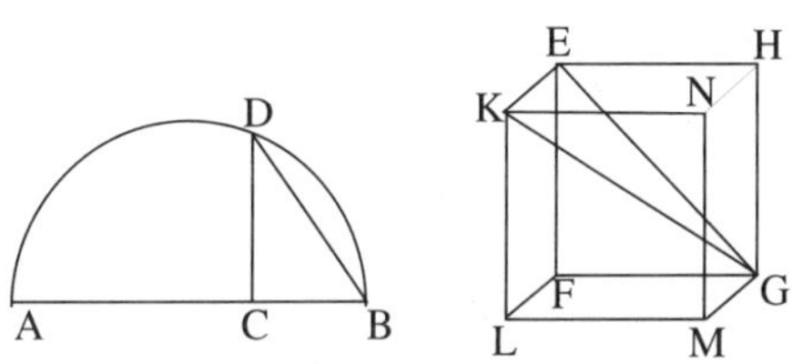

AB = D
AC = 2 CB
⇒ **sq. AB = 3 sq. BD**
EF, FG, *etc.* = s = BD
EK, HN, *etc.* = EF = s
⇒ sq. EG = 2 sq. s
⇒ sq. KG = 3 sq. s = 3 sq. BD
⇒ KG = AB = D

Fig. 8.1.5. *El.* XIII.15. Construction of a cube inscribed in a given sphere.

A cube FN is constructed with the side s = DB (Fig. 8.1.5, right), and the diagonals EG, KG are drawn. It is shown that a sphere with the diameter KG will pass through all the vertices of the cube. On the other hand,

sq. EG = sq. GF + sq. FE = 2 sq. EF = 2 sq. s, and
sq. KG = sq. GE + sq. EK = 3 sq. EF = 3 sq. s = 3 sq. DB = sq. AB = sq. D.

Therefore, KG = D, so that the cube is inscribed in the given sphere, and sq. D = 3 sq. s. Therefore, if the diameter of a sphere is expressible, then also the edge of an inscribed *cube* is expressible.

The analysis that must have preceded this synthetic construction is again simple and obvious, and will not be repeated here.

In ***El.* XIII.16**, the synthetic construction of a regular *icosahedron* inscribed in a given sphere begins with the 'setting out' of the diameter AB of the given sphere (Fig. 8.1.6, right), cut at the point C so that $AC = 4\ CB$. It is shown that the square on the diameter of the sphere is five times the square on the radius of the circle circumscribed around the pentagonal base of the top pyramid. Therefore it follows from XIII.11 that the edge of an icosahedron inscribed in a sphere is a *minor*, if the diameter of the sphere is an *expressible* straight line.

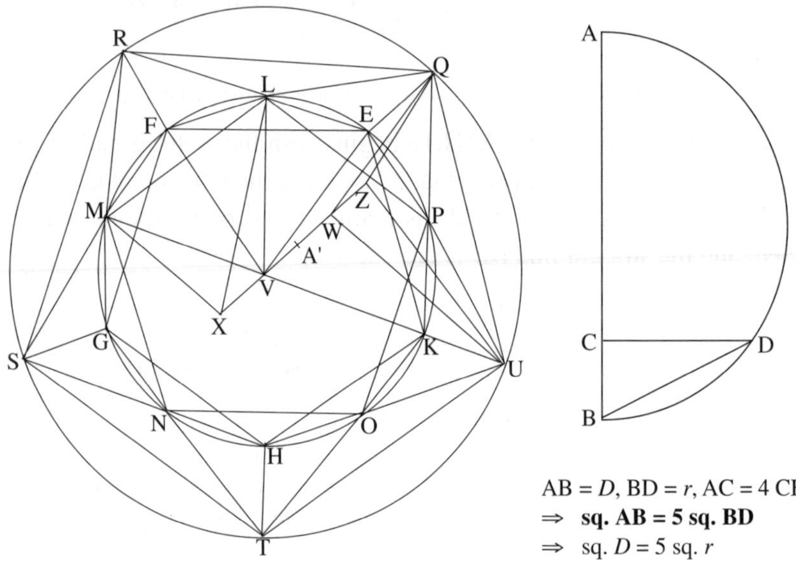

AB = D, BD = r, AC = 4 CB
⇒ **sq. AB = 5 sq. BD**
⇒ sq. D = 5 sq. r

Fig. 8.1.6. *El.* XIII.16. Construction of an icosahedron inscribed in a given sphere.

The complicated details of the synthetic construction in *El.*XIII.16 will not be discussed here. Instead, the form of the *analysis* that must have preceded the synthesis will be demonstrated. This analysis makes use of notations in the style of metric algebra, with reference to a simpler diagram (Fig. 8.1.7), based on the diagram in Heath, *ETBE 3* (1956), 487.

The analysis starts with the "top" of the icosahedron, a pyramid with a regular pentagon as base, and with five equilateral triangles as inclined faces. Let s be the edge of a face of the icosahedron, which is also the side of one of the equilateral triangular faces and the side of the pentagon, let r_5 be the radius of a circle circumscribed around the pentagon, and let h be the height of the pentagonal pyramid. Then

sq. h = sq. s – sq. r_5 = sq. s_{10}, so that $h = s_{10}$ [XIII.10]

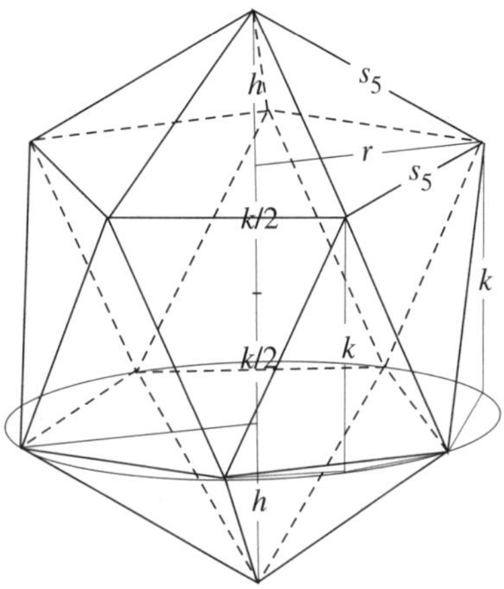

sq. h = sq. s_5 – sq. r_5
⇒ [XIII.10]
$h = s_{10}$

sq. k = sq. s_5 – sq. s_{10}
⇒ [XIII.10]
$k = r_5$

$k + h = r_5 + s_{10}$
⇒ [XIII.9]
$k + h$ cut in extreme and mean ratio
⇒ [XIII.3]
sq. $D/2$ = q. $(k/2 + h)$ = 5 sq. $k/2$ = 5 sq. $r_5/2$
⇒
sq. D = 5 sq. r_5, $D = k + 2\,h = r_5 + 2\,s_{10}$

Fig. 8.1.7. The missing analysis in *El.* XIII.16.

According to the diagram in Fig. 8.1.7, the "bottom" of the icosahedron is an inverted pentagonal pyramid, with its pentagonal base a certain distance k directly under the pentagonal base of the top pyramid, but rotated a tenth of a full circle. Therefore, the vertical projections of the vertices of the upper pentagonal base onto the plane of the lower pentagonal base can be identified with the five vertices of a decagon, situated halfway between the vertices of the lower pentagon. It follows that

sq. k = sq. s – sq. s_{10}, = sq. r_5, so that $k = r_5$ [*El.* XIII.10]

Consequently, the sphere circumscribed around the icosahedron has the diameter

$D = k + 2\,h = r_5 + 2\,s_{10}$.

It is clear from a look at the characteristic triangle for a pentagon in Fig. 7.3.2, left, that $s_{10}/(2\,r) = (s_5/2)/d_5 = \phi/2$. Therefore [XIII.9], the sum $k + h = r_5 + s_{10}$ is cut in extreme and mean ratio, with $k = r_5$ as the greater part. Consequently,

sq. $D/2$ = sq. $(k/2 + h)$ = 5 sq. $k/2$ = 5 sq. $r_5/2$, so that sq. D = 5 sq. r_5 [*El.* XIII.3]

The precise relationship between the edge s of the icosahedron and the diameter D of the circumscribed sphere is described by the equations

s = sqs. (5 – sqs. 5)/2 · D / sqs. 5 = sqs. (5 – sqs. 5)/10 · D (*cf.* Fig. 7.3.1)

(showing that s is a *minor* with respect to D), and

D = sqs. (5 + sqs. 5)/10 · s · sqs. 5 = sqs. (5+ sqs. 5)/2 · s (*cf.* Fig. 7.5.1)

In ***El.* XIII.17**, finally, is constructed a regular *dodecahedron* inscribed in a given sphere. The basic idea of the construction is the observation that *suitably selected diagonals of the 12 pentagonal faces of an dodecahedron can be identified with the 12 edges of an inscribed cube* (Fig. 8.1.8). It is shown that *the cube and the icosahedron can be inscribed in the same sphere*. Therefore, since the edge of the cube is expressible with respect to the diameter of the circumscribed sphere [*El.* XIII.15], it follows that also the diagonal of the pentagonal face is expressible. Now, the diagonal of a pentagon is cut in extreme and mean ratio, with the greater of the two parts equal to the side of the pentagon [*El.* XIII.9]. Moreover, the greater part of a straight line cut in extreme and mean ratio is an apotome with respect to the whole straight line [*El.* XIII.6]. It follows from a combination of these results that the edge of a dodecahedron inscribed in a given sphere is an *apotome* with respect to the diameter of the sphere.

The complicated details of the synthetic construction in *El.* XIII.17 will be discussed later. First, the form of the *analysis* that must have preceded the synthesis will be demonstrated here. This analysis makes use of notations in the style of metric algebra, with reference to a simpler diagram (Fig. 8.1.8), based on the diagram in Heath, *ETBE 3* (1956), 499.

In the diagram in Fig. 8.1.8, one of the diagonals of a pentagonal face of a dodecahedron is shown to coincide with one of the edges of an inscribed cube. Three orthogonal planes divide the cube into eight smaller cubes, all with the edge $d/2$, where d is the length of a diagonal in the pentagon. Now, it is easy to see that all those vertices of a pentagonal face of the dodecahedron, which are not simultaneously vertices of the inscribed cube, have the same distance to the nearest face of the cube. Let that common distance be called a. The size of a can be computed as follows, through *an application of the diagonal rule in three dimensions*:

sq. s = sq. a + sq. $d'/2$ + sq. $d/2$ ⇒ [*El.* XIII.4] sq. s = sq. a + 3 sq. $s/2$
⇒ sq. a = sq. $s/2$ ⇒ $a = s/2$.

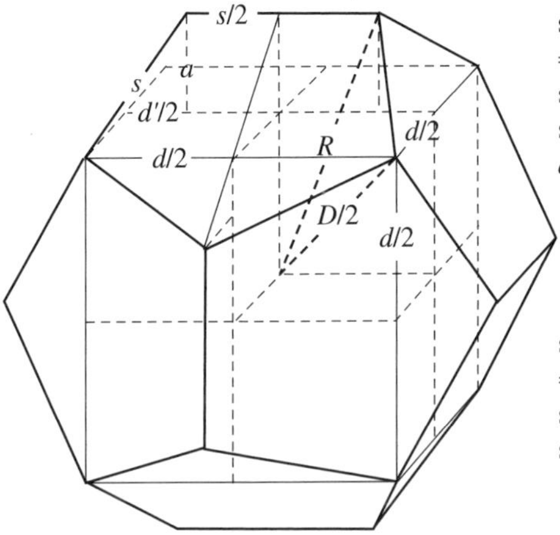

sq. s = sq. a + sq. $d'/2$ + sq. $d/2$
⇒ [XIII.4]
sq. s = sq. a + 3 sq. $s/2$
sq. a = sq. $s/2$
$a = s/2$

sq. R = sq. $s/2$ + sq. $(s + d)/2$
⇒ [XIII.4]
sq. R = 3 sq. $d/2 = D/2$
sq. 2 R = 3 sq. d = sq. D

Fig. 8.1.8. The missing analysis in *El.* XIII.17.

Next, the distance R from the centre of the inscribed cube to one of the vertices of the icosahedron which are not also a vertex of the cube can be computed as follows:

sq. R = sq. $s/2$ + sq. $(s + d)/2$ ⇒ [*El.* XIII.4] sq. R = 3 sq. $d/2$.

(Indeed, if s and d are the side and the diagonal of a regular pentagon, then the sum $s + d$ is cut in extreme and mean ratio, with d as the greater part.) On the other hand, *another application of the diagonal rule in three dimensions* shows that if $D/2$ is half the interior diagonal of the cube, then

sq. $D/2$ = sq. $d/2$ + sq. $d/2$ + sq. $d/2$ = 3 sq. $d/2$ (*cf. El.* XIII. 15)

Therefore, $R = D/ 2$. Consequently, D/2 is the distance from the centre of the inscribed cube to all vertices of the dodecahedron, whether they coincide with a vertex of the cube or not. This means that the cube and the dodecahedron are inscribed in the same sphere, and that if D is the diameter of the sphere and d the diagonal of a pentagonal face of the dodecahedron, then sq. D = 3 sq. d. Consequently, if D is expressible then d is also expressible. On the other hand, if the diagonal d is expressible, then the side s of the pentagon is an apotome [*El.* XIII.6]. Since the side of a pentagonal face of the dodecahedron is also an edge of the dodecahedron, it follows that each edge of a dodecahedron is an *apotome* with respect to the diameter of the circumscribed sphere.

Explicitly,

$s = (\text{sqs. } 5 - 1)/2 \cdot D \,/\, \text{sqs. } 3 = (\text{sqs. } 15 - \text{sqs. } 3)/6 \cdot D$

and

$D = (\text{sqs. } 5 + 1)/2 \cdot s \cdot \text{sqs. } 3 = (\text{sqs. } 15 + \text{sqs. } 3)/2 \cdot s.$

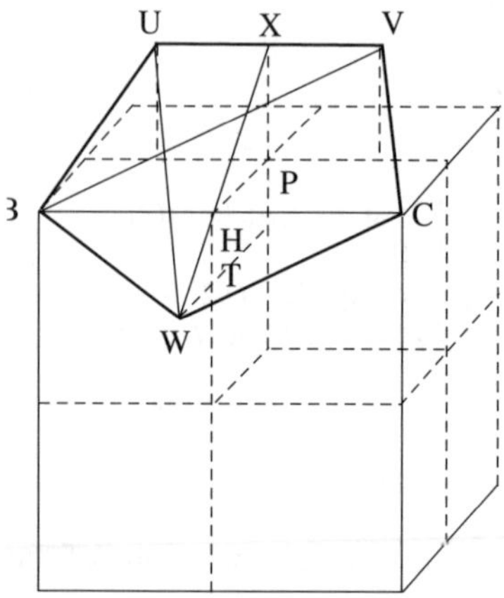

XHW is a straight line segment because
XP : PH = $s/2 : d/2 = d'/2 : s/2$ = HT : TW.

sq. BV = sq. $s/2$ + sq. $(s + d)/2$ + sq. $d/2$
⇒ [XIII.4]
sq. BV = 4 sq. $d/2$ = sq. d, BV = d.

sq. UW = sq. $s/2$ + sq. $d/2$ + sq. $(s + d)/2$
⇒ [XIII.4]
sq. UW = 4 sq. $d/2$ = sq. d, UW = d.

Fig. 8.1.9. *El.* XIII.17. Construction of a dodecahedron inscribed in a given sphere.

In the synthetic proof of *El.* XIII.17, it is notable that Euclid makes a somewhat uncongenial use of the two propositions *El.* VI.32 and *El.* XIII.7. The first of these propositions is used to prove that after the explicit construction of an equilateral pentagon as in Fig. 8.1.9 below, *the pentagon is in one plane.* However, *El.* VI.32 is more general than what is needed for the purpose. Since the triangles XPH and HTW are right triangles all that is needed to show that XHW are in a right line (and that therefore the pentagon is in one plane) is the observation that $s/2 : d/2 = d'/2 : s/2$. This relation, follows from the fact that the sum $d = s + d'$ is cut in extreme and mean ratio, with s as the greater part.

El. XIII.7 says that "If three angles of an equilateral pentagon are equal, then the pentagon is equiangular". Euclid starts by showing, by use of an application of the diagonal rule in three dimensions, that BV = BC (see again Fig. 8.1.9), and that therefore the angle BUV is equal to the angle BWC. Then also the angle CVU must be equal to the angle CWB. Next, he makes use of *El.* XIII.7 to show that all five angles in the pentagon are equal. However, he could equally well have shown, by another use of the diagonal rule in three dimensions, that UW = BC, from which it follows that the angle UBW is equal to the angle CWB. And so on.

Conclusion

In the detailed investigation above of *Elements* XIII.13-17, Euclid's synthetic constructions are complemented by analytic procedures expressed in the style of metric algebra. The unexpected result of the investigation is that all the procedures needed for the computation of various crucial parameters of the five regular polyhedrons probably was within the competence of Old (and Late?) Babylonian mathematicians, with the possible exception of the drawing of accurate diagrams, such as the ones in Figs. 8.1.6-9. (Known OB drawings of three-dimensional objects are of notoriously poor quality. However, nothing is known about the quality of corresponding drawings presumably made by Late Babylonian mathematicians.) Only the completely superfluous use of *El.* VI.32 and *El.* XIII.7 in the proof of *El.* XIII.17 is beyond the horizon of Babylonian mathematics!

Note that it is now known that the kind of application of *the diagonal rule in three dimensions* which plays such a prominent role in the construction of a dodecahedron in *El.* XIII.17 is documented in a mathematical cuneiform text. That text is the object of discussion in Sec. 8.2 below.

8.2. MS 3049 § 5. The Inner Diagonal of a Gate

MS 3049 (Friberg, *RC* (2007), Sec. 11.1) is a small fragment of a large cuneiform mathematical recombination text. It is either late OB or post-Old-Babylonian (Kassite). According to a post-script, which luckily is preserved on what remains of the reverse of the clay tablet, the text originally contained 6 problems for circles (of which one is preserved on the obverse of the fragment), 5 problems for squares, 1 for a triangle(?), 3 for 'brick molds', and 1 for an 'inner diagonal of a gate' (preserved on the reverse of the fragment). Here is a translation of the text of the preserved last problem:

MS 3049 § 5, literal translation	explanation
If the inner cross-over (diagonal) of a gate he shall do,	Compute the inner diagonal of a gate Its height is

5 cubits, 25, and 10 fingers, 1 40,	5 cubits = ;25 ninda and
the height of the gate.	10 fingers = ;01 40 ninda
x x x x x the table,	According to the x x x table
to this one 20 and *x x x x* enter,	and x x x x x it follows that
then this 26 40, 8 53 20 *the width*, and	when ;26 40 n. is the height, then
6 40, the thickness of the wall, you see.	;08 53 20 n. is the width, and
26 40, the height of the wall, let eat itself,	;06 40 n. the thickness of the wall
then 11 51 06 40 you see.	sq. ;26 40 = ;11 51 06 40
8 53 20, the width of the gate, let eat itself,	sq. ;08 53 20
then 1 19 … 44 26 40 you see.	= ;01 19 00 44 26 40
6 40, the thickness of the wall, let eat itself,	sq. ;06 40
then 44 26 40 you see.	= ;00 44 26 40
Heap them, 13 54 34 14 26 40 you see.	The sum of the squares is
Its likeside let come up,	;13 54 34 14 26 40
then 28 53 20 you see	The square side is ;28 53 20
(for) the gate of height 26 40.	(the diagonal) for a gate of height 26 40
So you do it.	Compute like this

Note that the problcm text is *not* accompanied by any illustrating diagram. Anyway, the statement of the problem is clear:

What is the 'inner diagonal' of a 'gate' with the height ;26 40 ninda?

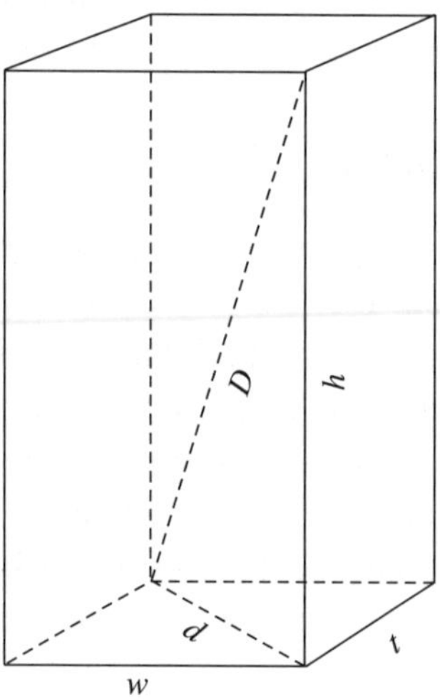

(a = ;02 13 20 n. = 1/27 n.)

t = 3 a = ;06 40 n.
w = 4 a = ;08 53 20 n.
d = 5 a (the bottom diagonal)
h = 12 a = ;26 40 n.
D = 13 a = ;28 53 20 n.

sq. t + sq. w = sq. d
sq. d + sq. h = sq. D
$\Rightarrow$
sq. t + sq. w + sq. h = sq. D

Fig. 8.2.1. MS 3049 § 5. Computation of the inner diagonal of a gate.

The solution procedure begins with the consultation of some mysterious mathematical table, according to which a gate with the height h = ;26

40 n. (= 5 cubits 10 fingers) has the width w = ;08 53 20 n. (= 1 cubit 23 1/3 finger), and the 'thickness' (which is also the thickness of the wall in which the gate is situated) t = ;06 40 n. (= 1 cubit 10 fingers). Unfortunately, the text is broken just where it presumably describes what kind of mathematical table is meant here and how it is used.

It is clear, anyway, that the inner diagonal D of the gate is computed through an application of *the diagonal rule in three dimensions*:

sq. D = sq. h + sq. w + sq. t = ;11 51 06 40 + ;01 19 00 44 26 40 + ;00 44 26 40
= ;13 54 34 14 26 40 (sq. n.) = sq. (;28 53 20 n.).

The counting with cubits and fingers and "many-place" sexagesimal numbers as in this text is typical for many Babylonian mathematical texts, and reveals that *the primary purpose of Babylonian education in mathematics was to teach the students to master the complexities of counting with sexagesimal numbers and with various traditional measures.*

Actually, in the present case, the appearance of many-place sexagesimal numbers is an example of a deliberately introduced difficulty hiding the underlying simplicity of the data. Indeed, it is easy to check that

t	= ;06 40 n.	= 3 · ;02 13 20 n.	where	;02 13 20 = 1/27
w	= ;08 53 20 n.	= 4 · ;02 13 20 n.		
h	= ;26 40 n.	= 12 · ;02 13 20 n.		
D	= ;28 53 20 n.	= 13 · ;02 13 20 n.		

The number quartet 3, 4, 12, 13 has the interesting property that

sq. 3 + sq. 4 + sq. 12 = sq. 13.

This "diagonal quartet" is constructed through composition of the two simple "diagonal triples" 3, 4, 5 and 5, 12, 13, in the following way:

sq. 3 + sq. 4 = sq. 5, and sq. 5 + sq. 12 = sq. 13 ⇒ sq. 3 + sq. 4 + sq. 12 = sq. 13.

In geometric terms, the computation in MS 3049 § 5 can be explained as follows (see Fig. 8.2.1): First the square of the "bottom diagonal" d of the gate is computed by use of the diagonal rule in two dimensions, as

sq. d = sq. t + sq. w.

Then the square of the inner diagonal of the gate is computed through a second application of the diagonal rule in two dimensions, as sq.

D = sq. d + sq. h.

It still remains to explain how the square side of the many-place sexagesimal number sq. D = ;13 54 34 14 26 40 was computed. An answer to this interesting question will be suggested in Sec. 16.7 below.

8.3. The Weight of an Old Babylonian Colossal Copper Icosahedron

MS 3876 (Friberg, *RC* (2007), Sec. 11.3) is a mathematical problem text written with very small cuneiform signs on a clay tablet of a most unusual format. Also some of the mathematical terminology in this text is unusual and quite difficult to interpret. It is likely that the text is Kassite, meaning post-Old-Babylonian, from the last half of the second millennium BC. This makes the text unique of its kind. Also the mathematical content of the text is highly unusual, as will be shown below.

Here is a tentative translation of the most important part of the text, written on the lower half of the obverse of the clay tablet (Fig. 8.3.1):

MS 3876 # 3, literal translation	explanation
x x x x the city wall x, a horn-figure,	? An icosahedron
copper behind.	made of copper(?)
x the copper x (is) what?	Compute the weight of the copper
At the rim (periphery?) an arc (a ball?)	A ball(?)
of gaming-piece-fields you see, then	of equilateral triangles.
what xx you take, then	Take one of them (?)
your ground (area) y*ou see*.	and compute its area
Heap them, then that ground	Add the areas
for 1 horn-figure, the copper behind (it), x,	and compute the weight of the copper
So you do (it).	Do it like this:
The reciprocal of 6, the constant,	Compute the reciprocal of the
resolve, then 10.	constant 6, it is 10
Steps of 1 30, the front of the city wall, 15.	1/6 of 1 30, the circumference (?) = 15
3 cubits each (the sides of)	s = ;15 ninda = 3 cubits is each side of
one gaming-piece are equal.	each equilateral triangle
If 3 cubits each (the sides of) a gaming-piece	If 3 cubits is the side of each
are equal, the volume (is) what?	equilateral triangle, what is the volume?
Half (of) 15, the front, break, then 7 30.	$s/2$ = ;15/2 = ;07 30
7 30 steps of 15, the second front,	$s \cdot s/2$ = ;07 30 n. · ;15 n.
1 52 30, the halved.	= ;01 52 30 sq. n. = (sq. s)/2
14 03 45, its eighth tear off, then	1/8 · (sq. s)/2 = ;00 14 03 45
1 38 26 15 (is) the ground (of)	(1 – 1/8) (sq. s)/2 = ;01 38 26 15 =
one gaming-piece-field that you see.	A (the area of one equilateral triangle)

The gaming-piece-fields how many?	How many equilateral triangles?
From 6, the constant, 1 tear off,	The constant 6, minus 1
then 5 (is) the remainder.	= 5
To 4 repeat (it), then 20.	$(6 - 1) \cdot 4 = 5 \cdot 4 = 20$
20 gaming-piece-fields.	There are 20 equilateral triangles
1 38 26 15 to 20 repeat, then 32 48 45,	$20\,A = 20 \cdot$;01 38 26 15 sq. n.
1/2 šar 2 2/3 gín 26 1/4 barleycorns.	= ;32 48 45 sq. n. = ···
For 1 metal-covered horn-figure,	How much copper in 1
what is its copper?	icosahedral shell?
32 48 45 times 2 (is) 1 05 37 30,	$V = 20\,A \cdot$ (1 finger)
1 gín 16 1/2 1/4 barleycorns	= ;32 48 45 sq. n. · ;02 cubit
and half 1/4 barleycorns (is) its volume.	= ;01 05 37 30 sq. n. · c. = ···
1 05 37 30, its volume,	$V \cdot$ 1 12 00 talents / sq. n. · c,
steps of 1 12, the constant of the copper, (is)	the density of copper
1 18 45, the copper.	= 1 18;45 talents of copper
Instead of its volume,	Instead of counting with volume
1 cubit each the square side *x x*	measure: The [weight of a sheet of]
for 1 metal-covered horn-figure,	1 sq. cubit [· 1 finger] of copper
1 talent, the copper in *x x x x x x x x x,*	is 1 talent
in this copper *x x x x x x x x x.*	

The exercise begins with *the statement of the problem*, which is quite obscure due to damage to the text and the use of the previously unknown terms ‘city wall’(?) and ‘horn figure’(?). Then follows *a brief description of the solution procedure* to be used, namely to consider a ‘ball(?) of ‘gaming-piece fields’(?), to compute the area of each such field, to sum the individual areas, and finally to compute the weight of the copper needed.

The solution procedure itself begins with the multiplication of the length(?) 1 30 of the ‘city wall’(?) with the reciprocal 1/6 = ;10 of the constant. The result is ‘15’, which is immediately explained as ;15 ninda = 3 cubits. (Remember that 1 ninda = 12 cubits = appr. 6 meters.)

In the next step of the solution procedure, the area A of a ‘gaming-piece figure’ with each side equal to 3 cubits is computed as

$$A = (1 - 1/8) \cdot s \cdot s/2 = (;01\ 52\ 30 - ;00\ 14\ 03\ 45) \text{ sq. n.} = ;01\ 38\ 26\ 15 \text{ sq. n.}$$

Evidently, therefore, gán.za.na ‘gaming-piece field’ is a Sumerian name for ‘equilateral triangle’ (*cf.* Sec. 7.7 above), possibly because the profile of some kind of gaming-piece may look like an equilateral triangle.

Fig. 8.3.1. MS 3876. Computation of the weight of a huge copper icosahedron.

The number *N* of such gaming-piece fields is computed as

$N = (6-1) \cdot 4$, where 6 is the 'constant' mentioned before.

The total area of *N* gaming-piece fields is then

$N \cdot A = 20 \cdot$;01 38 26 15 sq. n. = ;32 48 45 sq. n.

In *Old Babylonian area measure notations*, this total area is equal to

1/2 šar 2 2/3 gín 26 1/4 barleycorn, where 1 šar = 1 sq. n., 1 gín = 1/60 šar, *etc.*

The last part of the computation begins by repeating the question

'How much copper is needed to cover 1 'horn figure'?

The computation begins by computing a volume *V*:

V = '32 48 45' · '2' = '1 05 37 30' = 1 gín 16 1/2 1/4 1/2 · 1/4 barleycorns.

Now, in *Old Babylonian volume measure notations*,

1 šar = 1 sq. n. · 1 cubit, 1 gín = 1/60 šar, 1 barleycorn = 1/180 gín.

Therefore, the computed volume must be equal to

V = 1 gín 16 1/2 1/4 1/8 barley corns = ;01 05 37 30 sq. n. · 1 cubit.

Since A = ;32 48 45 sq. n., it follows that

'2' = ;02 cubit = 1/30 cubit = 1 finger (= appr. 1 2/3 cm).

In other words, V is computed as *the total area 20 · A times 1 finger.*

In the last step of the computation, a new constant is mentioned:

'1 12' the constant of copper.

Apparently, this constant is explained as follows in the badly damaged last few lines of the text:

1 talent (= 60 minas = appr. 30 kilograms) is the weight of
a square sheet of copper measuring 1 sq. cubit · 1 finger.

Therefore,

The weight of 1 volume-šar = 1 sq. n. · 1 cubit of copper is
12 · 12 · 30 talents = 1 12 00 talents.

Consequently, the weight W of all the copper covering the 'horn-figure' is

W = V · 1 12 00 talents / volume-šar = ;01 05 37 30 · 1 12 00 talents = 1 18;45 talents.

All the steps of the complicated "metro-mathematical" computation in MS 3876 # 3 have now been explained. It still remains to be explained what is meant by the Sumerian term gán.si 'horn-figure'. The horn-figure in the text appears to be covered by a 'ball'(?) of (6 – 1) · 4 equilateral triangles made of copper, each with the side 3 cubits (1 1/2 meters), and 1 finger (1 2/3 cm) thick, together weighing 1 18;45 talents (2,350 kg.).

A reasonable conjecture is that the 'horn-figure' is an *icosahedron*, a regular polyhedron with 20 equilateral faces. The strange computation of 20 as (6 – 1) · 4 can then be explained as follows (Fig. 8.3.2, left):

The OB construction of an icosahedron in MS 3876 # 3 begins with *a regular hexagon* bounded by a 'city wall' of length 1;30 ninda (9 meters). The hexagon is divided into 6 equilateral triangles, each with the side ;15 ninda = 3 cubits. One of the triangles is removed, so that 6 – 1 = 5 equilateral triangles remain. Then 3 equilateral triangles is added to each one of the 5 equilateral triangles, so that 6 – 1 = *5 chains* are formed, *each chain composed of 4 equal equilateral triangles*. The 5 chains together contain (6 – 1) · 4 = 20 equilateral triangles, and when the chains are folded in the appropriate way, an icosahedron is formed (Fig. 8.3.2, right).

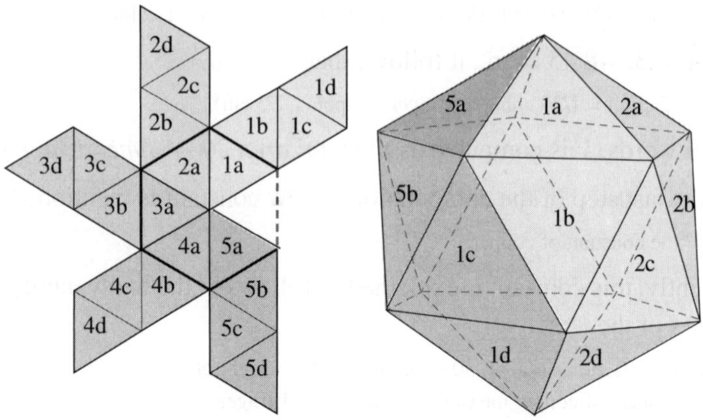

Fig. 8.3.2. Construction of an icosahedron by folding 5 chains of 4 equilateral triangles.

The discussion above of MS 3876 # 3 suggests that Babylonian mathematicians knew how to compute *the area of (the outer shell of) an icosahedron*. In view of the OB mathematicians' well known habit of computing the volumes of all kinds of solid figures (*cf.* Chapter 9 below), it is therefore also extremely likely that they tried to compute *the volume of an icosahedron*. That they may have been successful if they ever tried to do that was mentioned above, in the Conclusion to Sec. 8.1.

Chapter 9

Elements XII and Pyramids and Cones in Babylonian Mathematics

9.1. Circles, Pyramids, Cones, and Spheres in *Elements* XII

Areas and volumes are never explicitly mentioned in *Elements* XII, or anywhere else in the *Elements*. Yet, the main feature of *Elements* XII is the use of the *method of exhaustion*, based on *El.* X.1, in order to prove that

Circles are to one another as the squares on the diameters — *El.* XII.2
Pyramids which are of the same height and have triangular bases are to one another as the bases — *El.* XII.5
Any (circular) *cone* is a third part of the (circular) cylinder which has the same base with it and equal height — *El.* XII.10
Spheres are to one another in the triplicate ratio of their respective diameters — *El.* XII.18

Of particular interest in the present connection (comparison with Babylonian mathematics) are the propositions *El.* XII.3-7, all dealing with *triangular pyramids*. Their contents will be outlined briefly below, in intentionally modernized form:

El. XII.3. A dissection of a triangular pyramid by planes through the midpoints of its edges

Every *triangular pyramid* can be cut (by three planes through the midpoints of the six edges) into *two pyramids* of equal volumes, similar to the whole pyramid, and *two wedges* (triangular prisms) of equal volumes (but not similar to each other).
The combined volume of the two wedges is greater than half the volume of the whole pyramid.

The way in which a given triangular pyramid is dissected, according to *El.* XII.3, is shown in Fig. 9.1.1 below. Let the lengths of the edges of the given pyramid be *a, b, c, d, e, f.* Then the lengths of the edges of the two sub-pyramids cut off by two of the planes through five mid-points of the edges of the given pyramid are, in each case, *a*/2, *b*/2, *c*/2, *d*/2, *e*/2, *f*/2. Clearly, *the two sub-pyramids have all edges equal and parallel*. Therefore, they *are similar and "equal"*. (Actually, they are *congruent* and so, with any reasonable definition of volume, have the same volume.)

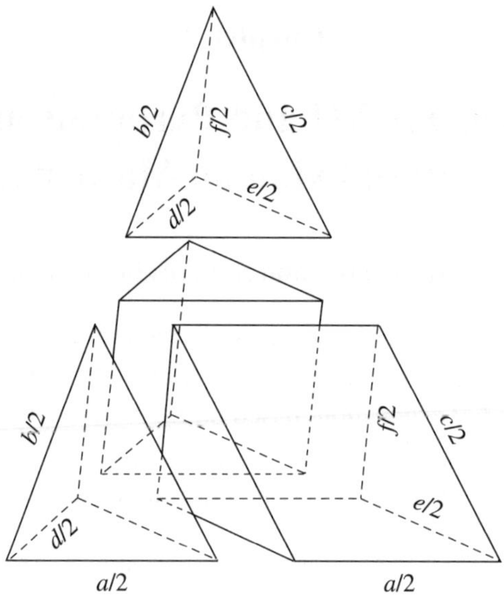

Fig. 9.1.1. A triangular pyramid dissected as in *El.* XII.3.

The two wedges (triangular prisms) remaining after the removal of the two sub-pyramids from the given pyramid are *both equal to one half of a parallelepipedal solid* with edges of lengths *a*/2, *e*/2, *f*/2, parallel to the edges with those lengths of the sub-pyramids. One of the wedges is formed by cutting the parallelepipedal solid with a plane through the two edges of length *a*/2 (see again Fig. 9.1.1), while the other wedge is formed by cutting a similar and equal parallelepipedal solid with a plane through the two edges of length *f*/2. Therefore, the two wedges have the same volume but are not similar. (Euclid proves that the two wedges are "equal" by reference to the strangely formulated *ad hoc* proposition *El.* XI.39.)

Finally, since each one of the two wedges is greater (in volume) than each one of the two pyramids, it follows that the two wedges together are greater (in volume) than half the original pyramid.

***El.*. XII.4. Corresponding dissections of two triangular pyramids of**

the same height.

Let two triangular pyramids of the same height both be cut into two pyramids and two wedges as in *El.* XII.3. Then *the combined volumes of the two wedges* in each pyramid separately *are proportional to the areas of the bases of the pyramids.*

***El.*. XII.5. The volumes of triangular pyramids of the same height are proportional to the areas of their bases.**

This proposition is proved by means of the *exhaustion method* of Eudoxus. Suppose that the two sub-pyramids produced by the dissection described in *El.* XII.3 are in their turn dissected in the same way. The result is two new, smaller wedges and two new, smaller sub-pyramids. The process can be repeated until the combined volumes of all the sub-pyramids becomes arbitrarily small. If there are two pyramids of the same height, and if the successive regular dissections are carried out *in tandem*, then, as in *El.* XII.4, after each step of the algorithm the combined volumes of all the produced wedges in each pyramid separately are proportional to the areas of the bases. Therefore, it can be shown, by an ingenious argument, that the ratio between the volumes of the pyramids can be neither greater nor smaller than the ratio between the areas of their bases.

***El.*. XII.6. The volumes of *polygonal* pyramids of the same height are proportional to the areas of their bases.**

***El.*. XII.7. Every triangular prism can be cut (by two planes through four of the six vertices) into three triangular pyramids (not similar to each other). The three sub-pyramids have, two by two, equal heights, and bases of equal areas. Therefore, the volume of each one of them is one third of the volume of the triangular prism.**

A *triangular prism* is a solid figure bounded by two parallel and congruent triangles and three parallelograms (*El.* XI.Def.13). In Fig. 9.1.2 below, the two bounding triangles both have sides of lengths a, b, c and the three parallelograms have sides of lengths 1) a, d, 2) b, d, 3) c, d. Let the prism be cut by a plane through one of the sides of length a and through the opposite vertex of the prism. The plane cuts two of the bounding parallelograms along their diagonals, and it divides the prism into two pyramids, one with the same triangular base as the prism, the other with the parallelogram with sides of lengths a, d as a base. Let the second pyramid, in its turn, be cut by a plane through the diagonal of its base and through the opposite vertex of the pyramid. As a result, this second pyramid is divided into two triangular pyramids. These two pyramids (the one to the left and the one in the middle in Fig. 9.1.2) have bases of the same area (half the area of the parallelogram with sides of lengths *a*, *d*), and the same height. Therefore, they have the equal volumes [*El.* XII.6]. On the other hand, in Fig. 9.1.2, the triangular pyramids in the middle and the one to the right also have bases of the same area (half the area of the parallelogram with sides of lengths *c*, *d*) and the same height. therefore, they too have equal volumes. This observation concludes the proof of the proposition.

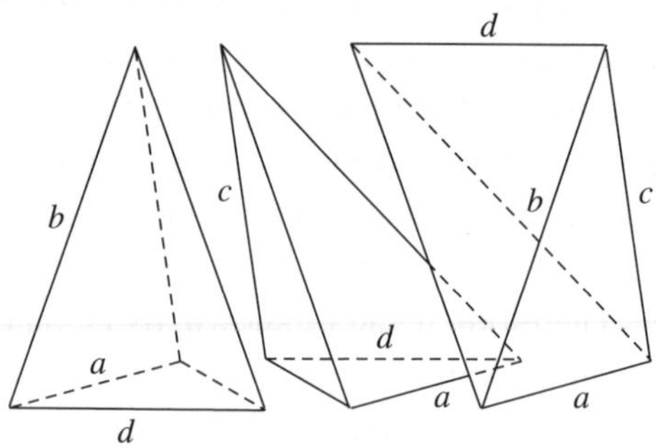

Fig. 9.1.2. A triangular prism dissected as in *El.* XII.7.

9.2. Pre-literate Plain Number Tokens from the Middle East in the Form of Circular Lenses, Pyramids, Cylinders, Cones, and Spheres

The basic object of *Elements* XII is (from a modern point of view) the computation of the area of a circle and the volumes of (triangular) pyramids, of (circular) cylinders or cones, and of spheres. It is interesting to note that it is well documented that *circles, pyramids, cylinders, cones, and spheres* were well known long before the time of the Greeks, although in a completely different setting. The arguments below supporting this proposition are borrowed from extensive accounts of related matters in Friberg, *OLZ* 89 (1994) and *RC* (2007), Appendix 4, Sec. A4 i.

In ***Before Writing, Vol. 1*** (1992), Schmandt-Besserat gave a detailed account of her revolutionary theory about the crucial role played by small clay-figures, so called "tokens", in the prehistory of writing. According to this theory, there were seven essential steps in the early development of writing as a tool for accounting and communication: 1) the appearance in various parts of the Middle East around 8000 BCE, that is at the time of the agricultural revolution, of six types of "plain tokens", small geometric objects in baked clay (*circular disks, tetrahedrons, cylinders, cones, spheres, and ovoids*), probably used as counters; 2) in the late fifth and early fourth millennia, the gradual introduction of additional types and subtypes of tokens, so called "complex tokens"; also the occasional use of perforations, allowing groups of tokens to be strung together; 3) around the middle of the fourth millennium, at the time of the first cities and beginning state formation, an explosive proliferation of the repertory of complex tokens at a limited number of sites (mainly Susa in Iran, Uruk in Iraq, and Habuba Kabira in Syria), probably in order to represent many new kinds of products from the city workshops; 4) the invention of "spherical envelopes", containing (mostly) plain tokens and often impressed with cylinder seals, sometimes for good measure also marked on the outside with more or less schematic representations of the tokens inside; 5) for a short while, around 3300 BCE, the use of "impressed tablets", instead of, or together with, spherical envelopes and yielding the same kind of information, the number signs on these first clay tablets being imitations of the previously used tokens; 6) the invention of writing on clay tablets, with a large inventory of sometimes pictographic but most often abstract signs, of which, apparently, the latter in some cases were two-dimensional representations of the complex tokens they replaced; 7) the complete disappearance of tokens from (almost) all excavated sites after the invention of writing.

For the history of number notations, the spherical envelopes are particularly important. Their importance derives from two hypothetical situations: Either the content of an envelope constitutes an account of a single *disbursement or delivery*, in which case the enclosed tokens record a number in a single system of number tokens. Or else, the content of an envelope constitutes a record of a single *transaction*, in which case the enclosed tokens record two numbers in two separate systems of number tokens, and there exists some simple mathematical relation between the two numbers.

There is no reason to doubt that the (mostly) plain tokens enclosed in spherical envelopes belonged to a small number of *pre-literate* systems of number tokens, very much similar to the now well known *proto-literate* systems of number notations impressed on clay tablets. The following (extremely) tentative and partial interpretation of the numerical meaning of plain tokens in spherical envelopes was suggested in Friberg, *op. cit.*:

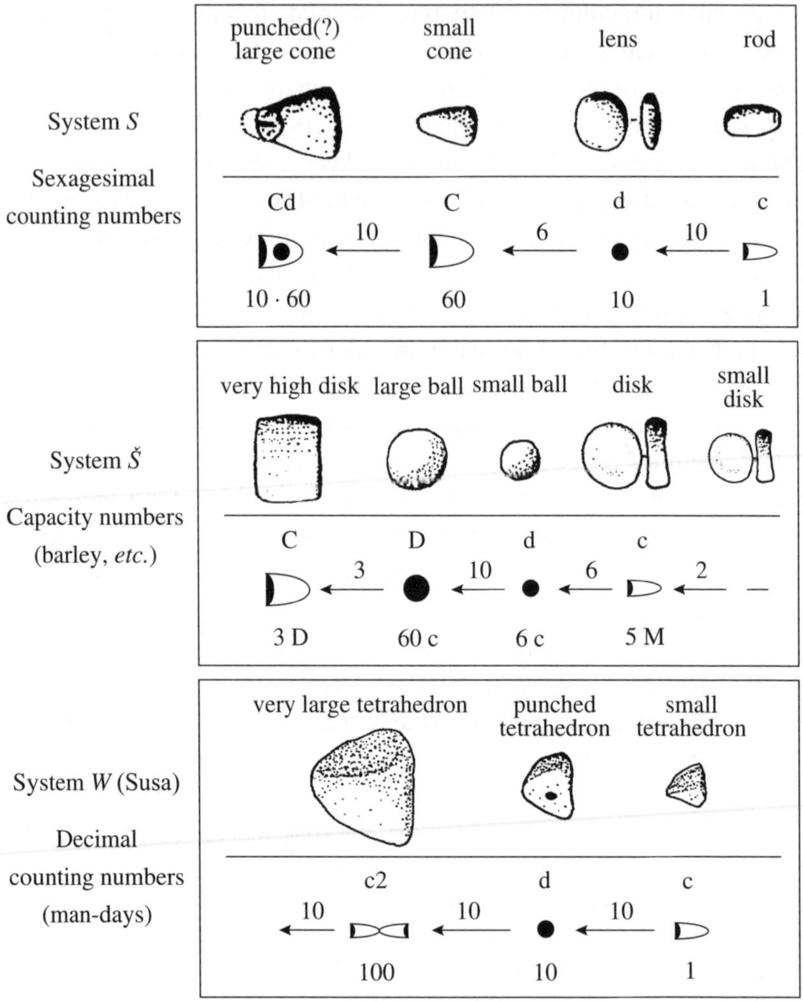

Fig. 9.2.1. Factor diagrams for parallel(?) pre- and proto-literate systems of numbers.

In the first of the three registers above, for instance, the *factor diagram for the proto-literate system of sexagesimal counting numbers* (used on clay tablets in Iran and Iraq near the end of the third millennium BCE) shows that units were written with small oblong punch signs (c), while tens were written with small circular punch signs (d), sixties with large oblong punch signs (C), and ten-times-sixties with large oblong punch signs with a small circular punch sign inside it (Cd). The corresponding pre-literate number signs seems to have been a "rod" (cylinder), a "lens" (circle), a small cone, and a large cone with a punch mark, respectively.

Here is an example of the kind of arguments that can be used in an effort to explain the numerical values of the plain tokens. The spherical envelope **MS 4632** (Friberg, *RC* (2007), Sec. A4 i) has turned out to contain 5 cones, 1 large and 8 small balls, and 3 non-plain tokens:

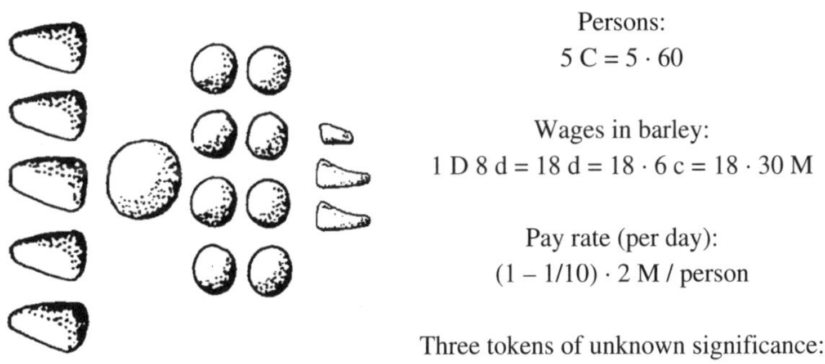

Fig. 9.2.2. The contents of the spherical envelope MS 4632 (courtesy P. Damerow).

With the interpretations of the numerical meaning of the plain tokens suggested above, the contents of MS 4632 seems to be a record of 18 d = 108 c of barley being paid out to 5 · 60 = 300 workers. The corresponding daily wage rate, close to 2/5 c of barley per person, agrees well with what is known about pay rates mentioned in proto-literate cuneiform texts.

Note that the use of plain tokens in the form of cones, pyramids (tetrahedrons), balls (spheres), *etc.*, does not mean that people in the Middle East long before the invention of writing were interested in solid geometry. The simple explanation is instead that cones, tetrahedrons, balls, *etc.* are very easy to fabricate out of lumps of clay by rolling and squeezing.

9.3. Pyramids and Cones in OB Mathematical Cuneiform Texts

An extensive discussion of occurrences of pyramids and cones in ancient Babylonian, Egyptian, Greek, Chinese, and Indian mathematical texts can be found in Friberg, *PCHM* 6 (1998), and *UL* (2005), Sec.4.8 g.

Selected passages from that discussion are reproduced below.

The volume and grain measure of a ridge pyramid

***TMS* 14** (Friberg, *UL* (2005), Sec. 1.5 f) is an OB mathematical cuneiform text from Susa (western Iran), which remained a mystery for more than 60 years after it was excavated (1933). The reason why it was not understood is that it starts by mentioning the object it is concerned with, a 'granary'. It was assumed that the shape of a granary would be a cylinder with a domed top, as in a well known depiction of a granary on a seal imprint from an archaic layer in Susa. The data given in the text of *TMS* 14 clearly did not fit this description of a granary. A renewed analysis of the text revealed that the form of the granary is instead a "ridge pyramid" of the type shown in Fig. 9.3.1 below, formed like a roof sloping uniformly from a ridge towards the ground at the sides and at the ends.

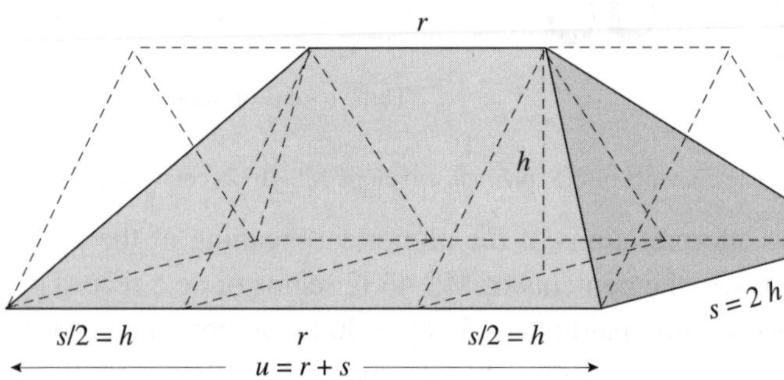

$f = 1$ c./c. $\Rightarrow$ $s = 2\,h$ and $V = r \cdot \text{sq.}\, h + 2 \cdot (1 - 1/3) \cdot h \cdot \text{sq.}\, h$

Fig. 9.3.1. *TMS* 14. An Old Babylonian problem for a ridge pyramid.

***TMS* 14**, literal translation	explanation
A granary, as much as 14 24 is the volume.	A granary. V = 14 24 volume-šar
3, reeds, is the height.	h = 3 ninda [f = 1 cubit / 1 cubit]
As much as 14 24 being the volume,	
length, front and ridge what do I set?	l, s, r = ?
You: The opposite of 12 of the depth,	1 ninda = 12 cubits
release, 5 you see,	$\Rightarrow$ 1 cubit = ;05 ninda
5 to 14 24, the volume,	14 24 volume-šar (sq. ninda · cubit)
raise, then 1 12 you see.	= 1 12 cubic ninda = V*
3, reeds, the upper length, square, 9 you see.	sq. h = sq. 3 ninda = 9 sq. ninda

The 9 to 3 of the height return, 27 you see.	$h \cdot$ sq. h = 27 cubic ninda
From 1, the normal step,	
20 of volume, a third of	1 – 1/3
what <to>$^?$ the normal volume,	
you added, tear off, 40 you see.	= 1 – ;20 = ;40
The 40, since 2 fronts of the granary,	There are 2 ends of the ridge pyramid
to 2 repeat, then 1$^!$ 20 you see.	2 · (1 – 1/3) = 2 · ;20 = ;40
The 1 20 to 27 raise, then 36 you see.	2 · (1 – 1/3) · h · sq. h = 36 cubic ninda
The 36 from 1 12 tear off, 36 you see.	V^* – 36 cubic ninda = 36 cubic ninda
Return.	
3, the height, square, 9 you see.	sq. h = sq. 3 ninda = 9 sq. ninda
The opposite of 9 break off, 6 40 you see.	1/9 = ;06 40
6 40 to 36 raise, 4 you see.	36 cubic ninda / 9 sq. ninda = 4 ninda
4 is the ridge.	r = 4 ninda
3, the height, since	h = 3 ninda, and
in a cubit a cubit,	f = 1 cubit / 1 cubit
3 double, 6 you see. 6 is the front.	⇒ $s = 2\,h$ = 6 ninda
The 6 to 4, the ridge, heap, 10 you see.	$r + s$ = 6 ninda + 4 ninda = 10 ninda
10 *is the length.*	l = 10 ninda
[… … …]	[(l + r/2)/3 = 6 ninda]
12 to 3, the height, raise, then 36 you see.	h = 3 ninda = 36 cubits.
36 to 24 raise, then	(l + r/2)/3 · h = 36 cubits · 24 ninda
14 24 you see, the volume.	= 14 24 volume-šar = V
14 24, the volume, to	
8, the storing number of the granary,	V · 8 00 00 sìla / volume-šar
raise, then 1 55 12 you see.	= 1 55 12 00 00 sìla
23 gur$_7$(?) 2 24 gur of barley … …	= 23 02 24 gur of barley …

Briefly, what this text means is the following: Consider a *ridge pyramid* like the one in Fig. 9.3.1. Suppose it is known that its volume V is 14 24 volume-šar (sq. ninda · 1 cubit), and that its height is 3 ninda. Suppose also that the uniform slope of the roof is 1:1. What are then the long and short sides of the base (u and s), and what is the length of the ridge (r)?

The first step of the solution procedure is to divide the volume 14 24 sq. ninda · 1 cubit by 12 (cubits per ninda), expressing it in the new form 1 12 sq. ninda · ninda (cubic nindas). Next, the cube of height h is computed; it is sq. $h \cdot h$ = 27 cubic nindas. The awkwardly worded passage of the text which then follows can be interpreted as describing the computation of the volume 2 · (1–1/3) · sq. $h \cdot h$ = 36 sq. cubic nindas. What this means is that the volumes of the rectangular pyramids at the two ends of the ridge pyramid (see again Fig. 9.3.1) are computed as *two thirds of the volumes of the two wedges (triangular prisms) containing them.* Indeed, since the uniform slope of the roof of the ridge pyramid is 1:1, each end pyramid has the height h and the sides h and $2h$. Therefore, the volume of each containing wedge is equal to sq. $h \cdot h$, and the combined volume of the two end pyramids is indeed equal to $2 \cdot (1–1/3) \cdot$ sq. $h \cdot h$.

In the next step of the procedure, the volume of the central wedge is computed as the given volume of the whole ridge pyramid, diminished by the volumes of the two end pyramids, that is as (1 12 – 36) sq. ninda· ninda = 36 sq. ninda· ninda. On the other hand, the volume of this central wedge is equal to $s \cdot r \cdot h/2 = r \cdot$ sq. h, since $s = 2h$. Therefore, the ridge r is equal to the volume of the central wedge divided by sq. h. In other words, r = (36 sq. ninda· ninda.)/(9 sq. ninda) = 4 ninda. In addition, $s = 2\ h = 6$ ninda and $l = r + s = 10$ ninda.

In the second part of the text, the obtained result is *verified*, in that the given value of h and the computed values of l, r, s are used to compute the volume V, which, of course, again is equal to 14 24 volume-šar. This value for the volume is then multiplied by the "storing number"

c = 8 00 00 sìla/volume-šar, with 1 sìla = somewhat less than 1 liter.

The final result is the "grain measure" of the granary:

$C = c \cdot V =$ 1 55 12 00 00 sìla.

BM 96954+BM 102366+SÉ 93 (Friberg, *PCHM* 6 (1996), Robson, *MMTC* (1999), Appendix 3; Friberg, *UL* (2005), Sec. 4.8 g) is a text composed of three fragments of a large clay tablet. As shown by the outline below of the clay tablet and its table of contents, it is a recombination text which has "whole or truncated pyramids and cones" as its dominating topic.

§ 1 of the recombination text may have consisted originally of 13 mathematical exercises, all dealing with a certain *ridge pyramid,* similar to the one treated in *TMS* 14 (Fig. 9.3.1 above). § 3 consists of 3 exercises dealing with various kinds of prisms. The content of each solid appearing in §§ 1 and 3 is expressed, not in terms of its *volume V*, but in terms of its "grain measure" $C = c \cdot V$, where c is the new storing number

c = 1 30 gur/šar = 7 30 00 sìla/volume-šar.

The equation used in § 1 for the volume of a ridge pyramid is

$V = (l + r/2) \cdot s \cdot h/3$, with l, s, r, h as in Fig. 9.3.1.

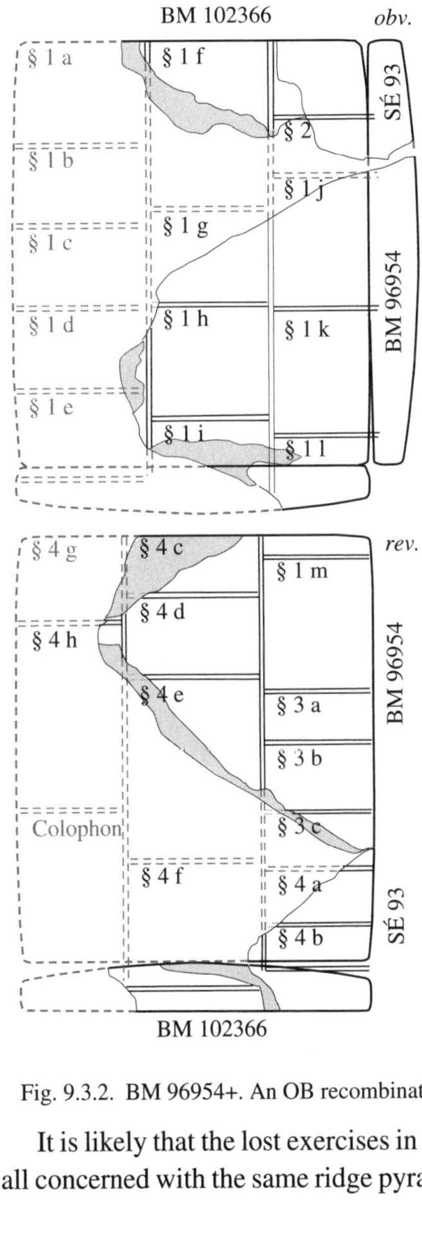

BM 96954+102366+SÉ 93.

Contents:

§§ 1 a-e:
Basic equations for the parameters of a ridge pyramid(?).

§ 1 f:
The content of a ridge pyramid truncated at mid-height.

§§ 1 g-i:
Equations for the parameters of a ridge pyramid.

§ 2:
The content of a square pyramid(?).

§ 1 j-m:
Equations for the parameters of a ridge pyramid.

§§ 3 a-c:
The contents of various prisms.

§§ 4 a-d:
Equations for the parameters of a cone.

§ 4 e:
The volume of a cone truncated at mid-height.

§ 4 f, h:
Problems for a cone truncated near the top.

Fig. 9.3.2. BM 96954+. An OB recombination text dealing with pyramids and cones.

It is likely that the lost exercises in the first column of BM 96954+ were all concerned with the same ridge pyramid as the one in §§ 1 g - 1 m. If this conjecture is correct, the following series of simple questions may have been asked there:

§ 1 a:	l, s, r, h	given	$C = ?$
§ 1 b:	s, r, h, C	given	$l = ?$
§ 1 c:	l, r, h, C	given	$s = ?$
§ 1 d:	l, s, h, C	given	$r = ?$
§ 1 e:	l, s, r, C	given	$h = ?$

The remaining exercises in § 1, except § 1 f, deal with linear or rectangular-linear *systems of equations* for the parameters of the ridge pyramid. Being a recombination text, BM 96954 is somewhat chaotically organized, so that § 1 f deals instead with a *truncated* ridge pyramid:

The grain measure of a ridge pyramid truncated at mid-height

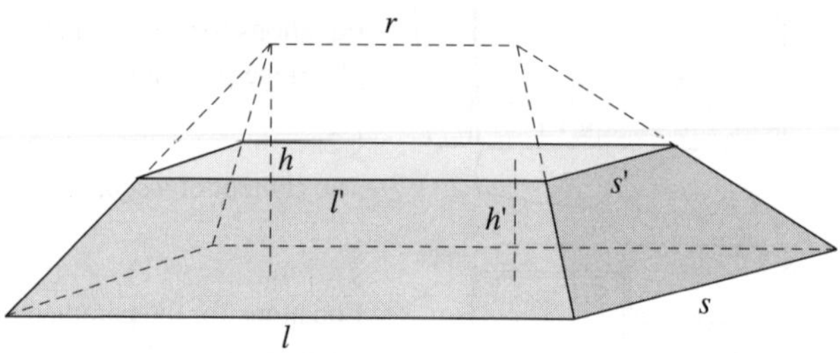

$l = 10$ ninda, $s = 6$ ninda, $r = 4$ ninda	$l' = 7$ ninda, $s' = 3$ ninda,
$h = 48$ cubits, $h - h' = 24$ cubits	$h' = 24$ cubits
$V = 19\ 12$ šar, c = 1 30 gur/šar	$V' = 15\ 36$ šar
$C = 28\ 48\ 00$ gur	$C' = 23\ 24\ 00$ gur

Fig. 9.3.3. BM 96954 § 1 f. A ridge pyramid truncated at mid-height.

In BM 96954+ § 1 f, the ridge pyramid common to all the other exercises in § 1 is *truncated at mid-height* (Fig. 9.3.3). Here is the text of § 1 f, of which only the first part is preserved:

BM 96954+ § 1 f, literal translation	explanation
A granary.	A granary
10 the length, 6 the front, 4 the ridge,	$l = 10$ ninda, $s = 6$ ninda, $r = 4$ ninda
28 48 <gur of barley>,	Grain measure $C = 28\ 48\ 00$ gur
48 the height, 24 I went down.	$h = 48$ cubits, $h - h' = 24$ cubits
The transversal(s) and the barley are what?	$l', s', C' = ?$
You: The opposite of 48 the height release,	1/48

1 15 you see.	= ;01 15
1 15 to *6, what* the length is more than	$l - r = 6$ ninda
the ridge, raise, 7 30 you see.	$f = (l - r)/h =$;07 30 (1/8) ninda/cubit
7 30 to 24 raise, 3 you see.	$f \cdot (h - h') = 3$ cubits
3 from 10 the length *tear off, 7 you see,*	$l - f \cdot (h - h') = 7$ cubits
7 the transversal.	$= l'$
3 from 6 the front tear off, *3 you see, the* ···	$s - f \cdot (h - h') = 3$ cubits [= s']
··· ··· ··· ··· raise, 1 you see. ··· ··· ···	$[l \cdot s] = 1\ 00$ sq. ninda. ··· ··· ···

Although only the first part of the text of § 1 f is preserved, it is clear that the equation for the volume of the truncated ridge pyramid must have been expressed in terms of its linear parameters, the length l and front s at the base, the "upper length" l' and "the upper front" s' at mid-height, and the lower height h'. (Unless, of course, the volume of the truncated ridge pyramid was computed as the volume of the whole ridge pyramid minus the volume of the small upper ridge pyramid.) The volume V of the truncated ridge pyramid is easily seen to be equal to the volume V_c of a central rectangular prism, plus the volumes $2 \cdot V_l$ and $2 \cdot V_s$ of four wedges along the sides, and the volumes $4 \cdot V_p$ of four square pyramids in the corners:[17]

$$V = V_c + 2 \cdot V_l + 2 \cdot V_s + 4 \cdot V_p$$
$$= l' \cdot s' \cdot h + l' \cdot (s - s') \cdot h/2 + s' \cdot (l - l') \cdot h/2 + (l - l') \cdot (s - s') \cdot h/3$$
$$= \{(l \cdot s + l' \cdot s') + (l \cdot s' + l' \cdot s)/2\} \cdot h/3.$$

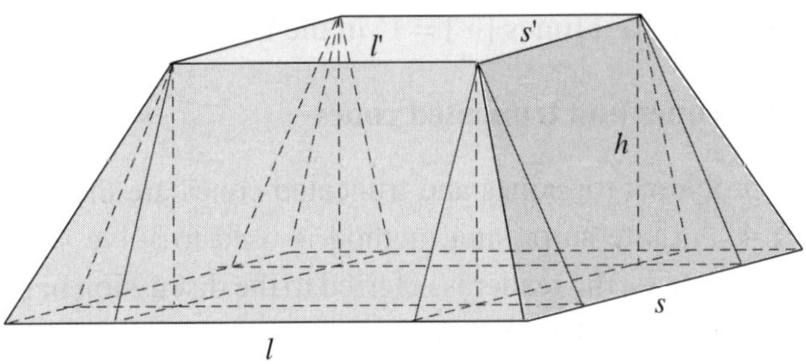

Fig. 9.3.4. Dissection of a truncated ridge pyramid.

However, before this equation can be used to compute the volume of

17. In the case when $l' = r$ and $s' = 0$, that is in the case of a whole ridge pyramid, this equation is reduced to $V = \{(l + r/2) \cdot s \cdot h/3$. In the case when $l = s$ and $l' = s' = t$, that is in the case of an ordinary truncated square pyramid, the equation is reduced to the well known equation $V = (\text{sq. } s + s \cdot t + \text{sq. } t) \cdot h/3$. *Cf.* the discussion of the Egyptian hieratic mathematical exercise *P.Moscow* # 14 in Friberg, *UL* (2005), Sec. 2.2 d.

the truncated ridge pyramid, the values of u' and s' must be known. As a matter of fact, the preserved first part of BM 96954 § 1 f is devoted to the computation of these values. The first step of the computation is to find the *combined feed* for the two ends of the ridge pyramid:

$2 \cdot f = (l - r)/h = (10 - 4)$ ninda · 1/(48 cubits) = 6 · ;01 15 n./c. = ;07 30 n./c.

The double feed is multiplied by the height of the truncated ridge pyramid:

$2 \cdot f \cdot h' =$;07 30 n./c. · 24 c. = 3 n.

So much smaller are the upper length and the upper front than the lower length and the lower front, respectively. Therefore,

$l' = l - 3$ n. = 10 n. – 3 n. = 7 ninda, $s' = s - 3$ n. = 6 n. – 3 n. = 3 ninda.

Inserting these computed values into the equation for the volume of the truncated ridge pyramid, one obtains the following result (unfortunately not present in the preserved part of the text):

$V = \{(10 \cdot 6 + 7 \cdot 3) + (10 \cdot 3 + 7 \cdot 6)/2\}$ sq. ninda · 24 cubits /3
= 1 57 sq. ninda · 8 cubits = 15 36 šar,
$C = c \cdot V =$ 1 30 gur/šar · 15 36 šar = 23 24 00 gur

Note that the first step in this computation of the volume V would be to compute the product of l and s as $10 \cdot 6 = 1$ (00). This proposed first step of the computation agrees well with the only preserved part of the calculation of V, which is '[···] times [···] = 1' in the last preserved line of § 1 f.

Problems for cones and truncated cones

Various problems for cones and truncated cones are the object of **BM 96954 +, § 4**. A quite surprising method is used to solve some of those problems. For details, the reader is referred to the discussion in Friberg, *UL* (2005), Sec. 4.8 g. No other examples are known of Babylonian mathematical texts dealing with cones.

9.4. Pyramids and Cones in Ancient Chinese Mathematical Texts

The fifth chapter in *Jiu Zhang Suan Shu*

The famous ***Jiu Zhang Suan Shu***, *'Nine Chapters on the Mathematical Art'* (Vogel, *NBAT* (1968); Shen, Crossley, and Lun, *NCMA* (1999)) is one of the oldest, and probably the most important, Chinese mathematical classic to have survived to the present day. It was assembled into one book not later than in the middle of the Eastern Han Dynasty (25–220 A.D.). The fifth chapter of *Jiu Zhang Suan Shu* has the misleading title 'Construction Consultations'. It is ostensibly devoted to the discussion of the amount of manpower needed, under various enumerated circumstances, for the excavation or building of various constructions in the form of more or less familiar types of (mostly) rectilinear solids. In this respect it calls to mind similarly dressed problems in a number of OB mathematical exercises. Nevertheless, it is obvious that the main emphasis is laid on *the computation of the volumes of various kinds of pyramids and cones, or related types of solids.* The text of the chapter is organized in a strikingly systematic way. The brief survey below of the contents of the chapter will make this fact clear.

In the first six problems, V.2-7, the object considered is a trapezoidal prism, which trivially has a known volume, since it is composed of one rectangular block and two wedges. In the next six exercises, V.8-13, are computed the volumes of cubes and cylinders, of truncated pyramids and cones, and of full pyramids and cones. Then, in V.14-16, the volumes are computed of the solids that result when a rectangular block is cut into two wedges, and each wedge into two pyramids.

V. 2-4. A wall, a dike. V. 5-7. A trench, a moat, a canal $V = (s + t)/2 \cdot h \cdot l$	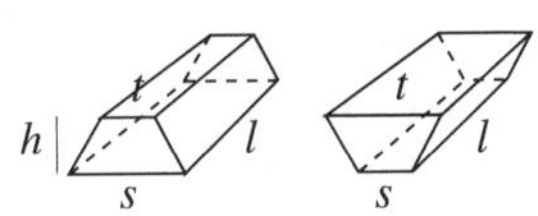
V. 8. A square fort $V = \text{sq. } s \cdot h$ V. 9. A round fort $V = \text{sq. } a \cdot h/12 \quad (\pi \approx 3)$ V. 10. A square pavilion $V = (\text{sq. } s + s \cdot t + \text{sq. } t) \cdot h/3$ V. 11. A round pavilion $V = (\text{sq. } a + a \cdot b + \text{sq. } b) \cdot h/36$ V. 12. A square needle $V = \text{sq. } s \cdot h/3$ V. 13. A round needle $V = \text{sq. } a \cdot h/36$	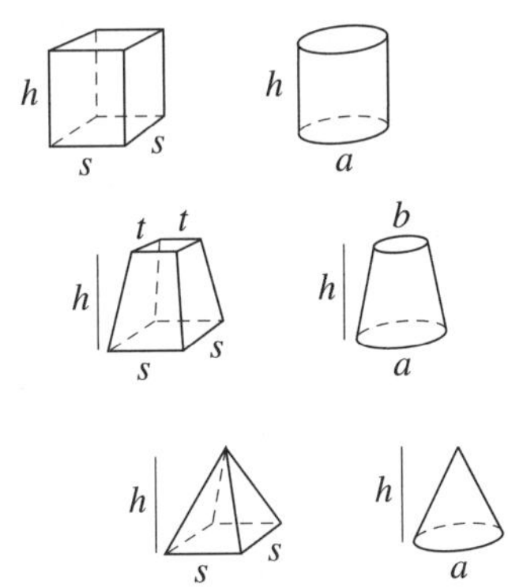

V. 14. A moat-wall, *qian du*
$V = l \cdot s \cdot h/2$
V. 15. A male horse, *yang ma*
$V = l \cdot s \cdot h/3$
V. 16. A turtle's bone, *bie nao*
$V = l \cdot s \cdot h/6$

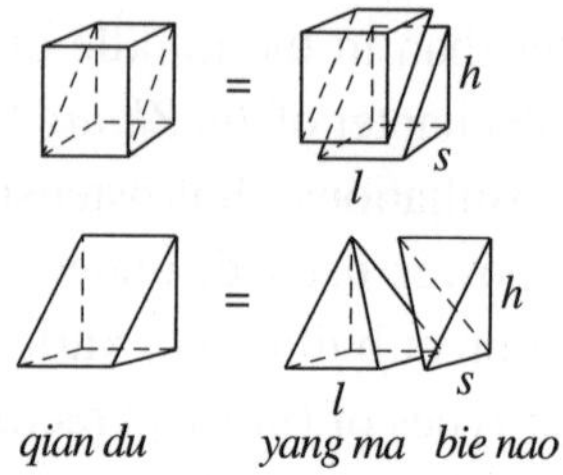

Fig. 9.4.1. Volumes of solid figures computed in *Jiu Zhang Suan Shu*, V.2-16.

More complicated types of solids are considered in V.17-20, a trapezoidal wedge in V.17, a ridge pyramid in V.18, and a truncated ridge pyramid in V.19. In V.20, the lining of a tapering well is considered as an astonishing example of a *curved* truncated pyramid, which can be treated in the same way as the non-curved variant.

In V.23-25, finally, the expressions for the volumes of a cone, a half-cone, and a quarter-cone, all in terms of the circumference of the base, are contrasted with each other.

V. 17. A drain
$V = (a + b + c) \cdot l \cdot h/6$

V. 18. A cut grass ridge
$V = (2\,l + r) \cdot s \cdot h/6$
V. 19. A cut grass overhang
$V = \{(2l + l') \cdot s + (2\,l' + l) \cdot s'\} \cdot h/6$

V. 20. A bent moat
$V = \{(2a + a') \cdot t + (2a' + a) \cdot t'\} \cdot h/6$
a, a' = lower, upper middle arcs
t, t' = lower, upper widths

V. 23. Cereal piled on the floor
$V = \text{sq. } a \cdot h/36$
V. 24. Cereal piled against a wall
$V = \text{sq. } a \cdot h/18$
V. 25. Cereal piled in a corner
$V = \text{sq. } a \cdot h/9$

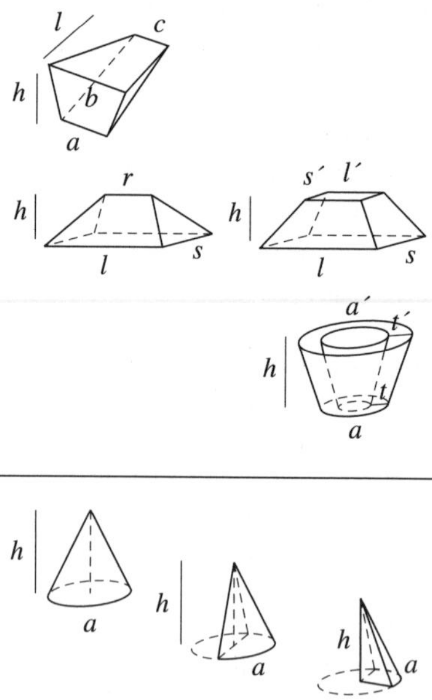

Fig. 9.4.2. Volumes of solid figures computed in *Jiu Zhang Suan Shu*, V.17-25.

All the rules in *Jiu Zhang Suan Shu* V.2-25 for the computation of the volume of pyramids, cones, and related objects are correct.

The reason why *Jiu Zhang Suan Shu*, Chapter V is mentioned here is its obvious close relationship with OB mathematics. There are OB counterparts to the exercises *Jiu Zhang Suan Shu* V.2-15 and V.17-19 in the mathematical recombination texts BM 96954+BM 102366+SÉ 93 (Fig. 9.3.2 above), BM 85194, BM 85196, and BM 96957+VAT 6598, all from the ancient Mesopotamian city Sippar (see Friberg, *PCHM* 6 (1996) § 1.5).[18] In addition, the juxtaposition of the *qian du*, the *yang ma* and the *bie nao* in *Jiu Zhang Suan Shu* V.14-16 (Fig. 9.4.1) is reminiscent of the computation of the volume of the end pyramids in *TMS* 14 (Fig. 9.3.1 above), and the computation of the 'bent moat' in *Jiu Zhang Suan Shu* V.20 (fig. 9.4.2) has an (imperfect) parallel in the computation of the volume of a ring-wall in BM 85194 # 3. It is also worth noting that the computations of the volumes of a circular cylinder, a truncated circular cone, and a full circular cone in *Jiu Zhang Suan Shu* V.9, 11, 13 are all based on the OB form of the rule for the computation of the area of a circle, A = sq. a /12, where a is the circumference of the circle.

Thus, if there ever existed an OB well organized *theme text* with computations of volumes of pyramids, cones, and related objects, as it main topic (and there almost certainly did), it is reasonable to suspect that it was organized very much like the fifth chapter of *Jiu Zhang Suan Shu*.

Even older than *Jiu Zhang Suan Shu* is the recently published work ***Suan Shu Shu*** (written on 190 bamboo strips dated to the second century BCE; see Cullen, *SSS* (2004)). It is interesting that the section 'Shapes and Volumes' (group 12) of *Suan Shu Shu* contains the following exercises: parallels to *Jiu Zhang Suan Shu* V. 17-19 in *Suan Shu Shu* ## 55-57, and parallels to *Jiu Zhang Suan Shu* V.13, 11, 9 in *Suan Shu Shu* ## 58-60.

For the reasons mentioned above, it seems to be justified to draw the conclusion that the rules for the computation of volumes of pyramids, cones, and related objects, which had been discovered by OB mathematicians, became part of a common mathematical tradition in large parts of the ancient world, a tradition which ultimately spread all the way to China

18. Some of the mentioned recombination texts are somewhat chaotically organized, and some of the computations of volumes of solid objects are only rough approximations.

and was still alive in the second century BCE.[19 20 21]

Liu Hui's commentary to *Jiu Zhang Suan Shu,* Chapter V.

The expressions given in *Jiu Zhang Suan Shu,* Chapter V, for the volumes of various kinds of pyramids and cones or related solids are all correct. However, as far as is known, justifications for these expressions, in the form of careful derivations from supposedly known facts were first given in the commentary to *Jiu Zhang Suan Shu* written by Liu Hui (the third century A.D.). In Wagner's article *HM* 6 (1979), which is a continuation of his unpublished master's thesis (1975), are discussed in full detail Liu Hui's correct derivations of the expressions for the volumes of the *fang ting* (a truncated pyramid) and of the pyramids *yang ma* and *bie nao*.

In this connection, Wagner mentions that the solution to Hilbert's Third Problem (Dehn (1900), Jessen (1939)) confirms that it is *not possible* generally, and particularly not in the case of a regular tetrahedron, to compute the area of a pyramid by means of a finite number of dissections and rearrangements, that is without the use of infinitesimal calculus.

Liu Hui's derivation of the correct expressions for the volumes of the square pyramid *yang ma* and of the triangular pyramid *bie nao* is closely related to the method used in *Elements* XII.3-9 (in the case of an *arbitrary* triangular pyramid). Thus, Liu Hui starts his derivation by considering two pyramids, one *yang ma* and one *bie nao*, which can be fitted together to form one *qian du* (a triangular wedge; see Fig. 9.4.1 above). He wants to show that *the volume of the yang ma is twice the volume of the bie nao*, because if that is so, then the volumes of the *yang ma* and the *bie nao* must be 2/3 and 1/3, respectively, of the known volume of the wedge they form

19. *Cf.* the discussion of rules for the computation of volumes of pyramids and cones in Heron's work ***Metrica II*** (Heath, *HGM* 2 (1981), 331 *ff.*), and in the Greek-Egyptian mathematical papyrus ***P.Vindobonensis G. 1996***, 1st century? (Friberg, *UL* (2005), Sec. 4.8).

20. *Cf.* also the discussion in Friberg, *PCHM* 6 (1996) § 1.4, of similar rules in the Indian mathematical work *Brāhmasphuṭasiddhānta* by Brahmagupta (628).

21. It is particularly interesting that in ***Brāhmasphuṭasiddhānta*** **VIII.50-51** there are rules for the computation of the volumes of *conical piles of grain* resting *on the floor, against the side of a wall, in a corner, or on the outside of a corner*, all in terms of the height and the circumference a of the base: $V = h \cdot$ sq. $(a/6)$, $V = h \cdot$ sq. $(a/6) \cdot 2$, $V = h \cdot$ sq. $(a/6) \cdot 4$, $V = h \cdot$ sq. $(a/6) \cdot 4/3$. These rules are parallels to *JZSS* V.23-25 (Fig. 9.4.2).

together. He imagines that, in a first step, the two pyramids are cut by planes through the midpoints of their edges, one into five smaller pieces, the other into four, as shown in Fig. 9.4.3 below.

Liu Hui suggests that the once dissected pyramids should be thought of as composed of half-size wooden building blocks, two small *wedges* and two small *bie nao*s, all colored red, in the case of the *bie nao*, and one small *cube*, two small *wedges* and two small *yang mas*, all colored black, in the case of the *yang ma*. Since the combined volume of the small black cube and the small black wedges is known and is twice the combined volume of the small red wedges, it remains to be shown that the volume of the two half-size *yang mas* is twice the volume of the two half-size *bie nao*s. In a second step of the algorithm, these half-size pyramids, in their turn, are dissected, and new red and black cubes and wedges of known volumes are removed, and so on. At each step of the algorithm, the volume of the remaining pyramids is less than the volume of the pieces just removed. Ultimately, the volume of the remaining pyramids will be negligible, and the stated goal will be reached.

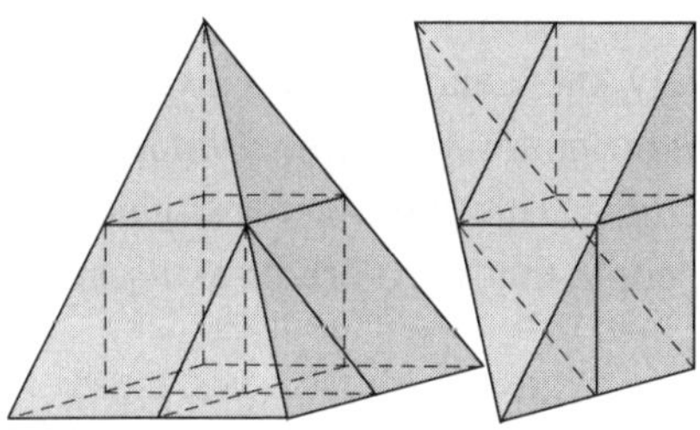

Fig. 9.4.3. Liu Hui's dissection of a *yang ma* and a *bie nao*.

9.5. A Possible Babylonian Derivation of the Volume of a Pyramid

As shown above, in Sec. 9.3, *the first ones to seriously consider non-trivial solid figures like pyramids, cones, and related objects were the Old Babylonian mathematicians* (or possibly their Sumerian predecessors). Moreover, *correct expressions for volumes of such solid figures were known to the Babylonians*. Even *the fundamental and non-trivial idea of cutting a triangular prism by a plane into a triangular and a square pyra-*

mid seems to have been their invention. (See Fig. 9.3.1.)

It remains to find out how it was possible for Babylonian mathematicians to *find* the correct expressions for the volumes of whole and truncated pyramids and to *prove*, to their own satisfaction, that the expressions are correct. Simple answers to these questions will be suggested below:

1) The idea to consider pyramids and cones must have come naturally to a people that constantly dug ditches, canals, and water reservoirs, built temple platforms and step pyramids, and heaped up grain in their granaries. Besides, easily fabricated small *pyramids, cylinders, cones, and spheres* of clay were in constant use as tokens for counting in Mesopotamia and surrounding regions for several millennia before the invention of writing (Sec. 9.2 above), which shows that pyramids and cones were familiar objects even long before the time of the Babylonians.

2) It is clear from the form of many geometrical entries in Old Babylonian "tables of constants" and from the solution procedures in many mathematical cuneiform problem texts that Babylonian mathematicians intuitively knew and routinely exploited the "scaling rule for plane figures" that *the areas of similar plane figures are proportional to the squares of their sides*. It is also clear that Babylonian mathematicians in a similar way intuitively knew and routinely exploited the "scaling rule for solid figures" that *the volumes of similar solid figures are proportional to the cubes of their edges*. This three-dimensional scaling rule must have been obvious to a people that used bricks as its most important building material.

Now consider the dissection of a triangular prism used by Liu Hui (Sec. 9.4 above) for his derivation of the volumes of the *yang ma* and the *bie nao,* by a method which is very close to the method used in *Elements* XII.3-5 and 7 (Sec. 9.1 above). Thus, consider, as in Fig. 9.4.3, a prism which has a cross-section in the form of a right triangle. (This is one half of a rectangular wedge such as the one at both ends of the ridge pyramid in Fig. 9.3.1). Let W be the volume of the wedge. Imagine that the wedge is dissected by means of three mutually orthogonal planes through the midpoints of its edges, two vertical and one horizontal (Fig. 9.5.1 below). Then, according to the scaling rule for solid figures, the wedge is divided into 4 wedges similar to itself, each with the volume $W/8$, and two rectangular blocks, each with the volume $W/4$.

Dissect the wedge further by a slanting plane through the upper left vertex and the lower right edge. The result is that the original triangular prism is divided into two pyramids, one rectangular (*cf.* the *yang ma*), the other triangular (*cf.* the *bie nao*). Let their unknown volumes be called R and T, respectively. The rectangular pyramid, in its turn, is divided by the mentioned orthogonal planes into two small rectangular pyramids similar to itself, two wedges, and one rectangular block. According to the scaling rule for solid figures, each one of the small rectangular pyramids has the volume $R/8$, each one of the two wedges has the volume $W/8$, and the rectangular block has the volume $W/4$. Similarly, the triangular pyramid is divided by the orthogonal planes into two triangular pyramids similar to it, each with the volume $T/8$, and two wedges, each with the volume $W/8$. Therefore, the unknown volumes R and T satisfy the equations

$$R = 2 \cdot R/8 + 2 \cdot W/8 + 1 \cdot W/4 = R/4 + W/2,$$
$$T = 2 \cdot T/8 + 2 \cdot W/8 = T/4 + W/4.$$

To find this out would be well within the competence of a Babylonian mathematician, who would also be able to solve the linear equations for R and T, finding immediately that

$$R = 4/3 \cdot W/2 = 2/3 \cdot W, \quad \text{and} \quad T = 4/3 \cdot W/4 = 1/3 \cdot W.$$

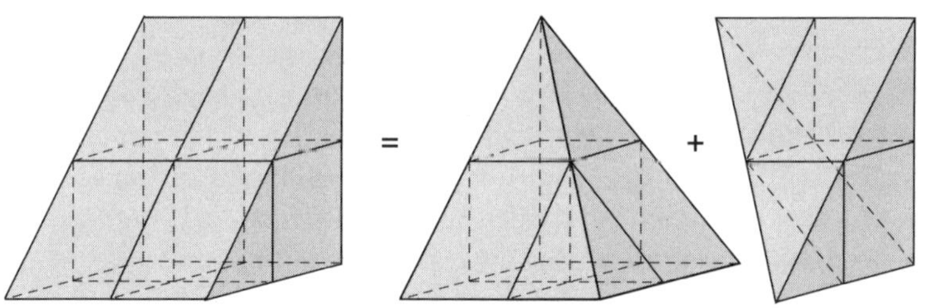

$W =$ $4 \cdot W/8 + 2 \cdot W/4$

$R =$ $2 \cdot R/8 + 2 \cdot W/8 + 1 \cdot W/4$ $\Rightarrow R = 2/3 \cdot W$

$T =$ $2 \cdot T/8 + 2 \cdot W/8$ $\Rightarrow T = 1/3 \cdot W$

Fig. 9.5.1. Possibly the way in which the Babylonians found the volume of a pyramid.

The conjecture that Babylonian mathematicians found the volume of a pyramid in the way suggested above (Fig. 9.5.1) is supported by two circumstances: 1) the fact that in *TMS* 14 (Fig. 9.3.1) the volumes of the end pyramids of a ridge pyramid are computed as 1 – 1/3 of the volume of a wedge, and 2) the fact that in BM 96954+ § 1 f the volume is computed of a ridge pyramid truncated at mid-height, while in BM 96954+ § 4 e the volume is computed of a circular cone truncated at mid-height. (It was probably intuitively clear that the ratio between the volume of a circular cone and the volume of a circumscribed square pyramid is equal to the ratio between the area of a circle and the area of a circumscribed square.)

Consequently, it is likely that Old Babylonian mathematicians did indeed both find and prove, *according to their own standards, and without any kind of infinitesimal calculus*, correct expressions for the volumes of a large variety of pyramids, cones, and related solids! Thus, of the various solid figures appearing in *Elements* XII, only the sphere seems to have been totally outside the scope of Babylonian mathematics.

Chapter 10

El. I.43-44, *El.* VI.24-29, Data 57-59, 84-86, and Metric Algebra

The most recent edition of Euclid's *Data* (in the sense of *Givens*) is Taisbak (2003), complete with the Greek text, an English translation, and extensive commentaries, based on geometric interpretations.

In his Preface, Taisbak explains why he felt he had to write the book:

> "After reading it (a modern translation of the *Data*) one is left in the same bewilderment as that already expressed by the ancient commentators *Pappus* and *Marinus*: What is all this really about?"
>
> "The most recent commentary is by Clemens Thaer (Data 1962), a very brief and succinct interpretation in modern algebraic jargon ··· . Needless to say (at least after you have read even part of *my* tale) I disagree with him at almost all points, so much so that one would think we were reading completely different texts."

In the Introduction, the nature of the *Data* is explained as follows:

> "In the *Data*, Euclid proves deductively that if some items are given, some other items are also given, *into the bargain* so to speak."
> "··· an essential feature of the *Data*: the Givens hang together in chains, the purpose of any proposition being to produce more links to them."

Taisbak also cites Wilbur Knorr's description of the *Data* (*ATGP* 1986):

> "The *Data* is a complement to the *Elements*, recast in a form more serviceable for the analysis of problems. ··· the subject matter overlaps that of the *Elements* ··· Indeed, only in rare instances does the *Data* present a result without a parallel in the *Elements*."

The 15 definitions and 94 propositions in the *Data* are divided by Taisbak into 14 chapters, of which the following ones are particularly interesting from the point of view of Babylonian mathematics (metric algebra):

Chapter 8. Application of areas I	*Data* 57-61
Chapter 12. Application of areas II	*Data* 84-85
Chapter 13. Intersecting hyperbolas. Zeuthen's conjecture	*Data* 86

10.1. *El.* I.43-44 & *Data* 57: Parabolic Applications of Parallelograms

Three of the propositions in *Elements* I have close associations with Babylonian metric algebra. One of them, I.47 (the diagonal rule) was discussed in Chapter 2 above. The other two are ***El.* I. 43-44**:

***El.* I.43**

In any parallelogram the complements of the parallelograms about the diagonal are equal to one another.

***El.* I.44**

To a given straight line to apply, in a given rectilineal angle, a parallelogram equal to a given triangle.

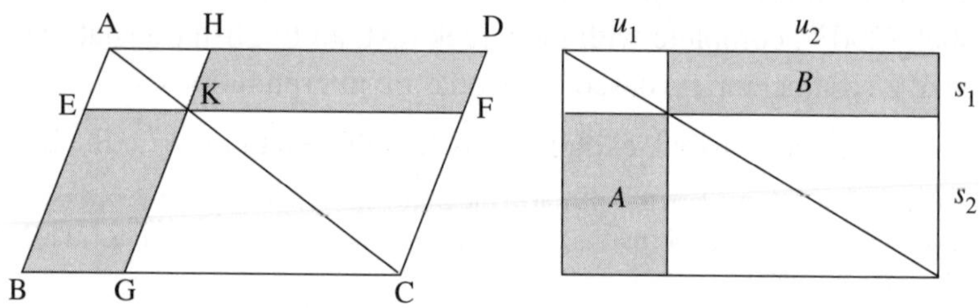

Fig. 10.1.1. *El.* I.43. Complements about the diagonal are equal (in area).

The proof of ***El.* I.43** is simple. Let a parallelogram ABCD be divided by a diagonal and two lines parallel to the sides of the parallelogram and intersecting in a point K on the diagonal, as in Fig. 10.1.1, left. Then

ABC = ACD, AEK = AHK, and KFC = KGC.

Subtracting AEK and KFC from ABC or AHK and KGC from ACD, one finds, as wanted, that what remains is the same in both cases:

BGKE = DHKF.

Parallelograms were never considered in Babylonian mathematics, so the (hypothetical) Babylonian counterpart to the diagram in *El.* I. 43 would be the divided *rectangle* in Fig. 10.1.1, right, *divided by the diagonal into two right triangles*. Now, one of the basic tools of Babylonian geometry was the intuitively obvious "right sub-triangle rule", according to which every line parallel to one of the sides of a right triangle cuts off a right sub-triangle *with the same ratio between the sides* as that in the whole triangle. (This is, of course, the Babylonian counterpart to ***El.* VI.24**.) In the case of the divided rectangle in Fig. 10.1.1, right, this means that, for instance, by the rule of three,

$$s_2 = u_2 \cdot s_1 / u_1.$$

Therefore, clearly,

$$u_1 \cdot s_2 = u_2 \cdot s_1,$$

which means that the areas *A*, *B* of the opposite sub-rectangles in Fig. 10.1.1, right, must be equal. (This kind of manipulation with equations was done routinely by Babylonian mathematicians.)

The idea behind the proof of *El.* I.44 is to first construct a parallelogram, say DHKF in Fig. 10.1.1, left, in the given angle, equal (in area) to the given triangle, and such that one of its sides, say FK, is in a straight line with KE, the given straight line.[22] Next, the point A is constructed as the intersection of a continuation of DH with a line through E parallel to FD, the point C as the intersection of a continuation of AK with a continuation of DF, the point B as the intersection of a continuation of AE with a line through C parallel to DH, and the point G as the intersection of a continuation of HK with BC. Then it follows from *El.* I.43 that BGKE is equal (in area) to DHKF, *etc.*

From the point of view of metric algebra, the proof of *El.* I.44 can be explained as follows: Let *A* be the given area, let *u* be the given length, and set $s = A/u$. Then

$$u \cdot s = u \cdot A/u = B.$$

In this way, the complicated geometric proof[23] of *El.* I.44 is replaced by a trivial metric algebra proof (still essentially geometric).

Actually, the problem *to find the length of the second side of a rectangle when the area of the rectangle and the length of one side are known* was considered by Mesopotamian mathematicians long before the time of the Greeks. See the discussion of "metric division exercises" from the Old Akkadian (or Sargonic) period in Mesopotamia (ca. 2340-2200 BCE) in Friberg, *CDLJ* (2005-2) §§ 2-3; *RC* (2007), Appendix 6, Sec. A6 c. One such exercise is ***DPA* 39** (Fig. 10.1.2 below). The brief text of that exercise, or rather assignment, since no answer is given, states that the length

---

22. For some obscure reason, the diagram in I.44 is not identical with the diagram in I.43.
23. Taisbak (*op. cit.*, 152), cites Heath's opinion (*ETBE* 1, 342) that "This proposition will always remain one of the most impressive in all geometry ···".

(of a rectangle) is 4(60) 3 ninda (a common length unit), and the area is 1 iku (a common area unit) = 100 square ninda. What is then the front (the short side of the rectangle)?

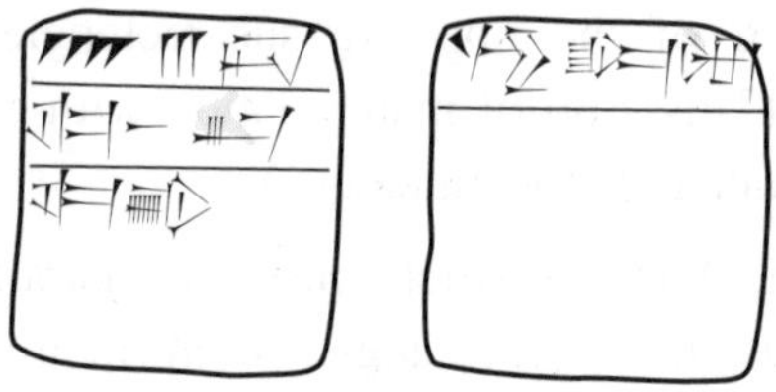

4(60) 3 the length.
(What is) the front
(if) the area is 1 (iku)?
Its front is ····
to be found.

Fig. 10.1.2. *DPA* 39. An Old Akkadian metric division exercise.

In the Old Akkadian period in Mesopotamia, the area of a rectangle was computed as the length times the front. Therefore the problem stated in DPA 39 can be expressed as the rectangular-linear system of equations

$u \cdot s = A = 1(60)\ 40$ square ninda (= 100 square ninda), $u = 4(60)\ 3$ ninda (= 243 ninda).

An *Old Babylonian school boy* 500 years later would have known the numerical answer to this problem right away, namely that

$s = A/u = 1\ 40 \cdot$ igi 4 03,

where igi 4 03 is a sexagesimal number such that 4 03 · igi 4 03 = some power of 60.

Here 4 03 = 4(60) 3 = 243 is a so called "regular sexagesimal number", by which it is meant that it is *a sexagesimal number with no other prime factors than 2, 3, and 5*. Because 2, 3, and 5 are also the only prime factors in the sexagesimal base 60, a given sexagesimal number is a factor in some power of 60 if and only if it is a regular sexagesimal number. Consequently, if n is a given sexagesimal number, then *there exists as a finite sexagesimal number igi n (the reciprocal of n) such that n · igi n = some power of 60 if and only if n is a regular sexagesimal number.*

Now, look at the particular case when n = 4 03 = 243. Since 243 = 3^5, the number 4 03 is a regular sexagesimal number, and (in floating sexagesimal numbers)

igi 4 03 = igi 3^5 = (igi 3)^5 = 20^5 = 14 48 53 20.

Therefore, the OB school boy would have found the answer

$s = A/u = 1\ 40 \cdot 14\ 48\ 53\ 20 = 24\ 41\ 28\ 53\ 20.$

If he was able to make a correct estimate of how big the answer reasonably ought to be, he could then interpret this floating sexagesimal number as

s = approximately ;24 40 ninda = 4 2/3 cubits 8 fingers.

(Indeed, 1 cubit = 1/12 ninda = ;05 ninda, and 1 finger = 1/30 cubit.)

An *Old Akkadian school boy*, on the other hand, would have had to solve the problem in *DPA* 39 in some other way, since sexagesimal numbers in place value notation and, in particular, sexagesimal fractions had not yet been invented in the Old Akkadian period. He could, conceivably, proceed as follows: Knowing that 1 ninda = 12 cubits, 1 cubit = 30 fingers, and 1 finger = 6 barleycorns, he would have

A = 1 iku = 1 ninda · 100 ninda
= 3 ninda · 33 ninda 4 cubits
= 9 ninda · 11 ninda 1 cubit 10 fingers
= 27 ninda · 3 ninda 8 cubits 13 fingers 2 barleycorns
= 1(60) 21 ninda · 1 ninda 2 cubits 24 fingers 2 2/3 barleycorns
= 4(60) 3 ninda · 4 2/3 cubits 8 fingers 2/3 and 1/3 of 2/3 barleycorns.

Therefore, the answer is that if a rectangle has the area 1 iku and length of the long side is 4(60) 3 ninda, then the length of the short side is 4 2/3 cubits 8 fingers 2/3 and 1/3 of 2/3 barleycorns.

The example shows that metric division is in principle much simpler but in actual practice sometimes much more complicated than the purely geometric procedure in *El.* I.44!

Interestingly, Euclid offers an alternative to *El.* I.44 in his ***Data* 57**, illustrated by the diagram in Fig. 10.1.3, left, below. Fig. 10.1.3, right, illustrates an interpretation of *Data* 57 in terms of metric algebra.

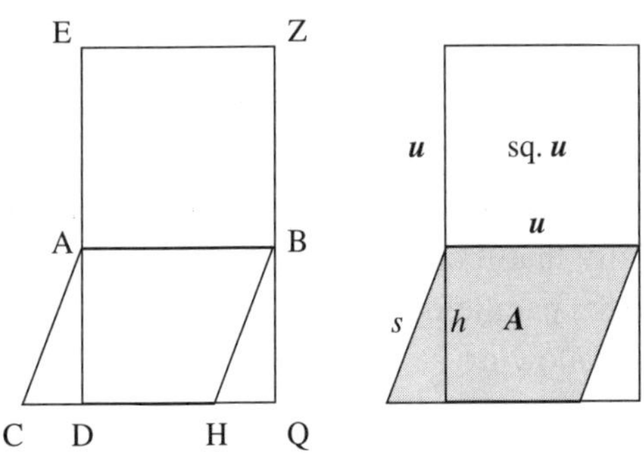

sq. $\boldsymbol{u}$: $\boldsymbol{A}$ = sq. $\boldsymbol{u}$: $\boldsymbol{u} \cdot h = \boldsymbol{u} : h$

$s : \boldsymbol{u} = (\boldsymbol{s} : \boldsymbol{h}) \cdot (h : \boldsymbol{u})$

$s = (\boldsymbol{s} : \boldsymbol{h}) \cdot (\boldsymbol{A}$: sq. $\boldsymbol{u}) \cdot \boldsymbol{u}$

Fig. 10.1.3. *Data* 57. Application of a given parallelogram to a given straight line.

***Data* 57**	explanation
If a given (parallelogram) is applied to a given (straight line) in a given angle, the width of the applied (parallelogram) is given.	
For let the given (parallelogram) AH	The area AH = A (given).
have been applied to a given (straight line) BA	The length BA = u (given).
in the given angle CAB.	The angle CAB is given.
I say that CA is given.	Then the length CA = s is given.
For let the square EB have been described on AB.	Construct the square EB on AB.
Then EB is given.	Then the area of EB = sq. u is given.
And let EA, ZB, CH have been produced to D, Q.	Draw ED, ZQ, CQ.
And since each of EB and AH is given,	
therefore the ratio EB : AH is given.	EB : AH = sq. $u : A = r_1$.
And HA is equal to AQ.	The area HA = the area AQ.
Hence the ratio EB : AQ is given,	EB : AQ = EB : AH = r_1.
so that the ratio EA : AD is given.	EA : AD = EB : AQ = r_1.
And EA = AB. Hence the ratio	EA = AB.
BA : AD is given.	$u : h$ = AB : AD = EA : AD = r_1.
And since angle CDA is given,	Angle CDA is given, and
of which angle DAB is given,	angle DAB is right. Therefore,
the remaining angle CAD is given.	angle CAD = CDA – DAB is given.
And angle CDA is also given, for it is right.	Angle CDA is right. Therefore,
Hence triangle ACD is given in form.	the form of triangle ACD is given.
Therefore the ratio CA : AD is given.	CA : AD = $s : h = r_2$ is given.
And the ratio DA : AB is given.	DA : AB = $h : u = 1/r_1 = r_3$.
Hence the ratio CA : AB is also given.	CA : AB = $s : u = r_2 \cdot 1/r_1 = r_4$.
And BA is given. Therefore	
AC is also given, and it is the width	$s = (s : u) \cdot u = r_4 \cdot u$.
of the applied (figure).	

From the point of view of metric algebra, the idea behind the proof in *Data* 57 is the following (see again Fig. 10.1.4, right): The area A of a parallelogram is known, as well as the length u of one side, and the angle of one vertex of the parallelogram. That one of the angles of the parallelogram is known means, essentially that the ratio s/h is known, where s is (length of) the unknown side of the parallelogram and h the (length of) the unknown height. Therefore, the following values can be computed:

1) sq. u, 2) u/h = sq. $u\,/A$, 3) $s/u = (s/h)\,/\,(u/h)$, 4) $s = (s/l) \cdot u$.

A Babylonian mathematician would, of course, have computed h directly as A/l, but for a Greek mathematician it was more correct to begin by expressing sq. $u : A$ as the ratio between *two magnitudes of the same kind.*

In the explanation above, the successive "givens" are interpreted as a sequence of computed values or ratios between values, where the latter are arbitrarily called $r_1, r_2, \cdots$. A parallel phenomenon is a well known trick in modern mathematics, where one conveniently calls all "constants" that appear in the course of a computation algorithm $C_1, C_2, \cdots$, sometimes even just C, instead of bothering to keep track of their actual values.

10.2. *El.* VI. 28 & *Data* 58. Elliptic Applications of Parallelograms

El. VI.28

To a given straight line to apply a parallelogram equal to a given rectilineal figure and deficient by a parallelogrammic figure similar to a given one; thus the given rectilineal figure must not be greater than the parallelogram described on the half of the straight line and similar to the defect.

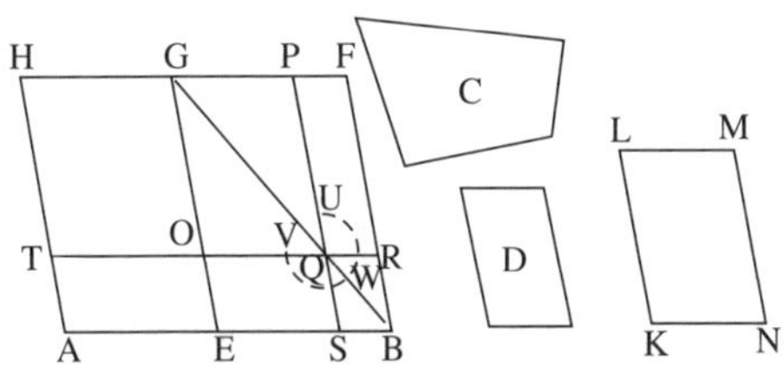

Fig. 10.2.1. *El.* VI.28. Elliptic application of a parallelogram of given area.

In the diagram associated with ***El.* VI.28**, AB is the given straight line, C the given rectilineal figure, and D the given parallelogrammic figure. It is required to construct a parallelogram ST (or SQTA) along a part AS of AB, such that the area of ST is equal to the given area of C, and such that the 'defect' BRQS is similar to the given parallelogram D. The given area of C is not allowed to be greater than the area of the parallelogram EF, which is similar to D and constructed on EB, the half of AB.

The procedure begins with the construction of EF along EB and the similar parallelogram AG along AE, the other half of AB. Then, if AG equals C (in area), AG is the wanted parallelogram. If not, then BG = AG is greater than C (in area). In that case, let the parallelogram KM (KLMN) be constructed so that it is similar to D and equal (in area) to the excess by which BG is greater than C. This is possible according to *El.* VI.25 (below, Sec. 10.4), which teaches how to construct a figure similar to one given rectilineal figure and equal (in area) to another given rectilineal figure.

Next, GQ (GOQP), a copy of KM, is placed in the upper left corner of the similar parallelogram BG, and the "figure" (*schema*) is described, by which is meant that the lines PS, OR, and GB are drawn. Obviously then, the "gnomon" which is the difference between BG and GQ is equal (in area) to D. On the other hand, the gnomon is equal in area to the parallelogram TS, because PR is equal to OS (in view of *El.* I.44) and OS + QB is equal to TE. Therefore, TS is the wanted parallelogram.

***Data* 58**

If a given (parallelogram) is applied to a given (straight line) deficient by a form given in form, the widths of the defect are given.

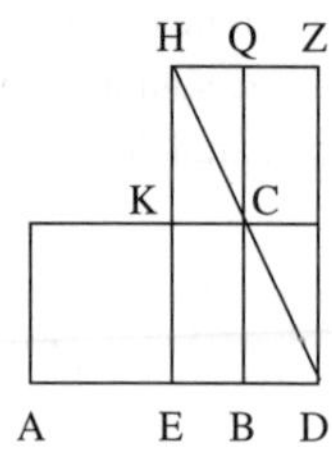

Fig. 10.2.2. *Data* 58. The widths of the defect are given.

The proof of *Data* 58 begins with the assumption that a parallelogram AC of given area has already been applied to a straight line AB of given length in such a way that it is deficient by a parallelogram DC of given form. This can have been done by use of *El.* VI.28.

The construction that was used for the proof of *El.* VI.28 is then repeated, up to the point where the 'figure' is described. It is noted that the parallelogram EZ is given both 'in form', since it is similar to DC, and 'in magnitude', since, in addition, the side ED is half of the given straight line AD. On the other hand, according to *El.* VI.28, EZ = AC + KQ, and AC is given. Therefore, KQ is also given, at least in magnitude, but it is also given in form since it is similar to DC, according to *El.* VI.24, Euclid's counterpart to the Babylonian "right sub-triangle rule".

Since the parallelogram KQ is now given in both form and in magnitude, its sides are also given (*Data* 55, below, Sec. 10.4). Therefore, KC is given, and so is EB = KC. Since ED is given, BD is also given, and since DC is given in form, BC is given. Consequently, the sides of the defect are given, as claimed in the proposition.

From the point of view of metric algebra, the construction problem posed in *El.* VI.28 is *a Babylonian basic quadratic equation of type B4c:*

$p \cdot s - \text{sq. } s = R.$

(See Sec. 1.1 above.) Indeed, in Fig. 10.2.1, let

(the lengths of) AB = $a\,p$, SB = $a\,s$, BR = $b\,s$, and (the area of) ASQT = A.
and let h be the height of the parallelogram ABRT.

Then

$a\,p \cdot b\,s - a\,s \cdot b\,s = A \,/\, r$, where $r = h \,/\, a\,b\,s.$

This equation can immediately be reduced to

$p \cdot s - \text{sq. } s = R$, where $R = A \,/\, a\,b\,r.$

(*Cf.* Taisbak *SC* 16 (2003).) From this point of view, the appearance of a parallelogrammic defect in *El.* VI.28 instead of simply a square defect is a meaningless complication of the situation.

In *El.* VI.28, it is shown that there is a geometric solution to every *elliptic application problem* corresponding to a quadratic equation of type B4 c (under the obvious assumption that the given area of the applied parallelogram is not too big). The method used is *synthetic and constructive*. In the closely related proposition *Data* 58, on the other hand, it is shown by use of an *analytic* method, that once a solution to the elliptic application problem has been found, the sides of the defect can be computed. [24]

Essentially, what is shown in *Data* 58 is that the solution to the basic quadratic equation of type B4c is of the form

$s = p/2 - \text{sqs. }(\text{sq. } p/2 - R)$, with the silent assumption that $R \leq \text{sq. } p/2.$

10.3. *El.* VI. 29 & *Data* 59. Hyperbolic Applications of Parallelograms

El. VI.29

To a given straight line to apply a parallelogram equal to a given rectilineal figure and exceeding by a parallelogrammic figure similar to a given one.

---

24. *Cf.* the following remark in Taisbak *ED* (2003), 155: "··· some of us would think that the 'unknowns' to be proved given (in *Data* 58) were the sides of the applied parallelogram, and not those of that phantom, *the deficient parallelogram*. However, the givenness of the sides (of the applied parallelogram) is postponed till Dt 85 ···". It is clear that the persons referred to as "some of us" have not fully appreciated why *Data* 58-59 and *Data* 84-85 are two separate pairs of propositions. See the continued discussion of the matter in Sec. 10.4.

***Data* 59**

If a given (parallelogram) is applied to a given (straight line) exceeding by a figure given in form, the widths of the excess are given

It is obvious that *El.* VI.29 and *Data* 59 are direct counterparts to *El.* VI.28 and *Data* 58, only with excesses instead of defects in the applications of parallelograms. From the point of view of metric algebra, *El* VI.29 and *Data* 59 can be explained as being concerned with geometric solutions to *quadratic equations of the basic type B4a*:

$$\text{sq. } s + q \cdot s = P.$$

(See again Sec. 1.1 above.) The proofs of *El.* VI.29 and *Data* 59 are, with the necessary modifications, repetitions of the proofs of *El.* VI.28 and *Data* 58. See Fig. 10.3.1 below, where the first diagram shows the metric algebra interpretation of the proof of *Data* 59, while the third diagram shows the metric algebra interpretation of the proof of *Data* 58.

The diagram in the middle in Fig. 10.3.2 shows how a quadratic equation of type B4b can be reduced to an equation of type B4a through what is essentially a simple change of variable. Therefore, this case, too, can be taken care of by the procedure in *Data* 59.

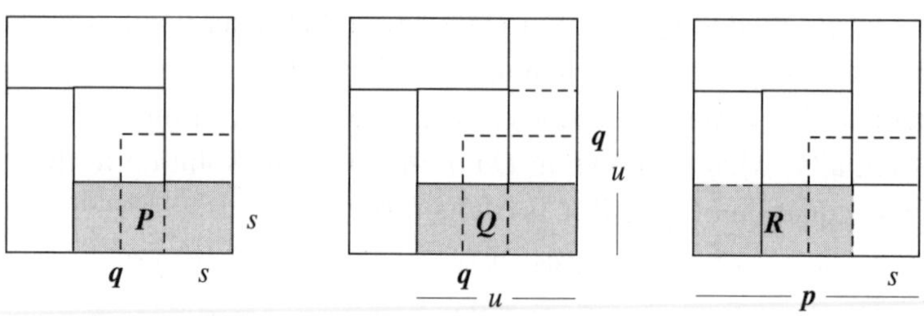

El. VI.29 & *Data* 59	(*El.* VI.29 & *Data* 59)	*El.* VI.28 & *Data* 58
B4a : sq. $s + \boldsymbol{q} \cdot s = \boldsymbol{P}$	B4b : sq. $u - \boldsymbol{q} \cdot u = \boldsymbol{Q}$	B4c : $\boldsymbol{p} \cdot s$ – sq. $s = \boldsymbol{R}$
sq. $(s + \boldsymbol{q}/2)$ = sq. $\boldsymbol{q}/2 + \boldsymbol{P}$	sq. $(u - \boldsymbol{q}/2)$ = sq. $\boldsymbol{q}/2 + \boldsymbol{Q}$	sq. $(\boldsymbol{p}/2 - s)$ = sq. $\boldsymbol{p}/2 - \boldsymbol{R}$

Fig. 10.3.1. Metric algebra interpretations of the proofs of *El.* VI.28-29 and *Data* 58-59.

10.4. *El.* VI.25 and *Data* 55

A crucial part of the arguments in *El.* VI.28-29 and *Data* 58-59 is the application of ***El.* VI.25**, or ***Data* 55**, for *the construction of a parallelogram of given form and magnitude* (for instance KM or OP in Fig. 10.2.1).

***El.* VI.25**

To construct one and the same figure similar to a given rectilineal figure and equal to another rectilineal figure.

In the proof it is assumed that the given form is, for instance, a triangle ABC. (See Fig. 10.4.1 below.) A parallelogram BE equal to ABC is then applied to BC, and a parallelogram CM of the given magnitude is applied to CE [*El.* I.44-45]. Next, GH is constructed as a mean proportional to BC [*El.* VI.13 or II.14] and CF, and KGH is constructed on GH similar and similarly situated to ABC [*El.* VI.18].

Then it follows that as BC is to CF, that is, as the parallelogram BE to the parallelogram EF, so is the triangle ABC to the triangle KGH [*El.* VI.19, Por., see below]. Therefore, since BE = ABC and EF = D, it follows that KGH = D, so that KGH has the desired properties.

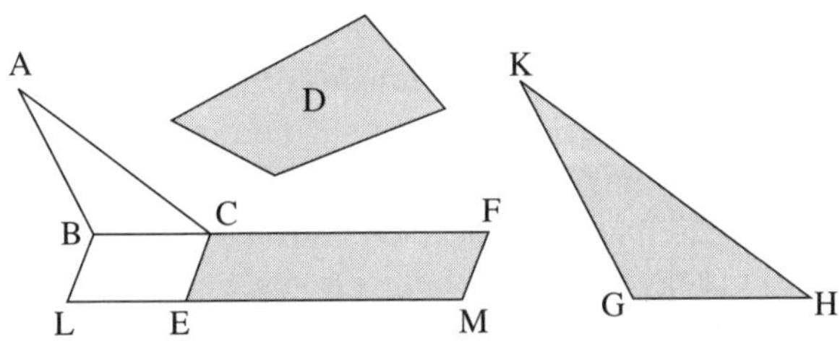

Fig. 10.4.1. *El.* VI.25. Construction of a figure of given form and magnitude.

El. VI.19, Por. which plays a crucial role in the construction in *El.* VI 25 is a corollary to *El.* VI.19:

***El.* VI.19**

Similar triangles are to one another in the duplicate ratio of the corresponding sides.

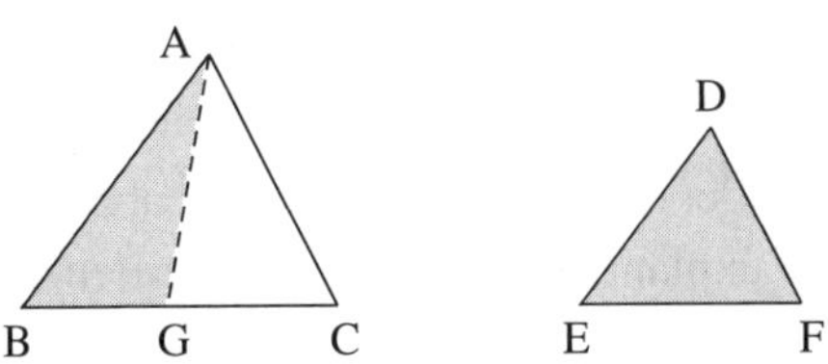

Fig. 10.4.2. *El.* VI.19. Similar triangles are in the duplicate ratio of corresponding sides.

The proof of *El.* VI.19 begins (see Fig. 10.4.2) by constructing BG as the third proportional to BC and EF, so that as BC is to EF, so is EF to BG.

Then, if ABC and DEF are similar, as AB is to DE, so is BC to EF, and EF to BG. Therefore, in the triangles ABG and DEF, the sides about the equal angles are reciprocally proportional, and it follows that ABG = DEF [*El.* VI.15].

Consequently, ABC is to DEF as ABC is to ABG, that is, as BC is to BG. Since BG is the third proportional to BC and CF, BC is to BG in the duplicate ratio of what BC is to EF. Which was to be proved.

The mentioned corollary to *El.* VI.19 says that

> If three straight lines are proportional, then, as the first is to the third, so is the figure described on the first to that which is similar and similarly described on the second.

In Euclid's *Data*, the two propositions ***Data* 54-55** are closely related to *El.* VI.19 and *El.* VI.25. As pointed out by Taisbak, *ED* (2003), 116-118. 143, Data 55 "will play a dominant role in the theory of 'application of areas' (Dt 58 and 59) by way of *Elements* VI.25, which according to Dt 55 has a 'given' solution."

***Data* 54**

If two forms given in form have a given ratio to one another, I say that their sides will also have a given ratio to one another.

***Data* 55**

If a figure is given in form and in magnitude, its sides will also be given in magnitude.

The metric algebra counterpart to the statement in *El.* VI.19 that "similar triangles are to one another in the duplicate ratio of the corresponding sides", and to the related statement in *Data* 54, is that *the areas of similar triangles are proportional to the squares of (the lengths of) corresponding sides.* This is a special case of the following more general "OB quadratic similarity rule":

> The areas of similar figures bounded by straight lines and/or circular arcs are proportional to the squares of (the lengths of) corresponding sides or arcs in the figures.

This intuitively understood rule is frequently applied in OB mathematical texts, and is the explanation for some of the items in the well known OB mathematical "tables of constants". Take, for instance, the following items from the OB table of constants *TMS* 3 (= BR), mentioned in Sec. 6.2 above:

5 igi.gub *šà* gúr	5, constant of the arc	BR 2
26 40 igi.gub *šà a-pu-sà-am-mi-ki*	26 40, constant of the lyre-window	BR 22

The first of these items means that the area of a circle is equal to ;05 (1/12) *times the square of the whole arc* (circumference) of the circle. The second item means that the area of a concave square is equal to ;26 40 (4/9) times the square of the length of any one of the circular arcs bounding the figure. (See, for instance, Fig. 6.2.6 above.)

Note that the complicated proof of *El.* VI.19 can be replaced by the following more straightforward argument in the style of metric algebra:

> The quadratic similarity rule is trivially true for a right triangle with the base b, the height h and the area A, since $A = b/2 \cdot h = (b/h)/2 \cdot$ sq. h. Consequently, it is also true for an arbitrary triangle, since any triangle can be understood as either the sum or the difference of two right triangles, glued together along a common height.

The metric algebra counterpart to the statement in *Data* 55: "If a (rectilineal) figure is given in form and in magnitude, its sides will also be given in magnitude" is the understanding that *if the area and the "constant" of a figure are known, then the sides of the figure can be computed.* In particular, if A = c sq. s, where A and c are known, then s can be computed as the square side of 1/c $\cdot A$.

An explicit solution to a "form and magnitude problem" by use of metric algebra is given in ***P.Moscow* # 17** (Friberg, *UL* (2003), Sec. 2.2 c), a metric algebra exercise in an Egyptian hieratic mathematical text:

> ***P.Moscow* # 17**
>
> Method of calculating a triangle. If it is said to you:
> A triangle of 20_s in field, and as for what you set as length you have 3' $\overline{15}$ as width.
> You double the 20_s, it makes 40. You count with 3' $\overline{15}$ to find 1. It makes 2 2' times.
> You count 40 times 2 2', it makes 100. You count the corner (square side), it makes 10.
> Look, this 10 is the length.
> You count 3' $\overline{15}$ of 10, it makes 4. Look, this 4 is the width.
> You have found correctly.

What this means is that if l is the length, w the width and A the area of a triangle, and $f = w/l$ the ratio between the sides of the triangle, then

$$A = l \cdot w/2 = f/2 \cdot \text{sq. } l.$$

In the given example, when A = 20 *setat* (with 1 *setat* = 1 sq. *khet*, and 1 *khet* = 100 cubits), and f = 3' $\overline{15}$ (= 2/5), it follows from this equation that

sq. $l = 1/f \cdot 2\,A = 1\,/\,3'\,\overline{15} \cdot 40 = 2\,2' \cdot 40 = 100$ (*setat*), so that
$l = 10$ (*khet*), and $w = 3'\,\overline{15} \cdot 10 = 4$ (*khet*).

The steps of the computation are repeated in a diagram in *P.Moscow* # 17.

An OB explicit solution to a form and magnitude problem is **IM 121613 # 1** (Friberg, *op. cit.*), an exercise in a large theme text.

IM 121613 # 1

2/3 of the length (is) the front. 1 èše (is) a field I built.
Length and front (are) what? You:
1, the length, (and) 40, its 2/3, let eat each other, then 40, the false field, you see.
The opposite of 40, the false field, resolve, to 10, the true field, raise, then 15 you see.
The equalside of 15 let come up, 30 you see.
30 to 1 and 40, your numbers, always raise, then 30, the length, 20, the front, you see.
Such is your doing.

In this exercise, the given ratio between the front and the length of a rectangle is $f = 2/3$, and the given area is A = 1 èše (= 10 00 sq. ninda). The solution procedure is an example of the application of the OB "rule of false value" (see Friberg, *RlA 7* (1990), Sec. 5.7 d).

First, it is assumed that a tentative 'false' length is $u' = 1\ 00$. The corresponding 'false' front is $s' = f \cdot u' = 40$, and the corresponding 'false' area

$$A' = f \cdot \text{sq. } u' = 2/3 \cdot 1\ 00\ 00 = 40\ 00.$$

The ratio between the 'true' and the 'false' area is the "quadratic correction factor"

$$\text{sq. } u \,/\, \text{sq. } u' = A \,/\, A' = 10\ 00 \text{ (sq. ninda)} / 40\ 00 = 1/4 \text{ (sq. ninda)} = ;15 \text{ (sq. ninda)}.$$

This is the square of the "linear correction factor"

$$u \,/\, u' = \text{sqs. } (A \,/\, A') = 1/2 \text{ (ninda)} = ;30 \text{ (ninda)}.$$

Therefore, the true length and the true front are

$$u = 1\ 00 \cdot ;30 \text{ (ninda)} = 30 \text{ (ninda)}, \quad \text{and} \quad s = 40 \cdot ;30 \text{ (ninda)} = 20 \text{ (ninda)}.$$

It is interesting to compare this solution to the form and magnitude problem in IM 121613 # 1 by use of the OB rule of false value with Euclid's proof of the *form and magnitude* proposition ***Data* 55**:

***Data* 55, proof**

Let the (rectilineal) figure A be given in form and magnitude;
I say that its sides are given in magnitude.
Let the straight line BC have been set out given in position and in magnitude,
and on BC let D have been described similar and similarly situated to A.
Then D is given in form.
And since on the straight line BC given in magnitude
the given form D has been described, therefore D is given in magnitude.
And A is given; therefore the ratio A : D is given. And A is similar to D;

therefore the ratio EZ : BC is given [*Data* 54 or *El.* VI.19].
And BC is given; therefore EZ is given.
And the ratio ZE : EH is given; therefore EH is given.
For the same reason each of the other sides is also given in magnitude.

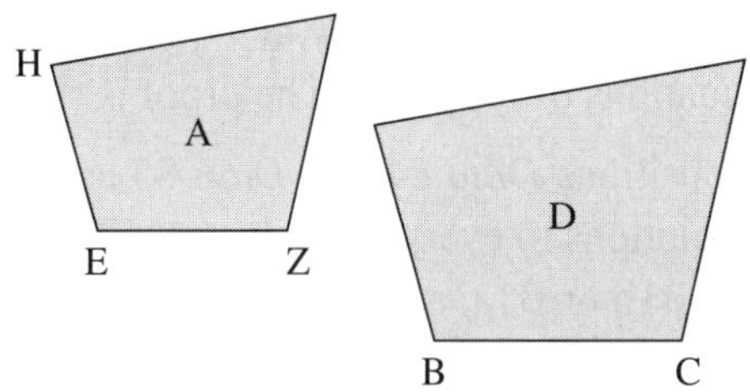

Fig. 10.4.3. *Data* 55. A form and magnitude proposition.

It is easy to see that *Euclid's proof of the form and magnitude proposition Data 55 step by step corresponds to the solution of the form and magnitude problem IM 121613 #1 by an application of the rule of false value!*

10.5. *Data* 84-85. Rectangular-Linear Systems of Equations

In Sec. 1.2 above, it was suggested that the purpose of the two propositions *El.* II.2-3, never adequately explained before, may have been to show that basic *quadratic equations* of the Babylonian types B4a-c can be replaced by equivalent basic *rectangular-linear systems of equations* of the Babylonian types B1a-b (described in Sec. 1.1), which then in their turn can be solved by use of *El.* II.5-6.

This hypothesis is strongly supported by the evidence of a lemma in *Elements* X preceding the crucial propositions *El.* X.17-18 and of a pair of propositions in Euclid's *Data*. All three are accompanied by diagrams resembling the diagram illustrating *El.* II. 3 (Fig. 1.2.3 above).

Lemma *El.* X. 16/17

If to any straight line there be applied a parallelogram deficient by a square figure, the applied parallelogram is equal to the rectangle contained by the segments of the straight line resulting from the application.

***Data* 84**

If two straight lines contain a given figure in a given angle, and one of them is greater than the other by a given (straight line), each of them will be given, too.

***Data* 85**

If two straight lines contain a given figure in a given angle, and their sum is given, each of them will be given.

The lemma says, essentially, quite explicitly, that a basic quadratic equation of type B4c can be replaced by an equivalent basic rectangular-linear system of equations of type B1a. The proof is straightforward.

The pair of propositions *Data* 84 and *Data* 85 assert, essentially, that there exist unique solutions to every given basic rectangular-linear system of equations of type B1b or B1a, respectively.

The proof of *Data* 85, for instance, goes as follows:

***Data* 85, proof**

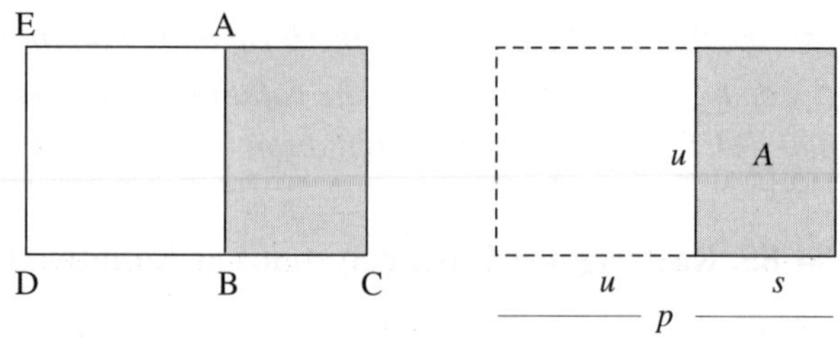

Fig. 10.5.1. *Data* 85. Existence of a solution to a rectangular-linear system of equations.

For let two straight lines AB, BC contain the given area AC in the given angle ABC, and let the sum ABC be given. I say that each of AB, BC is given.
For, let CB have been produced to D, and let BD have been laid out equal to AB, and through D let DE be drawn parallel to BA, and let AD have been completed.
And, since DB is equal to BA, and the angle ABD is given,
because its adjacent angle is given, too, EB is given in form.
And since the sum ABC is given, and AB is equal to B, DC is given, then,
since the given AC has been applied to the given DC deficient by the given form EB, the widths of the defect are given [Data 58].
Therefore AB and BD are given; but the sum ABC is also given;
therefore the remainder BC is also given; therefore each of AB, BC is given.

In terms of metric algebra, *Data* 85 states that if there exists a solution u, s to a basic rectangular-linear system of type B1a:

$$u \cdot s = A, \quad u + s = p,$$

then the solution is uniquely determined.

The proof of the proposition begins by explicitly showing that if there exists a solution u, s to a given system of equations of type B1a, then u is also a solution to a corresponding basic quadratic equation of type B4c:

$p \cdot u - \text{sq.}\ u = A.$

In view of *El.* VI. 28 and *Data* 58 (see Sec. 10.2 above), there exists a unique solution u to every such equation, provided that $A \leq \text{sq.}\ p/2$. Then also $s = p - u$ is uniquely determined.

With only small modifications, the arguments in the proof of *Data* 85 can be used to prove also the *existence* of a solution to every rectangular-linear system of equations of type B1a (when $A \leq \text{sq.}\ m/2$). It is likely, however, that the purpose of the propositions *Data* 85 and *Data* 58 together was not theoretical but practical, namely *to indicate the essential steps in an actual computation of a solution to a system of type B1a.*

10.6. *Data* 86. A Quadratic-Rectangular System of Equations

A particularly intriguing proposition in Euclid's *Data* is ***Data* 86**, extensively discussed in Taisbak, *ED* (2003), Chapter 13:

***Data* 86**

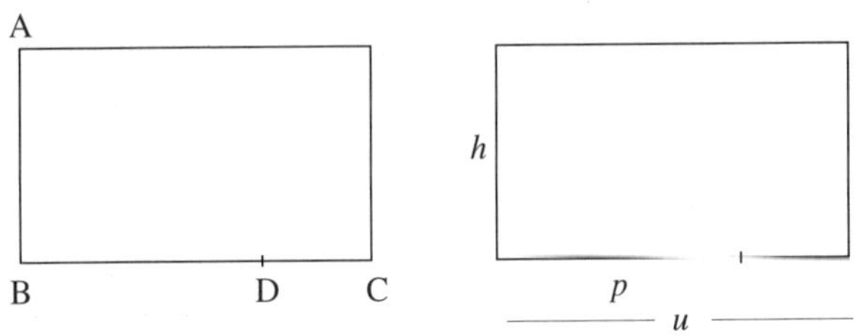

Fig. 10.6.1. *Data* 86. A quadratic-rectangular system of equations.

If two straight (lines) contain a given field in a given angle and if in power one is by a given greater than the other, in ratio, also each of them will be given.

Neither the diagram accompanying the proposition (Fig. 10.6.1, left), nor the wording of the statement itself, is very helpful. Nevertheless, it is clear from the proof what the statement means. In the terminology of metric algebra, the proposition states that if the sides u and s of a parallelogram satisfy a system of equations of the following type

$s : h = r_1, \quad u \cdot h = A, \quad (\text{sq.}\ u - B) : \text{sq.}\ s = r_2,$

where A and B are given areas, and r_1, r_2 given ratios, then u and s will be uniquely determined. Here the equation $s : h = r_1$ is a way of expressing the condition that the parallelogram with the sides u, s and the height h against u is contained in a given angle (see Fig. 10.6.2 below), although the diagram in the text shows only a rectangle.

The mentioned system of equations can be rewritten in the form

$$s = r_1 \cdot h, \quad u \cdot h = A, \quad \text{sq. } u - r_2 \cdot \text{sq. } s = B.$$

Consequently, the given system of equations for u, s can be replaced by a quadratic-rectangular system of equations of the following type for u, h:

$$\text{sq. } u - r_3 \cdot \text{sq. } h = B, \quad u \cdot h = A\,,$$

where r_3 is a new given ratio. Now, recall that in Sec. 5.4 above, a system of equations of the following type

$$\text{sq. } p + \text{sq. } q = S, \quad p \cdot q = P$$

was called a *basic quadratic-rectangular system of equations of type B5*. Systems of type B5 appear in *Elements* X and in OB mathematical cuneiform texts. Systems of equations of a similar type, only with a minus sign instead of a plus sign, can be called "basic quadratic-linear systems of equations of type B6", although no such systems of equations were previously known to appear in either Greek or Babylonian mathematics. Nevertheless, it is still motivated to call the system of equations for u and h mentioned above, with its arbitrary given ratio r_3, a *quadratic-rectangular system of equations of type B6*.

Five different ways of solving basic quadratic-rectangular systems of equations of type B5, documented in *Elements* X or in OB mathematical texts, are exhibited in the form of diagrams in Figs. 5.4.1-2 above. Of the five methods for systems of type B5, only two will work also for a system of type B6, namely the ones used in *El.* X.54, 57 and BM 13901 # 12.

Here is how method of *El.* X.54 would work in the case of a "modified" system of equations of type B6 like the one in *Data* 86 (with $s = h$):

$$\text{sq. } u - r \cdot \text{sq. } s = B, \quad u \cdot s = A\,.$$

First choose a straight line e and set

$$\text{sq. } u = a \cdot e, \quad \text{sq. } s = b \cdot e, \quad A = v/2 \cdot e, \quad B = w \cdot e.$$

Then u, s is a solution to the original system if a, b is a solution to

$$a - r \cdot b = w, \quad a \cdot b = \text{sq. } v/2.$$

This system of equations for a, b can, in its turn, be replaced by a basic system of equations of type B1b for the modified pair a, $r \cdot b$:

$$a - r \cdot b = w, \quad a \cdot (r \cdot b) = r \cdot \text{sq. } v/2.$$

The solution to this system of equations is, of course,

$$a = \text{sqs. } \{\text{sq. } w/2 + r \cdot \text{sq. } v/2\} + w/2, \quad b = [\text{sqs. } \{\text{sq. } w/2 + r \cdot \text{sq. } v/2\} - w/2] \,/\, r.$$

Consequently,

$$u = a \cdot e = [\text{sqs. } \{\text{sq. } w/2 + r \cdot \text{sq. } v/2\} + w/2] \cdot e = [\text{sqs. } \{1 + r \cdot \text{sq. } (2\,A/B)\} + 1] \cdot B/2,$$

and

$$s = b \cdot e = [\text{sqs. } \{1 + r \cdot \text{sq. } (2\,A/B)\} - 1] \,/\, r \cdot B/2.$$

In the proof of *Data* 86, however, Euclid demonstrates an entirely different way of solving a (modified) system of equations of type B6. Euclid's alternative method is illustrated by the diagram in Fig. 10.6.2 below, essentially identical with Fig. 86.4 in Taisbak, *ED* (2003):

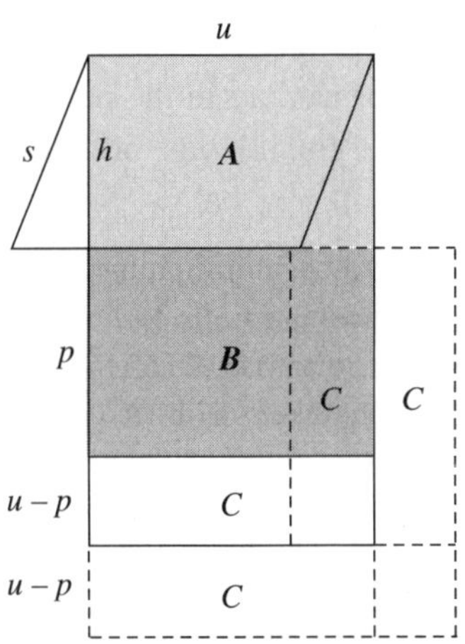

Assume that $s : h$ ***is given***, $u \cdot h$ $= \boldsymbol{A}$ ***is given***
(sq. $u -$ $\boldsymbol{B}$) : sq. s ***is given***
Let $\boldsymbol{B} = u \cdot p$
Then $h : p$ *is also given*
(sq. $u -$ $u \cdot p$) : sq. p *is also given*
sq. $(2\,u -$ $p)$: sq. p *is also given*
$(2\,u -$ $p) : p$ *is also given*
$u : p$ *is also given*
Since $\boldsymbol{B} =$ $u \cdot p$ *is given*
sq. p *is also given*
p *is also given*
u, h, s *are also given*

Fig. 10.6.2. *Data* 86. The procedure in terms of metric algebra.

In terms of metric algebra, Euclid's procedure works in the following way: Consider, in the simple case when $h = s$, the system of equations

$$\text{sq. } u - B = r \cdot \text{sq. } s, \quad u \cdot s = A.$$

Interpret B as the (area of) a rectangle with the sides u and p. Then

$$\text{sq. } u - u \cdot p = r \cdot \text{sq. } s, \text{ and } \; s \,/\, p = u \cdot s \,/\, u \cdot p = A/B.$$

Consequently, the given quadratic-rectangular system of equations for u, s

can be replaced by the following quadratic equation for u, p:

$$\text{sq. } u - u \cdot p = C, \quad \text{with} \quad C = r' \cdot \text{sq. } p, \quad \text{and} \quad r' = r \cdot \text{sq. } A/B.$$

By use of *El.* II.8, for instance, this equation can be transformed into

$$\text{sq. } (2\,u - p) = 4\,C + \text{sq. } p = (4\,r' + 1) \cdot \text{sq. } p.$$

Therefore,

$$2\,u - p = \text{sqs. } (4\,r' + 1) \cdot p, \quad \text{so that} \quad u = r'' \cdot p, \quad \text{with} \quad r'' = \{\text{sqs. } (4\,r' + 1) + 1\}/2.$$

This means that the original equations for u and p have been reduced to a rectangular-linear system of equations of the following simple kind:

$$u \cdot p = B, \quad u = r'' \cdot p.$$

The solution to this system of equations is, of course, that

$$r'' \cdot \text{sq. } p = B, \quad \text{so that} \quad p = \text{sqs. } (B\,/\,r'') \quad \text{and} \quad u = r'' \cdot p = r'' \cdot \text{sqs. } (B\,/\,r'').$$

When p and u have been determined in this way, it is easy to find also the value of $s = A/B \cdot p$.

The more general case when $s/h = r_1$ can be treated similarly.

The proposed explanation above of Euclid's procedure in the proof of *Data* 86 operates somewhat carelessly with ratios. Euclid was, of course, much more careful, which is evident from the full text below. See Fig. 10.6.1, left, for the meaning of the notations in the left column.

***Data* 86, proof**	explanation
For. let the two straight lines AB, BC contain	Let AB = s, BC = u
the given field AC in the given angle ABC,	$s : h = r_1$ (given)
and let the square on CB be by a given	$u \cdot h = A$ (given)
greater than in ratio to the square on BA.	(sq. $u - B$ (given)) : sq. $s = r_2$ (given)
I say that each of AB, BC is also given.	Then s and u are also given
———————	
For ··· let the given rectangle CBD	Set B = rect. CB, BD = $u \cdot p$
have been subtracted.	
Then the ratio of the remaining rect. DCB	Then (sq. $u - u \cdot p$) : sq. h
to the square on AB is given.	= sq. $r_1 \cdot r_2 = r_3$
And since rect. ABC is given, and	$u \cdot h : u \cdot p$
rect. CB, BD is also given, therefore	$= A : B$
the ratio rect. AB, BC : rect. CBD is given.	$= r_4$
But rect. ABC : rect. CB, BD :: AB : BD.	$u \cdot h : u \cdot p = h : p$
Therefore the ratio AB : BD is also given:	Therefore, $h : p = r_4$, and
Hence the ratio sq. AB : sq. BD is also given.	sq. h : sq. p = sq. $r_4 = r_5$
The ratio sq. AB : rect. BCD is given.	sq. h : (sq. $u - u \cdot p$) = $1/r_3 = r_6$

Therefore rect. BCD: sq. DB is also given.	$(\text{sq. } u - u \cdot p) : \text{sq. } p = r_5/r_6 = r_7$
Hence the ratio 4 rect. BCD : sq. BD is given,	$4\,(\text{sq. } u - u \cdot p) : \text{sq. } p = 4\,r_7 = r_8$
and the ratio	$\{4\,(\text{sq. } u - u \cdot p) + \text{sq. } p\} : \text{sq. } p$
(4 rect. BCD + sq. BD) : sq. BD is given,	$= r_8 + \text{sq. } p : \text{sq. } p = r_9$
But 4 rect. BCD + sq. BD	$4\,(\text{sq. } u - u \cdot p) + \text{sq. } p$
= sq. (BC + CD). Hence the ratio	$= \text{sq. } (2\,u - p)$
sq. (BC + CD) : sq. BD is given.	$\text{sq. } (2\,u - p) : \text{sq. } p = r_9$
And so the ratio (BC + D) : BD is given.	$(2\,u - p) : p = \text{sqs. } r_9 = r_{10}.$
Therefore, *synthenti*,	
the ratio 2 CB : BD is given, so that	$2\,u : p = r_{10} + p : p = r_{11}$
the ratio of CB alone to BD is given.	$u : p = r_{11}/2 = r_{12}$
But CB : BD :: rect. CBD : sq. BD.	$u : p = u \cdot p : \text{sq. } p$
Therefore the ratio rect. CBD : sq. BD is given.	$u \cdot p : \text{sq. } p = r_{12}$
Rect. CB, BD is given.	$u \cdot p = B$
Therefore, sq. BD is also given,	$\text{sq. } p = B/r_{12} = c_1$
and so BD is given,	$p = \text{sqs. } c_1 = c_2$
so that BC, too, is given,	$u = r_{12} \cdot c_2 = c_3$
because the ratio CB : BD is given.	since $u : p = r_{12}$
[And the ratio AB : BD is given.]	$[h : p = r_4]$
[Therefore AB is given.]	$[h = r_4 \cdot c_2 = c_4.]$
And the parallelogram AC is given,	$u \cdot h = A$ (given)
and the angle B is given.	$s : h = r_1$ (given)
Therefore AB is also given.	$s = r_1 \cdot c_4 = c_5$
Therefore each of AB and BC is given.	$s = c_5,\ u = c_3$

The translation above of the text of *Data* 86 follows closely the translation given by Taisbak in *ED* (2003), with the important exception that the text has been broken up into sentences or parts of sentences so that *each step of the procedure is written on a separate line of the translation.* This way of representing a difficult ancient mathematical text makes it easy to follow the course of the arguments and to juxtapose the explanations as in the right column above.

Breaking up the text mass in this way also makes the repetitive nature of the text stand out, with the phrase "is given" finishing each step of the procedure. It is a striking observation that *Babylonian mathematical texts are often of precisely the same repetitive nature, with each step of a complicated computation finishing with a fixed phrase*. See, for instance, IM 121613 (Friberg, *UL* (2005), 73; Sec. 10.4 above), an OB mathematical text where the fixed phrase is 'you see' (*ta-mar*). VAT 7532 (*op. cit.*, 118) is another OB mathematical text with the fixed phrase 'it gives' (in.sì), and BM 34800 (*op. cit.*, 34), is a Late Babylonian mathematical text with another fixed phrase meaning 'it gives' (ì.mu).

Comparing the repeated use of the intentionally vague phrase 'is given' in *Data* 86 with the difficulty of keeping track of the changing values of the successive ratios r_1, r_2, ··· in the explanation, one begins to understand how ingenious the style of Euclid's *Data* really is. *The persistent use of the phrase 'is (also) given' makes it possible to go through all the steps of an algorithmic computation without being bothered by obscuring details.*

10.7. Zeuthen's Conjecture: Intersecting Hyperbolas

Interpretations of *Data* 86 as an example of the application of "geometric algebra" was suggested already by Tannery in his paper "De la solution géometrique des problèmes du second degrée avant Euclide" (1882) (see Saito *HS* 28 (1985), fn. 27), and by Zeuthen in "Sur la reforme qu'a subie la mathématique de Platon à Euclide, et grâce à laquelle elle est devenue science raisonnée" (1917) (see Taisbak *ED* (2003), fn. 153).

In his paper, Zeuthen also makes the interesting conjecture that

> "The proposition (*Data* 86) may be said to deal with the givenness of the intersection of two hyperbolas having the same two given straight lines for conjugate diameters and asymptotes, respectively" (Taisbak, *op. cit.*, 212, 223-224).

Apparently without knowing about Zeuthen's conjecture, Saito (*op. cit.*, 50-53; *CHGM* (2004), 160-161) made a similar interpretation of *Data* 86. In Saito's diagram, reproduced in Fig. 10.7.1 below, I and II are the two hyperbolas, and P their point of intersection.

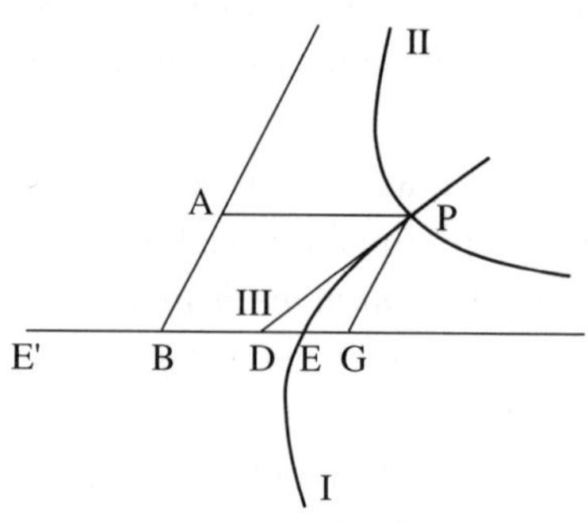

I: (sq. AP – sq. BE) : sq. PG is given

II: AP · PG is given

(III: sq. BE = BG · BD = AP · BD)

⇒

AP and PG are also given

Fig. 10.7.1. *Data* 86 interpreted as a proposition concerning intersecting hyperbolas.

The diagram shows that *Data* 86 can be interpreted as saying that if two given hyperbolas intersect each other in a point P, then also the abscissa AP and the ordinate PG of P are given, so that P is given in position. In addition, as observed by Saito, the crucial trick of setting $B = u \cdot p$ can be explained as simply an application of Appolonius' *Conics I* 37 (a), giving the equation for the position of the point D, the point where the diameter is intersected by the tangent to hyperbola I at the point P!

10.8. A Kassite Series Text with Modified Systems of Types B5 and B6

YBC 4709 (Neugebauer, *MKT 1* (1935), 412-420) is one of several known medium size clay tablets with mathematical "series texts". A series text is *an extremely compressed theme text*, typically with about 50 closely related exercises, ending with a colophon (a subscript). According to Neugebauer (*QSB 3* (1934-36), 113), the writing style shows that the series texts are post-Old-Babylonian, possibly Kassite.

In the case of YBC 4709, the subscript is the following:

55 hand tablets (assignments). 5th clay tablet (in a certain series).

The 55 exercises can be divided into 15 paragraphs. Here are two of them:

	YBC 4709 § 1 a-c, literal translation	explanation
a	The field (is) 1 èše.	$A = 1$ èše = 10 00 sq. ninda
	The length times 3 repeat, equalsided.	sq. $(3\,u)$
	The field of the front add, then 2 21 40.	+ sq. s = 2 21 40
b	Times 2 repeat, add, then 2 28 20.	+ 2 sq. s = 2 28 20
c	The field of the front tear off, then 2 08 20.	– sq. s = 2 08 20

	YBC 4709 § 15 a-c, literal translation	explanation
a	The field (is) 1 èše.	$A = 1$ èše = 10 00 sq. ninda
	The front times 3 repeat,	
	as much as the length over the front is beyond	
	add, then equalsided.	sq. $\{3\,s + (u - s)\}$
	The field of the length add, then 1 36 40.	+ sq. u = 1 36 40
b	Times 2 repeat, add, then 1 51 40.	+ 2 sq. u = 1 51 40
c	The field of the length (and) the front add then 1 43 20.	+ (sq. u + sq. s) = 1 43 20

The question in YBC 4709 § 1 a can be interpreted as *a quadratic-rectangular system of equations of type B5*:

§ 1 a $\quad$ sq. (3 u) + sq. s = 2 21 40, $\; u \cdot s = A$ = 10 00 (sq. n.)

The question in § 1 b is a *slightly modified system of type B5*:

§ 1 b $\quad$ sq. (3 u) + 2 sq. s = 2 28 20, $\; u \cdot s = A$ = 10 00 (sq. n.)

The question in YBC 4709 § 1 c, on the other hand, is *a quadratic-rectangular system of equations of type B6*:

§ 1 c $\quad$ sq. (3 u) – sq. s = 2 08 20, $\; u \cdot s = A$ = 10 00 (sq. n.)

The questions in YBC 4709 § 15 can all be reduced to *modified quadratic-rectangular* systems of equations of type B5:

§ 15 a $\quad$ sq. {3 s + (u – s)}+ sq. u = 1 36 40, $\; u \cdot s = A$ = 10 00 (sq. n.)
$\Rightarrow$ $\quad$ 2 sq. u + 4 sq. s = 1 36 40, $\; u \cdot s = A$ = 10 00 (sq. n.)

§ 15 b $\quad$ sq. {3 s + (u – s)} + 2 sq. u = 1 51 40, $\; u \cdot s = A$ = 10 00 (sq. n.)
$\Rightarrow$ $\quad$ 3 sq. u + 4 sq. s = 1 51 40, $\; u \cdot s = A$ = 10 00 (sq. n.)

§ 15 c $\quad$ sq. {3 s + (u – s)} + (sq. u + sq. s = 1 43 20, $\; u \cdot s = A$ = 10 00 (sq. n.)
$\Rightarrow$ $\quad$ 2 sq. u + 5 sq. s = 1 43 20, $\; u \cdot s = A$ = 10 00 (sq. n.)

It is likely that the author of YBC 4709 intended all the problems in §§ 1 and 15 to be solved by use of a method closely related to the solution method for a similar problem in the OB text BM 13901 # 12 (Fig. 5.4.1, bottom).[25] [26] In the case of § 15 c, for instance, the trick is to set

2 sq. $u = a$, $\quad$ 5 sq. $s = b$.

In this way, the mentioned system of type B5 for the pair u, s is reduced to a simpler system of type B1a for the pair a, b:

$a + b$ = 1 03 20, $\; a \cdot b$ = 10 · sq. 10 00 = 16 40 00 00.

The solution to this system of equations can be found in the usual way. It turns out to be a, b = 30 00, 33 20. Therefore, u, s = 30, 20. This, by the way, is also the solution to all the other 55 problems in YBC 4709.

---

25. Compare with the suggested solution procedure for the triangle division problem *TMS* 18 in Sec. 11.2 e below.

26. Note that of the five methods to solve a quadratic-rectangular system of type B5 shown in Figs. 5.4.1-2, only the method used in BM 13901 # 12 and the related method used in *El.* X. 54, 57 work also in the case of *modified* systems of type B5, like the ones in YBC 4709 § 1 b and § 15 a-c. As observed above, it is also only those two methods of the five shown in Figs. 5,4,1-2 that work in the case of a quadratic-rectangular system of equations *of type B6*, like the one in *Data* 86, modified or not. Reversely, the method used in *Data* 86 works also in the case of a *modified* quadratic-rectangular system of equations of type B5.

Chapter 11

Euclid's Lost Book *On Divisions* and Babylonian Striped Figures

The Greek text of Euclid's book ***On Divisions*** is lost. The only Greek references to it can be found in Proclus' commentary to Euclid's *Elements*. In 1851, Woepcke published a French translation of an abstract of *On Divisions* composed by the 10th-century Persian geometer al-Sijzī. Al-Sijzī reproduced the statements of all the 36 propositions in *On Divisions*, but the solution procedures for only four of them (## 19, 20, 28, 29).

In 1915, Archibald published a reconstruction (in English) of *On Divisions*, complete with procedures for all the problems, based on Woepcke's translation and on possible traces of the work in **Leonardo Pisano's *Practica Geometriae*** (ed. Boncompagni 1862). Finally, in 1993, Hogendijk made available the Arabic text translated by Woepcke, now with an English translation, and a couple of briefer related Arabic manuscripts. In the discussion below the numbering and the translations of the propositions in *On Divisions* follow Archibald. Also some of the reconstructed solution procedures are borrowed from Archibald.

The figures divided in various ways in *On Divisions* are *triangles, trapezoids, parallelograms, a circle, and a circle segment attached to the base of a triangle*. The figures are divided in two or several parts, the parts being either equal or in given ratios. The dividing lines are parallel to a side of the figure or to each other, or drawn from a vertex, or from a point inside the figure, outside the figure, or on a side of the figure.

Old Babylonian parallels exist mainly to division problems in *On Divisions* concerned with *triangles or trapezoids divided by lines parallel to the base*. On the other hand, there are quite a few known OB examples of division problems *without counterparts* in Euclid's book. All problems in *On Divisions* with OB parallels will be discussed below, plus a few others of particular interest.

11.1. Selected Division Problems in *On Divisions*

OD 1-2, 30-31. To divide a triangle by lines parallel to the base

OD 1

To divide a given triangle into two equal parts by a line parallel to its base.

OD 2

To divide a given triangle into three equal parts by two lines parallel to its base.

OD 30

To divide a given triangle into two parts by a line parallel to its base, such that the ratio of one of the two parts to the other is equal to a given ratio.

OD 31

To divide a given triangle by lines parallel to its base into parts which have given ratios to one another.

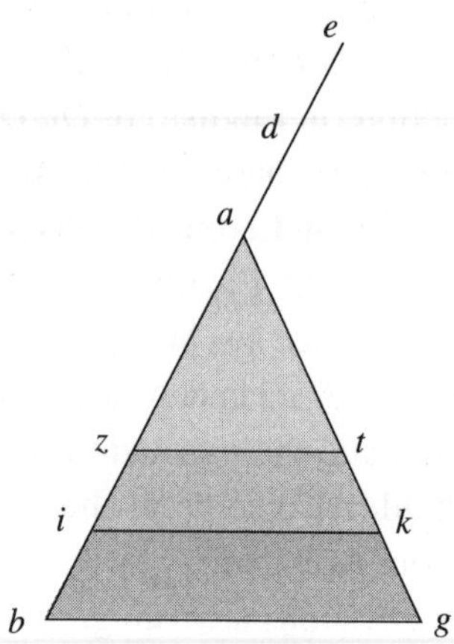

$ad = 1/3\ ba$	$ae = 2/3\ ba$
$ba : az = az : ad$	$ba : ai = ai : ae$
$\Rightarrow$ [*El.* VI.19]	$\Rightarrow$ [*El.* VI.19]
$abg : azt$	$abg : aik$
$=$ sq. ba : sq. az	$=$ sq. ba : sq. ai
$=$ sq. $ba : (ba \cdot ad)$	$=$ sq. $ba : (ba \cdot ae)$
$= ba : ad = 3 : 1$	$= ba : ae = 3 : 2$
$\Rightarrow\ azt = 1/3\ abg$	$\Rightarrow\ aik = 2/3\ abg$

Fig. 11.1.1. *On Divisions* 2. To divide a triangle by two parallels in three equal parts.

Procedure for *OD* 2 (in *Practica Geometriae*): In the triangle *abg*, the side *bg* is extended to *d* and *e*, with *ba* = 3 *ad* and *ad* = *de*. The points *z* and *i* are constructed so that *az* is the mean proportional between *ba* and *ad*, and *ia* the mean proportional between *ba* and *ae* (Fig. 11.1.1), and the parallels *zt* and *ik* are drawn. Then it follows from *El.* VI.19: "Similar triangles are to one another in the duplicate ratio of the corresponding sides" that *abg* : *azt* = *ba* : *ad* = 3 : 1 and *abg* : *aik* = *ba* : *ae* = 3 : 2. Hence the triangle *abg* has been divided in three equal parts, as required.

OD 3. To bisect a triangle by a line through a point on a side

***OD* 3**

To divide a given triangle into two equal parts by a line drawn from a given point situated on one of the sides of the triangle.

If the given point *d* is the midpoint on the side *bg* (see Fig. 11.2.1), then the line through *d* and the opposite vertex *a* solves the problem. If not, and if *d* is between *b* and the midpoint *e*, first the line *da* is drawn, then *ez* parallel to da, and *zd* is joined. The triangle *adz* is then equal to the triangle *ade*, and if *abd* is added to both, it becomes evident that the quadrilateral *abdz* is equal to the triangle *abe* which is half of *abg*.

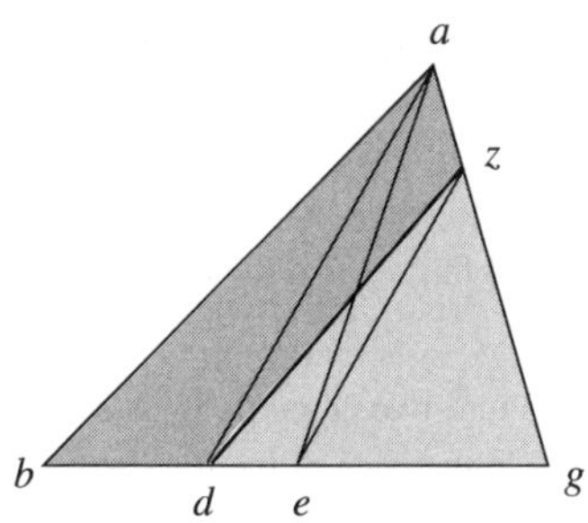

Fig. 11.1.2. *On Divisions* 3. To bisect a triangle by a line through a given point on a side.

OD 4-5. To divide a trapezoid by lines parallel to the base

***OD* 4**

To divide a given trapezoid into two equal parts by a line parallel to its base.

***OD* 5**

To divide a given trapezoid into three equal parts by lines parallel to its base.

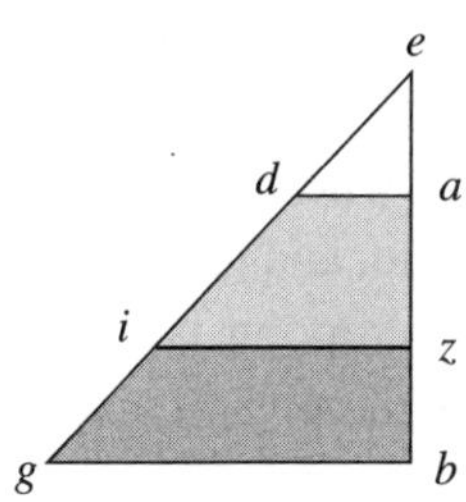

sq. *ze* = (sq. *eb* + sq. *ea*)/2
⇒
2 *ezi* = *ebg* + *ead*,
ezi = *zbgi* + *ead*,
azid = *zbgi*

Fig. 11.1.3. *On Divisions* 4. To bisect a trapezoid by a line parallel to its base.

If *abgd* is the given trapezoid (Fig. 11.1.3), the sides *gd* and *bd* are extended until they meet in the point *e*. The point z is constructed so that

sq. *ze* = (sq. *eb* + sq. *ea*)/2.

Then 2 sq. *ze* = sq. *cb* + sq. *ea*, and it follows (by *El*. VI.19 as in the procedure of *OD* 2) that 2 times *ezi* is equal to the sum of *ebg* and *eda*. If first *ezi* and then *eda* are subtracted from both sides of this equation, it follows that *azid* = *zbgi*, as required.

In *OD* 5 it is shown that a similar procedure can be used for the construction of two lines parallel to the base of a trapezoid, cutting the trapezoid into *three equal parts*.

OD 8, 12. To bisect a trapezoid by a line through a point on a side

OD 8

To divide a given trapezoid into two equal parts by a straight line drawn from a given point situated on the longer of the sides of the trapezoid.

OD 12

To divide a given trapezoid into two equal parts by a straight line drawn from a point which is not situated on the longer side of the trapezoid.

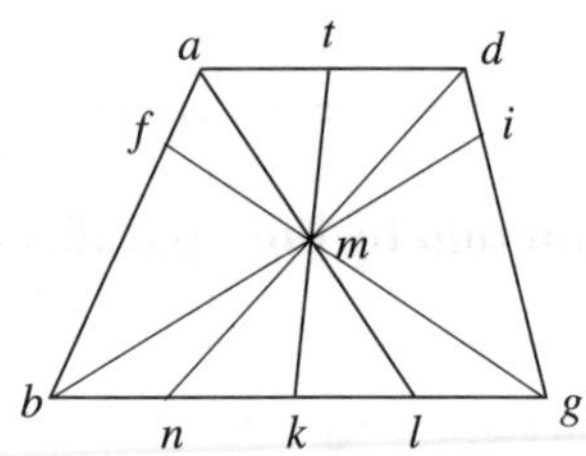

Fig. 11.1.4. *On Divisions* 8. To bisect a trapezoid with a line through a point on a side.

The solution procedure in Leonardo's *Practica Geometriae* is divided into a number of cases. The first step in the procedure (abbreviated here) is to draw the diagram in Fig. 11.1.4, left, where *t* and *k* are the midpoints of the parallel sides *ad* and *bg* of a trapezoid, and *m* the midpoint of *tk*. The simplest case is, of course, when the given point is *t* or *k* and the *tk* the dividing line. Three other cases are when the given point is 1) on *ad* or *nl*, 2) on *bn* or *lg*, or 3) on *ab* or *dg*.

An example of the first case is shown in Fig. 11.1.5, left, where the dividing line passes through the given point *p* and the midpoint *m* on *tk*.

This is a simple generalization of the case when the given point is the midpoint *t* on *ad*. It is clear that *pmq* is the required dividing line since the triangles *ptm* and *qkm* are equal so that the quadrilateral *pdgq* is equal to the quadrilateral *tdgk*, which is one half of the given trapezoid.

An example of the third case is shown in Fig. 11.1.5, right, where *zi* is a parallel to the base bisecting the given trapezoid (*OD* 4). If the given point *k* lies on the side *ab*, then first the line *ki* is drawn, then the line *zh* parallel to *ki*, and finally the line *kh*, which is the required dividing line. This is a simple generalization of the situation in *OD* 4, since it is clear that the triangles *zhk* and *zhi* are equal, and that consequently the quadrilateral *khgb* is equal the sub-trapezoid *zigb*, half the given trapezoid, plus and minus two equal triangles.

The third case, when the given point is on *bn* or *lg* in Fig. 11.1.4, left, is a simple generalization of the situation in *OD* 3 (see above).

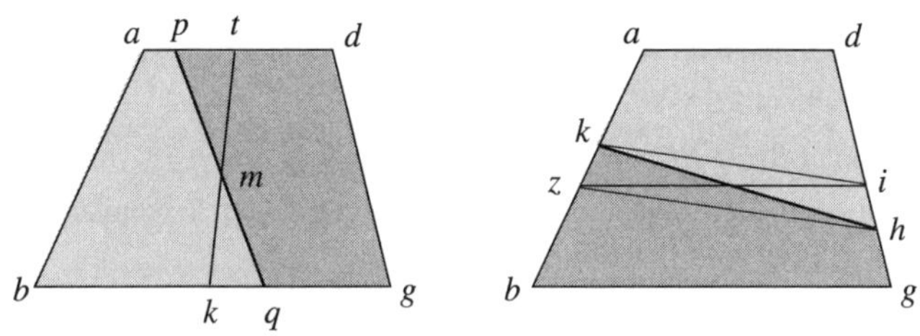

Fig. 11.1.5. *On Divisions* 8. The solution in two different cases.

In ***OD* 9**, the slightly more general case is considered when the dividing line cuts off a certain fraction of the given trapezoid.

OD 19-20. To divide a triangle by a line through an interior point

OD 19

To divide a given triangle into two equal parts by a line which passes through a point situated in the interior of the triangle.

OD 20

To cut off a certain part from a given triangle by a line drawn from a given point situated in the interior of the triangle.

These are two of the four problems provided with explicit solution procedures in al-Sijzī's Arabic manuscript (Hogendijk, *VM* (1993)). In the explanation below, the notations used are the same as in Fig. 11.1.6, right.

***OD* 19, solution**	explanation
We draw from point *D* a line parallel to *BG*, namely *DE*.	$DE = t$ is drawn parallel to b
We apply to *DE* an area equal to half $AB \cdot BG$.	$BT = m$
let it be $TB \cdot ED$.	$m \cdot t = r \cdot a \cdot b \quad (r = 1/2)$
We apply to line *TB* a parallelogram	$BH = p,\ BE = s$
equal to $BT \cdot BE$, deficient from its	Find p a s the solution to the equation
completion by a square area;	$m \cdot p - \text{sq.}\, p = m \cdot s$
let the applied area be $BH \cdot HT$.	or $p \cdot (m - p) = m \cdot s$
We join line *DH* and we extend it towards *Z*.	Draw the line *HDZ*
I say that the line *DHZ* has been drawn such as to divide triangle *ABG* into two equal parts, namely *HBZ*, *HZGA*.	It is the required dividing line
Proof of this:	Proof:
$TB \cdot BE$ is equal to $TH \cdot HB$, so the ratio of	$m \cdot s = (m - p) \cdot p$ (by assumption)
BT to *TH* is equal to the ratio of *HB* to *BE*.	$\Rightarrow\ m : (m - p) = p : s$
Separando, the ratio of *TB* to *BH* is also equal to the ratio of *BH* to *HE*.	$\Rightarrow\ m : p = p : (p - s)$
But the ratio of *BH* to *HE* is equal to the ratio of *BZ* to *ED*.	$p : (p - s) = q : t$ (similar triangles)
Thus the ratio of *TB* to *BH* is equal to the ratio of *BZ* to *ED*.	$\Rightarrow\ m : p = q : t$
So $TB \cdot ED$ is equal to $BH \cdot BZ$.	$\Rightarrow\ m \cdot t = p \cdot q$
But $TB \cdot ED$ is equal to half of $AB \cdot BG$,	
and the ratio of $BH \cdot BZ$ to $AB \cdot BG$ is equal to	Hence, $m \cdot t = r \cdot a \cdot b \quad (r = 1/2)$
the ratio triangle *HBZ* to triangle *ABG*,	$\Rightarrow\ p \cdot q : a \cdot b = r$
because the angles of point *B* are common.	$\Rightarrow$ [by *Data* 66]
So triangle *HBZ* is half of triangle *ABG*.	cut off triangle : given triangle
so triangle *ABG* has been divided into	$= p \cdot q : a \cdot b = r$
two equal parts, namely *BHZ*, *AHZG*.	.

There follows a discussion of the extreme case when the points *H* and *A* coincide. The preceding problem *OD* 18 contains an oblique reference to *El.* VI.28 and a vague discussion of the condition for the existence of a solution to a quadratic equation of type B4 c. The solution procedures in *OD* 19 and *OD* 20 are essentially the same, but the one in *OD* 19 (the case when the ratio r between the cut off triangle and the whole triangle is equal to 1/2) is more detailed.

The wholly synthetic solution arguments in *OD* 19-20 must have been preceded by an *analysis*, which is not provided. It is easy to restore the missing analysis, for instance as follows:

In Fig. 11.1.6 below, right, a, b are two sides of the given triangle, while s, t are the ordinate and abscissa of the given point, parallel to a and b. Suppose that a triangle has already been constructed, r times smaller than the given triangle, with sides of lengths p and q along a and b, respectively, and with its third side passing through the given point D.

Then it follows from a similarity argument that

$$p / (p - s) = q / t,$$

and since the areas of the two triangle are to each other in the ratio r,

$$p \cdot q / a \cdot b = r.$$

(See, for instance, the simple proof of ***Data* 66**, which says that

> If a triangle has a given angle, the rectangle contained by the lines that contain the given angle has a given ratio to the triangle.)

Consequently, the pair p, q must satisfy the equations

$$p \cdot q = r\, a \cdot b, \quad t \cdot p = q \cdot (p - s).$$

If both sides of the second equation are multiplied by p / t, the result is the following *quadratic equation for p*:

$$\text{sq.}\, p = m \cdot (p - s), \quad \text{where} \quad m = r\, a \cdot b / t.$$

This equation for p is where the synthetic solution procedure begins.

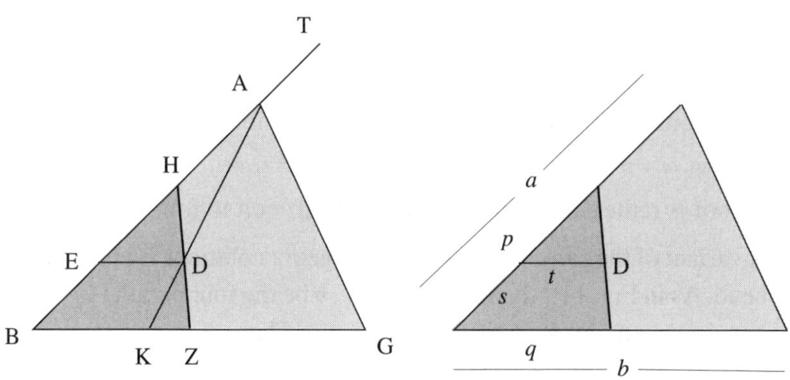

Fig. 11.1.6. *On Divisions* 19. To cut off part of a triangle by a line through a given point.

OD 32. To divide a trapezoid by a parallel in a given ratio

***OD* 32**

To divide a given trapezoid by a line parallel to its base into two parts such that the ratio of one of these parts to the other is equal to a given ratio.

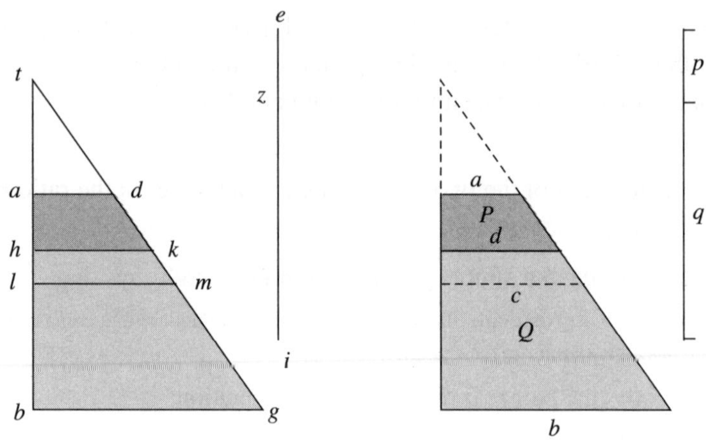

Fig. 11.1.7. *On Divisions* 32. To divide a trapezoid by a parallel in a given ratio.

The solution procedure for this problem in *Practica Geometriae* begins by extending the non-parallel sides of the given trapezoid until they meet in the point t (Fig. 11.1.7, left). The parallel sides are called ad and bg, the dividing line parallel to ad and bg is called hk, and an auxiliary parallel line is called lm. The given ratio is called $ez : zi$. Then, without any preceding analysis, the solution to the problem is claimed to be given by the following equations determining the position of the point h:

$$\text{sq.}\, tl : \text{sq.}\, at = zi : ez, \quad \text{and} \quad \text{sq.}\, ht : (\text{sq.}\, bt + \text{sq.}\, tl) = ez : ei.$$

The proof is rather long-winded and quite difficult to follow.

A variant of the same proof in metric algebra notations is easier to comprehend. As in Fig. 11.1.7, right, let a, d, c, b be the four parallel lines, from the top down, and let $p : q$ be the given ratio. Then

$$ta : th : tl : tb = a : d : c : b.$$

Therefore, the equations for the solution in *OD* 32 can be reformulated as

$$\text{sq.}\, c : \text{sq.}\, a = q : p, \quad \text{and} \quad \text{sq.}\, d : (\text{sq.}\, b + \text{sq.}\, c) = p : (p + q).$$

Since a, b, p, and q are given, the first of these equations causes c to be known, and when c is known, the second equation causes d to be known.

Now, it follows from the first equation (in view of *El.* V.18) that also

$$\text{sq.}\,a : (\text{sq.}\,c + \text{sq.}\,a) = p : (p + q).$$

Combining this result with the second equation, one gets that

$$\text{sq.}\,d : (\text{sq.}\,b + \text{sq.}\,c) = p : (p + q) = \text{sq.}\,a : (\text{sq.}\,c + \text{sq.}\,a).$$

Consequently also (in view of *El.* V.19)

$$(\text{sq.}\,d - \text{sq.}\,a) : (\text{sq.}\,b - \text{sq.}\,a) = p : (p + q).$$

Note that now the auxiliary straight line $c = lm$ has been eliminated from the equation! Finally, also (in view of *El.* V.17)

$$(\text{sq.}\,d - \text{sq.}\,a) : (\text{sq.}\,b - \text{sq.}\,d) = p : q.$$

This is the required result, since (with the notations in Fig. 11.1.7)

$$P : Q = (\text{sq.}\,d - \text{sq.}\,a) : (\text{sq.}\,b - \text{sq.}\,d)$$

The mentioned solution procedure in *OD* 32 is completely synthetic; the necessary preceding analysis is missing. However, reading the synthetic procedure backwards, *one can easily restore the missing analysis*. The obvious point of departure for the analysis is the equation

$$\mathbf{P : Q = (sq.\,d - sq.\,a) : (sq.\,b - sq.\,d) = p : q} \qquad (*)$$

In this equation, the unknown (length of the) dividing line d appears in two places. The equivalent equation (in view of *El.* V.18)

$$(\text{sq.}\,d - \text{sq.}\,a) : (\text{sq.}\,b - \text{sq.}\,a) = p : (p + q)$$

is simpler in this respect, as the unknown d appears only once. In the next step, the term sq. d – sq. a is to be replaced by sq. d alone. For this purpose, the auxiliary length c is introduced, satisfying the equation

$$\mathbf{sq.\,c : sq.\,a = q : p} \qquad (**)$$

Then (in view of *El.* V.18)

$$(\text{sq.}\,d - \text{sq.}\,a) : (\text{sq.}\,b - \text{sq.}\,a) = p : (p + q) = \text{sq.}\,a : (\text{sq.}\,c + \text{sq.}\,a).$$

Hence, finally (in view of *El.* V.19),

$$\mathbf{sq.\,d : (sq.\,b + sq.\,c) = p : (p + q)} \qquad (***)$$

In this way, the relatively complicated equation (*) for the unknown length d can be replaced the simpler equation (**) for c and the equally simple equation (***) for d.

Summary. In the preceding section, the following problems from Euclid's *On Divisions* were discussed:

OD 1-2	To divide a *triangle* by parallels to the base in 2 or 3 equal parts.
OD 3	To divide a triangle by a line through a point on a side in 2 equal parts.
OD 4-5	To divide a *trapezoid* by 2 or 3 parallels to the base in equal parts.
OD 8, 12	To divide a trapezoid by a line through a point on a side in 2 equal parts.
OD 19-20	To divide f a triangle by a line through an interior point in 2 (equal) parts.
OD 32	To divide a trapezoid by a parallel to the base in a given ratio.

Of these problems, all the ones where the dividing lines are *parallel* to a side have Babylonian parallels (see below). The ones where the dividing line passes *through a given point on a side* are simple generalizations of problems with Babylonian parallels. Only the problems where the dividing line passes *through a given point in the interior* of the figure are not closely related to any Babylonian problems.

Note, by the way, that *the notion of an arbitrarily given point seems to have been completely unknown in Babylonian mathematics*.

The problems in *On Divisions* not discussed in Sec.1 1.1 above are:

OD 6-7	To divide a *parallelogram* by a line through a point on a side.
OD 9	To divide a trapezoid in a given ratio by a line through a point on a side.
OD 10-13	To divide a parallelogram by a line through an exterior point.
OD 14-17	To divide a *quadrilateral* in equal parts by a line through a given point.
OD 26-27	To divide a triangle in a given ratio by a line through an exterior point.
OD 28	To divide a figure composed of *a triangle and a circle segment*.
OD 29	To cut off a third, fourth, fifth, of a *circle* by two parallel chords.
OD 30-31	To divide a triangle in given ratios by parallels to the base in parts.
OD 33	To divide a trapezoid in given ratios by parallels to the base.
OD 34-36	To divide a quadrilateral in equal ratios by lines through a given point.

Most of these problems are not related to any known Babylonian mathematical problems. In particular, *the notion of an arbitrary parallelogram seems to have been completely unknown in Babylonian mathematics*.

11.2. Old Babylonian Problems for Striped Triangles

11.2 a. Str. 364 § 2. A model problem for a 3-striped triangle

Metric algebra problems for triangles or trapezoids divided into two or several "stripes" by *transversals parallel to the front* was a quite popular topic in OB mathematics. An interesting first example of a text with problems of this type is the well organized theme text **Str. 364**, probably from Uruk in southern Mesopotamia (Neugebauer, *MKT 1* (1935), 248; photo *MKT 2* (1935), pl. 11).

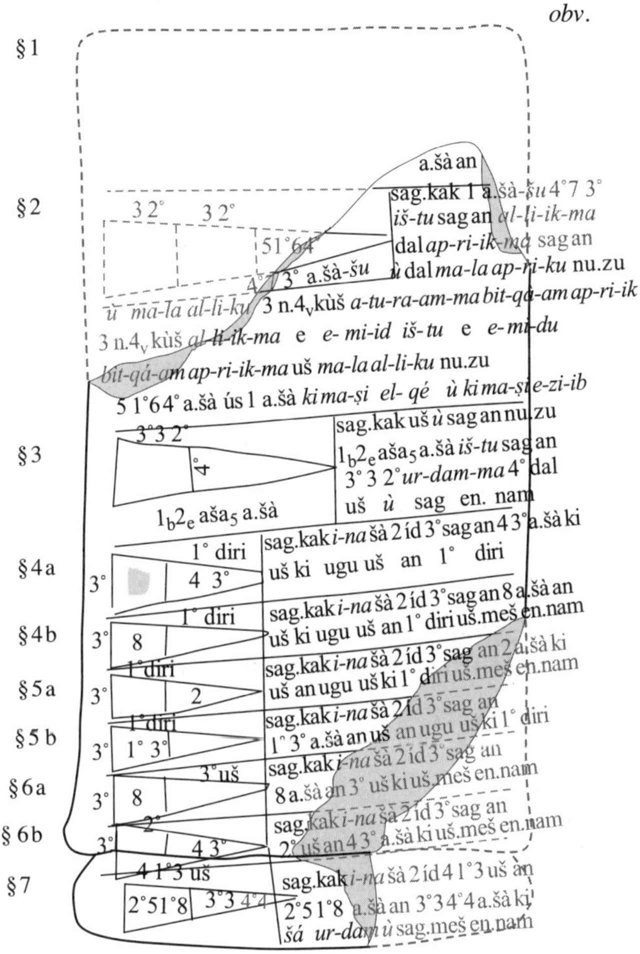

Fig. 11.2.1. Str. 364 *obv*. Metric algebra problems for 3- and 2-striped triangles.

As shown by the hand copy of the obverse of Str. 364 in Fig. 11.2.1 above, the first problem on the clay tablet is lost. The second problem, **Str. 364 § 2**, is only partly preserved. Nevertheless, a likely reconstruction of the text of § 2, as well as of the associated diagram, is presented below.

Str. 364 § 2

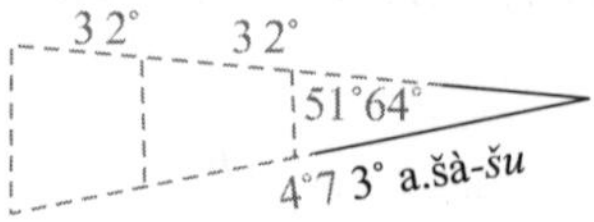

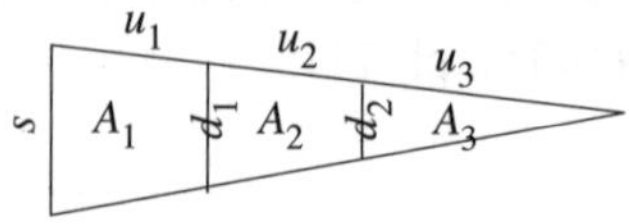

A peg-head. Fie*ld, 47 30*. From the upper front *I went,* a transversal I laid across.
The upper front and the transversal as much as I laid across I do not know,
but as much as I went, 3 ninda 4 cubits. I returned, and an opening I laid across.
3 ninda 4 cubits I went and a dike I installed. From the dike that I installed,
an opening I laid across, but the length as much as I went I do not know.
5 16 40 the next field. Fields, how much did I take and how much did I leave?

The pretense in this unusually explicit problem text is that a triangular field is divided, for irrigation purposes(?), into three canals(?) separated from each other by dikes. (There is no other known mathematical cuneiform text using similar terminology.) The two partial lengths u_1 = 3;20, u_2 = [3;20], and the areas $A = A_1 + A_2 + A_3$ = [47] 30 and A_3 = 5 16;40 are known. (See the notations used in the metric algebra diagram above, to the right.) No solution procedure is given in the text, but the remaining parameters for the divided field can be computed easily, one at a time, as below.

First, according to *the OB quadratic similarity rule for triangles*,

sq. $s = A / A_3 \cdot$ sq. $d_2 = 9$ sq. d_2 so that $s = 3\ d_2$.

Next, according to *the OB area rule for trapezoids*,

$s + d_2 = 2\ (A_1 + A_2) / (u_1 + u_2) = 2 \cdot (47\ 30 - 5;16\ 40) / 6;40 = 12;40$.

Therefore,

$4\ d_2 = 12;40$ so that $d_2 = 3;10$ and $s = 9;30$.

Then, according to *the OB area rule for triangles*,

$u_3 = 2A_3 / d_2 = 2 \cdot 5;16\ 40 / 3;10 = 3;20$.

This means that

$u_1 = u_2 = u_3 = 3;20$.

Therefore, obviously,

$d_1 = 2\ d_2 = 6;40$ and $A_1 = 5\ A_3 = 26;23\ 20$, $A_2 = 3\ A_3 = 15;50$.

Note also that the 'feed' f for the triangle can be computed as follows by use of *the OB linear similarity rule for triangles*:

$f = s / u = s / (u_1 + u_2 + u_3) = 9;30 / 10 = ;57$.

Apparently, § 2 (and the lost § 1) were chosen as easy introductory problems, illustrating both the area rules and the linear and quadratic similarity rules. § 2 is also a clear demonstration of what appears to have been an "OB given parameters rule":

> *In a triangle divided into n stripes* by transversals parallel to the front, there are n partial lengths, n partial areas, and n fronts (if the transversals are counted as fronts). These 3 n parameters are related to each other by n area equations and $n - 1$ similarity equations. Therefore, *the values for only n + 1 of the 3 n parameter can be given arbitrarily.*
> In particular, 3 parameters can be given for a triangle divided into 2 stripes (a "2-striped triangle"), 4 for a triangle divided into 3 stripes (a "3-striped triangle"), and so on.
> In Str. 364 § 2, $n = 3$ and $n + 1 = 4$.

11.2 b. Str. 364 § 3. A quadratic equation for a 2-striped triangle

An interesting series of intimately connected metric algebra problems for 2-striped triangles on the obverse of Str. 364 begins with **Str. 364 § 3**. All the problems are illustrated by diagrams (see Fig. 11.2.1 above).

> **Str. 364 § 3**
> A peg-head. The length and the upper front I do not know. 1 bùr 2 èše is the field.
> From the upper front 33 20 I went down, then 40 the transversal.
> Length and front are what?

In this example there are $n = 2$ stripes and $n + 1 = 3$ given values:

$u_a = 33;20$ (ninda), $d = 40$ (ninda), $A = 1$ bùr 2 èše = 50 00 sq. ninda.

(The bùr and the èše were OB area measures equal to 30 00 and 10 00 square ninda, respectively.).

No explicit solution procedures are given for the exercises in Str. 364. It is, however, definitely worthwhile to try to reconstruct the intended Babylonian solution procedures. Only in this way can one truly appreciate the mathematical sophistication of the OB mathematician who composed the systematically arranged series of problems in Str. 364 §§ 3-9, and the deep insight he must have had into the geometric and algebraic aspects of the problems he devised.

In Str. 364 § 3, an equation for the unknown front s in terms of the given quantities u_a, d, and A can be obtained through a combined application of the linear and quadratic similarity rules for a striped triangle. Thus, according to the *quadratic similarity rule*,

sq. $s = f \cdot 2\,A$, where $f = s/u$ is the (unknown) 'feed' for the triangle.

According to the *linear similarity rule*,

$s - d = f \cdot u_a$.

From these two equations together it follows that

sq. s / 2 $A = (s - d)/u_a$ (= f)

or

sq. $\mathbf{s = 2\,r \cdot (s - d)}$, where $r = A / u_a$ = 50 00 / 33;20 = 1 30.

This is a *quadratic equation for s*, which can be solved in the usual way by *a completion of the square*

$2\,r \cdot s$ – sq. $s = 2\,r \cdot d$ ⇒ sq. $(r - s)$ = sq. $r - 2\,r \cdot d$.

After a second completion of the square one then finds that[27]

sq. $\mathbf{(r - s)}$ + sq. $\mathbf{d}$ = sq. $\mathbf{(r - d)}$.

Therefore,

sq. (1 30 – s) + sq. 40= sq. 50.

Consequently,[28]

1 30 – s = 30 so that s = 1 00, and $u = 2\,A / s$ = 1 40.

Note that the computation above surprisingly showed that $r - d, d, r - s$ is a *diagonal triple*. Indeed,

$r - d, d, r - s$ = 50, 40, 30 = 10 · (5, 4, 3).

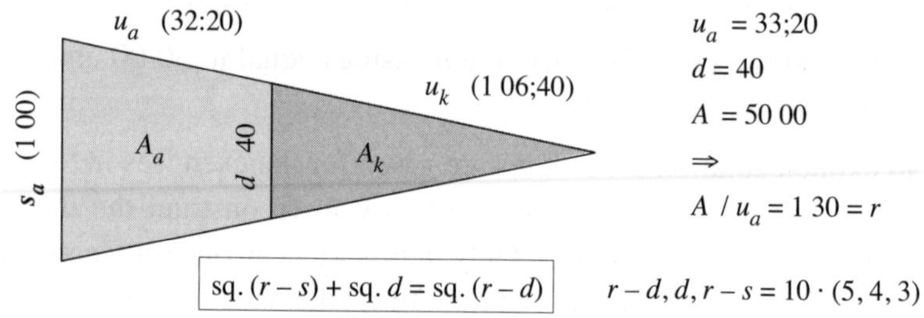

Fig. 11.2.2. Str. 364 § 3. The trick of making two completions of squares.

27.Here, the idea of making *a second completion of the square* in a situation of this kind is borrowed from the explicit solution procedure in **MS 3052** § 1 c, an interesting OB problem text dealing with a clay wall with a triangular cross section (see Friberg, *RC* (2007), Sec. 10.2).

28. The alternative s – 1 30 = 30 (see Neugebauer, *MKT 1* (1935), 255) must be rejected, because it leads to the solution s = 2 00, u = 1 40 00 / 2 00 = 50, with u less than s. In Babylonian mathematical texts, the length u is always greater than the front s.

The remaining parameters are easily calculated, for instance as follows:

u_k = 1 30 – 33;20 = 1 06;40, $f = s/u$ = 1 00 / 1 40 = 3/5 = ;36,
$u_a : u_k$ = 33;20 : 1 06 40 = 1 : 2, $A_a : A_k$ = (sq. 3 – sq. 2) : sq. 2 = 5 : 4.

11.2 c. Str. 364 §§ 4-7. Quadratic equations for 2-striped triangles

Str. 364 §§ 4-7 is a cleverly organized series of problems for 2-striped triangles, all leading to quadratic equations. It is easy to reconstruct the missing solution procedures.

Str. 364 § 4 a

A peg-head. Inside it two canals. 30 the upper front, 4 30 the lower field.
The lower field is 10 beyond the upper field.

The diagram illustrating this problem shows a 2-striped triangle with the given values

s = 30, A_k = 4 30, $u_k - u_a$ = 10.

The unknown values are those for the transversal d, the partial lengths u_a and u_k, and the upper area A_a. According to *the linear similarity rule*,

$d = f \cdot u_k$ and $s - d = f \cdot u_a$.

Subtracting here the terms of the second equation from those of the first equation one gets the new equation

$2\,d - s = f \cdot (u_k - u_a)$.

In addition, according to *the quadratic similarity rule*,

sq. $d = f \cdot 2\,A_k$.

A combination of the two equations above shows that

sq. d / 2 A_k = $(2\,d - s) / (u_k - u_a)$ (= f).

This quadratic equation for the transversal d can be reformulated as

2 $r \cdot d$ – sq. d = $r \cdot s$, where $r = 2\,A_k / (u_k - u_a)$ = 9 00 / 10 = 54.

Hence, *after two completions of squares*, as in Str. 364 § 3,

sq. ($r - d$) + sq. s/2 = sq. ($r - s$/2) so that sq. ($r - d$) = sq. 39 – sq. 15 = sq. 36.

Therefore, in this case, the triple $r - s/2, r - d, s/2$ is *a diagonal triple*, with

$r - s/2, r - d, s/2$ = 39, 36, 15 = 3 · (13, 12, 5).

Consequently, the solution to the problem in Str. 364 § 4 a is that

$d = r - 36$ = 54 – 36 = 18, f = sq. d / 2 A_k = ;36 (= 3/5), $u_k = d / f$ = 30,

$u_a = 30 - 10 = 20,\ \ u_a : u_k = 2 : 3,\ \ A_a : A_k = 16 : 9,\ \ A_a = 8\ 00.$

The next problem is a simple variant of § 4 a:

Str. 364 § 4 b

A peg-head. Inside it two canals. 30 the upper front, 8 (00) the upper field. The lower field is 10 beyond the upper field. The lengths are what?

This time, the quadratic equation for the transversal *d* has the form

$$(\text{sq.}\,s - \text{sq.}\,d)\,/\,2\,A_a = (2\,d - s)\,/\,(u_k - u_a)\quad (= f).$$

The equation can be reformulated as

$$\textbf{sq.}\,\boldsymbol{d} + \mathbf{2}\,\boldsymbol{r}\cdot\boldsymbol{d} = \textbf{sq.}\,\boldsymbol{s} + \boldsymbol{r}\cdot\boldsymbol{s},\quad \text{where}\quad r = 2\,A_a\,/\,(u_k - u_a) = 16\ 00\,/\,10 = 1\ 36.$$

After a completion of the square, the equation is reduced to

$$\text{sq.}\,(d + r) = \text{sq.}\,s + r\cdot s + \text{sq.}\,r = 3\ 36\ 36 = \text{sq.}\ 1\ 54.$$

Therefore, $d = 1\ 54 - 1\ 36 = 18$, *etc.*, as in § 4 a.

Str. 364 § 5 a

A peg-head. Inside it two canals. 30 the upper front, 2 (00) the lower field. The upper length is 10 *beyond* the lower length. The lengths are what?

In this case, the upper length is greater than the lower length. The equation for *d* is modified accordingly (compare with the equation in § 4 a):

$$\text{sq.}\,d\,/\,2\,A_k = (s - 2\,d)\,/\,(u_a - u_k)\quad (= f).$$

Consequently, the quadratic equation for *d* becomes

$$\textbf{sq.}\,\boldsymbol{d} + \mathbf{2}\,\boldsymbol{r}\cdot\boldsymbol{d} = \boldsymbol{r}\cdot\boldsymbol{s},\quad \text{where}\quad r = 2\,A_k\,/\,(u_a - u_k) = 4\ 00\,/\,10 = 24.$$

After two completions of squares, the equation is reduced to the form

$$\textbf{sq.}\,(\boldsymbol{r} + \boldsymbol{d}) + \textbf{sq.}\,\boldsymbol{s}/\mathbf{2} = \textbf{sq.}\,(\boldsymbol{r} + \boldsymbol{s}/\mathbf{2})\quad \text{so that}\quad \text{sq.}\,(r + d) = \text{sq.}\ 39 - \text{sq.}\ 15 = \text{sq.}\ 36.$$

Therefore, in this case, $r + s/2,\ r + d,\ s/2$ is a diagonal triple, with

$$r + s/2,\ r + d,\ s/2 = 39, 36, 15 = 3\cdot(13, 12, 5).$$

It follows that

$$d = 36 - 24 = 12,\ \ f = \text{sq.}\,d\,/\,2\,A_k, = ;36\ (= 3/5),\ \ u_k = d\,/\,f = 20,$$
$$u_a = 20 + 10 = 30,\ \ u_a : u_k = 3 : 2,\ \ A_a : A_k = 21 : 4,\ \ A_a = 10\ 30.$$

Str. 365 § 5 b has the same relation to § 4 b as § 5 a has to § 4 c.

The text of **Str. 364 § 6 a** is lost, but the illustrating diagram is perfectly preserved. The diagram shows a 2-striped triangle in which

$$s = 30,\ \ A_a = 8\ (00),\ \ u_k = 30.$$

Then

$(s + d) \cdot u_a = 2\,A_a, \quad (s - d) / u_a = d / u_k.$

Consequently, d is the solution to the following quadratic equation:

$(\text{sq.}\, s - \text{sq.}\, d) = 2\,A_a \cdot d / u_k.$

Equivalently,

$\textbf{sq.}\, \boldsymbol{d + r \cdot d} = \textbf{sq.}\, s,$ where $r = 2\,A_a / u_k = 16\;00 / 30 = 32.$

After a completion of the square, this equation is reduced to

$\textbf{sq.}\, \boldsymbol{(d + r/2)} = \textbf{sq.}\, s + \textbf{sq.}\, \boldsymbol{r/2} = \text{sq.}\, 30 + \text{sq.}\, 16 = \text{sq.}\, 34.$

Therefore, in § 6 the triple $d + r/2, s, r/2$ is a diagonal triple, with

$d + r/2, s, r/2 = 34, 30, 16 = 2 \cdot (17, 15, 8).$

It follows that $d = 18$, *etc*., precisely as in § 4.

Also the text of **Str. 364 § 6 b** is lost, while the illustrating diagram is perfectly preserved. The diagram shows a 2-striped triangle in which

$s = 30, \quad A_k = 4\;30, \quad u_a = 20.$

Then,

$d \cdot u_k = 2\,A_k, \quad d / u_k = (s - d) / u_a.$

In this case, the quadratic equation for the transversal d is

$\text{sq.}\, d = 2\,A_k \cdot (s - d) / u_a.$

Equivalently,

$\textbf{sq.}\, \boldsymbol{d + r \cdot d = r \cdot s},$ with $r = 2\,A_k / u_a = 9\;00 / 20 = 27.$

After two completions of squares, this equation becomes

$\textbf{sq.}\, \boldsymbol{(d + r/2)} + \textbf{sq.}\, s = \textbf{sq.}\, \boldsymbol{(s + r/2)}$ so that $\text{sq.}\, (d + r/2) = \text{sq.}\, 43;30 - \text{sq.}\, 30 = \text{sq.}\, 31;30.$

Therefore, in § 7 the triple $s + r/2, d + r/2, s$ is a diagonal triple, with

$s + r/2, d + r/2, s = 43;30, 31;30, 30 = 1;30 \cdot (29, 21, 20).$

It follows that $d = 18$, *etc*., again precisely as in § 4.

The last exercise on the obverse is **Str. 364 § 7**.

The diagram shows a 2-striped triangle in which

$A_a = 25\;18, \quad A_k = 33\;[44], \quad u_a = 4\;13.$

In spite of the complicated form of the given numbers, the solution procedure is uncomplicated. (*Cf.* the solution procedure in the case of the related exercise Str. 364 § 2, above.) Note that

$A_a = 25\;18 = 6 \cdot 4\;13$ and $A_k = 33\;44 = 8 \cdot 4\;13$ (where $4\;13 = 23 \cdot 11$).

Therefore, according to *the OB quadratic similarity rule*

$\text{sq.}\, s = (A_a + A_k) / A_k \cdot \text{sq.}\, d = (6 + 8) / 8 \cdot \text{sq.}\, d = 7 / 4 \cdot \text{sq.}\, d.$

Consequently,

$s = \text{sqs.}\, 7 / 2 \cdot d,$ where $\text{sqs.}\, 7 =$ appr. $8/3 = 2;40$.

Also, according to *the OB area rule for trapezoids*,

$s + d = 2\, A_a / u_a = 12.$

Consequently,

$s = 4\,(7 - 2\ \text{sqs.}\, 7) =$ appr. $20/3 = 6;40$ and $d = 8\,(\text{sqs.}\, 7 - 2) =$ appr. $16/3 = 5;20$.

And so on. Note that this exercise has *intentionally* been made more complicated than necessary, by arbitrarily introducing the scale factor 4 13 = 23 · 11, and by choosing the data so that the numbers in the answer are non-rational, being expressed in terms of the square side of 7.

11.2 d. Str. 364 § 8. Problems for 5-striped triangles

On the reverse of Str. 364 (Fig. 11.2.3 below) there are the well preserved texts of three problems, § 8a-c, illustrated by diagrams showing a triangle divided into 5 stripes. Hence there can be *6 arbitrarily given values* for each such divided triangle. In all three cases, two of the given values are

$A_1 = 18\ 20$ and $A_2 = 15\ (00)$.

In § 8 a and § 8 c, two further given values are

$s_1 - s_2 = 13;20,\quad s_2 - s_3 = 13;20.$

In § 8 c, which alone of the three problems on the reverse will be discussed below, the last two given values are

$A_4 = 13;20$ and $(s_5 + s_6)/2 = 26;40$.

Str. 364 § 8 c

A peg-head. Inside it 5 canals. The upper field 18 20, the 2nd field 15, the 3rd field I do not know, the 4th field 13 20, at half 26;40, the 5th field I do not know. The upper front over the transversal is 13;20 beyond, transversal over transversal is 13;20 beyond. The fields and the lengths and the transversals are what?

There are 15 parameters for this 5-striped triangle, 5 partial areas, 5 partial lengths, and 5 parallels. Of these 15 parameters, 3 are explicitly given:

$A_1 = 18\ 20,\quad A_2 = 15\ (00),$ and $A_4 = 13\ 20$.

There are also 3 equations that are explicitly given:

$s_1 - s_2 = 13;20$, $s_2 - s_3 = 13;20$, and $(s_4 + s_5)/2 = 26\ 40$.

In addition to these 6 arbitrarily imposed conditions, there are the usual 5 area equations and 4 similarity equations for a 5-striped triangle. Altogether, there are *12 equations that have to be satisfied simultaneously by 12 unknowns* namely 3 partial areas, 4 partial lengths, and 5 parallels. This looks like quite a formidable system of equations to be solved.

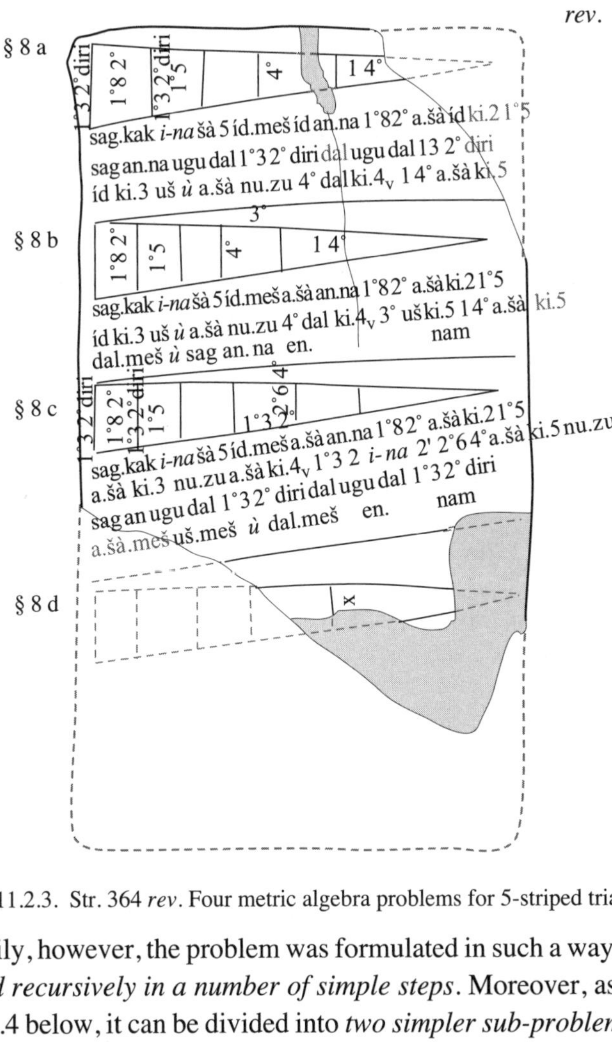

Fig. 11.2.3. Str. 364 *rev*. Four metric algebra problems for 5-striped triangles.

Happily, however, the problem was formulated in such a way that it can be *solved recursively in a number of simple steps*. Moreover, as shown in Fig. 11.2.4 below, it can be divided into *two simpler sub-problems*, one for a 2-striped trapezoid, and one for a 3-striped triangle.

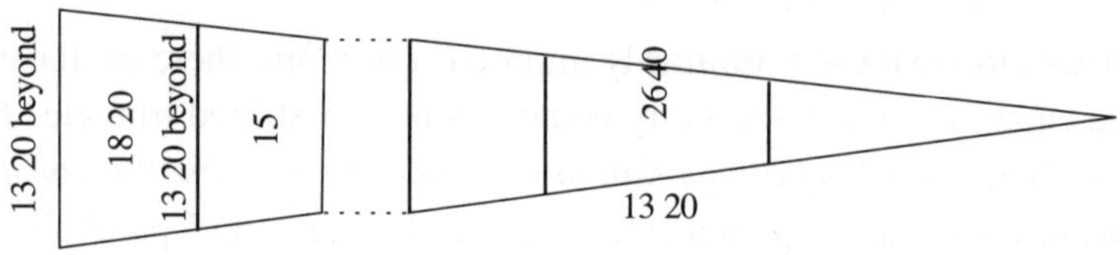

Fig. 11.2.4. Str. 364 § 8 c. A system of simultaneous equations for 12 unknowns.

The first sub-problem is concerned with a 2-striped (parallel) trapezoid, where the two partial areas and two differences are given:

$A_1 = 18\ 20,\ \ A_2 = 15\ (00),$ and $s_1 - s_2 = 13;20,\ \ s_2 - s_3 = 13;20.$

(Note that since a 2-striped trapezoid can be interpreted as part of a 3-striped triangle, there can be 3 + 1 = 4 arbitrarily imposed conditions on the parameters of such a trapezoid.)

Str. 364 § 8 c, proposed reconstruction of the intended solution procedure, part 1

Step 1: *An application of the linear similarity rule for triangles* shows that

$u_1 / u_2 = (s_1 - s_2)/(s_2 - s_3) = 13;20 / 13;20 = 1,$ so that $u_1 = u_2.$

Step 2: *Two applications of the area rule for trapezoids* show that

$A_1 - A_2 = \{(s_1 + s_2) - (s_2 + s_3)\}/2 \cdot u_1,$ because $u_1 = u_2.$

Then also

$A_1 - A_2 = \{(s_1 - s_2) + (s_2 - s_3)\}/2 \cdot u_1,$ so that $3\ 20 = 13;20 \cdot u_1$

Consequently,

$u_1 = u_2 = 3\ 20 / 13;20 = 15.$

Step 3: *A renewed application of the area rule for trapezoids* shows that

$(s_1 + s_2)/2 = A_1 / u_1 = 18\ 20 / 15 = 1\ 13;20,$ while

$(s_1 - s_2)/2 = 6;40.$

Consequently,

$s_1 = 1\ 13;20 + 6;40 = 1\ 20,\ \ s_2 = 1\ 13;20 - 6;40 = 1\ 06;40,$

and

$s_3 = 1\ 06;40 - 13;20 = 53;20.$

Step 4: *A renewed application of the linear similarity rule* shows that

$f = (s_1 - s_2)/u_1 = 13;20 / 15 = 8/9 = ;53\ 20.$

The second sub-problem is concerned with a 3-striped triangle, for which the following 4 values are given:

$A_4 = 13\ 20,\ \ (s_4 + s_5)/2 = 26;40,$

$s_3 = 53;20$ (by step 3), $f = 8/9 = ;53\ 20$ (by step 4).

Str. 364 § 8 c, proposed reconstruction of the intended solution procedure, part 2

Step 5: An *application of the area rule for trapezoids* shows that
$u_4 = A_4 / (s_4 + s_5)/2$ = 13 20 / 26;40 = 30.

Step 6: *An application of the linear similarity rule for triangles* shows that
$(s_4 - s_5)/2 = f \cdot u_4 /2$ = ;53 20 · 15 = 13;20.
Consequently,
s_4 = 26;40 + 13;20 = 40, s_5 = 26;40 – 13;20 = 13;20.

Step 7: *Two renewed applications of the linear similarity rule* show that
$u_3 = (s_3 - s_4) / f$ = (53;20 – 40) / ;53 20 = 15, $u_5 = s_5 / f$ = 13;20 / ;53 20 = 15.

Step 8: Two renewed *application of the area rule* show that
A_3 = (53;20 + 40)/2 · 15 = 11 40, A_5 = 13;20 /2 · 15 = 1 40.

Combining all the results in Steps 1-8, one finds that

the 5 partial lengths are 15, 15, 15, 30, 15, in the ratios 1 : 1 : 1 : 2 : 1.
the 5 parallels are 1 20, 1 06;40, 53;20, 40, 13;20, in the ratios 6 : 5 : 4 : 3 : 1,
the 5 partial areas are 18 20, 15, 11 40, 13 20, 1 40, in the ratios 11 : 9 : 7 : 8 : 1
the whole length of the triangle is 1 30, and the whole area is 1 00 00.

11.2 e. *TMS* 18. A cleverly designed problem for a 2-striped triangle

In the case of the triangle in Str. 364 §§ 4 and 6, the partial lengths 20, 30 are in the ratio 2 : 3, the partial areas 8 00 and 4 30 are in the ratio 16 : 9, and the front and the transversal 30 and 18 are in the ratio 5 : 3. This particular 2-striped triangles occurs also in at least one other OB metric algebra problem (discussed below). The reason for the apparent popularity of this 2-striped triangle may have been that not only u_a = 30, u_k = 20, are *regular* sexagesimal numbers, but so are also the difference $u_a - u_k$ = 10 and the sum $u_a + u_k$ = 50. Furthermore, there is the fortuitous circumstance that $A : A_a : A_k$ = 25 : 16 : 9 = sq. 5 : sq. 4 : sq. 3.

***TMS* 18** (Bruins and Rutten, *TMS* (1961)) is a fragment of a small clay tablet from the ancient city Susa (Western Iran) with a single OB mathematical exercise. It is one of the further texts where the mentioned 2-striped triangle appears. The statement of the problem in *TMS* 18 is preceded by a small drawing of a divided triangle. In Fig. 11.2.5 below, there is a larger drawing of the divided triangle. The values within brackets were probably computed in the course of the solution procedure, of which, however, the larger part is destroyed. Actually, the lower two-thirds, or so, of the clay tablet are lost

***TMS* 18**, literal transliteration	explanation
The lower length to the upper length *frame, 10,*	$u_k \cdot u_a = 10\ (00)$
the upper field to the lower field frame, 36,	$A_a \cdot A_k = 36\ (00\ 00)$
the *upper front* frame, the transversal frame,	sq. s_a + sq. d
sum 20 24.	= 20 24
You:	Procedure:
36 that field with field was framed to 4 go,	$4 \cdot A_a \cdot A_k$
then 2 24 you see.	$= 4 \cdot 36 = 2\ 24$
The opposite of 10 that descent with descent	$1/\ u_k \cdot u_a = 1/10$
was framed resolve, 6 you see.	= ;06
2 24 to 6 raise, then 14 24 you see.	$4 \cdot A_a \cdot A_k\ /\ u_k \cdot u_a = 2\ 24\ /\ 10 = 14\ 24$
14 24 frame, 3 27 21 *36* you see.	sq. 14 24 = 3 27 21 36
To 2 raise (it),	$2 \cdot$ sq. 14 24 =
6 54 43 12 you see.	6 54 43 12
Return. 14 24 to 2 rai*se, 28 48 you see.*	$2 \cdot 14\ 24 =$ *28 48*
....	....
1 10	
.... 30 you see	$u_k = 30$
..... *to* ..... raise, then 20 you see.	$u_a = 20$

Only the beginning and the last couple of lines of the solution procedure for *TMS* 18 are preserved. Luckily, enough of the solution procedure is preserved to confirm the correctness of the explanation below.[29]

Given in *TMS* 18 are the product $u_k \cdot u_a = 10\ (00)$ of the partial lengths, the product $A_a \cdot A_k = 36\ (00\ 00)$ of the partial areas, and the sum of the squares of the front and the transversal, sq. s_a + sq. $d = 20\ 24$.

The partial areas of the divided triangle can be expressed as follows:

$$A_a = u_a \cdot (s_a + d)/2, \quad A_k = u_k \cdot d/2.$$

Therefore, the division in the first step of the solution procedure in *TMS* 18 can be explained as the computation of

$$4 \cdot A_a \cdot A_k\ /\ u_k \cdot u_a = (s_a + d) \cdot d = 14\ 24.$$

This equation, together with the third of the three given equations leads to

29. The badly flawed explanation proposed by Bruins and Rutten (*TMS* (1961)) in the original publication of the text was based on an mistaken restoration of a broken part of the statement and on a lacking understanding of the conventions of OB geometry.

the following system of equations for the unknowns s_a and d:

$(s_a + d) \cdot d = R = 14\ 24, \quad \text{sq.}\, s_a + \text{sq.}\, d = S = 20\ 24.$

This system of equations seems to have been reduced in the following way to a more familiar-looking system of equations:

$\text{sq.}\,(s_a + d) + 2\ \text{sq.}\, d = S + 2\,R = 49\ 12, \quad (s_a + d) \cdot d = R = 14\ 24.$

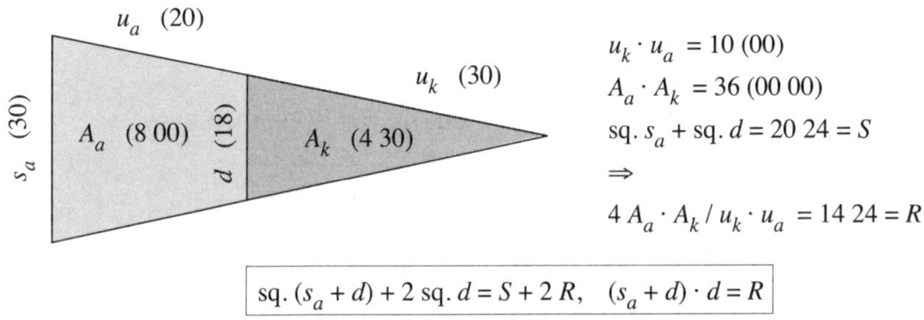

Fig. 11.2.5. *TMS* 18. A cleverly designed metric algebra problem for a divided triangle.

This is *a modified quadratic-rectangular system of equations of type B5* for the pair of unknowns $s_a + d$, d. (*Cf.* Figs. 5.4.1-2 above.) It is not a "basic" system of this type, since the coefficient for sq. d in the first equation is 2, not 1. In this respect, the modified system above of type B5 is similar to the modified system of type B6 with which, apparently, *Data* 86 is concerned. See Sec. 10.6 above.

It is likely that the OB author of *TMS* 18 solved the indicated system of type B5 by use of a method closely related to the better preserved solution method for a similar problem in the OB text BM 13901 # 12 (Sec. 5.4 above). Thus, he probably set

$\text{sq.}\,(s_a + d) = a, \quad 2\ \text{sq.}\, d = b.$

In this way he could reduce the mentioned system of type B5 for the pair $s_a + d$, d to a simpler system of type B1a for the pair a, b:

$a \cdot b = 2\ \text{sq.}\, R = 6\ 54\ 43\ 12, \quad a + b = S + 2\,R = 49\ 12.$

(Note that both 2 sq. B and 2 B are computed in the preserved beginning of the solution procedure in *TMS* 18!)

The system of equations for a, b can then be solved in the usual way:

$(a + b)/2 = 49\ 12 / 2 = 24\ 36, \quad \text{sq.}\,(a + b)/2 = 10\ 05\ 09\ 36,$
$\text{sq.}\,(a - b)/2 = 10\ 05\ 09\ 36 - 6\ 54\ 43\ 12 = 3\ 10\ 26\ 24, \quad (a - b)/2 = 13\ 48,$

$a = 24\ 36 + 13\ 48 = 38\ 24, \quad b = 24\ 36 - 13\ 48 = 10\ 48.$

Consequently,

$\text{sq.}\,(s_a + d) = 38\ 24$ and $2\ \text{sq.}\,d = 10\ 48$, so that
$s_a + d = 48, \quad d = 18$, and $s = 30$.

Now, when s_a and d are known, u_{a} and u_{k} can be computed as follows, by use of what may be called the "form and magnitude rule":

$u_{\text{k}} / u_{\text{a}} = d / (s_a - d) = 18/12 = 1;30, \quad u_{\text{k}} \cdot u_{\text{a}} = 10\ 00$, hence
$\text{sq.}\,u_{\text{k}} = 10\ 00 \cdot 1;30 = 15\ 00, \quad u_{\text{k}} = 30, \quad u_{\text{a}} = 30 / 1;30 = 20.$

The final computation of $A_{\text{a}} = 8\ 00$ and $A_{\text{k}} = 4\ 30$ is then easy.

11.2 f. MLC 1950. An elegant solution procedure

MLC 1950 (Neugebauer and Sachs, *MCT* (1945) text Ca) is a single problem text, probably from Uruk. The partial lengths of a 2-striped triangle are 20, 30 as in the case *TMS* 18 and Str. 364 §§ 4, 6, but the triangle is narrower, with a feed *f* of only ;24 (2/5), instead of ;36 (3/5).

In this problem, the front and the transversal of the divided triangle are called in Sumerian sag an 'the upper front' and sag ki 'the lower front'. The solution procedure is short and elegant, making use of a surprising equation for the half-difference $(s_{\text{a}} - s_{\text{k}})/2$. (Here s_{a} and s_{k} are suitable notations for sag an and sag ki.)

MLC 1950, literal translation — explanation

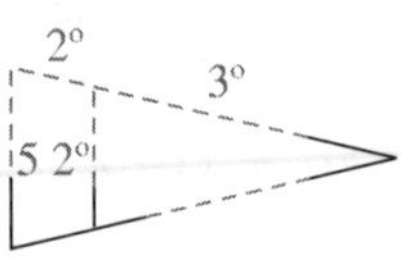

MLC 1950, literal translation	explanation
A peg-head.	A triangle
20 ninda the *upper* length, 5 20 its field,	$u_{\text{a}} = 20$ n., $A_a = 5\ 20$
30 ninda the [·····].	$u_{\text{k}} = 30$ n.
The upper front and the lower front are what?	$s_{\text{a}}, s_{\text{k}} = ?$
You in your doing (it).	Procedure:
The opposite of 20 resolve, 3 you see.	$1 / u_{\text{a}} = 1/20 = ;03$
3 to 5 20 raise, then 16.	$A_a / u_{\text{a}} = 5\ 20 \cdot ;03 = 16$
16 to the upper front and [·····].	????
30 the length to 2 repeat, 1,	$2 \cdot u_{\text{k}} = 2 \cdot 30 = 1\ 00$
and (with) 20 the upper descent sum (it), 1 20.	$u_{\text{a}} + 2 \cdot u_{\text{k}} = 1\ 20$
The opposite of 1 20 is 45,	$1 / 1\ 20 = ;00\ 45$

to 5 20 the field *raise it, then 4.*	$A_a / (u_a + 2 \cdot u_k) = 5\ 20 / 1\ 20 = 4$
4 to 16 add, from 16 tear off,	$16 + 4\ (= 20),\ 16 - 4\ (= 12)$
20 *the upper front, 12 the lower front.*	$s_a = 20,\ s_k = 12$

In metric algebra notations:

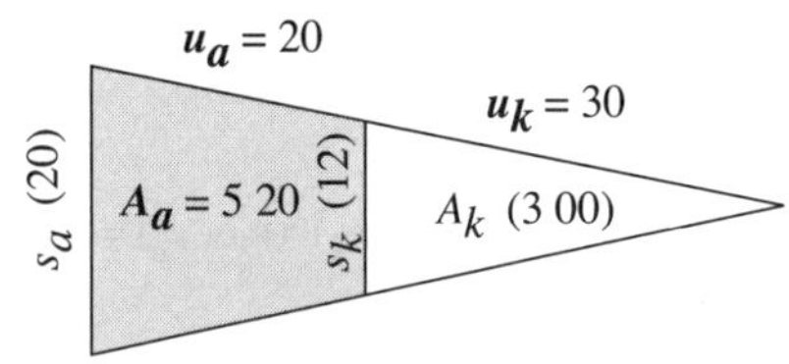

$(s_a + s_k)/2 = \mathbf{A_a} / \mathbf{u_a} = 16$
$(s_a - s_k)/2 = \mathbf{A_a} / (\mathbf{u_a} + 2\ \mathbf{u_k}) = 4$

Fig. 11.2.6. MLC 1950. An elegant solution procedure.

Given are the 'upper area' $A_a = 5\ 20$, and the two 'descents' or 'lengths' $u_a = 20, u_k = 30$. The first, easy step of the solution procedure is to compute

$$(s_a + s_k)/2 = A_a / u_a = 5\ 20 / 20 = 16.$$

(Here u_a and A_a are the known values of the upper length and the upper area.) The second step is more unexpected, with the computation of

$$(s_a - s_k)/2 = A_a / (u_a + 2\ u_k) = 5\ 20 / 1\ 20 = 4.$$

The latter equation can b explained as follows, for instance: Let f be the 'feed' of the divided triangle, the ratio of the front to the length. Then, by *the OB linear similarity rule*,

$$s_a = f \cdot (u_a + u_k), \quad s_k = f \cdot u_k \quad \text{and} \quad s_a - s_k = f \cdot u_a.$$

Then also[30]

$$s_a + s_k = f \cdot (u_a + 2\ u_k) \quad \text{and consequently} \quad \mathbf{(s_a - s_k) / (s_a + s_k) = u_a / (u_a + 2\ u_k)}.$$

Therefore, as in the second part of the solution procedure in MLC 1950,

$$\mathbf{(s_a - s_k)/2} = (s_a + s_k)/2 \cdot u_a / (u_a + 2\ u_k) = \mathbf{A_a / (u_a + 2\ u_k)}.$$

11.2 g. VAT 8512. Another cleverly designed problem

VAT 8512 (Høyrup, *LWS* (2002), 234-238) is a problem text with a single exercise closely related to the theme of Str. 364 § 4. In VAT 8512 is considered a striped triangle with a single transversal. With the usual

30. The argumentation here is, of course, related to the manipulation of proportions in Euclid's *Elements* V, for instance in ***El.* V. 19**, which says, essentially, that if $a : b = c : d$, where c and d are parts of a and b, then also $(a - c) : (b - d) = a : b$.

notations, the given parameters are

$$s = 30 \text{ n.}, \quad u_k - u_a = 20 \quad A_a - A_k = 7\ (00).$$

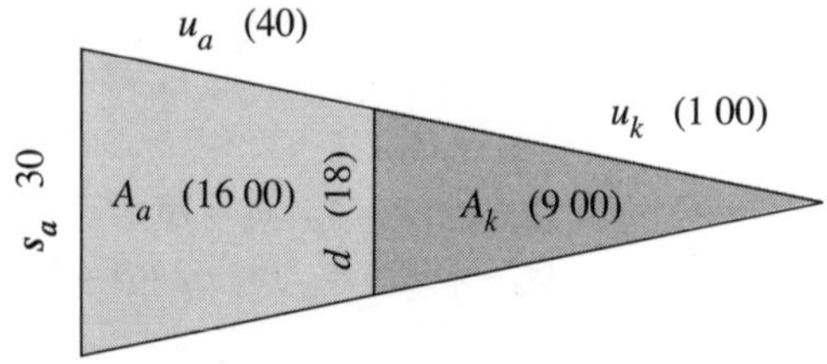

$s = 30$

$u_k - u_a = 20$

$A_a - A_k = 7\ (00)$

$\Rightarrow$

$(A_a - A_k)/(u_k - u_a) = 21 = r$

$\text{sq.}\,(d + r) = \{\text{sq.}\,(s + r) + \text{sq.}\,r\}/2 \qquad s + r, d + r, r = 3 \cdot (17, 13, 7)$

Fig. 11.2.7. VAT 8512. Another clever metric algebra problem for a 2-striped triangle.

In metric algebra notations, the first step of the explicit solution procedure is the computation of the transversal, here called *pirkum* 'crossline', as

$$d = \text{sqs.}\,[\{\text{sq.}\,(s + r) + \text{sq.}\,r\}/2] - r, \quad \text{where} \quad r = (A_a - A_k) / (u_k - u_a) = 7\ 00 / 20 = 21.$$

This solution formula may have been found as follows, by use of the *linear and quadratic similarity rules*: If f is the 'feed' for the triangle, then

$$s - d = f \cdot u_a \quad \text{and} \quad d = f \cdot u_k \quad \Rightarrow \quad 2d - s = f \cdot (u_k - u_a).$$

(*Cf.* the discussion of MLC 1950 in Sec. 11.2 f, in particular footnote 27.) Similarly,

$$\text{sq.}\,s - \text{sq.}\,d = f \cdot 2\,A_a \quad \text{and} \quad \text{sq.}\,d = f \cdot 2\,A_k \quad \Rightarrow \quad \text{sq.}\,s - 2\ \text{sq.}\,d = f \cdot 2\,(A_a - A_k).$$

Combining these two results, one finds that

$$\text{sq.}\,s - 2\ \text{sq.}\,d = 2\,r \cdot (2d - s), \quad \text{where} \quad r = (A_a - A_k) / (u_k - u_a).$$

Hence, the value of d can be found as the solution to the quadratic equation

$$\mathbf{sq.}\,\boldsymbol{d} + \mathbf{2}\,\boldsymbol{r} \cdot \boldsymbol{d} = (\mathbf{sq.}\,\boldsymbol{s} + \mathbf{2}\,\boldsymbol{r} \cdot \boldsymbol{s})/\mathbf{2}.$$

After two completions of squares, this equation can be written in the form

$$\mathbf{sq.}\,(\boldsymbol{d} + \boldsymbol{r}) = \{\mathbf{sq.}\,(\boldsymbol{s} + \boldsymbol{r}) + \mathbf{sq.}\,\boldsymbol{r}\}/\mathbf{2}, \quad \text{with} \quad r = (A_a - A_k) / (u_k - u_a).$$

(More about this below, in Sec. 11.3 b.) With the given numerical values,

$$r = 21 \quad \text{and} \quad \text{sq.}\,(d + 21) = (\text{sq.}\,51 + \text{sq.}\,21)/2 = (43\ 21 + 7\ 21)/2 = 25\ 21 = \text{sq.}\,39.$$

Hence,

$$d + 21 = 39 \quad \text{so that} \quad d = 18.$$

The remaining unknown values are computed as follows:

$f = (A_a - A_k) / (1/2 \cdot \text{sq.}\, s - \text{sq.}\, d) = 7\ 00 / 2\ 06 = 3;20\ (= 10/3).$

(See the equation above for sq. s – 2 sq. d.) Hence,

$u_a = f \cdot (s - d) = 3;20 \cdot (30 - 18) = 40,$
$A_a = (s + d)/2 \cdot u_a = 16,$
$u_k = u_a + (u_k - u_a) = 40 + 20 = 1\ 00,$
$A_k = d/2 \cdot u_k = 9 \cdot 1\ 00 = 9\ 00.$

Note that the triangle in VAT 85 12 is again, as the triangles in Str. 364 §§ 4, 6, and in *TMS* 18, a 2-striped triangle in which the partial lengths are in the ratio 2 : 3, and the partial areas in the ratio 16 : 9.

11.2 h. YBC 4696. A series of problems for a 2-striped triangle

YBC 4696 (Neugebauer, *MKT 2* (1935) 60-64, *MKT 3* (1937) pl. 4) is a *series text* like YBC 4709 (Sec. 10.8 above) with 52 metric algebra problems for the 2-striped triangle known from Str. 364 §§ 4 and 6.

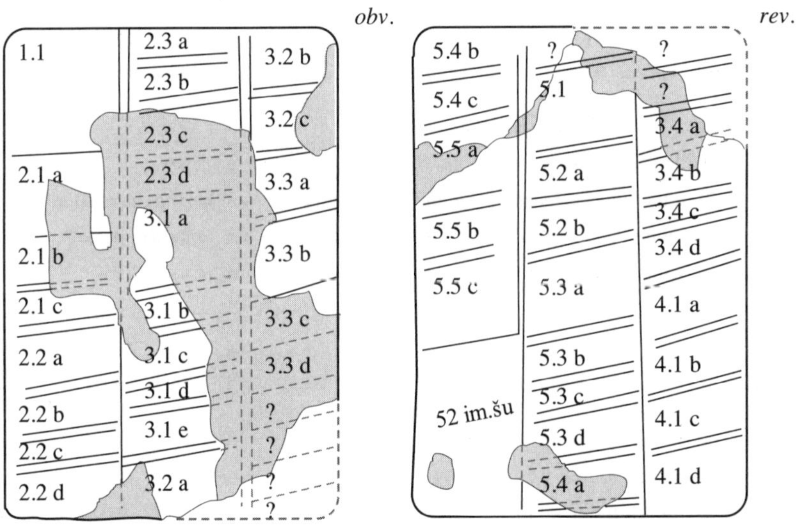

Fig. 11.2.8. YBC 4696. A series text with problems for a 2-striped triangle.

	YBC 4696, literal translation	explanation
1.1	A peg-head, 50 n. the length, 30 n. the front.	u = 50 n., s = 30 n.
	Inside it 2 canals.	2 stripes

	20 n. the upper descent, 30 n. the lower descent.	$u_a = 20$ n., $u_k = 30$ n.
	The fields of the 2 canals are what?	A_a and A_k = ?
	8 the upper field, 4 30 the lower field.	$A_a = 8\ (00)$, $A_k = 4\ 30$
2.1 a	A peg-head, 50 n. the length.	$u = 50$ n.
	Inside it 2 canals.	2 stripes
	20 n. the upper descent, *30 n. the lower descent.*	$u_a = 20$ n., $u_k = 30$ n.
	Front over transversal 12 n.	$s - d = 12$ n.
2.1 b	Half the front and 3 n. is the *transversal.*	$s/2 + 3$ n. $= d$
2.1 c	A 3rd of *the front and 8 n.* is the transversal.	$s/3 + 8$ n. $= d$
2.2 a	A 3rd of front over transversal to the front add, 34.	$s + (s - d)/3 = 34$
2.2 b	Times 2 repeat, (to) the front add, 38.	$s + 2\ (s - d)/3 = 38$
2.2 c	(From) the front tear off, 26.	$s - (s - d)/3 = 26$
2.2 d	Times 2 repeat, (from) the front tear off, 22.	$s - 2\ (s - d)/3 = 22$
2.3 a	(To) the transversal add, 22.	$d + (s - d)/3 = 22$
2.3 b	Times 2 repeat, (to) the transversal add, 26.	$d + 2\ (s - d)/3 = 26$
2.3 c	(From) the transversal tear off, 14.	$d - (s - d)/3 = 14$
2.3 d	*To 2 repeat, (from) the transversal tear off, 10.*	$d - 2\ (s - d)/3 = 10$
3.1 a	A peg-head, *50 n. the length.*	$u = 50$ n.
	Inside it *2 canals.*	2 stripes
	20 n. the *upper descent*, 30 *n. the lower descent.*	$u_a = 20$ n., $u_k = 30$ n.
	The upper field is 8. *The lower field is* what?	$A_a = 8\ (00)$. A_k = ?
3.1 b	The front to the *upper field* add, 8 *30.*	$A_a + s = 8\ 30$
3.1 c	Times 2 repeat, add, 9.	$A_a + 2\ s = 9\ (00)$
3.1 d	Tear off, 7 *30.*	$A_a - s = 7\ 30$
3.1 e	Times 2 repeat, tear off, 7.	$A_a - 2\ s = 7\ (00)$
3.2 a	The field of the front (to) the *field* of the upper canal *add*, 23.	$A_a +$ sq. $s = 23\ (00)$
3.2 b	The field of the front times 2 *repeat*, (to) the upper field add, *38.*	$A_a + 2$ sq. $s = 38\ (00)$
3.2 c	The field of the front beyond the upper field, 7.	sq. $s - A_a = 7\ (00)$
3.3 a	The front and the field of the front (to) the field of the upper canal add, 23 30.	$A_a + s +$ sq. $s = 23\ 30$
3.3 b	The front times 2 repeat and the field of the front (to) the field of the upper canal add, 24	$A_a + 2\ s +$ sq. $s = 24\ (00)$

.....	....	....
3.4 a	The transversal to the field of the upper canal add, 8 18.	$A_a + d = 8\ 18$
3.4 b	The transversal times 2 repeat, add, 8 36.	$A_a + 2\,d = 8\ 36$
3.4 c	Tear off, 7 42.	$A_a - d = 7\ 42$
3.4 d	Times 2 repeat, tear off, 7 24.	$A_a - 2\,d = 7\ 24$
4.1 a	The field of the transversal (to) the field of the lower! canal add, 9 54.	A_k + sq. $d = 9\ 54$
4.1 b	Times 2 repeat, add, 15 18.	A_k + 2 sq. $d = 9\ 54$
4.1 c	The field of the transversal over the lower field, 54 beyond.	sq. $d - A_k = 54$
4.1 d	The field of the transversal to 2 repeat, over the lower field, 6 18 beyond.	2 sq. $d - A_k = 6\ 18$
5.1	A peg-head, *50 n. the length,* 30 n. the front.	$u = 50$ n., $s = 30$ n.
	Inside it *2 canals*.	2 stripes
	20 n. the *upper* descent, 30 n. the *lower* descent.	$u_a = 20$ n., $u_k = 30$ n.
	The field of the upper canal over the lower field, 3 30 beyond.	$A_a - A_k = 3\ 30$
5.2 a	Half the upper field and 30 šar, the lower field.	$1/2 \cdot A_a + 30 = A_k$
5.2 b	A 3rd of the upper field and 1 50, the lower field.	$1/3 \cdot A_a + 1\ 50 = A_k$
5.3 a	A 7th of the upper field over the field of the lower canal beyond (to) the field of the upper canal add, 8 30.	$A_a + 1/7 \cdot (A_a - A_k) = 8\ 30$
5.3 b	Times 2 repeat, add, 9.	$A_a + 2 \cdot 1/7 \cdot (A_a - A_k) = 9$
5.3 c	Tear off, 7 30.	$A_a - 1/7 \cdot (A_a - A_k) = 7\ 30$
5.3 d	Times 2 repeat, tear off, 7.	$A_a - 2 \cdot 1/7 \cdot (A_a - A_k) = 7$
5.4 a	*(To) the field of the lower canal* add, *5*.	$A_k + 1/7 \cdot (A_a - A_k) = 5$
5.4 b	(From) the field of the lower canal tear off, 4.	$A_k - 1/7 \cdot (A_a - A_k) = 5$
5.4 c	Times 2 repeat, tear off, 3 30.	$A_k - 2 \cdot 1/7 \cdot (A_a - A_k) = 5$
5.5 a	*A peg-head,* 30 n. the front. *Inside it* 2 canals.	$s = 30$
	Field over field 3 30 beyond.	$A_a - A_k = 3\ 30$
5.5 b	Half the upper field and 30 šar the lower field.	A_a + 30 šar $= A_k$
5.5 c	A third of the upper field and 1 50 the lower field.	$1/3 \cdot A_a + 1\ 50 = A_k$

The fields of the canals are what?

Col. 52 hand tablets. 52 assignments

In all the 52 problems on YBC 4696, the partial heights of the 2-striped triangle are given, in each case as $u_a = 20$ and $u_k = 30$. A third given value is specified in each sub-paragraph.

No solution procedures are given in the text. However, the obvious way of solving the stated problems is to first find the values of the pair s, d, and then compute the partial areas A_a and A_k by use of the area rules for trapezoids and triangles.

In view of the linear similarity rule and the assumption that always $u_a = 20$ and $u_k = 30$, it is clear that

$(s - d)/20 = d/30$, so that $3\,(s - d) = 2\,d$, or simply $3\,s = 5\,d$.

This is *a linear equation for s and d common to all the 52 assignments*. In the problems in § 2, the third given condition is *another linear equation* for s and d. Therefore, all the problems in § 2 can be interpreted as *systems of linear equations* for the two unknowns s and d. In § 2.1 a, for instance, the equations for s and d are

$3\,s = 5\,d$, $s - d = 12$, so that $5\,s - 5\,d = 5 \cdot 12 = 1\ 00$, and consequently
$2\,s = 1\ 00$, $s = 30$, and $d = 18$.

In § 3.1, the third condition is an equation for A_a and s. However, since $A_a = (s + d)/2 \cdot u_a = (s + d)/2 \cdot 20$, all such equations are again *linear equations for s and d*. Therefore, also the problems in § 3.1 can be reduced to *systems of linear equations* for the two unknowns s and d.

As a matter of fact, all the problems in YBC 4696 can be reduced to systems of linear equations, except the problems in §§ 3.2-3 and § 4.1, which can instead be reduced to *quadratic-linear systems* for s and d. The solution is, in all the separate cases, that $s, d = 30, 18$.

11.2 i. MAH 16055. A table of diagrams for 3-striped triangles

It follows from the quadratic similarity rule that if s, d are the front and the transversal of a 2-striped triangle, then

$A_a : A_k = (\text{sq.}\ s - \text{sq.}\ d) : \text{sq.}\ d.$

Therefore, in particular,

$A_a = A_k \;\Rightarrow\; \text{sq.}\ s - \text{sq.}\ d = \text{sq.}\ d \;\Rightarrow\; \text{sq.}\ s = 2\ \text{sq.}\ d \;\Rightarrow\; s = \text{sqs.}\ 2 \cdot d.$

If a triangle divided in this way had appeared in an OB mathematical text, it is likely that the approximation sqs. 2 = appr. 1;24 (7/5), or the slightly more accurate approximation sqs. 2 = appr. 1;25 (17/12), would have been used. However, no such text is known.

On the other hand, two examples are known of OB mathematical texts where *a 3-striped triangle is divided so that the first and the third of the partial areas are equal*. The two texts are **MAH 16055** in this section and IM 43996 in the next section.

If the front and the two transversals of a 3-striped triangle are s, d_1, d_2, and if the partial are A_1, A_2, A_3, then clearly

$$A_1 = A_3 \quad \Rightarrow \quad \text{sq.}\, s - \text{sq.}\, d_1 = \text{sq.}\, d_2.$$

A particularly interesting case is, of course, when s, d_1, d_2 are *integers* satisfying the mentioned condition, because that means that they form *a diagonal triple*. In the diagram in Fig. 11.2.9 below is shown the simplest case of this type, the case when

$$s : d_1 : d_2 = 5 : 4 : 3.$$

Then the partial lengths are in the ratios

$$u_1 : u_2 : u_3 = (5 - 4) : (4 - 3) : 3 = 1 : 1 : 3,$$

and the partial areas in the ratios

$$A_1 : A_2 : A_3 = (25 - 16) : (16 - 9) : 9 = 9 : 7 : 9.$$

If the total length u is "normalized", in the sense that $u = 1\ 00$, then it follows that the partial lengths are 12, 12, and 36.

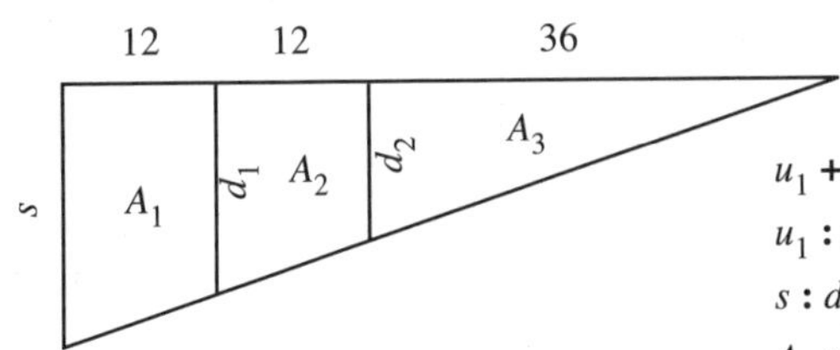

$$u_1 + u_2 + u_3 = 1\ 00$$
$$u_1 : u_2 : u_3 = 1 : 1 : 3$$
$$s : d_1 : d_2 = 5 : 4 : 3$$
$$A_1 : A_2 : A_3 = 9 : 7 : 9$$

Fig. 11.2.9. MAH 16055. A 3-striped triangle with $A_1 = A_3$.

The clay tablet **MAH 16055** (Bruins, *ND* (1961) 11-14) is a "geometric table text", with drawings of 5 different 3-striped triangles on the obverse, and 5 on the reverse. All 10 of the 3-striped triangles are of the type shown in Fig. 11.2.9 above, with the 3 partial lengths equal to 12, 12, 36.

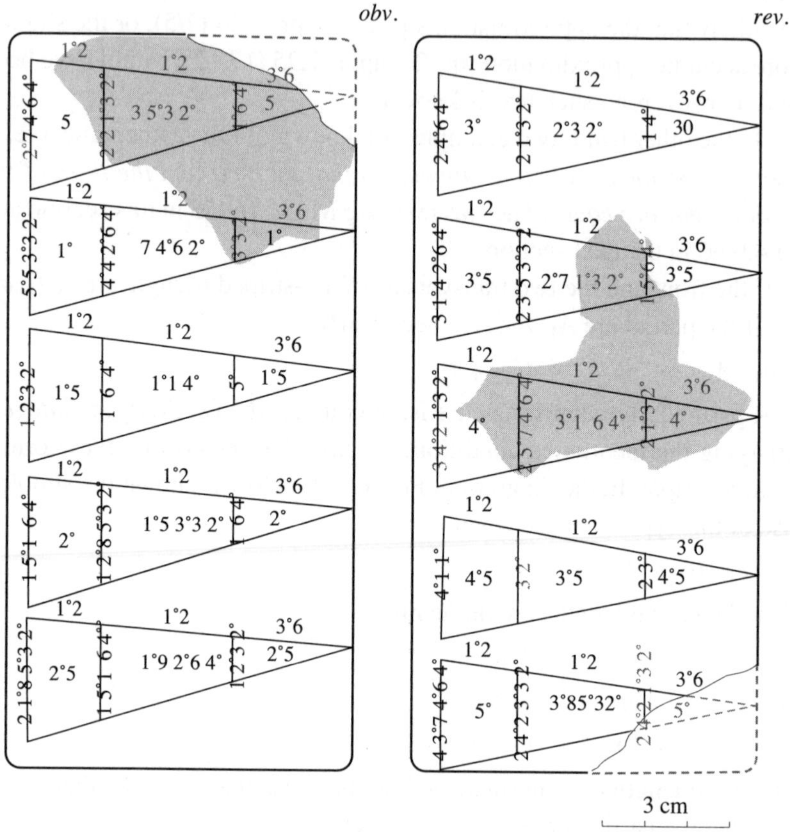

Fig. 11.2.10. MAH 16055. A table of 10 diagrams for 3-striped triangles.

The idea behind the construction of the data for the 3-striped triangles in MAH 16055 (Fig. 11.2.10 above) is simple, namely that

$A_1 = A_3 = n \cdot 5\;00$, for $n = 1, 2, \cdots, 10$, n being the number of the diagram.

Consequently, the second partial area is

$A_2 = 7/9 \cdot A_3 = n \cdot 5\;00 = n \cdot 3\;53;20$, for $n = 1, 2, \cdots, 10$.

The second transversals are also easily computed:

$d_2 = 2\,A_3 \,/\, u_3 = n \cdot 5\;00 \cdot 2/36 = n \cdot 16;40$, for $n = 1, 2, \cdots, 10$.

Consequently,

$d_1 = 4/3 \cdot s_3 = n \cdot 22;13\;20$,
$s = 5/3 \cdot s_3 = n \cdot 27;46\;40$, for $n = 1, 2, \cdots, 10$.

These are also the values indicated in the 10 diagrams.

11.2 j. IM 43996. A 3-striped triangle divided in given ratios

IM 43996 was published by Bruins in *Sumer* 9 (1953), with a photo in Bruins, *CCPV 1* (1964), Part 3, pl. 2. It is an OB square hand tablet with a geometric assignment on the obverse in the form of a diagram, showing a 3-striped triangle with its transversals and some associated numbers.

The given numbers are the three partial areas, the lengths of the two transversals, and the first partial length:

$A_1 = 9\ 22;30,\ \ A_2 = 20\ 37;30,\ \ A_3 = 10\ 00,\ \ d_1 = 17;30,\ \ d_2 = 10,\ \ u_1 = 30.$

The two remaining segments of the length were apparently first given, too, but were then erased by the tip of a finger (the teacher's?), probably an indication that the numbers should be found again by the student. (*Cf.* the discussion in Friberg, *UL* (2005), Sec. 2.1 of a similar assignment in the Egyptian hieratic mathematical text *P.Rhind* # 53 a.)

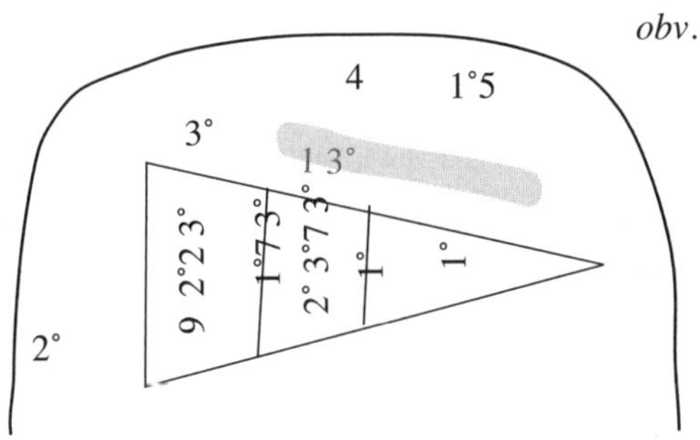

Fig. 11.2.11. IM 43996. A 3-striped triangle with the partial areas in the ratios 1 : 2 : 1.

It is easy to find a solution to the stated (overdetermined) problem. The first step can be to compute the third partial length and the 'feed' of the triangle as follows:

$u_3 = 2\ A_3\ /\ d_2 = 20\ 00\ /\ 10 = 2\ 00,$ and $f = d_2\ /\ u_3 = 10\ /\ 2\ 00 = 1\ /\ 12 = ;05.$

Then it follows that

$u_2 = (d_1 - d_2)\ /\ f = 7;30\ /\ ;05 = 1\ 30,$ and $s = d_1 + u_1 \cdot f = 17;30 + 30 \cdot ;05 = 20.$

There is more to say about the actual *construction* of the problem. The given first and second partial areas, 9 22;30 and 20 37;30, are relatively close to two round area numbers, 10 00 and 20 00. Suppose that some OB mathematician ad the intention *to construct a 3-striped triangle with the three partial areas in the ratios 1 : 2 : 1*. He would then understand that for this to happen it was necessary to let

$$(\text{sq.}\, s - \text{sq.}\, d_1) : (\text{sq.}\, d_1 - \text{sq.}\, d_2) : \text{sq.} d_2 = 1 : 2 : 1.$$

Hence,

$$\text{sq.}\, d_1 - \text{sq.}\, d_2 = 2\ \text{sq.}\, d_2 \quad \text{so that}$$
$$\text{sq.}\, d_1 = 3\ \text{sq.}\, d_2, \quad \text{and} \quad \text{sq.}\, s - \text{sq.}\, d_1 = \text{sq.}\, d_2$$

so that

$$\text{sq.}\, s = \text{sq.}\, d_1 + \text{sq.} d_2 = 4\ \text{sq.}\, d_2.$$

Therefore, it was also necessary to let

$$s : d_1 : d_2 = 2 : \text{sqs.}\, 3 : 1.$$

Now, it is known that the standard OB approximation to sqs. 3 was 1;45 (= 7/4). Therefore, the author of the problem would be led to choose his given values for the 3-striped triangle so that

$$s : d_1 : d_2 = 8 : 7 : 4 \quad \text{and, consequently,} \quad u_1 : u_2 : u_3 = 1 : 3 : 4.$$

In agreement with these conditions, he chose to set

$$u_1, u_2, u_3 = 30,\ 1\ 30,\ 2\ 00, \quad \text{so that} \quad u = u_1 + u_2 + u_3 = 4\ 00, \quad \text{and} \quad s = 20.$$

(Note the numbers 4 and 20 written close to the upper and left margin of the clay tablet.) He seems then to have computed the feed f of the triangle as $s/u = 20 \cdot 1/u = 20 \cdot {;}00\ 15 = {;}05$. (Note the number 15 written near the number 4 close to the upper margin.) With this value for the feed, he could rapidly compute also

$$d_2 = f \cdot u_3 = {;}05 \cdot 2\ 00 = 10, \quad \text{and}$$
$$d_1 = f \cdot (u_2 + u_3) = {;}05 \cdot 3{;}30 = 17{;}30.$$

The corresponding values for the partial areas would then be

$$A_1 = 30 \cdot (20 + 17{;}30)/2 = 9\ 22{;}30,$$
$$A_2 = 1\ 30 \cdot (17{;}30 + 10)/2 = 20\ 37{;}30,$$
$$A_3 = 10 \cdot 2\ 00/2 = 10\ 00.$$

Therefore, the construction of all the given values in IM 43996 can be explained as a consequence of an effort to *try* to let the three partial areas of a 3-striped triangle be proportional to 1, 2, 1.

11.3. Old Babylonian Problems for 2-Striped Trapezoids

11.3 a. IM 58045, an Old Akkadian problem for a bisected trapezoid

IM 58045 (Friberg, *RlA 7* (1990), Sec. 5.4 k; Fig. 11.3.1 below) is a round hand tablet from the Old Akkadian period in Mesopotamia, c. 2340-2200 BCE. It is by its find site in a collapsed house in the ruins of the ancient city Nippur firmly dated to the reign of the king Šarkallišarri. There is drawn on it a trapezoid with a transversal line parallel to the upper and lower fronts of the trapezoid. The lengths of all four sides of the trapezoid, but not the length of the transversal, are indicated in the diagram.

The indicated common length of the two long sides of the trapezoid is u = 2 'reeds', which is as much as 12 cubits (= 1 ninda), since 1 reed = 6 cubits. The given lengths of the two parallel fronts are m = 3 reeds – 1 [cubit] = 17 cubits and n = 1 reed 1 cubit = 7 cubits, respectively. It is likely that the area of the trapezoid was meant to be computed by use of the "false area rule" as

A = (2 reeds + 2 reeds)/2 · {(3 reeds – 1 cubit) + (1 reed 1 cubit)}/2
= 2 reeds · 2 reeds = 1 sq. ninda = 1 šar.

The circumstance that the area of the trapezoid, computed in this way, is *a conspicuously area round number* suggests that the hand tablet is *a mathematical assignment, rather than some surveyor's sketch of a field.*

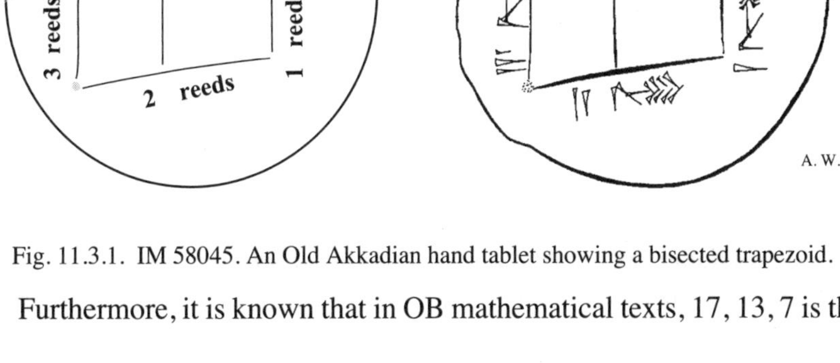

Fig. 11.3.1. IM 58045. An Old Akkadian hand tablet showing a bisected trapezoid.

Furthermore, it is known that in OB mathematical texts, 17, 13, 7 is the most frequently occurring example of a "transversal triple", with the interesting property that the area of a trapezoid with the fronts 17 and 7 is "bisected", that is, *divided in two equal parts* by a transversal of length 13. (See Friberg, *RlA* 7 (1990), Sec. 5.4 k.) Generally, the OB "trapezoid bisection rule" says that a trapezoid with the fronts s_a and s_k is divided in two parts of equal area by a transversal d parallel to the fronts and satisfying the "trapezoid bisection rule"

$$\text{sq. } d = (\text{sq. } s_a + \text{sq. } s_k)/2.$$

It is likely that the trapezoid bisection rule was known already in the Old Akkadian period, and that the teacher who handed out IM 58045 as an assignment had the intention that his students should compute the length d of the transversal in the trapezoid as follows:

$$\begin{aligned} \text{sq. } d &= (\text{sq. } s_a + \text{sq. } s_k)/2 = \{\text{sq. (3 reeds – 1 cubit) + sq. (1 reed 1 cubit)}\} \\ &= \text{5 sq. reeds – 2 reeds} \cdot \text{cubit + 1 sq. cubit} \\ &= \text{4 sq. reeds + 4 reeds} \cdot \text{cubit + 1 sq. cubit} \\ &= \text{sq. (2 reeds 1 cubit) = sq. } d \quad \text{so that} \quad d = \text{2 reeds 1 cubit.} \end{aligned}$$

Note that in the diagram on IM 58045 all lengths are given in the form of traditional length numbers, measured in reeds and cubits. This is in contrast to drawings of trapezoids in OB mathematical texts, where lengths normally are given in the form of *abstract* (sexagesimal) numbers, always thought of as multiples of the main length unit, the ninda. An OB school boy, living 500 years after the Old Akkadian period, would have computed the transversal of the trapezoid in Fig. 11.3.1 (essentially) as follows:

s_a = 1 ninda 5 cubits = 1;25, s_b = 1/2 ninda 1 cubit = ;35,

sq. d = (sq. s_a + sq. s_k)/2 = (sq. 1;25 + sq. ;35)/2 = (2;00 25 + ;20 25)/2 = 1;10 25,

d = sqs. 1;10 25 = 1;05 = 1 ninda 1 cubit.

It is an interesting question how the trapezoid bisection rule can have been discovered originally by some mathematician living in Mesopotamia in the Old Akkadian period or maybe even earlier. The discovery can, of course, have been made in a number of ways, but one particularly intriguing possibility is that it happened as follows:

In *decimal* numbers, the squares of 7, 13, and 17 are 49, 169, and 289, which cannot be said to be very interesting. In contrast to this, a Sumerian or Old Akkadian mathematician constructing *a table of areas of squares with sides measured in* ninda and counting with *sexagesimal* numbers may have made the interesting observation that [31]

sq. (7 ninda) = 49 sq. ninda = 49 šar,
sq. (13 ninda) = 2(60) 49 sq. ninda = 2(60) 49 šar,
sq. (17 ninda) = 4(60) 49 sq. ninda = 4(60) 49 šar.

The observation may have led him to draw a diagram of the following kind, with three "concentric (and parallel) squares":

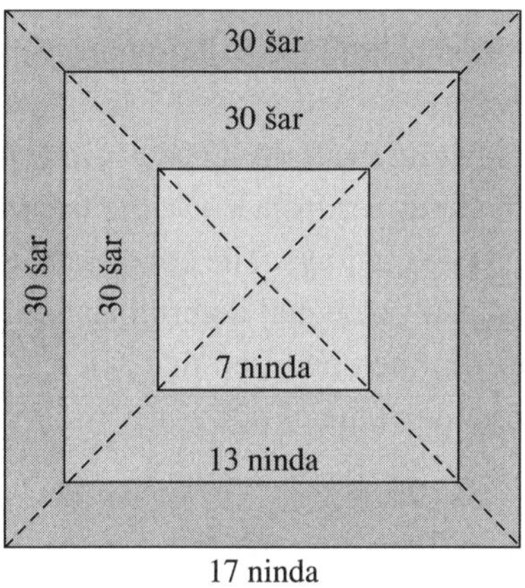

sq. (17 ninda) = 4(60) 49 šar
sq. (13 ninda) = 2(60) 49 šar
sq. (7 ninda) = 49 šar

{sq. (17 ninda) + sq. (7 ninda)}/2
= sq. (13 ninda)

Fig. 11.3.2. How the transversal triple 17, 13, 7 may have been discovered.

11.3 b. VAT 8512, interpreted as a problem for a bisected trapezoid

The OB problem for a 2-striped triangle in VAT 8512 was discussed above, in Sec. 11.2 g. The given parameters in that problem are

s = 30 n., $u_k - u_a$ = 20 n, $A_a - A_k$ = 7 (00) sq. n.

and the first step of the solution procedure is the following computation:

---

31. Three Mesopotamian tables of areas of squares are known at present, all from the third millennium BCE. One of them, ***OIP 14*, 70** (Friberg, *CDLJ* (2005/2) § 4.9; *RC* (2007), Fig. A1.4), is a table of areas of squares with sides measured in cubits. It is a Sumerian table text from the Early Dynastic period IIIb, which preceded the Old Akkadian period. Another such table text, **VAT 12593** (Nissen/Damerow/Englund, *ABK* (1993), Fig. 119; Friberg, *RC* (2007), Fig. 6.3) is a table of areas of squares with sides measured in tens or sixties of the ninda. This table text is even older, from the Early Dynastic period IIIa. A third text of a similar kind is **CUNES 50-08-001** (Early Dynastic IIIb), a combined table of areas of large and small squares (Friberg, *op. cit.*, Figs. A.7.1-2).

d = sqs. [{sq. (s + r) + sq. r}/2] – r, where $r = (A_a - A_k) / (u_k - u_a) = 7\ 00 / 20 = 21$.

The equation shows that the (length of) the transversal d is

d = sqs. {(sq. 51 + sq. 21)/2} – 21 = sqs. (25 21) – 21 = 39 – 21 = 18.

An interesting interpretation was suggested by Gandz (1948) and Huber (1955). See the references to these authors in Høyrup, *LWS* (2002), 234-238. According to Gandz and Huber, the equation for d shows that

$s + r, d + r, r = 51, 39, 21 = 3 \cdot (17, 13, 7)$

is a transversal triple. Therefore, the idea behind the solution procedure in VAT 8512 may have been the following.

The 2-striped triangle with the front s and the transversal r is interpreted as a part of *a bisected trapezoid* with the upper front $s + r$, the transversal $d + r$, and the lower front r (see Fig. 11.3.3 below). The condition that the partial areas of the bisected trapezoid shall be equal determines the value of the extension r, and when r is known, the condition that $s + r, d + r, r$ shall be a transversal triple determines the value of d.

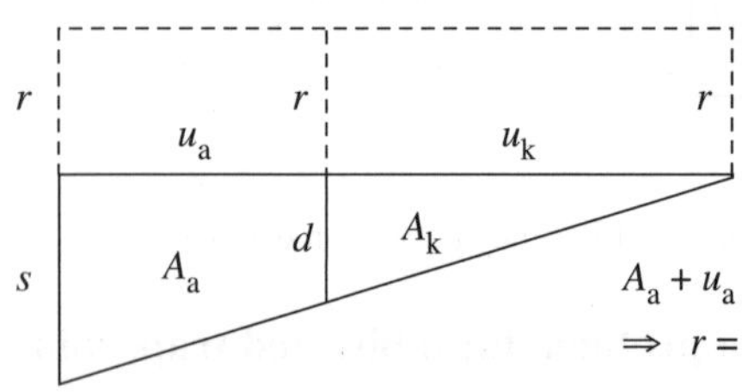

$s = 30$,

$A_a - A_k = 7\ 00$,

$u_k - u_a = 20$

$A_a + u_a \cdot r = A_k + u_k \cdot r$

$\Rightarrow\ r = (A_a - A_k)/(u_k - u_a) = 21$

sq. $(d + r)$ = {sq. $(s + r)$ + sq. r}/2

$s + r = 51,\ r = 21 \Rightarrow d + r = 39,\ d = 18$

Fig. 11.3.3. VAT 8512. The Gandz/Huber interpretation of the solution procedure.

11.3 c. YBC 4675. A problem for a bisected quadrilateral

As will be shown below, the basic ideas of a bisected trapezoid was generalized in an amazing number of ways by OB mathematicians. One interesting example is the problem in **YBC 4675** (Høyrup, *LWS* (2002), 244-249), a single problem text from the ancient Mesopotamian city Larsa, where the usual bisected trapezoid is replaced by a very long and thin *quadrilateral*, and where, consequently, in the solution procedure the accurate area rule for trapezoids has to be replaced by the more or less inaccurate "false area rule" for quadrilaterals. Here is the statement of the problem in YBC 4675:

YBC 4675, literal translation | explanation

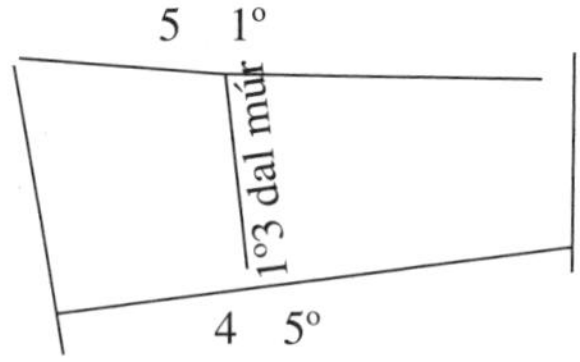

YBC 4675, literal translation	explanation
If a field of 'length-eats-length',	A long quadrilateral
the 1st length 5 10, the 2nd length 4 50,	$u' = 5\ 10,\ u'' = 4\ 50$
the upper front 17, the lower front 7,	$s_a = 17,\ s_k = 7$
its field 2 bùr.	$A = 2$ bùr $= 1\ 00\ 00$ sq. ninda
1 bùr each, the field in two I divided.	$A_a = A_k = 1$ bùr $= 30\ 00$ sq. ninda
The middle transversal, how much?	$d = ?$
The long length and the short length,	$u_a', u_a'', u_k', u_k'' = ?$
how much shall I set them,	
so that they border 1 bùr?	

The successive steps of the solution procedure in YBC 4675 are:

1.	$u = (u' + u'')/2 = (5\ 10 + 4\ 50)/2 = 5\ 00$	the average length
2.	$f = (s_a - s_k)/u = (17 - 7)\ /\ 5\ 00 = 1/30 = ;02$	the 'feed'
3.	$2f \cdot A_a = 2 \cdot ;02 \cdot 30\ 00 = 2\ 00$	silently understood
4.	$\text{sq.}\ d = \text{sq.}\ s_a - 2f \cdot A_a = \text{sq.}\ 17 - 2\ 00 = 2\ 49,\quad d = 13$	the transversal
5.	$u_a = A_a\ /\ (s_a + d)/2 = 30\ 00\ /\ 15 = 2\ 00$	the average upper length
6.	$g = (u' - u'')/(u' + u'') = (5\ 10 - 4\ 50)/(510 + 4\ 50) = ;02$	the 'obliquity factor'
7.	$u_a' = u_a + g \cdot u_a = 2\ 04,\quad u_a'' = u_a - g \cdot u_a = 1\ 56$	the upper lengths
8.	$u_k = A_k\ /\ (d + s_k)/2 = 30\ 00\ /\ 10 = 3\ 00$	the average lower length
9.	$u_k' = u_k + g \cdot u_k = 3\ 06,\quad u_k'' = u_k - g \cdot u_k = 2\ 54$	the lower lengths

Apparently, everything here was assumed to be at least *approximately correct*; thus, the upper and lower front were assumed to be approximately parallel, the average length was assumed to be approximately equal to the distance between the upper and the lower fronts, *etc*. (Actually, as was remarked in the original publication of YBC 4675 by Neugebauer and Sachs in *MCT* (1945), text B, the data were poorly chosen, since they violate the so called triangle inequality, so that there simply does not exist any quadrilateral with the given fronts and lengths!)

With these reservations in mind, most of the steps in the solution procedure make sense. Step 4, for instance, can be explained as follows, in view of the linear and quadratic similarity rules,

$$A_a = \{(s_a - d)/f\} \cdot (s_a + d)/2 = (\text{sq.}\ s_a - \text{sq.}\ d)/(2f).$$

In step 6, the term 'obliquity factor' is a conjectured translation of the word *arakarûm* in the text. The value of the obliquity factor is actually not computed in the text; it is introduced without any explanation. Nevertheless, its use in this text can reasonably be explained as follows: It was assumed that the 'longer' and 'shorter' upper lengths, and also the 'longer' and 'shorter' lower lengths are proportional to the 'longer' and 'shorter' lengths of the quadrilateral. More precisely, it was assumed that

$$u_a' / u' = u_a'' / u'' \quad \text{and} \quad u_k' / u' = u_k'' / u''.$$

It follows for the upper lengths, for instance, that

$$u_a' = g \cdot u' \quad \text{and} \quad u_a'' = g \cdot u'', \quad \text{for some (unknown) factor } g.$$

Then also

$$u_a' + u_a'' = g \cdot (u' + u'') \quad \text{and} \quad u_a' - u_a'' = g \cdot (u' - u''),$$

so that

$$g = (u_a' - u_a'') / (u_a' + u_a'') = (u' - u'') / (u' + u'') = \text{the } arakar\hat{u}m.$$

Similarly for the lower lengths. (The reasoning is a naive variant of the reasoning in the well known proportion theory in *Elements* V.)

It follows that, for instance, the first part of step 7 of the solution procedure in YBC 4675 can be explained as follows:

$$u_a + g \cdot u_a = (u_a' + u_a'')/2 + (u_a' - u_a'') / (u_a' + u_a'') \cdot (u_a' + u_a'')/2$$
$$= (u_a' + u_a'')/2 + (u_a' - u_a'')/2 = u_a'.$$

The discussion above shows that YBC 4675 is a splendid example of *the surprising mixture of naive and sophisticated arguments* that one can meet in Babylonian mathematics!

11.3 d. YBC 4608. A 2-striped trapezoid divided in the ratio 1: 3

The quadrilateral in YBC 4675 and the trapezoid in Gandz' and Hubert's explanation of VAT 8512 are both divided in two parts of equal area by a transversal d, and in both texts the triple s_a, d, s_k is proportional to the *transversal triple* 17, 3, 7, just as in the case of the trapezoid appearing in the Old Akkadian text IM 58045. As mentioned above, in a trapezoid divided in *two equal parts* by a transversal d parallel to the fronts, the triple s_a, d, s_k satisfies the *trapezoid bisection equation*

$$\text{sq.}\, s_a - \text{sq.}\, d = \text{sq.}\, d - \text{sq.}\, s_k \quad \text{or, equivalently,} \quad \text{sq.}\, s_a + \text{sq.}\, s_k = 2\ \text{sq.}\, d.$$

In a couple of examples discussed below, however, trapezoids are divided instead by a transversal parallel to the fronts *in two parts in a given ratio $P : Q$*. The "general trapezoid bisection equation" will then have a correspondingly modified form, so that

If $A_a : A_k = P : Q$, then

$(\text{sq.}\, s_a - \text{sq.}\, d)/P = (\text{sq.}\, d - \text{sq.}\, s_k)/Q$, or, equivalently,

$\boldsymbol{Q \cdot \text{sq.}\, s_a + P \cdot \text{sq.}\, s_k = (Q + P) \cdot \text{sq.}\, d}$.

This is *an indeterminate quadratic equation* for the triple s_a, d, s_k, very much like the diagonal equation for the sides of a right triangle. If arbitrary rational values are prescribed for two of the parameters in the triple, then in the general case, the third parameter will be a square side. However, as the next example will show, *Old Babylonian mathematicians had found a way to construct rational triples satisfying the general trapezoid bisection equation*. *Cf.* the discussion in Sec. 3.3 above of the OB generating rule for the construction of rational triples satisfying the diagonal equation.

YBC 4608 (Neugebauer and Sachs, *MCT* (1945), text D) is an OB single problem text from the ancient Mesopotamian city Uruk. Here is the statement of the problem in that text:

YBC 4608, literal translation	explanation
An ox-face.	A trapezoid
The field in *two I divided*,	divided in two parts
42 11 15 the lower canal,	$A_k = 42\ 11;15$
14 03 45 the upper canal,	$A_a = 14\ 03;45$
the 5th-part of the lower canal the upper canal,	$u_a = 1/5 \cdot u_k$
52 30 its transversal.	$d = 52;30$
The upper front and the lower front are what?	s_a and s_k = ?

Note that here the given partial areas are in the ratio

$A_a : A_k = 14\ 03;45 : 42\ 11;15 = 1 : 3.$

The successive steps of the solution procedure in YBC 4608 are:

1. $u_a{}^* = 1\ 00,\ \ u_k{}^* = 5\ 00$	false lengths in the ratio 1 : 5
2. $A = A_a + A_k = 14\ 03;45 + 42\ 11;15 = 56\ 15$	the total area
3. $u = u_k{}^* + u_a{}^* = 5\ 00 + 1\ 00 = 6\ 00$	the false total length
4. $s_a{}^* + s_k{}^* = 2\ A\ /\ u^* = 2 \cdot 56\ 15\ /\ 6\ 00 = 18;45$	the sum of the false fronts
5. $s_a{}^* + d^* = 2\ A_a\ /\ u_a{}^* = 28;07\ 30$	the sum of the false upper front and the false transversal
6. $d^* - s_k{}^* = (s_a{}^* + d^*) - (s_a{}^* + s_k{}^*)$ $= 2\ A_a\ /\ u_a{}^* - 2\ A\ /\ u^* = 28;07\ 30 - 18;45 = 9;22\ 30$	the difference between the false transversal and false lower front
7. $(d^* + s_k{}^*)/2 = A_k\ /\ u_k{}^* = 42\ 11;15\ /\ 5\ 00 = 8;26\ 15$	the half-sum of the false transversal and the false lower front
8. $d^* = A_k\ /\ u_k{}^* + (A_a\ /\ u_a{}^* - A\ /\ u^*)$ $= 8;26\ 15 + 4;41\ 15 = 13;07\ 30$	the false transversal
9. $c = d\ /\ d^* = 52;30\ /\ 13;07\ 30 = 4$	the correction factor
10. $u_a = u_a{}^*\ /\ c = 1\ 00\ /\ 4 = 15$	the upper length
11. $u_k = u_k{}^*\ /\ c = 5\ 00\ /\ 4 = 1\ 15$	the lower length
12. $s_a + d = 2\ A_a\ /\ u_a = 2 \cdot 14\ 03;45\ /\ 15 = 1\ 52;30$	the sum of the upper front and the transversal
13. $s_a = (s_a + d) - d = 1\ 52;30 - 52;30 = 1\ 00$	the upper front
14. $s_a + s_k = 2\ A\ /\ u = 1\ 52\ 30\ /\ 1\ 30 = 1\ 15$	the sum of the fronts
15. $s_k = (s_a + s_k) - s_a = 1\ 15 - 1\ 00 = 15$	the lower front

(In the text of YBC 4608 all values are in the form of *relative* sexagesimal numbers in place value notation without zeros, *etc*. In the explanation above, the most likely *absolute* values of the sexagesimal numbers in the text are indicated, always in accordance with the observed rule that lengths and fronts regularly are measured in *tens or sixties* of the ninda.)

Note that the transversal triple in this example is

$s_a, d, s_k = 1\ 00, 52;30, 15 = 7;30 \cdot (8, 7, 2).$

Clearly this triple satisfies the general trapezoid bisection equation

$3 \cdot \text{sq.}\ s_a + 1 \cdot \text{sq.}\ s_k = 4 \cdot \text{sq.}\ d.$ Indeed,

$3 \cdot \text{sq.}\ 8 + 1 \cdot \text{sq.}\ 2 = 3 \cdot 1\ 04 + 1 \cdot 4 = 3\ 12 + 4 = 3\ 16,\ \ 4 \cdot \text{sq.}\ 7 = 4 \cdot 49 = 3\ 16.$

The solution procedure in YBC 4608 is an unusual variant of the OB rule of false value. The trick here is to *choose false values for the partial lengths in the prescribed ratio* 5 : 1 and then *compute the corresponding false value for the transversal*. It turns out that the computed false value for the transversal is 1/4 of the prescribed value. However, since also the partial areas of the trapezoid have prescribed values, the length of the trapezoid on one hand and the two fronts and the transversal on the other are *inversely proportional*. Therefore, since the computed "correction factor" for the transversal is 4, the corresponding correction factor for the partial lengths must be 1/4. This is why the "true" values for the partial lengths are computed as $1/4 \cdot 1\ 00 = 15$ and $1/4 \cdot 5\ 00 = 1\ 15$.

An inspection of the solution procedure in YBC 4608 reveals that OB mathematicians had found the following "generating rule for (rational) solutions to the trapezoid bisection equation":

Let u, u_a, u_k and A, A_a, A_k be given (rational) values for the whole or partial lengths and areas of a 2-striped trapezoid. Then it follows from the area rule for trapezoids that the triple s_a, d, s_k satisfies the following *system of linear equations*:

$(s_a + s_k)/2 = A / u$, $(s_a + d)/2 = A_a / u_a$ $(d + s_k)/2 = A_k / u_k$.

Then one finds by subtraction that

$(s_a - d)/2 = A / u - A_k / u_k$ and

$(d - s_k)/2 = A_a / u_a - A / u$.

Consequently,

$\mathbf{s_a} = A_a / u_a + A / u - A_k / u_k$,

$\mathbf{d} = A_k / u_k + A_a / u_a - A / u$,

$\mathbf{s_k} = A / u + A_k / u_k - A_a / u_a$.

11.3 e. Str. 367. A 2-striped trapezoid divided in the ratio 29 : 51

Str. 367 (Høyrup, *LWS* (2002), 239-244) is a single problem text from Uruk, just like TBC 4608 above. It is another problem for a 2-striped trapezoid. Here is the statement of the problem:

Str. 367, literal translation	explanation
An ox-face. Inside it two canals.	A 2-striped trapezoid
13 03 the upper field, 22 57 field 2.	$A_a = 13\ 03, A_k = 22\ 57$
The 3rd-part of the lower length in the upper length.	$u_a = 1/3 \cdot u_k$
What the upper front over the transversal is beyond	$(s_a - d) + (d - s_k) = 36$
and the transversal over the lower front is beyond heap,	
then 36. The lengths, the fronts, and the transversal are what?	$u_a, u_k, s_a, s_k, d = ?$

The given values in this problem are:

$A_a = 13\ 03$, $A_k = 22\ 57$, $u_a = 1/3 \cdot u_k$, and $(s_a - s_k) = (s_a - d) + (d - s_k) = 36$.

The solution procedure is another application of the rule of false value, where the trick this time is to *choose false values for the partial lengths in the prescribed ratio* 1 : 3 and then *compute the corresponding false value for the feed of the trapezoid in two different ways*. A comparison of the two results will then give the needed correction factor.

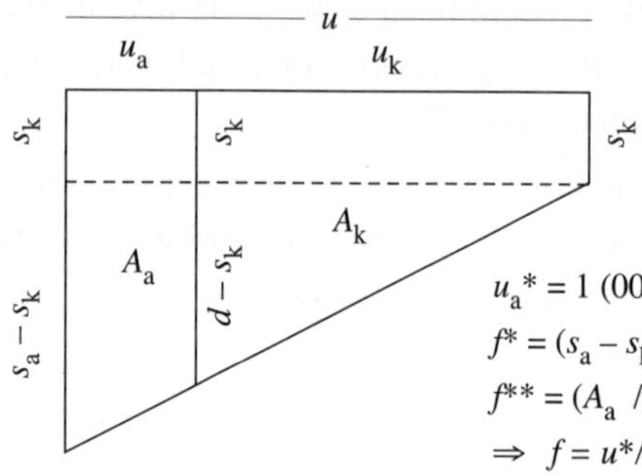

$A_a = 13\ 03\ (= 27 \cdot 29)$,
$A_k = 22\ 57\ (= 27 \cdot 51)$
$u_a = 1/3 \cdot u_k$
$s_a - s_k = 36$

$u_a^* = 1\ (00),\ \ u_k^* = 3\ (00),\ \ u^* = 4\ (00)$
$f^* = (s_a - s_k) / u^* = ;09$
$f^{**} = (A_a / u_a^* - A_k / u_k^*) / u^*/2 = ;02\ 42$
$\Rightarrow\ f = u^*/u \cdot f^*$ and $f = \text{sq.}\,(u^*/u) \cdot f^{**}$
$\Rightarrow\ u^*/u = f^{**} / f^* = ;02\ 42/;09 = ;18$
$\Rightarrow\ u_a = 1 / (u^*/u) \cdot u_a^* = 18,$ *etc.*

Fig. 11.3.4. Str. 367. A surprising application of the rule of false value.

The false values chosen for the lengths are

$u_a^* = 1\ (00),\ \ u_k^* = 3\ (00),$ hence $u^* = 4\ (00).$

In the first approach, the corresponding *false feed* is computed as

$f^* = (s_a - s_k) / u^* = 36 / 4\ (00) = ;09.$

Then it follows that

$s_a - d = f^* \cdot u_a^* = 9,\ \ d - s_k = f^* \cdot u_k^* = 27.$

In the second approach, another value for the false feed is computed as

$f^{**} = (s_a^* - s_k^*) / u^* = \{(s_a^* + d^*)/2 - (d^* + s_k^*)/2\} / u^*/2$
$= (A_a / u_a^* - A_k / u_k^*) / u^*/2 = 5;24 / 2\ (00) = ;02\ 42.$

The meaning of the next series of computations is less obvious:

$f^* / f^{**} = ;09 / ;02\ 42 = 3;20,\ \ 1 / 3;20 = ;18,\ \ ;18 \cdot 1\ (00) = u_a,\ \ ;18 \cdot 3\ (00) = u_k.$

What this means is, presumably, that, as is easy to check,

$f = u^*/u \cdot f^*$ but $f = \text{sq.}\,(u^*/u) \cdot f^{**}.$

Therefore,

$f^* / f^{**} = u^*/u$ so that $1 / (f^* / f^{**}) = u/u^*,$ and $u/u^* \cdot u_a^* = u_a,\ \ u/u^* \cdot u_k^* = u_k.$

In other words, f^{**}/f^* is the "correction factor" for the false values.

This far into the solution procedure, it is known that

$u_a = 18$, $u_k = 54$, hence $u = 18 + 54 = 1\ 12$, and
$A_a = 13\ 03$, $A_k = 22\ 57$, hence $A = 13\ 03 + 22\ 57 = 36\ 00$.

Since the trapezoid can be divided into a triangle of front $s_a - s_k = 36$ and a rectangle of front s_k, both with the length $u = 1\ 12$, it follows that

$A = 36\ 00 = 36 / 2 \cdot 1\ 12 + s_k \cdot 1\ 12$ so that $s_k = (36\ 00 - 21\ 36) / 1\ 12 = 12$.

Therefore

$s_a = 12 + 36 = 48$ and $d = 12 + 27 = 39$.

11.3 f. Ist. Si. 269. Five 2-striped trapezoids divided in the ratio 60 : 1

Ist. Si. 269 is a clay tablet from the ancient Mesopotamian city Sippar with 4 diagrams of 2-striped trapezoids on the obverse and 2 on the reverse. The hand copy in Fig. 11.3.5 below, with sexagesimal numbers in transliteration, is based on a hand copy with cuneiform numbers, courteously provided by V. Donbaz.

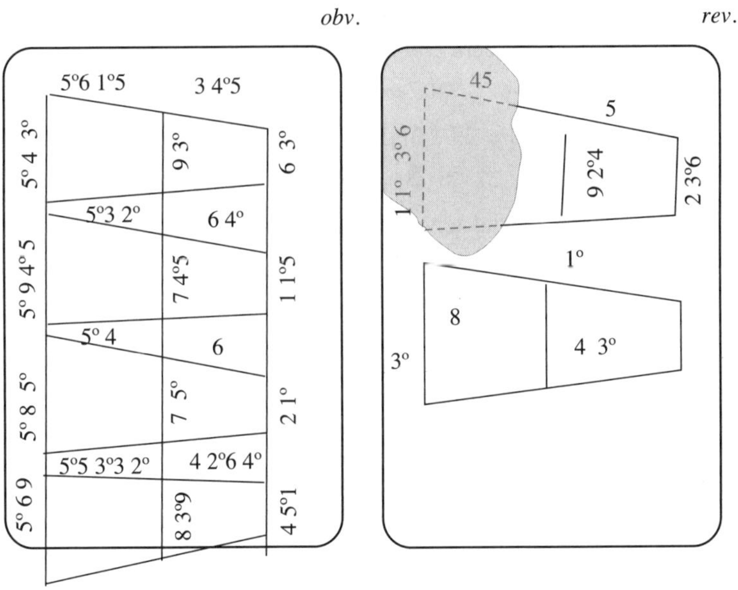

Fig. 11.3.5. Ist. Si. 269. Six 2-striped trapezoids with associated values.

The partial lengths, the two fronts, and the transversals are given for all the four trapezoids on the obverse and probably also for the first trapezoid on the reverse. The text may be either an assignment, where a student was supposed to compute the partial areas of all those trapezoids, or it may be the answer to an assignment, where a student has successfully computed some, or all, of the indicated values.

Already a preliminary analysis of the trapezoids on the obverse reveals some interesting features. The sum of the partial lengths is in all the four cases equal to '1':, probably meaning 1 00:

$$56;15 + 3;45 = 53;20 + 6;40 = 54 + 6 = 55;33\ 20 + 4;26\ 40 = 1\ 00.$$

Similarly, the sum of the two fronts is in all the four cases equal to 1 01:

$$54;30 + 6;30 = 59;45 + 1;15 = 58;50 + 2;10 = 56;09 + 4;51 = 1\ 01.$$

On the other hand, the four 2-striped trapezoids are *not* similar figures, since the upper lengths are various multiples of the lower lengths:

$$56;15 = 15 \cdot 3;45,\ \ 53;20 = 8 \cdot 6;40,\ \ 54 = 9 \cdot 6,\ \ 55;33\ 20 = 12;30 \cdot 4;26\ 40.$$

Note that the four factors 15, 8, 9, and 12;30 are not arbitrarily chosen numbers. Instead, thcy are small numbers n such that *both n and $n + 1$ are regular sexagesimal numbers* (numbers for which there exists a reciprocal sexagesimal number). Indeed,

$$15 + 1 = 16,\ \ 8 + 1 = 9,\ \ 9 + 1 = 10,\ \ \text{and}\ \ 12;30 + 1 = 13;30.$$

Here, for instance, 12;30 = 25/2 and 13;30 = 27/2, with the reciprocals

$$2 \cdot 12 \cdot 12 = 4\ 48\ \ \text{and}\ \ 2 \cdot 20 \cdot 20 \cdot 20 = 4\ 26\ 40.$$

It is likely that in the partial lengths were chosen, in all the four cases, so that both the partial lengths and the whole length would be regular sexagesimal numbers. More specifically, it is easy to see that

$$\begin{aligned} 56;\ 15, 3;45 &= (15/16, 1/16) \cdot 1\ 00,\\ 53;20 &= (8/9, 1/9) \cdot 1\ 00,\\ 54, 6 &= (9/10, 1/10) \cdot 1\ 00,\\ 55;33\ 20, 4;26\ 40 &= (25/27, 2/27) \cdot 1\ 00. \end{aligned}$$

Another idea behind the construction of the 2-striped trapezoids on the obverse of Ist. Si. 269 is revealed if the area and the partial areas of one of the trapezoids are computed. In the case of the first trapezoid, for instance,

$$\begin{aligned} A_a &= 56;15 \cdot (54;30 + 9;30)/2 = 56;15 \cdot 32 = 30\ 00,\\ A_k &= 3;45 \cdot (9;30 + 6;30)/2 = 3;45 \cdot 8 = 30. \end{aligned}$$

Thus, in this case (and also in the three other cases) the transversal divides the trapezoid in two parts in the ratio 1 00 : 1 (that is, 60 : 1).

It is possible, therefore, that the obverse of Ist. Si. 269 is some student's answer to an assignment of the following kind:

Construct four 2-striped trapezoids
with the partial areas always equal to 30 00 and 30,
and with the total length always equal to 1 00.

Take, for instance, the first example, where the student apparently chose

$u_a = 15 \cdot 1\ 00 / 16 = u_k = 56;15,$
$u_k = 1\ 00 / 16 = 3;45.$

Knowing u_a and u_k, he could then set up the equations for s_a, d, s_k:

$s_a + s_k = 2\ A / u = 1\ 01\ 00 / 1\ 00 = 1\ 01,$
$s_a + d = 2\ A_a / u_a = 1\ 00\ 00 / 56;15 = 1\ 04$
$d + s_k = 2\ A_k / u_k = 1\ 00 / 3;45 = 16.$

Note that division by $u = 1\ 00$, $u_a = 56;15$, and $u_k = 3;45$ is possible (without approximations) precisely because u, u_a and u_k all are regular.

This system of equations can be solved as follows:

$s_a + d + s_k = A / u + A_a / u_a + A_k / u_k = (1\ 01 + 1\ 04 + 16)/2 = 2\ 21 / 2 = 1\ 10;30,$
$s_a = (s_a + d + s_k) - (d + s_k) = 1\ 10;30 - 16 = 54;30\ (= A / u + A_a / u_a - A_k / u_k),$
$d\ = (s_a + d + s_k) - (s_a + s_k) = 1\ 10;30 - 1\ 01 = 9;30\ (= A_k / u_k - A / u + A_a / u_a),$
$s_k = (s_a + d + s_k) - (s_a + d) = 1\ 10;30 - 1\ 04 = 6;30\ (= A_a / u_a - A_k / u_k + A / u).$

These are the values actually recorded in and around the first trapezoid. The values associated with the remaining trapezoids on the obverse can be computed analogously.

Only three of the five values associated with the first trapezoid on the *reverse* are preserved. It is likely, however that these values were computed with departure from the values for the third trapezoid on the obverse by scaling the lengths by the scaling factor ;50 = 1 – 1/6 and the fronts and the transversal by the *reciprocal* scaling factor 1;12 = 1 + 1/5. Indeed,

5 = (1 – 1/6) · 6,
[45] = (1 – 1/6) · 54,
2;36 = (1 + 1/5) · 2;10,
9;24 = (1 + 1/5) · 7;50,
[1 10;36] = (1 + 1/5) · 58;50.

In this way, the first trapezoid on the reverse still has the partial areas

30 00, 30, and it still has the partial lengths in the ratio 9 : 1, but the total length is now 50 instead of 1 00 = 60.

The second trapezoid on the reverse of Ist. Si. 269 is inscribed with data for a new kind of problem. Interestingly, the given partial areas $A_{\text{a}} = 8\ 00$, $A_{\text{k}} = 4\ 30$ and the upper front $s_{\text{a}} = 30$ are the same as in the favorite kind of 2-striped triangle, the one appearing in *TMS* 18 and in Str. 364 §§ 4, 6. It is likely that this second problem on Ist. Si. 269, *rev*. is identical with the problem in Str. 364 § 4 a (Sec. 11.2 c above and Fig. 11.2.1), even if the number '10' in the diagram for the former text is a careless notation corresponding to the more correct '10 diri' in the diagram for the latter text!

11.3 g. The Bloom of Thymaridas and its relation to Old Babylonian generating equations for transversal triples

The **'Bloom' of Thymaridas**, an ancient Pythagorean, not later than the time of Plato, is mentioned by Iamblichus (the first half of the fourth century) in his book *On Nichomachus' Introductio Arithmetica* (see Heath, *HGM I* (1981), 94; Thomas *GMW 1* (1980), 139). The 'Bloom' is a rule for solving a certain kind of systems of linear equations. Its name suggests that the rule was well known and appreciated for its elegance.

The rule was stated as follows, in general terms, without the use of symbolic notation (here in the translation proposed by Thomas):

> "When any determined or undefined quantities amount to a given sum, and the sum of one of them plus every other is given, the sum of these pairs minus the first given sum is, if there are three quantities, equal to the quantity which was added to all the rest; if there are four quantities, one-half is so equal; if there are five quantities, one-third; if there are six quantities, one-fourth, and so on continually, there being always a difference of 2 between the number of quantities to be divided and the denomination of the part."

In modern notations, this rule says that a system of $n + 1$ linear equations of the type

$$d + s_1 + s_2 + \cdots + s_n = c, \quad d + s_1 = c_1, \quad d + s_2 = c_2, \cdots, d + s_{\text{n}} = c_n$$

has the solution

$$d = \{(c_1 + c_2 + \cdots + c_n) - c\} / (n - 1).$$

The proof is easy, since one gets by summation

$$(c_1 + c_2 + \cdots + c_n) - c = n \cdot d + (s_1 + s_2 + \cdots + s_n) - (d + s_1 + s_2 + \cdots + s_n) = (n - 1) \cdot d.$$

Now, consider again the OB system of equations for the fronts and transversal of a 2-striped trapezoid, in the simplified form

$s_a + s_k = c, \quad s_a + d = c_1, \quad d + s_k = c_2,$
where $c = 2\,A\,/\,u, \quad c_1 = 2\,A_a\,/\,u_a, \quad c_2 = 2\,A_k\,/\,u_k.$

Since one gets here by summation

$c + c_1 + c_2 = 2 \cdot (d + s_a + s_k),$

the mentioned system of equations can be reduced to the equivalent system

$d + s_a + s_k = (c + c_1 + c_2)/2\,, \quad d + s_a = c_1, \quad d + s_k = c_2.$

In this form, the OB system of equations for the fronts and transversal of a 2-striped trapezoid is clearly *identical with the Bloom of Thymaridas in the case when* $n = 2$. Accordingly,

$d = (c_1 + c_2) - (c + c_1 + c_2)/2 = (c_1 + c_2 - c)/2 = A_a\,/\,u_a + A_k\,/\,u_k - A\,/\,u.$

The obvious conclusion of this brief consideration is that the 'Bloom of Thymaridas' can be interpreted as a generalization to the case of an arbitrary n of the OB system of generating equations for the fronts and the transversal of a 2-striped trapezoid, as it appears in YBC 4608 and, indirectly, in Ist. Si. 269!

11.3 h. Relations between diagonal triples and transversal triples

There is an obvious connection between diagonal triples and transversal triples in the case of a *bisected* trapezoid ($A_a = A_k$). This can be shown either algebraically or geometrically. The algebraic proof is as follows (Vaiman, *SVM* (1961), 195):

If $\text{sq.}\, s_a + \text{sq.}\, s_k = 2 \cdot \text{sq.}\, d$, then $\text{sq.}\,(s_a + s_k)/2 + \text{sq.}\,(s_a - s_k)/2 = \text{sq.}\, d$,

and conversely

If $\text{sq.}\, u + \text{sq.}\, s = \text{sq.}\, d$, then $\text{sq.}\,(u + s) + \text{sq.}\,(u - s) = 2 \cdot \text{sq.}\, d$.

In other words,

If s_a, d, s_k is a transversal triple, then $d, (s_a + s_k)/2, (s_a - s_k)/2$ is a diagonal triple,
and if d, u, s is a diagonal triple, then $u + s, d, u - s$ is a transversal triple.

The geometric proof is just as simple. In Fig. 11.3.6 below, left, *a square band* viewed as a ring of four equal trapezoids *is divided in two parts of equal area* by a concentric square of side d, provided that s_a, d, s_k is a transversal triple. In Fig. 11.3.6, right, the same square band, now viewed as a ring of four equal rectangles, *is again divided in two parts of equal area* by a oblique square of side d, where d now is the diagonal of the rectangles. Therefore, *in both cases the area of the square of side d is the half-sum of the areas of the squares bounding the square band*. This is the geometric explanation of the "algebraic" identities mentioned above.

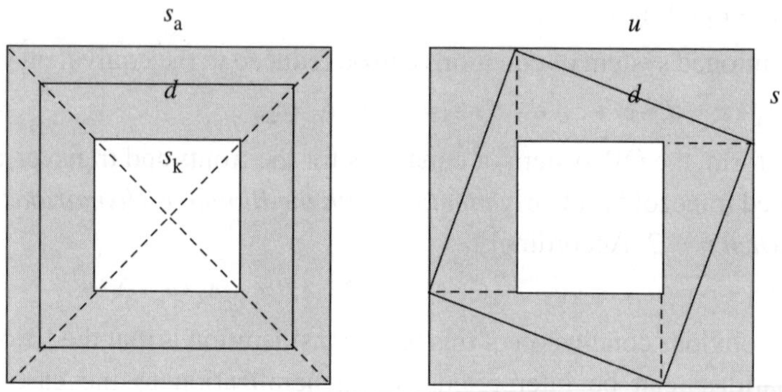

Fig. 11.3.6. Two ways of constructing a square halfway between two given squares.

Now, since there exists a simple relation between transversal triples and diagonal triples, it is clear that *there must also exist a simple relation between the generating rule for transversal triples* (Sec. 11. 3 d above) *and the generating rule for diagonal triples* (Sec. 3.3 above).

Indeed, consider the generating rule for diagonal triples, in the form

$$d : u : s = (\text{sq.}\, m + \text{sq.}\, n) : 2\, m \cdot n : (\text{sq.}\, m - \text{sq.}\, n), \quad \text{with} \quad n < m < n \cdot (\text{sqs.}\, 2 + 1).$$

(The stated restriction for the pair m, n ensures that $0 < s < u$.) Then also

$$u + s : d : u - s = \{2\, m \cdot n + (\text{sq.}\, m - \text{sq.}\, n)\} : (\text{sq.}\, m + \text{sq.}\, n) : \{2\, m \cdot n - (\text{sq.}\, m - \text{sq.}\, n)\}.$$

This is a corresponding generating rule for transversal triples. The same rule can be derived as follows from the equations in Sec. 11.3 d. Set

$$A_a = A_a = B, \quad \text{and} \quad u_a = n, u_k = m, \quad \text{where} \quad n < m < n \cdot (\text{sqs.}\, 2 + 1).$$

Then it follows that

$$\begin{aligned} &s_a : d : s_k \\ &\quad = \{2B/(m+n) - B/m + B/n\} : \{B/n - 2B/(m+n) + B/m\} : \{B/m - B/n + 2B/(m+n)\} \\ &\quad = \{2\, m \cdot n + (\text{sq.}\, m - \text{sq.}\, n)\} : (\text{sq.}\, m + \text{sq.}\, n) : \{2\, m \cdot n - (\text{sq.}\, m - \text{sq.}\, n)\}. \end{aligned}$$

11.4. Old Babylonian Problems for 3-and 5-Striped Trapezoids

Two OB hand tablets with metric algebra diagrams of 3-striped trapezoids are shown in Fig. 11.4.1 below.

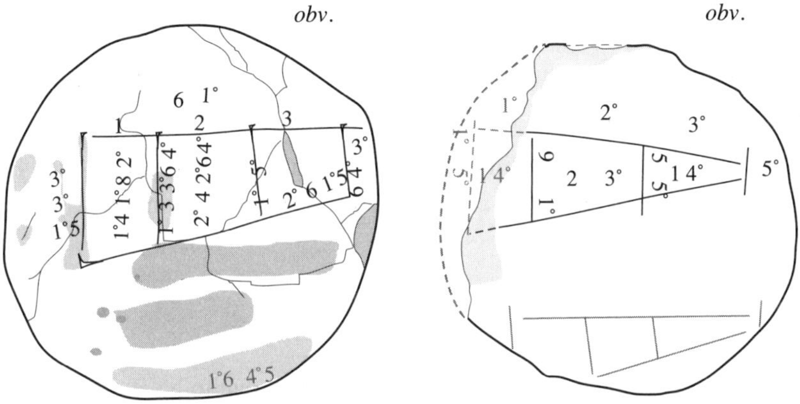

Fig. 11.4.1. Ash. 1922.168 and MS 3908. Hand tablets with 3-striped trapezoids.

In and around the trapezoid on **Ash. 1922.168** (Robson, *MMTC* (1999), 273) are recorded the following values:

u_1, u_2, u_3 = 1 (00), 2 (00), 3 (00),
s_1, s_2, s_3, s_4 = 15, 13;36 40, 10;50, 6;40,
A_1, A_2, A_3 = 14 18 20, 24 26 40, 26 15.

Similarly, in and around the trapezoid on **MS 3908** (Friberg, *RC* (2007), Sec. 8.1) are recorded the values

u_1, u_2, u_3 = 10, 20, 30,
s_1, s_2, s_3, s_4 = [10;50], 9 10, 5;50, ;50,
A_1, A_2, A_3 = [1] 40, 2 30, 1 40.

It is likely that both texts are answers to assignments, and that in each case the simplest of the recorded values are the ones that were given initially, while the more complicated values are the ones that were computed by the student. In Ash. 1922.168, the given values may have been

u_1, u_2, u_3 = 10, 20, 30, $A_1 = A_3$ = 1 40.

In MS 3908, on the other hand, the given values were probably instead

u_1, u_2, u_3 = 1 (00), 2 (00), 3 (00), s_1, s_4 = 15, 6;40.

In both cases, the partial lengths are to each other in the ratios 1 : 2 : 3, and there are altogether *5 given values*, which is precisely what is needed to make a 3-striped trapezoid fully determined.

With the mentioned given values, the problem in Ash. 1922.168 is quite simple, since it is clear that the 'feed' of the trapezoid is

$$f = (15 - 6{:}40) / (10 + 20 + 30) = 8;20 / 1\ 00 = ;08\ 20.$$

Therefore, three applications of the linear similarity rule will show that

$$s_2 = 15 - ;08\ 20 \cdot 10 = 15 - 1;23\ 20 = 13;36\ 40, \quad \text{and so on.}$$

In MS 3809, the following system of linear equations results from two applications of the area rule and two applications of the similarity rule:

$$s_1 + s_2 = 2 \cdot 1\ 40 /10 = 20,$$
$$s_3 + s_4 = 2 \cdot 1\ 40 / 30 = 6;40,$$
$$s_3 = s_1 - 3\ (s_1 - s_2) = 3\ s_2 - 2\ s_1,$$
$$s_4 = s_1 - 6\ (s_1 - s_2) = 6\ s_2 - 5\ s_1.$$

This system of linear equations can easily be shown to have the solution

$$s_1, s_2, s_3, s_4 = 10;50, 9;10, 5;50, ;50 = ;50 \cdot (13,11, 7, 1).$$

These are the values recorded on the hand tablet.

A diagram showing a 5-striped trapezoid precedes the problem text (of which most is lost) on the clay tablet **IM 31248** (Bruins, *Sumer* 9 (1953)). It is likely that here the given values were, just as in the case of Ash. 1922.168, the partial lengths, 3 (00), 1 (00), 3 (00), 1(00), 3 (00), and the upper and lower fronts, 45 and 1. With the 'feed' f equal to 44 / 11 (00) = ;04, the four transversals (33, 29, 17, 13) can be computed by repeated use of the similarity rule.

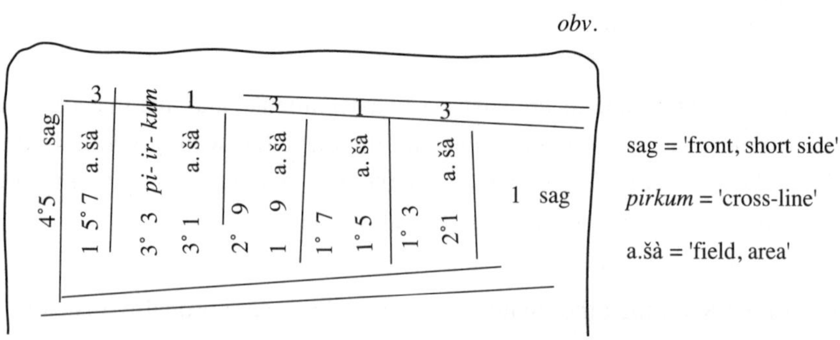

Fig. 11.4.2. IM 31248. An easy problem for a 5-striped trapezoid.

11.5. Erm. 15189. Diagrams for Ten Double Bisected Trapezoids

Bisected trapezoids with $s_a : d : s_k = 7 : 13 : 17$ and $A_a = A_k$ were discussed above in Secs. 11.3 a (IM 58045), 11.3 b (VAT 8512), and 11 3 c (YBC 4675).

With departure from this standard example, OB mathematicians generalized the idea of a bisected trapezoid in a variety of ways. (*Cf.* the survey in Friberg, *RlA 7* (1990), Sec. 5.4 k, and see the continued discussion of the topic in the sections below.) One interesting generalization is demonstrated by the geometric table text **Erm. 15189** (Vaiman, *EV* 10 (1955)), which displays 10 closely related examples of a "double bisected trapezoid", where in each case two bisected trapezoids are joined in such a way that they together form a new trapezoid.

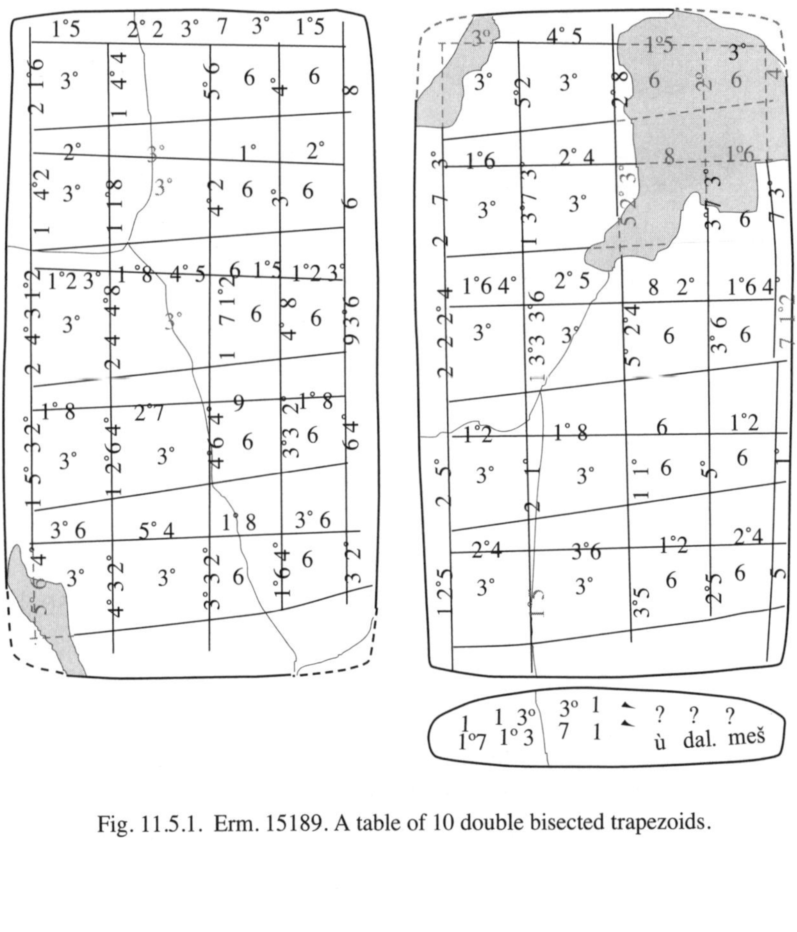

Fig. 11.5.1. Erm. 15189. A table of 10 double bisected trapezoids.

Fig. 11.5.2. Erm. 15189. Scale 1:2.
Photos: The State Ermitage Museum, St. Petersburg.

In example # 1 on the obverse of Erm. 15189, for instance, the first of the two joined bisected trapezoids has the partial areas 30, 30, the partial lengths 15, 22;30, and the fronts (actually two fronts and a transversal) 2 16, 1 44, 56. The second trapezoid has the partial areas 6, 6, the partial lengths 7;30, 15, and the fronts 56, 40, 8. Here

$$2\ 16, 1\ 44, 56 = 8 \cdot (17, 13, 7) \quad \text{and} \quad 56, 40, 8 = 8 \cdot (7, 5, 1).$$

All the other 9 examples are constructed in precisely the same way as # 1, so that in the first bisected trapezoid the partial areas are 30, 30 and the three fronts are equal to a multiple of the standard triple 17, 13, 7, while in the second bisected trapezoid the partial areas are 6, 6 and the three fronts are equal to a multiple of triple 7, 5, 1.

The situation is explained by a rudimentary diagram inscribed on the lower edge of the reverse of the clay tablet. Apparently, according to the diagram, the basic configuration is a double bisected trapezoid with the partial lengths 1 (00), 1 30, 30, 1 (00), and the fronts 17, 13, 7, <5> 1.

Compared to this basic configuration, the lengths of the trapezoid #1 in Erm. 15189 are scaled down by the factor 4 and the fronts are scaled up by the factor 8. However, it is more interesting to investigate the relations between # 1 and the remaining examples. Which are the relations between the lower front 8 in # 1 and the lower fronts 6, 9;36, 6;40, 3;20 in the remaining cases on the obverse, and the lower fronts 4, 7;30, 7;12, 10, 5 in the five cases on the reverse? The answer is that, in a way which is typical for OB mathematics, the lower fronts of the 10 trapezoids are

	8		4 =	8 · (1 – 1/2) =	1 · 8 / 2
6 =	8 · (1 – 1/4) =	3 · 8 / 4	7;30 =	8 · (1 – 1/16) =	15 · 8 / 16
9;36 =	8 · (1 + 1/5) =	6 · 8 / 5	7;12 =	8 · (1 – 1/10) =	9 · 8 / 10
6;40 =	8 · (1 – 1/6) =	5 · 8 / 6	10 =	8 · (1 + 1/4) =	5 · 8 / 4
3;20 =	1/2 · 6;40		5 =	1/2 · 10	

Here the pairs of numbers (1, 2), (3, 4), (4, 5), (5, 6), (9, 10), (15, 16) are regular sexagesimal "twins", just like the pairs (8, 9), (9, 10), (15, 16), (12;5, 13;5) used to construct the data for the divided trapezoids in Ist. Si. 269, *obv*. (Sec. 11.3 f). Thus, the idea behind this particular choice of scaling factors is to subtract *n*th parts where both *n* and *n* – 1 are regular sexagesimal numbers, or to add *n*th parts, where both *n* and *n* + 1 are regular!

Moreover, in Erm. 15189, *the scalings of the partial lengths is inversely proportional to the scaling of the fronts*, in order to ensure that the partial areas stay the same. Therefore, the lower lengths of the 10 trapezoids are

	15		30 =	15 · (1 + 1) =	2 · 15 / 1
20 =	15 · (1 + 1/3) =	4 · 15 / 3	16 =	15 · (1 + 1/15) =	16 · 15 / 15
12;30 =	15 · (1 – 1/6) =	5 · 15 / 6	16;40 =	15 · (1 – 1/9) =	8 · 15 / 9
18 =	15 · (1 + 1/5) =	6 · 15 / 5	12 =	15 · (1 – 1/5) =	4 · 15 / 5
36 =	2 · 18		24 =	2 · 12	

Note that the use of regular sexagesimal twins is imperative in this situation, in view of the following OB "reciprocity rule" (*cf.* Friberg, *UL* (2005), Sec. 3.1 f):

The reciprocal of $1 - 1/n$ is $1 + 1/(n - 1)$, and
the reciprocal of $1 + 1/n$ is $1 - 1/(n + 1)$.

Note also that it is not obvious why there should exist any double bisected trapezoid at all. The conditions for the existence of a double bisected trapezoid can be analyzed as follows.

There are two equations that the parameters for a double bisected trapezoid like the one in Fig. 11.5.3 below must satisfy. The first condition is that the lower front in the upper trapezoid must be equal to the upper front in the lower trapezoid. The second condition is that the 'feed' f must be the same for the two trapezoids.

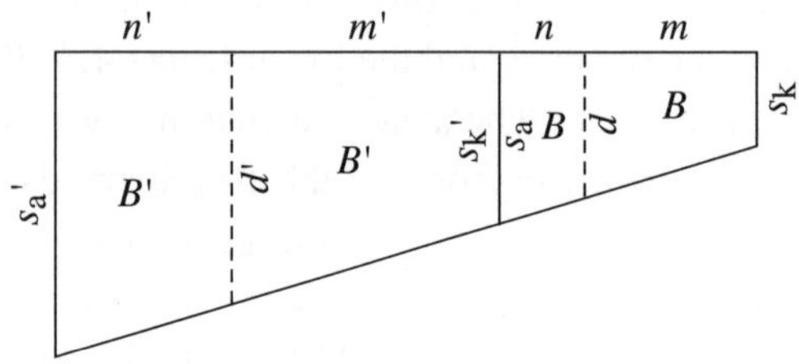

Fig. 11.5.3. The parameters for a double bisected trapezoid.

In view of the generating rule for (rational) transversal triples, these conditions can be expressed in the following way:

$$B'/m' - B'/n' + 2\,B'/(m'+n') = s_k' = s_a = 2\,B/(m+n) - B/m + B/n,$$

and

$$(2\,B'/n' - 2\,B'/m')/(m'+n') = f' = f = (2\,B/n - 2\,B/m)/(m+n).$$

Equivalently,

$$[B'/\{(m'+n')\cdot m'\cdot n'\}]\cdot\{2\,m'\cdot n' - (\text{sq.}\,m' - \text{sq.}\,n')\} = s_k'$$
$$= s_a = [B/\{(m+n)\cdot m\cdot n\}]\cdot\{2\,m\cdot n + (\text{sq.}\,m - \text{sq.}\,n)\},$$

and

$$[B'/\{(m'+n')\cdot m'\cdot n'\}]\cdot(m'-n') = f' = f = [B/\{(m+n)\cdot m\cdot n\}]\cdot(m-n).$$

Together, these two equations show that

$$\{2\,m'\cdot n' - (\text{sq.}\,m' - \text{sq.}\,n')\}/(m'-n') = \{2\,m\cdot n + (\text{sq.}\,m - \text{sq.}\,n)\}/(m-n),$$

or

$$\{2\,\text{sq.}\,n' - \text{sq.}\,(m'-n')\}/(m'-n') = \{2\,\text{sq.}\,m - \text{sq.}\,(m-n)\}/(m-n).$$

This single equation for the two unknowns m' and n' can be *arbitrarily complemented* by the second equation

$$m' - n' = m - n.$$

Then the first equation collapses to

$$\text{sq.}\,n' = \text{sq.}\,m \quad \text{so that} \quad n' = m.$$

What all this means is that if m, n, B are known parameters for a given bisected trapezoid, then the corresponding three parameters for a second bisected trapezoid which together with the given trapezoid forms a double bisected trapezoid can be computed as follows:

$n' = m, \ \ m' = n' + (m - n) = 2\,m - n,$ and $B' = B \cdot \{(m' + n') \cdot m' \cdot n'\}/\{(m + n) \cdot m \cdot n\}.$

Let, for instance, the given bisected trapezoid be the one with

$m, n, B = 2, 1, 6$

and, consequently,

$s_a, d, s_k = 7, 5, 1,$ and $m - n = 1.$

Then,

$n' = 2, \ \ m' = 3, \ \ B' = 6 \cdot (5 \cdot 3 \cdot 2) / (3 \cdot 2 \cdot 1) = 30,$

and, consequently,

$s_a, d, s_k = 17, 13, 7.$

Similarly, if

$m, n, B = 3, 2, 30,$

then

$m - n = 1, \ \ n' = 3, \ \ m' = 4, \ \ B' = 30 \cdot (7 \cdot 4 \cdot 3) / (5 \cdot 3 \cdot 2) = 1\ 24,$

and

$s_a, d, s_k = 31, 25, 17.$

Clearly it is possible, in this way, to construct *chains of bisected trapezoids*, 3, 4, or more. As will be shown by the example in the problem text AO 17264 (below), this fact was well known in Babylonian mathematics.

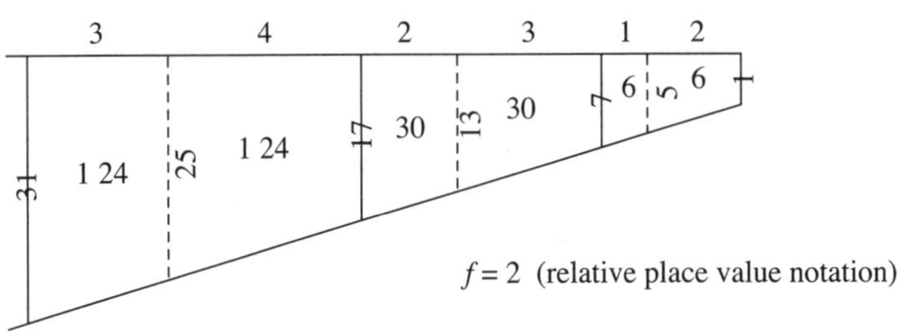

Fig. 11.5.4. A chain of 3 bisected (rational) trapezoids.

Note, by the way, that the simplest way to compute the consecutive transversal triples in a chain of bisected trapezoids is to compute the 'feed' f for the first trapezoid and then use the similarity rule. When, for instance, m, n, B = 2 (00), 1 (00), 6 (00), and s_a, d, s_k = 7, 5, 1 in the first bisected trapezoid, that (see Fig.11.5.4 above), then

$$f = (7-1) / (2\ 00 + 1\ 00) = ;02, \quad d' = 7 + ;02 \cdot 3\ 00 = 13, \quad s_a' = 13 + ;02 \cdot 2\ 00 = 17, \ etc.$$

11.6. AO 17264. A Problem for a Chain of 3 Bisected Quadrilaterals

AO 17264 (Neugebauer, *MKT 1* (135), 126-134) is a post-Old-Babylonian (Kassite) single problem text with a curiously corrupt problem for a threefold bisected quadrilateral. Here is the statement of the problem:

AO 17264, literal translation	explanation
A ····· .	A quasi-trapezoid
2 15 the upper length, 1 21 the lower length,	u = 2 15, v = 1 21
3 33 the upper front, 51 the lower front.	s_a = 3 33, s_k = 51
6 brothers. The oldest and the next one equal,	6 sub-fields, $A_1 = A_2$
3 and 4 equal, 5 and 6 equal.	$A_3 = A_4$, $A_5 = A_6$
The extents, the transversals, and the descents are what?	

Here, as in the case of the bisected quadrilateral in YBC 4675 (Sec. 11. 3 c) the introduction of a quadrilateral instead of a trapezoid is an intentional but quite meaningless complication of the problem. The corresponding problem for a trapezoid would be the case *when the upper and lower fronts and the length of the trapezoid are given and when one asks for the division of the trapezoid into 6 parallel stripes* as in the example in Fig. 11.5.4 above. This appears to be an *under-determined* problem.[32] The text seems to be the result of a thoughtless teacher handing out a defective assignment and a smart but dishonest student handing in his faked answer.

Note that the given 'upper' and 'lower' lengths of the quadrilateral are so different in size that the quadrilateral would be unacceptably lopsided. Therefore, in the explanation below (with notations as in Fig. 11.6.1) the condition that the lower length should be given has been disregarded, and

---

32.Thus, Neugebauer writes (*MKT 1*, 130): "Man sieht sofort dass die Aufgabe in dieser Form unbestimmt ist ··· Es ist also klar, dass zur eindeutigen Lösung noch zwei weitere Bedingungen nötig sind ···".

it is assumed instead that the figure is a trapezoid.

The solution in the text proceeds (essentially) in the following steps:

1. $s_2 = (s_1 + s_4 + v / u) / (u + v)/2 = 26\ 40;36 / 1\ 48 = 2\ 27\ (= 3 \cdot 49)$
2. $s_3 = s_2 - (u - v) = 2\ 27 - 54 = 1\ 33\ (= 3 \cdot 31)$
3. $\text{sq. } d_1 = (\text{sq. } s_1 + \text{sq. } s_2)/2 = 9\ 18\ 09, \quad d_1 = 3\ 03\ (= 3 \cdot 1\ 01)$
4. $\text{sq. } d_2 = (\text{sq. } s_2 + \text{sq. } s_3)/2 = 4\ 12\ 09, \quad d_2 = 2\ 03\ (= 3 \cdot 41)$
5. $\text{sq. } d_3 = (\text{sq. } s_3 + \text{sq. } s_4)/2 = 1\ 33\ 45, \quad d_3 = 1\ 15\ (= 3 \cdot 25)$
6. $n' = u \cdot (s_1 - d_1)/(s_1 - s_4) = ;11\ 06\ 40 \cdot 2\ 15 = 25$
7. $m' = u \cdot (d_1 - s_2)/(s_1 - s_4) = ;13\ 20 \cdot 2\ 15 = 30$
6. $n = u \cdot (s_2 - d_2)/(s_1 - s_4) = ;08\ 53\ 20 \cdot 2\ 15 = 20$
7. 'you make the 3 remaining descents like the preceding ones'

Apparently, the student who handed in this answer to the assignment knew beforehand that part of the answer should be that

$$s_a, d_1, s_2, d_2, s_3, d_3, s_4$$
$$= 3 \cdot (1\ 11, 1\ 01, 49, 41, 31, 25, 17) = 3\ 33, 3\ 03, 2\ 27, 2\ 03, 1\ 33, 1\ 15, 51$$

Therefore, in steps 1 and 2, he brashly invented meaningless combinations of the given data which gave the correct values for s_2 and s_3. However, from there on he proceeded correctly.

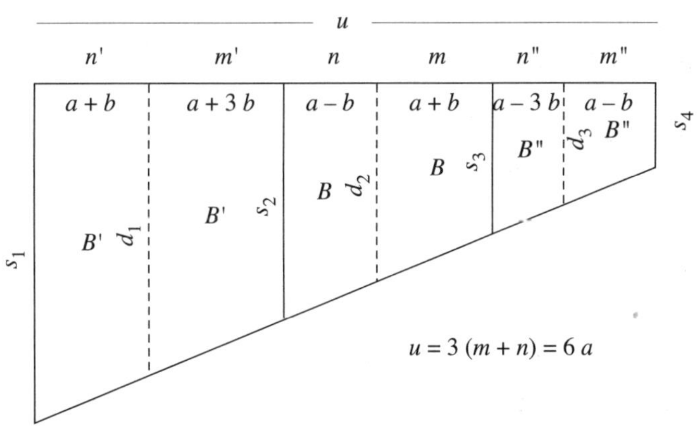

Fig. 11.6.1. A chain of three bisected trapezoids. The general case.

It is a remarkable fact that, in spite of appearances, the problem is well posed, in the following sense:

Suppose that in Fig. 11.6.1 the chain of three bisected trapezoids has been constructed with departure from the bisected trapezoid in the middle, with the parameters m, n, B, and that *the added bisected trapezoids to the left and right satisfy the condition, mentioned above, that*

$$m' - n' = m - n = m'' - n''.$$

Then all the parameters of the chain of bisected trapezoids are uniquely determined as soon as the values of u, s_1, *and* s_4 *are given.*

It is clear that if the values of m, n, and B can be found, then the other parameters can be computed. Therefore, what is needed is 3 equations for the 3 unknowns m, n, B. The first of these equations is easy to find. Indeed, recall that (see Sec. 11.5 above):

$$n' = m, \quad m' = n' + (m - n) = 2\,m - n, \quad \text{and}$$
$$B' = B \cdot \{(m' + n') \cdot m' \cdot n'\}/\{(m + n) \cdot m \cdot n\},$$

and similarly for the parameters of the trapezoid to the right. Therefore,

$$u = (n' + m') + (n + m) + (n'' + m'') = (3\,m - n) + (m + n) + (3\,n - m) = 3\,(m + n).$$

If, for the sake of symmetry, the pair m, n is called $a + b, a - b$, then

$$m, n = a + b, a - b \;\Rightarrow\; m' = a + 3\,b, n' = a + b, \quad \text{and} \quad m'' = a - b, n'' = a - 3\,b.$$

Then also

$$u = 3\,(m + n) = 6\,a, \quad \text{so that} \quad a = u\,/\,6.$$

Next, in view of the equations above for n', m', B', together with the generating rule for (rational) transversal triples

$$s_1 = \{2B'\,/\,(m' + n') \cdot m' \cdot n'\} \cdot \{2\,m' \cdot n' + (\text{sq.}\,m' - \text{sq.}\,n')\}$$
$$= \{2B\,/\,(m + n) \cdot m \cdot n\} \cdot \{2\,(a + 3\,b) \cdot (a + b) + \text{sq.}\,(a + 3\,b) - \text{sq.}\,(a + b)\}$$
$$= \{2B\,/\,(m + n) \cdot m \cdot n\} \cdot 2\,(\text{sq.}\,a + 6\,a \cdot b + 7\,\text{sq.}\,b).$$

Similarly, then

$$s_4 = \{2B\,/\,(m + n) \cdot m \cdot n\} \cdot 2\,(\text{sq.}\,a - 6\,a \cdot b + 7\,\text{sq.}\,b).$$

Consequently,

$$s_1 \cdot (\text{sq.}\,a - 6\,a \cdot b + 7\,\text{sq.}\,b) = s_4 \cdot (\text{sq.}\,a + 6\,a \cdot b + 7\,\text{sq.}\,b).$$

Since s_1 and s_2 are known, this is a quadratic equation for a/b, and since $a = u/6$, both a and b are now known. The values of the two interior fronts and the three transversals can then be computed by use of similarity.

For a numerical example, choose

$$s_1 = 3\;33 = 3 \cdot 1\;11, s_4 = 51 = 3 \cdot 17, \text{and } u = 2\;15$$

as in AO 17264 (disregarding the given value $v = 1\;21$, which transforms the trapezoid into a distorted quadrilateral). Then

$a = 2\ 15\ /\ 6 = 22;30$

and

$1\ 11\ (\text{sq.}\ t - 6\ t + 7) = 17\ (\text{sq.}\ t + 6\ t + 7),\quad \text{where}\quad t = a/b.$

This quadratic equation can be simplified to

$9\ \text{sq.}\ t + 1\ 08 = 1\ 28\ t.$

The solution can be found in the usual way. One finds that

$\text{sq.}\ (3\ t - 14;40) = \text{sq.}\ 12;20\quad \text{so that}\quad 3\ t = 14;40 + 12;20 = 27,\quad \text{and}\quad t = 9.$

Therefore,

$a = 22;30\quad \text{and}\quad b = 22;30\ /\ 9 = 2;30,$

and consequently

$a + b, a + 3\ b, a - b, a + b, a - 3\ b, a - b = 25, 30, 20, 25, 15, 20.$

Then, finally, by similarity,

$f = (3\ 33 - 51)\ /\ 2\ 15 = 6\ /\ 5 = 1;12$

$\Rightarrow$

$s_2 = 3\ 33 - 55 \cdot 1;12 = 2\ 27,\quad s_3 = 2\ 27 - 45 \cdot 1;12 = 1\ 33,$

and so on.

It is likely that it was in this way that the *author of the problem* intended it to be solved. If it was, then this Kassite (post-Old-Babylonian) problem and its intended solution mark one of the high points of Babylonian mathematics. In this connection it is worth pointing out that the only other known Kassite mathematical problem text is MS 3876 (Sec. 8.3 above), which with its computation of the weight of an icosahedron marks another high point of Babylonian mathematics. Therefore, perhaps, Høyrup was too pessimistic with regard to the level of Kassite mathematics when he wrote in his *LWS* (2002), 387, the following words about AO 17264:

> "· · · what the text offers is a mock solution · · · a piece of sham mathematics — all that remains of the stringency and creativity of the Old Babylonian mathematical school is the higher level on which the fraud is perpetrated."

In the terminology of modern mathematics, the problem stated (and incorrectly solved) in AO 17264 is the oldest known example of a "boundary value problem", where the initial and final values are known for a sequence of numbers generated by a recursive procedure.

11.7. VAT 7621 # 1. A 2 · 9-striped trapezoid

In OB examples of divided trapezoids, the lengths of the transversals are, as a rule, *exactly determined sexagesimal numbers*. An exception from this rule is **VAT 7621** (Thureau-Dangin, *TMB* (1938), 99).

VAT 7621, literal translation	explanation

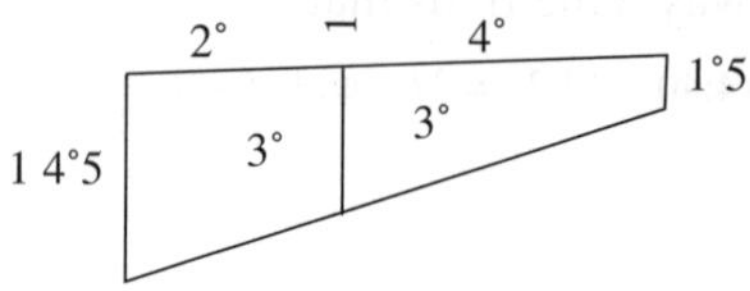

# 1	Two are they, 9 each their sons. The upper bùr in nine divide, and the lower bùr in nine divide, and (each) soldier show him his stake.	Divide each part of the 2-striped trapezoid in the diagram into 9 equal parts. Find the partial lengths.
# 2	To 3 men divide equally, and (each) soldier show him his stake.	Divide into 3 equal parts. Find the partial lengths.
# 3	....	....

As shown in the diagram, the given upper and lower fronts of a bisected trapezoid are s_a, s_k = 1 45 = 15 · 7, and 15 = 15 · 1. The two partial areas are both equal to 30 (00) sq. ninda = 1 bùr. Therefore, the transversal is d = 15 · 5 = 1 15. The normalized length of the trapezoid, 1 00 ninda, is divided in two parts, 20 and 40, in the ratio 1: 2.

An unusual feature of problem # 1 is that the two sub-trapezoids are further divided in nine parts each, allotted to the 2 · 9 sons of two (men). Thus, each son gets 1/9 bùr = 1/3 èše = 2 iku. To 'show each soldier his stake' means, as usual, to determine the positions of the transversals separating the 18 lots from each other. This can be done by use of a method suggested by Parker, *JEA* 61 (1975) as an explanation of the given length numbers in the Egyptian demotic mathematical papyrus ***P.Heidelberg 663* # 2**. (See the discussion of *P.Heidelberg 663* in Friberg, *UL* (2005), Sec. 3.7.)

Let the sum of the first n sons' lots, counted from the left, be the trapezoid with the fronts s_a, d_n, and the length $(s_a - d_n)/f$, where f is the inclination of the sloping side of the trapezoid. Let the corresponding area be A_n.

Then (*cf.* YBC 4675, Sec. 11.3 c above)

$$A_n = (s_a + d_n)/2 \cdot (s_a - d_n)/f = (\text{sq.}\ s_a - \text{sq.}\ d_n)/(2f), \quad \text{so that} \quad \text{sq.}\ d_n = \text{sq.}\ s_a - 2f \cdot A_n.$$

In VAT 7621, s_a = 1 45, f = (1 45 – 15)/1 00 = 1;30, and A_n = 1 00 00/18 · n = 3 20 · n. Hence, the length d_n of the n-th transversal can be computed as

$$d_n = \text{sqs.}\ (\text{sq.}\ s_a - 2f \cdot A_n) = \text{sqs.}\ (3\ 03\ 45 - 10\ 00 \cdot n), \quad \text{for} \quad n = 1, 2, \cdots, 18.$$

Once the transversals are known, it is easy to find also their positions.

The problem in VAT 7621 #2 seems to be to divide the same trapezoid into three stripes of equal area. The solution can be found as follows:

$$\text{sq.}\ d_1 = \text{sq.}\ s_a - 2f \cdot A_a = \text{sq.}\ 1\ 45 - 2 \cdot 1;30 \cdot 20\ 00 = 3\ 03\ 45 - 1\ 00\ 00 = 2\ 03\ 45,$$
$$\text{sq.}\ d_2 = \text{sq.}\ s_k + 2f \cdot A_k = \text{sq.}\ 15 + 2 \cdot 1;30 \cdot 20\ 00 = 3\ 45 + 1\ 00\ 00 = 1\ 03\ 45.$$

Hence,

$$d_1 = 15 \cdot \text{sqs.}\ 33 = \text{appr.}\ 1\ 26;15, \quad \text{and} \quad d_2 = 15 \cdot \text{sqs.}\ 17 = \text{appr.}\ 1\ 01;52.$$

Or, one can make use of the general trapezoid bisection equations:

$$\text{sq.}\ d_1 = (2\ \text{sq.}\ s_a + \text{sq.}\ s_k)/3 = \text{sq.}\ 15 \cdot (2 \cdot 49 + 1)/3 = \text{sq.}\ 15 \cdot 33,$$
$$\text{sq.}\ d_2 = (\text{sq.}\ s_a + 2\ \text{sq.}\ s_k)/3 = \text{sq.}\ 15 \cdot (49 + 2 \cdot 1)/3 = \text{sq.}\ 15 \cdot 17.$$

11.8. VAT 7531. Cross-wise striped trapezoids.

VAT 7531 ## 1-4 (Friberg, *UL* (2005), Sec. 3.7 c; cf. Fig. 1.12.7 above) are the only known OB examples of problems for "cross-wise striped trapezoids". In the trapezoids considered there, the lengths are parallel, not the fronts, and it is likely that the transversals are required to be orthogonal to the parallel lengths, as in the example shown in Fig. 11.8.1 below. (There are no explicit solution procedures in the text that can confirm this reasonable conjecture) Here is, for instance, the text of problem # 1.

VAT 7531 # 1, literal translation	explanation
2 35 50 the long length, 1 54 10 *the short length,*	u' = 2 35;50, u'' = 1 54;10
50 the upper front, 41 40 the low*er* front.	s_a = 50, s_k = 41;40
Its area, how much it is, find o*ut, then*	A = ?
to 3 brothers equal*ly divide it,*	Divided in 3 equal parts
and (each) soldier s*how him his stake*	Compute the partial lengths

In VAT 7531 # 1, the given figure can be interpreted as a trapezoid composed of a central rectangle and two flanking non-equal triangles. If the rectangle is removed, what remains is a rotated symmetric triangle, with two sides equal to 41;40 and the third side equal to 50. The height against the side 50 can be computed by use of the diagonal rule. It is 33;20, so that the triangle can be reinterpreted as the composition of two equal right triangles with the sides 8;20 · (3, 4, 5).

The height h against the side 41;40 can be computed as follows:

$$h \cdot 41;40/2 = A_{\text{triangle}} = 33;20 \cdot 50/2, \quad \text{so that} \quad h = 33;20 \cdot 50 \,/\, 41;40 = 40.$$

After h has been computed, the two components a and b of the base of the triangle can be computed by use of the diagonal rule. They are 30 and 11;40. Hence, the triangle has an alternative composition as two right triangles with the sides 10 · (3, 4, 5) and 1;40 · (7, 24, 25) joined together.

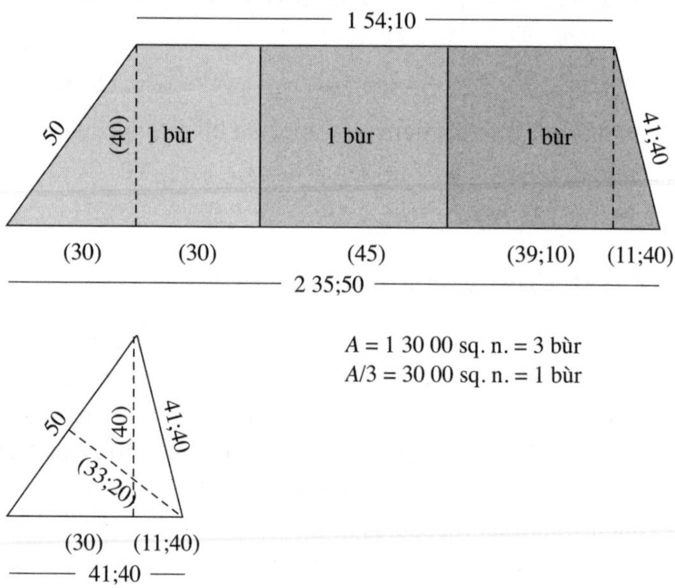

Fig. 11.8.1. VAT 7531 # 1. Three brothers sharing a trapezoidal field.

The aim of exercise # 1 is to divide the given trapezoid equally between three brothers. Now, it is clear that the area of the trapezoid is equal to

(2 35;50 + 1 54;10)/2 n. · 40 n. = 2 15 n. · 40 n. = 1 30 00 (sq. ninda) = 3 bùr.

One third of that area is 30 00 (sq. ninda) = 1 bùr, which is equal to the area of a rectangle with the length 45 n. and the height 40 n. Hence, the middle brother gets a central rectangle with these sides, while the first brother gets the left triangle plus a rectangle with the sides 30 and 40, and the third brother gets the right triangle plus a rectangle with the sides 39;10 and 40. In this way, each brother gets a field with the area 1 bùr.

11.9. *TMS* 23. Confluent Quadrilateral Bisections in Two Directions

***TMS* 23** (Bruins and Rutten (1961)) is a fragment from the middle of one side of a clay tablet. (See the proposed reconstruction in Fig. 11.9.1.)

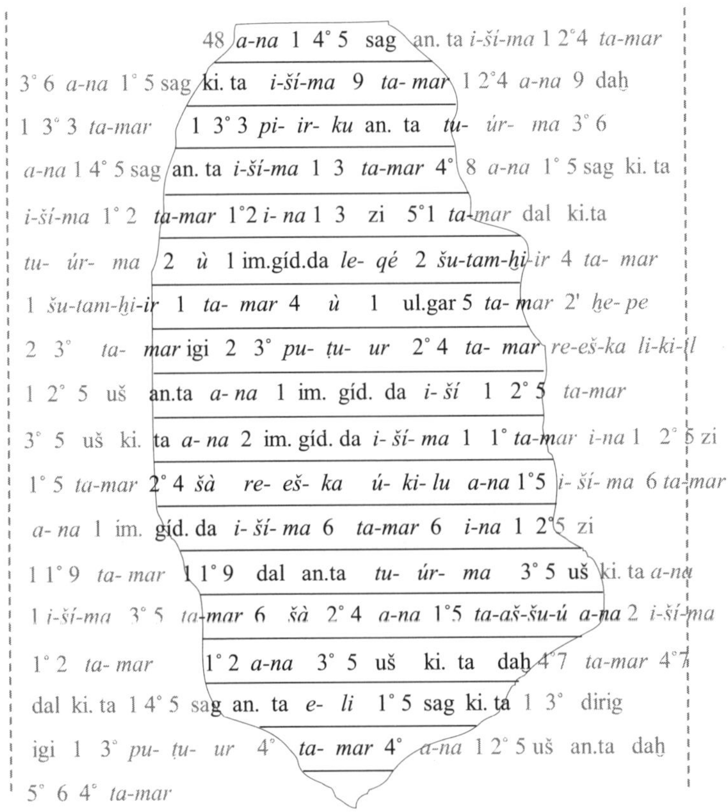

Fig. 11.9.1. *TMS* 23. Proposed reconstruction of part of the text.

The interpretation of *TMS* 23 in the original publication of the text was only partly successful. A more thorough analysis shows that *TMS* 23 is among the most important of all known OB mathematical texts. Here is the translation of the proposed reconstruction of the text:

TMS 23, literal translation	explanation
....	....
48 to 1 45, the *upper* front *raise, then 1 24 you see.*	;48 · s_{a} = ;48 · 1 45 = 1 24
36 to 15, the lower *front* raise, then 9 you see.	;36 · s_{k} = ;36 · 15 = 9
1 24 to 9 add, then 1 33 you see,	**;48 · s_{a} + ;36 · s_{k}**
1 33 the upper crossline	= 1 24 + 9 = 1 33 = $\boldsymbol{d_{\text{a}}}$
Turn around	Begin again
36 to 1 45, the upper *front,* raise, then 1 03 you see.	;36 · s_{a} = ;36 · 1 45 = 1 03
48 to 15, the lower front, raise, then 12 you see.	;48 · s_{k} = ;48 · 15 = 12
12 from 1 03 tear off, 51 you see, *the lower crossline.*	**;36 · s_{a} – ;48 · s_{k}** = 51 = $\boldsymbol{d_{\text{k}}}$
Turn around	Begin again
Take 2 and 1 from the table.	Choose the parameters 2 and 1
Make 2 equalsided, *4 you see.*	sq. 2 = 4
Make 1 equalsided, 1 you see.	sq. 1 = 1
4 and 1 heap, 5 you see.	sq. 2 + sq. 1 = 4 + 1 = 5
1/2 break 2 30 you see.	(sq. 2 + sq. 1)/2 = 2;30
The opposite of 2 30 resolve, 24 you see.	1 / 2;30 = ;24
Let it keep your head.	Remember this value!
1 25 the upper *length*, to 1 from the table,	1 · u_{a} = 1 · 1 25
raise, 1 25 *you see.*	= 1 25
35, the lower length, to 2, from the table,	2 · u_{k} = 2 · 35
raise, 1 10 you see.	= 1 10
From 1 25 tear off, 15 you see.	1 · u_{a} – 2 · u_{k} = 1 25 – 1 10 = 15
24 that held your head to 15 *raise, then 6 you see.*	(1 · u_{a} – 2 · u_{k}) · ;24 = 6
To 1, from the table, raise, then 6 you see.	1 · (1 · u_{a} – 2 · u_{k}) · ;24 = 6
6 from 1 25 *tear off, 1 19 you see,*	$\boldsymbol{u_{\text{a}}}$ **– 1 · (1 ·** $\boldsymbol{u_{\text{a}}}$ **– 2 ·** $\boldsymbol{u_{\text{k}}}$**) · ;24**
1 19 the upper transversal.	= 1 19 = $\boldsymbol{e_{\text{a}}}$
Turn around	Begin again
35, the *lower* length, *to 1 raise, then 35 you see.*	1 · u_{k} = 1 · 35 = 35
6, that of 24 that you raised to 15, to *2 raise,*	2 · (1 · u_{a} – 2 · u_{k}) · ;24 = 6 · 2
then 12 you see.	= 12
12 to 35, the lower length, add, *47 you see.*	$\boldsymbol{u_{\text{k}}}$ **+ 2 · (1 ·** $\boldsymbol{u_{\text{a}}}$ **– 2 ·** $\boldsymbol{u_{\text{k}}}$**) · ;24**
47, the lower transversal.	= 35 + 12 = 47 = $\boldsymbol{e_{\text{k}}}$
1 45, the upper *front*, over 15, the lower front,	$s_{\text{a}} - s_{\text{k}}$ = 1 45 – 15
is 1 30 beyond.	= 1 30
The opposite of 1 30 resolve, 40 you see.	1 / ($s_{\text{a}} - s_{\text{k}}$) = 1 / 1 30 = :00 40
40 to 1 25, the upper length, raise, then 56 40 you see.	u_{a} / ($s_{\text{a}} - s_{\text{k}}$) = ;56 40
....	....

The statement of the problem is not preserved, but apparently the object considered in this text is a quadrilateral with the 'upper front' s_{a} = 1 45 (=

15 · 7), the 'lower front' s_k = 15 (= 15 · 1), the 'upper length' u_a = 1 25 (= 5 · 17), and the 'lower length' u_k = 35 (= 5 · 7). See Fig. 11.9.2 below. Note that if the Sumerian/OB "quadrilateral area rule" had been used in this text (which it is not), the area of the quadrilateral would have been found to be a conspicuously round number:

$$A = (u_a + u_k)/2 \cdot (s_a + s_k)/2 = (1\ 25 + 35)/2 \cdot (1\ 45 + 15)/2 = 1\ 00 \cdot 1\ 00 = 1\ 00\ 00 \text{ (sq. n.)}$$

In the present case, the rule would yield a wildly inaccurate value.

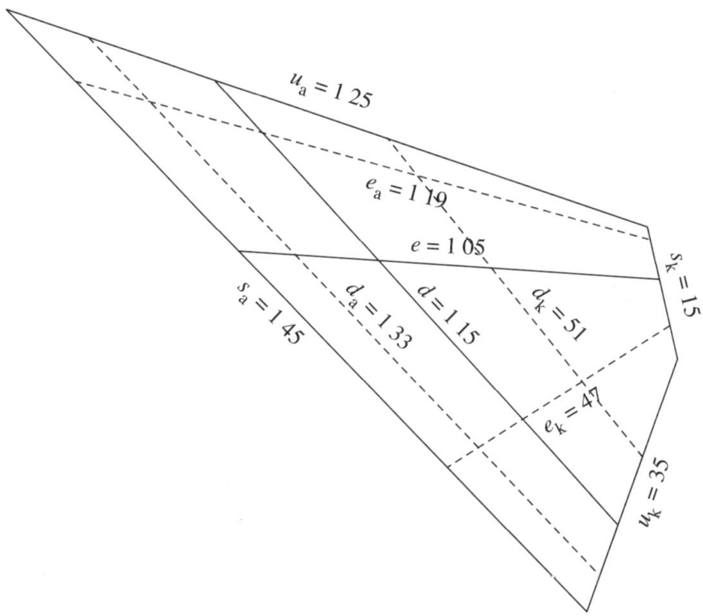

Fig. 11.9.2. *TMS* 23. Confluent bisections in two directions.

As a matter of fact, all the computations in this text are wildly inaccurate, because it is not taken into consideration that the quadrilateral is *not* a (parallel) trapezoid in two directions. Apart from that, however, the procedures in the text are correct, so one way of saving the situation is to interpret the text as dealing with *two separate trapezoids*, one where the 'fronts' are parallel and another where the 'lengths' are parallel.

With this amended interpretation of the text, the problem considered in the exercise can be explained as follows: Let s_a = 1 45 (= 15 · 7) and s_k = 15 (= 15 · 1) be given values for the upper and lower fronts of a (parallel) trapezoid. Then it is clear that the trapezoid is bisected by the transversal $d = 1\ 15$ (= $15 \cdot 5$), so that the triple s_a, d, s_k satisfies the equation

$$\text{sq. } s_a + \text{sq. } s_k = 2 \cdot \text{sq. } d.$$

The question is now if it is possible to find an 'upper' and a 'lower' transversal, d_a and d_k, in the same trapezoid, such that also the trapezoid with d_a and d_k as its upper and lower fronts is bisected by the transversal d. This phenomenon can be called "a confluent trapezoid bisection".

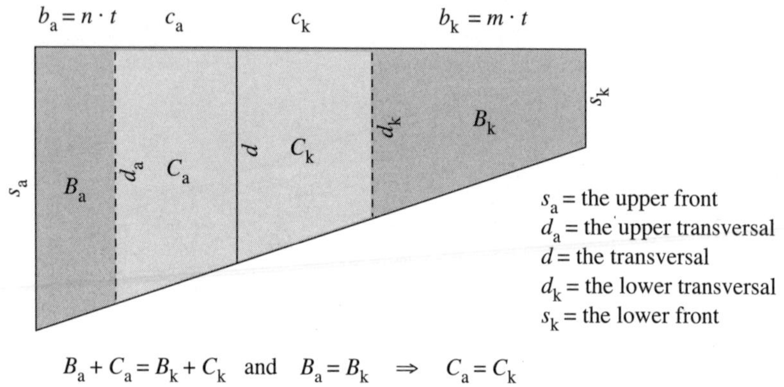

Fig. 11.9.3. A confluent trapezoid bisection.

The first part of the solution procedure in *TMS* 23 probably began with the computation of (the length of) the transversal d. That computation is now lost. Anyway, the solution procedure continues immediately with the computation of the upper and lower transversals as

$$d_a = ;48 \cdot s_a + ;36 \cdot s_k = ;48 \cdot 1\ 45 + ;36 \cdot 15 = 1\ 24 + 9 = 1\ 33,$$
$$d_k = ;36 \cdot s_a - ;48 \cdot s_k = ;36 \cdot 1\ 45 - ;48 \cdot 15 = 1\ 03 - 12 = 51.$$

Although this is not done in the text, it is easy to check that then, indeed,

$$\text{sq. } d_a + \text{sq. } d_k = \text{sq. } 1\ 33 + \text{sq. } 51 = 2\ 24\ 09 + 43\ 21 = 3\ 07\ 30 = 2 \cdot 1\ 33\ 45 = 2 \cdot \text{sq. } d.$$

Therefore, in the trapezoid in Fig. 11.9.3, $C_a = C_k$, as desired.

The explanation for this way of solving the problem comes in the second part of the solution procedure, which is much more explicit. The main idea is the following: The trapezoid in Fig. 11.9.3 is bisected by the transversal d. Therefore the areas $B_a + C_a$ and $B_k + C_k$ are equal. Hence, if also B_a and B_k are equal, it will follow that C_a and C_k are equal, as desired.

Suppose now that B_{a} and B_{k} are equal, and that, in addition,

$b_{\text{a}} : b_{\text{k}} = n : m$ where $b_{\text{a}}, b_{\text{k}}$ are partial lengths and m, n a given pair of integers.

(See again Fig. 11.9.3.) Then the upper and lower transversals d_{a} and d_{k} can be computed as the solutions to the following system of equations:

$$(s_{\text{a}} - d_{\text{a}}) / n = (d_{\text{k}} - s_{\text{k}}) / m, \quad n \cdot (s_{\text{a}} + d_{\text{a}}) = m \cdot (d_{\text{k}} + s_{\text{k}}).$$

(The first of these equations is a similarity equation, the second an area equation.) The system of equations can be solved in the following way:

Set $d_{\text{a}} = s_{\text{a}} - n \cdot t$ and $d_{\text{k}} = s_{\text{k}} + m \cdot t$ where t is a new unknown.

Then the first equation is satisfied. The second equation is also satisfied if

$n \cdot (2\, s_{\text{a}} - n \cdot t) = m \cdot (2\, s_{\text{k}} + m \cdot t)$ that is, if $n \cdot s_{\text{a}} - m \cdot s_{\text{k}} = (\text{sq.}\, m + \text{sq.}\, n)/2 \cdot t$.

Consequently,

$$t = (n \cdot s_{\text{a}} - m \cdot s_{\text{k}}) / (\text{sq.}\, m + \text{sq.}\, n)/2.$$

Thus, finally,

$$d_{\text{a}} = s_{\text{a}} - n \cdot t = s_{\text{a}} - n \cdot (n \cdot s_{\text{a}} - m \cdot s_{\text{k}}) / (\text{sq.}\, m + \text{sq.}\, n)/2, \quad \text{and}$$
$$d_{\text{k}} = s_{\text{k}} + m \cdot t = s_{\text{k}} + m \cdot (n \cdot s_{\text{a}} - m \cdot s_{\text{k}}) / (\text{sq.}\, m + \text{sq.}\, n)/2.$$

This OB "confluent bisection rule" is used explicitly in the second part of the solution procedure in *TMS* 23, where the upper and lower transversals e_{a} and e_{k} between the "parallels" u_{a} and u_{k} are computed as follows:

$$e_{\text{a}} = u_{\text{a}} - 1 \cdot t = u_{\text{a}} - 1 \cdot (1 \cdot u_{\text{a}} - 2 \cdot u_{\text{k}}) / (\text{sq.}\, 2 + \text{sq.}\, 1)/2 = 1\ 19, \quad \text{and}$$
$$e_{\text{k}} = u_{\text{k}} + 2 \cdot t = u_{\text{k}} + 2 \cdot (1 \cdot u_{\text{a}} - 2 \cdot u_{\text{k}}) / (\text{sq.}\, 2 + \text{sq.}\, 1)/2 = 47.$$

Thus, the solution procedure in the *second part* of *TMS* 23 makes use of the confluent bisection rule with $m, n = 2, 1$. Note, by the way, that

$$\text{sq.}\ 1\ 19 + \text{sq.}\ 47 = 1\ 46\ 01 + 36\ 49 = 2\ 20\ 50 = 2 \cdot 1\ 10\ 25 = 2 \cdot \text{sq.}\ 1\ 05.$$

(1 05 is the transversal in the bisected trapezoid with the parallels 1 25, 35.)

How is this confluent bisection rule related to the method used in the *first part* of the solution procedure in *TMS* 23? The answer to this question is that one can show, with a little bit of algebraic manipulation, that the equations above for the upper and lower transversals can be reduced to

$$d_{\text{a}} = s_{\text{a}} - n \cdot (n \cdot s_{\text{a}} - m \cdot s_{\text{k}}) / (\text{sq.}\, m + \text{sq.}\, n)/2$$
$$= \{(\text{sq.}\, m - \text{sq.}\, n)/2 \cdot u_{\text{a}} + m \cdot n \cdot u_{\text{k}}\} / (\text{sq.}\, m + \text{sq.}\, n)/2,$$
$$d_{\text{k}} = s_{\text{k}} + m \cdot (n \cdot s_{\text{a}} - m \cdot s_{\text{k}}) / (\text{sq.}\, m + \text{sq.}\, n)/2$$
$$= \{m \cdot n \cdot u_{\text{a}} - (\text{sq.}\, m - \text{sq.}\, n)/2 \cdot u_{\text{k}}\} / (\text{sq.}\, m + \text{sq.}\, n)/2.$$

In other words, *an alternative form of the confluent bisection rule* is that

$d_a = a \cdot s_a + b \cdot s_k,\quad d_k = b \cdot s_a - a \cdot s_k,$

where

$b = m \cdot n$ / (sq. m + sq. n)/2, and a = (sq. m – sq. n)/2 / (sq. m + sq. n)/2.

One recognizes here that *a and b are the short sides of a right triangle with the diagonal 1, given by a generating rule for diagonal triples*. In the special case when $m, n = 2, 1$, the equations for a and b show that

$b = 2 \cdot 1$ / (sq. 2 + sq. 1)/2 = 4/5 = ;48, and
a = (sq. 2 – sq. 1)/2 / (sq. 2 + sq. 1)/2 = 3/5 = ;36.

Therefore, *the alternative form of the confluent bisection rule is used in the first part of the solution procedure in TMS 23, with the same choice of the parameters m, n as in the second part of the solution procedure*.

This observation finally makes sense of the following cryptic remark at the beginning of the second part of the solution procedure:

2 *ù* 1 im.gíd.da *le-qé*	take 2 and 1 (from) the table

Here im.gíd.da 'long clay' or 'long clay tablet' is a Sumerian word which normally refers to a longish clay tablet inscribed with a mathematical table text, such as a single multiplication table, or a table of reciprocals. In the present context, it obviously refers to a mathematical table similar to but slightly different from Plimpton 322 (Sec. 3.3), namely *a table with separate columns for the parameters m, n and for the fronts and lengths of a number of right triangles satisfying the condition that the diagonal = 1*.

11.10. Erm. 15073. Divided Trapezoids in a Recombination Text

Erm. 15073 (Vaiman, *SVM* (1961), 232-244), is a large fraction of an OB mathematical recombination text. It is of the same kind as the more well known recombination texts BM 85194 and BM 85196 (both mentioned in Sec. 1.12 above) and BM 96954+ (see Sec. 93, in particular Fig. 9.3.2), all from the ancient Mesopotamian city Sippar. Most of the text on the reverse of Erm. 15073 is destroyed, but small parts remain of three exercises, all dealing with divided figures.

The final part of the first exercise in the leftmost column on the reverse (col. *vi*) contains the explicit computation of the partial lengths $u_a = 20$ and $u_k = 40$ of a bisected trapezoid with the upper front, the transversal, and the lower front equal to 35, 25, 5 = 5 · (7, 5, 1).

What remains of the second exercise in col. *vi* is only a diagram showing a quadrilateral with some associated numbers. However, the numbers are such that it is clear that this exercise is about a quadrilateral bisected in two directions, thus to some extent a parallel to *TMS* 23.

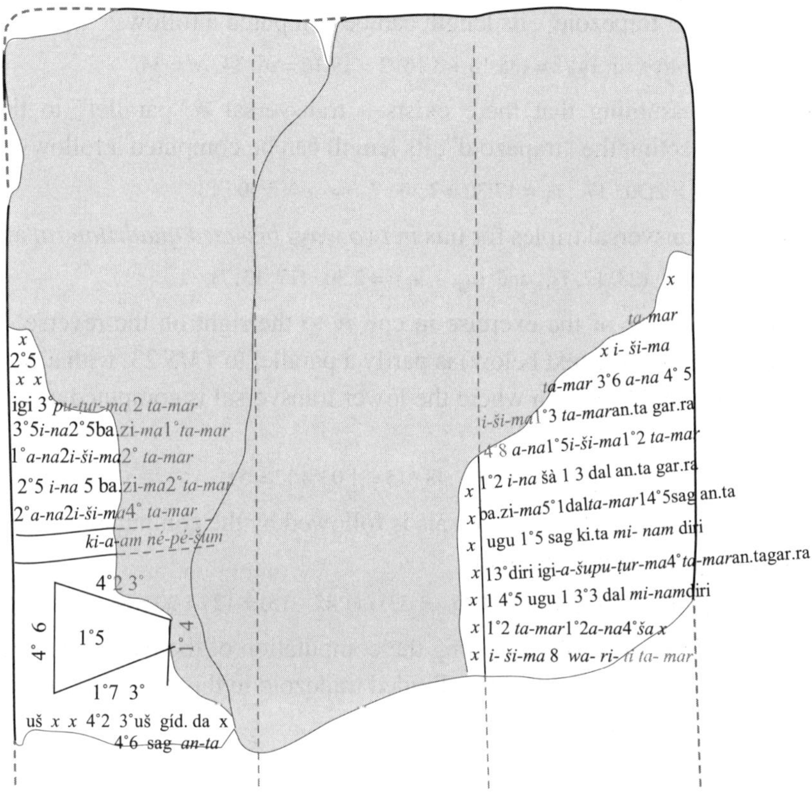

Fig. 11.10.1. Erm. 15073 *rev*. Remains of three exercises concerned with divided figures.

Erm. 15073 col. *vi* # 2, literal translation — explanation

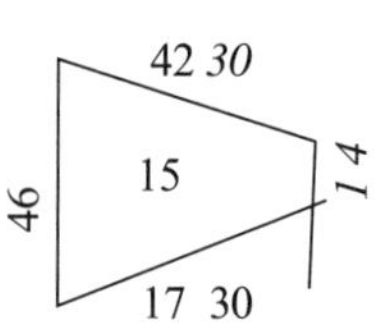

A · · · 42 30 the long length, — A trapezoid(?). u' =42;30

17 30 the short length, 46 the upper front — u'' = 17;30, s_a = 46

According to the "quadrilateral area rule", the area of the figure is

$A = (42;30 + 17;30)/2 \cdot (46 + 14)/2 = 30 \cdot 30 = 15\ (00).$

This number is recorded in the center of the diagram.

Assuming that there exists a transversal d "parallel" to the fronts and bisecting the "trapezoid", its length can be computed a follows:

$\text{sq. } d = (\text{sq. } 46 + \text{sq. } 14)/2 = (35\ 16 + 3\ 16)/2 = 19\ 16 = \text{sq. } 34, \quad d = 34.$

Similarly, assuming that there exists a transversal e "parallel" to the lengths, bisecting the "trapezoid", its length can be computed a follows:

$u_a = 42;30 = 2;30 \cdot 17, \quad u_k = 17;30 = 2;30 \cdot 7 \quad \Rightarrow \quad e = 2;30 \cdot 13.$

Thus, the transversal triples for this *in two ways bisected quadrilateral* are

$(s_a, d, s_k) = 2 \cdot (23, 17, 7), \quad \text{and} \quad (u_a, e, u_k) = 2;30 \cdot (17, 13, 7).$

What remains of the exercise in col. *iv* to the right on the reverse of Erm. 15073 (see the text below) is partly a parallel to *TMS* 23, with a confluent trapezoid bisection where the lower transversal is computed as follows:

$d_k = ;36 \cdot s_a - ;48 \cdot s_k = ;36 \cdot 1\ 45 - ;48 \cdot 15 = 1\ 03 - 12 = 51.$

The computation of the transversals is followed by the computation of the partial lengths, beginning with

$b_a = \{(s_a - d_a) / (s_a - s_k)\} \cdot 1 = (1\ 45 - 1\ 33) / (1\ 45 - 15) = 12 / 1\ 30 = 8.$

The rest of the exercise, including the computation of the partial areas, is lost. It is clear, anyway, that the divided trapezoid in this case was the one depicted in Fig. 11.10.2 below:

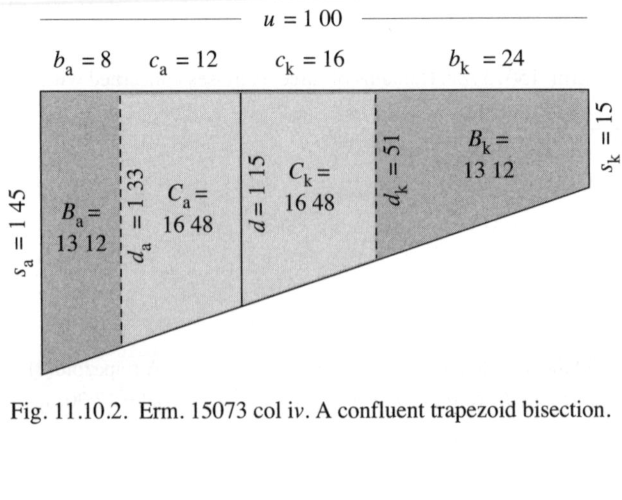

Fig. 11.10.2. Erm. 15073 col iv. A confluent trapezoid bisection.

Erm. 15073 col. *iv*, literal translation	explanation
. . . . you see	. . .
. . . . you see	. . .
. . . . you lift	. . .
. . . . you see.	. . .
36 to 1 45 lift, then 1 03 you see. Set it above.	$;36 \cdot s_a = ;36 \cdot 1\ 45 = 1\ 03$
48 to 15 lift, then 12 you see.	$;48 \cdot s_k = ;48 \cdot 15 = 12$
12 our from 1 03 that you set above	$\mathbf{;36 \cdot s_a - ;48 \cdot s_k} = 1\ 03 - 12$
tear off, then 51 the transversal you see.	$= 51 = \mathbf{d_k}$
1 45 the upper front over 15 the lower front	$s_a - s_k = 1\ 45 - 15$
is what beyond? 1 30 it is beyond.	$= 1\ 30$
its opposite resolve, then 40 you see. Set it.	$1/\ 1\ 30 = ;00\ 40$
1 45 over 1 33 the transversal is what beyond?	$s_a - d_a = 1\ 45 - 1\ 33 = 12$
12 you see. 12 to 40 that · · · lift,	$12 \cdot ;00\ 40$
then 8 the descent.	$= (s_a - d_a)\ /\ (s_a - s_k) = ;08$
. . .	. . .

The problems on the obverse of Erm. 15073 are not related to the division of figures problems on the reverse. The first problem appears to be a combined work norm exercise, while the remaining problems on the obverse are various kinds of volume computations. For details see Vaiman, *op cit*.

The importance of Erm. 15073 *rev*. is that it clearly demonstrates how readily mathematical ideas could spread from one Mesopotamian city to another in the OB period. (In the present case from Sippar(?) in central Mesopotamia to Susa far to the east.) This is evident in the case of the *confluent trapezoid bisections*, and even more so in the case of the *in two ways bisected quadrilaterals*, where a surprisingly lax attitude towards features apparently judged to be non-essential allows the treatment of decidedly unsymmetrical quadrilaterals as figures with two parallel fronts *and* two parallel lengths.

Fig. 11.10.3. Erm. 15073. Scale 3:5.
Photos: The State Ermitage Museum, St. Petersburg

Chapter 12

Hippocrates' Lunes and Babylonian Figures With Curved Boundaries

12.1. Hippocrates' Lunes According to Alexander

Hippocrates of Chios is the most famous of the Greek geometers of the (last part of the) 5th century BCE. In the *Summary* of Proclus, it is said that "he is the first of those mentioned as having compiled *Elements*" (Thomas, *GMW 1* (1980), 151. He also occupied himself with questions related to the quadrature of the circle, in particular his famous "quadratures of lunes" (Gr. *menískos* 'little moon'). Simplicius mentions, in his *Commentary on Aristotle's Physics* (van der Waerden, *SA* (1975), 131 *ff.;* Thomas, *op. cit.*, 235 *ff.;* Knorr, *ATGP* (1993), 29 *ff.*) two different sources allegedly dealing with Hippocrates' quadratures of lunes.

One of these sources is a statement by Simplicius' teacher Alexander of Aphrodisias. Two different quadratures of lunes described by Alexander are illustrated by the diagrams in Fig. 12.1.1 below, using metric algebra notations. The construction of the first diagram begins with an isosceles right triangle (a half-square) with the base p, the sides s and the area T. Three semicircles are applied to the sides of the triangle, a large semicircle to the base and smaller semicircles to the legs. In this way, two lunes are formed, both bounded on one side by a semicircle of diameter s and on the other side by a 1/4-circle of radius $p/2$. Clearly,

$$\text{sq. } p = 2 \text{ sq. } s.$$

Assuming it to be known that the area of a semicircle is proportional to the square of its diameter, it follows that *the area S_p of the large semicircle is equal to the sum $2\,S_s$ of the areas of the small semicircles*. On the other hand, *the figure formed by the two lunes and the large semicircle is the* *same as the figure formed by the right triangle and the two small semi-circles*. This means that

$S_p = 2\,S_s$ and $2\,L + S_p = T + 2\,S_s$, where L is the area of each lune.

Consequently,

$2\,L = T$ so that $L = T\,/\,2$ (= sq. $s/2$).

Therefore, a lune bounded by a semicircle and a 1/4-circle is squarable.

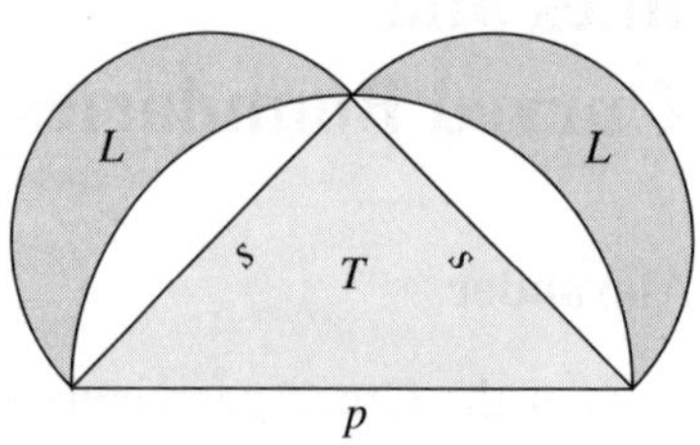

sq. p = 2 sq. s

$2\,L + S_p = T + 2\,S_s$

(S_p and S_s semicircles on p and s)

$\Rightarrow$

$S_p = 2\,S_s$

$2\,L = T = A_{triangle}$ (squareable)

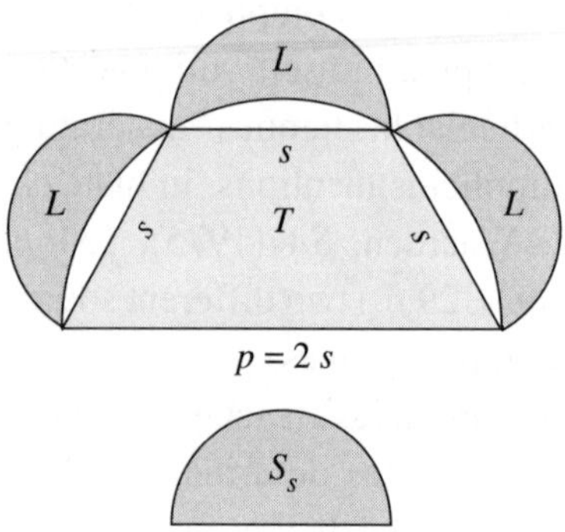

sq. p = 4 sq. s

$3\,L + S_p = T + 3\,S_s$

(S_p and S_s semicircles on p and s)

$\Rightarrow$

$S_p = 4\,S_s$

$3\,L + S_s = T = A_{trapezoid}$ (squareable)

Fig. 12.1.1. Hippocrates' lunes according to Alexander. Modern notations.

The next construction begins with one half of a regular hexagon with the side s the base p, and the area T. Four semicircles are applied to the sides of the trapezoid, a large semicircle to the base and smaller semicircles to the legs. In this way, three lunes are formed, all bounded on one side by a semicircle of diameter s and on the other side by a 1/6-circle of radius s. Clearly,

$p = 2\,s$, so that sq. p = 4 sq. s and, consequently, $S_p = 4\,S_s$.

On the other hand, *the figure formed by the three lunes and the large semicircle is the same as the figure formed by the trapezoid and the three small semicircles*. This means that

$S_p = 4\,S_s$ and $3\,L + S_p = T + 3\,S_s$, where L is the area of each lune.

Consequently,

$3\,L + S_s = T.$

Assuming that it is known that every rectilineal figure (a figure bounded by straight lines) is squareable (*cf.* Euclid's *Elements* II.14), it then follows that *3 times the area of a lune bounded by a semicircle and a 1/6-circle, and the area of a semicircle, all together, are equal to the area of a square*.

(In other words, the area of a circle is equal to the area of a square minus the area of 6 such lunes. Maybe the idea was that in this way the problem of the quadrature of the circle can be reduced to the problem of the quadrature of this kind of lune.)

12.2. Hippocrates' Lunes According to Eudemus

The other source used by Simplicius is a passage in the *History of Geometry* by Eudemus. The constructions of lunes in that passage are more sophisticated than the constructions mentioned by Alexander. Also the argumentation is more sophisticated. In particular, Eudemus says that

> "The quadratures of lunes, which seemed to belong to an uncommon class of propositions by reason of the close relationship to the circle, were first investigated by Hippocrates, and seemed to be set out in correct form; therefore we shall deal with them in length and go through them. He made his starting-point, and set out as the first of his theorems useful to his purpose, that similar segments of circles have the same ratios as the squares on their bases. And this he proved by showing that the squares on the diameters have the same ratios as the circles."

Unfortunately, Eudemus does not tell how Hippocrates proved that circles have the same ratios as their squares. (*Cf. El.* XII.2.) Neither does he explain under which circumstances segments of circles are similar to each other. Hippocrates' definition of similar segments was probably the same as Euclid's in *El.* III, Def. 11:

> "Similar segments of circles are those which admit equal angles."

What this means is made clear in *El.* III.33, where the angle admitted by a circle segment is seen to be the angle between the chord which is the base of the segment and the tangent to the circle at the endpoint of that chord. The situation is particularly simple in the case of semicircles, all of which can be interpreted as similar circle segments, since the angles that they admit are always right angles. Therefore, the constructions of lunes by means of similar circle segments (Fig. 12.2.1), can be viewed as direct generalizations of the constructions of lunes by means of semicircles (Fig. 12.1.1).

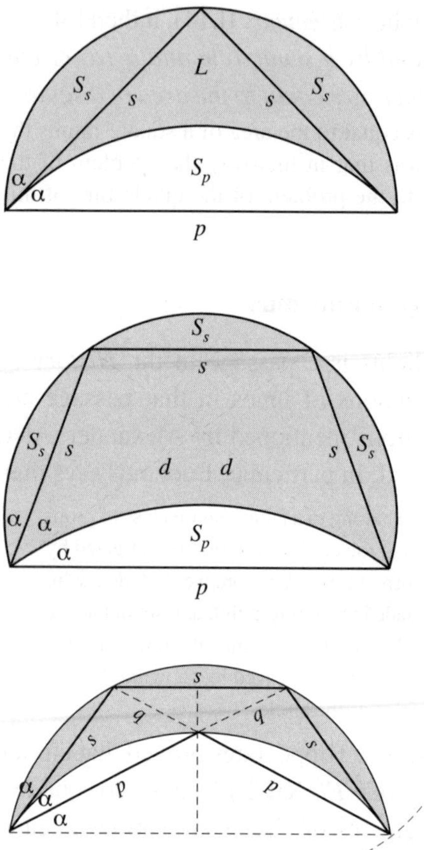

sq. $p = 2$ sq. s

$L + S_p = T + 2\,S_s$

(S_p and S_s circle segments on p and s)

$\Rightarrow$

$S_p = 2\,S_s$

$L = T = A_{half\text{-}square}$

sq. $p = 3$ sq. s

$L + S_p = T + 3\,S_s$

(S_p and S_s circle segments on p and s)

$\Rightarrow$

$S_p = 3\,S_s$

$L = T = A_{trapezoid}$

$d = p + q,\ d : s = s : q$

$\Rightarrow\ d \cdot q = \text{sq. } s,\quad d - q = p$

2 sq. $p = 3$ sq. s,

$L + 2\,S_p = T + 3\,S_s$

(S_p and S_s circle segments on p and s)

$\Rightarrow$

$2\,S_p = 3\,S_s$

$L = T = A_{trapezoid} - A_{triangle}$

Fig. 12.2.1. The three first of Hippocrates' lunes according to Eudemus. Modern notations.

The construction of the first of Hippocrates' lunes according to Eudemus starts, just like the construction of the first lune according to Alexander, with *a half-square T* of side *s* and base *p*, where then sq. *p* =

2 sq. s. A semicircle is circumscribed around this half-square, and on the base of the half-square is constructed a large circle segment S_p similar to the two small circle segments S_s with the base s which are cut off from the semicircle by the sides of the half-square. The common chord-tangent angle of the three segments is then half a right angle, and the large segment is tangent to the sides of the half-square. In the process, a lune is formed, bounded on one side by a semicircle with the diameter p and on the other side by the arc of a 1/4-circle segment with the base p. The area of the lune can be computed as follows:

$L = T + 2\,S_s - S_p$, where $2\,S_s = S_p$, so that $L = T$ = the area of the half-square.

The conclusion is that *a lune bounded on one side by a semicircle and on the other by a 1/4-circle is squareable*. Although it is not explicitly stated, it is also clear that *the chord-tangent angle of the outer arc of the lune is twice as large as the chord-tangent angle of the inner arc*.

The construction of the second of Hippocrates' lunes according to Eudemus starts with *a trapezoid T* with the sides and the smaller top all equal to s and with the base p, where sq. p = 3 sq. s. A circle segment with the base p is circumscribed around the trapezoid. It is shown as follows that *the circular arc of the segment is greater than that of a semicircle*:

Let d be (the length of) the diagonal of the trapezoid. Then sq. p = 3 sq. s
and sq. d > 2 sq. s because the angle opposite to d is obtuse
⇒ sq. p < sq. d + sq. s so that the angle opposite to p is acute.
Therefore the arc of the circumscribed segment is greater than that of a semicircle.

(Clearly, the arguments used here by Hippocrates are forerunners of Euclid's *Elements* II.12-13 and III.31.)

Next, a segment S_p is constructed on the base p, similar to the three small segments S_s with the base s which are cut off from the circumscribed segment by the trapezoid. A lune is then formed, bounded on one side by an arc greater than that of a semicircle and on the other side by the arc of the segment S_p. The area L of the lune can be computed as follows:

$L = T + 3\,S_s - S_p$, where $3\,S_s = S_p$, so that $L = T$ = the area of the trapezoid.

Therefore, *also this lune with an outer arc greater than that of a semicircle is squareable*. Although it is not explicitly stated, it is clear that *the chord-tangent angle of the outer arc of the lune is three times as large as the chord-tangent angle of the inner arc*. (*Cf.* Euclid's ***Elements* III.32**.)

The construction of the third of Hippocrates' lunes according to Eudemus is accompanied by a diagram which is excessively complicated because it wants to show the construction in full detail. The third diagram in Fig. 12.2.1 above is a somewhat simplified version of that diagram.

Essentially, the construction begins with a straight line of given length s, the top of the trapezoid in the diagram. A circle of diameter $2\,s$ is drawn with its center at one end point of the given straight line. Next a straight line of length p where sq. $p = 1\ 1/2 \cdot$ sq. s is constructed by use of *neusis* 'verging', with one end point on the circle, with the second end point on a perpendicular bisector of the given straight line, and such that its extension $p + q$ passes through the second end point of the given straight line.

Note that instead of using the *neusis* construction, Hippocrates could have computed the length $d = p + q$ of the diagonal as follows:

The triangle with the sides s, q, q is similar to the triangle with the sides d, s, s.
Therefore, $d : s = s : q$ so that $d \cdot q =$ sq. s.
Hence, d and q can be found as the solution to the rectangular-linear system of equations
$d \cdot q =$ sq. s, $d - q = p$.

A trapezoid with three sides equal to s and with the diagonal $p + q$ can now be constructed, as in the third diagram in Fig. 12.2.1. Finally, a circle segment is constructed, with the same base as the trapezoid and circumscribing the triangle formed by the base and the two straight lines of length p. The triangle cuts off two segments S_p of base p from this circle segment, at the same time as the trapezoid cuts off three segments S_s of base s from the circle segment circumscribing the trapezoid. It is clear that S_p and S_s are similar circle segments, since their chord-tangent angles are equal. (*Cf.* again *El.* III.32.) It is also clear that the angle at the base of the trapezoid is twice as big. (*Cf. El.* III.27.)

A lune is now formed with its outer and inner arcs equal to the arcs of the circle segments circumscribing the trapezoid and the triangle, respectively. The area L of the lune can be computed as follows:

Since S_p and S_s are (the areas of) similar circle segments with the base p and s, respectively, and since 2 sq. $p = 3$ sq. s, it follows that $2\,S_p = 3\,S_s$.
Let now $T =$ (the area of) the figure equal to *the trapezoid with the triangle torn off.*
Then $L = T + 3\,S_s - 2\,S_p = T$.

Since rectilineal figures are squareable (*cf.* again *El.* II.14), it follows that *the lune with its outer and inner arcs equal to the arcs circumscribing the trapezoid and the triangle, respectively, is squareable.* Although it is not explicitly stated, it is also clear that *the chord-tangent angle of the outer arc of the lune is 1 1/2 times as large as the chord-tangent angle of the inner arc*. Finally, it is can be shown that *the outer arc is less than the arc of a semicircle* (Thomas, *op. cit.*, 247, footnote a).

The fourth of Hippocrates' lunes according to Eudemus (see Fig. 12.2.2) is bounded on the outside by the arc of a 1/3-circle segment of base p and on the inside by the arc of a 1/6-circle segment S_p with the same base. The area L of the lune can be computed as

$L = T + 2\,S_s - S_p$, where T is a triangle with the sides p, s, s, with sq. p = 3 sq. s.

Since S_s and S_p are similar circle segments, it follows that

$S_p = 3\,S_s$ so that $L = T - S_s$.

This is a somewhat unsatisfactory result, so the construction continues as follows: A circle $C_{s'}$ is circumscribed around a regular hexagon $H_{s'}$ with the diameter 2 s', and with sq. s = 6 sq. s'. The hexagon then cuts of 6 circle segments $S_{s'}$ from the circumscribed circle, and it is clear that

$S_s = 6\,S_{s'}$ so that $L = T - 6\,S_{s'}$.

Then also

$L + (H_{s'} + 6\,S_{s'}) = T + H_{s'}$ so that $L + C_{s'} = T + H_{s'}$ is squareable.

Thus, the final conclusion is that *the area of the lune in Fig. 12.2.2 plus the area of the small circle is equal to the area of a square.* This is a result of the same kind as the one for Hippocrates' second lune according to Alexander. See the second diagram in Fig. 12.1.1 above.

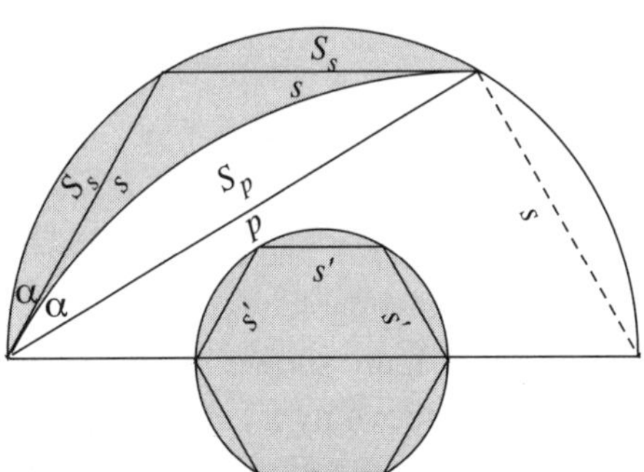

sq. p = 3 sq. s and sq. s = 6 sq. s'

$L + S_p = T + 2\,S_s$

$\Rightarrow$

$S_p = 3\,S_s$ and $S_s = 6\,S_{s'}$

$L + S_s = T = A_{triangle}$

$L + C = A_{triangle} + A_{small\ hexagon}$

(C circle of diameter 2 s')

Fig. 12.2.2. The fourth of Hippocrates' lunes according to Eudemus. Modern notations.

12.3. Some Geometric Figures in the OB Table of Constants BR

A well known sequence of entries in the OB mathematical table of constants BR = Bruins and Rutten, *TMS* (1961) text 3 was mentioned above in Sec. 6.2. In these entries are listed, in a systematic way, the main parameters for (in particular) the following plane geometric figures:

the 'arc'	(circle)	BR 2-4
the 'crescent'	(semicircle)	BR 7-9
the 'bow'	(bow-like figure)	BR 10-12
the 'boat field'	(boat-like figure, rhomb)	BR 13-15
the 'barleycorn field'	(thin double circle segment)	BR 16-18
the 'ox-eye'	(thick double circle segment)	BR 19-21
the 'lyre-window'	(concave square)	BR 22-24
the 'lyre-window of 3'	(concave triangle)	BR 25
the '5-front, 6-front, 7-front'	(regular polygons)	BR 26-28
the 'peg-head'	(equilateral triangle)	BR 29
the 'šár field'	(ring of right triangles)	BR 30
the 'divider of the square'	(diagonal of a square)	BR 31
the 'divider of the length-and-front'	(diagonal of a 1 00 × 45 rectangle)	BR 32

The rather obvious meaning of the parameters for the circle and the semicircle was discussed above in Sec. 6.2. Constants for regular polygons (including the equilateral triangle) were discussed in Sec. 7.4. The much less obvious meaning of the parameters for the geometric figures in entries 10-25 of the list above (among them the šár field; see Sec. 2.4) was explained successfully for the first time by Vaiman in *VDI* I:83 (1963). The discussion below is based on Vaiman's ideas.

12.3 a. BR 10-12. The 'bow field'

The constants listed for the **bow field** in BR 10-12 are

$A = 6\ 33\ 45, \quad d = 52\ 30, \quad p = 15.$

Here A, d, p are notations for the *arc*, the *transversal*, and the *crossline* of a given geometric figure, "normalized" in some suitable way.

Suppose that the bow field is the bow-like figure shown in Fig. 12.3.1 below, bounded below by a straight line, and above by a curved line.The curved line is composed of 1/3 of the perimeter of a circle in the middle and 1/6 of the perimeter of a circle at either end. If the length of the whole curved line is called a, then each 1/6 of the perimeter of the circle is equal to $a/4$. Therefore, the radius of the circle is $(3 \cdot a/4)/\pi$ = appr. $a/4$. Now, in a circle with the radius $a/4$ the side of an inscribed regular hexagon is also $a/4$, and the side of an inscribed equilateral triangle is sqs. $3 \cdot a/4$. Consequently, the *(longest) transversal* in the bow field is

$d = 2 \cdot$ sqs. $3 \cdot (3 \cdot a/4)/\pi$ = appr. $2 \cdot 7/4 \cdot a/4 = 7/8 \cdot a$ = ;52 30 $\cdot a$.

When $a = 1$ (00), this gives the listed value for the transversal d. Similarly, the (longest) crossline in the bow field, orthogonal to the transversal, is

$p = (3 \cdot a/4)/\pi$ = appr. $1/4 \cdot a$ = ;15 $\cdot a$.

When $a = 1$ (00), this gives the listed value for the crossline p.

As can be seen from the diagram in Fig. 12.3.1, *the area of the bow field is equal to the area of a triangle with the base d and the height p*! Consequently, the area of the bow field is

$A = d/2 \cdot p$ = sqs. 3 $\cdot$ sq. $\{(3 \cdot a/4)/\pi\}$ = appr. $7/4 \cdot 1/16 \cdot$ sq. a = ;06 33 45 $\cdot$ sq. a.

Again, when $a = 1(00)$, this gives the listed value for the parameter A.

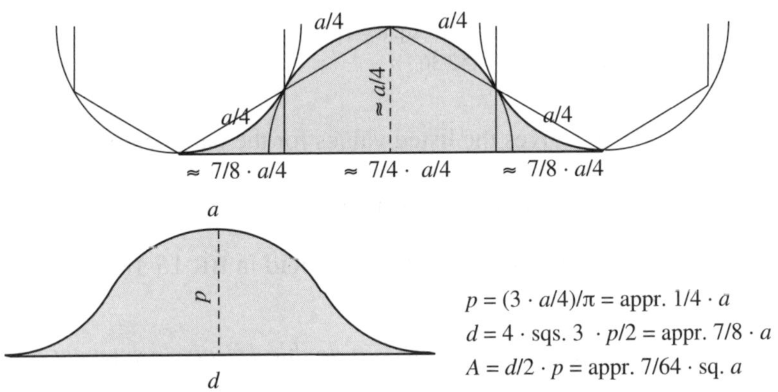

Fig. 12.3.1. BR 10-12. Parameters for the 'bow field'.

12.3 b. BR 13-15. The 'boat field'

The constants listed for the **boat field** in BR 13-15 are

A =13 07 30, d = 52 30, p = 30.

A probable connection with the bow field discussed above is obvious, since the area and the crossline for the bow field are exactly twice as large as the area and crossline for the bow field. Therefore, a reasonable interpretation is that the boat field is in some way equal to either a double bow field or *a double triangle* with the same transversal and crossline as the bow field. As shown in Fig. 12.3.2 below, the second alternative gives the most likely interpretation, namely that *the boat field is a rhomb*, or, more precisely, *a double equilateral triangle.*

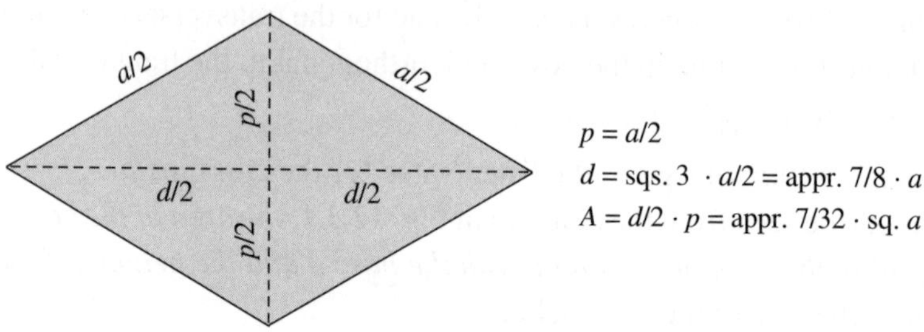

Fig. 12.3.2. BR 13-15. Parameters for the 'boat field', a double equilateral triangle.

Accordingly, the parameters for the boat field are

$A = 1/2 \cdot$ sqs. $3 \cdot$ sq. $a/2 =$ appr. $7/32 \cdot$ sq. $a = ;13\ 07\ 30 \cdot$ sq. a
$d =$ sqs. $3 \cdot a/2 =$ appr. $7/8 \cdot a = ;52\ 30 \cdot a$
$p = a/2 = ;30 \cdot a.$

When $a = 1(00)$, this gives the listed values for the parameters.

12.3 c. BR 16-18. The 'barleycorn field'

The constants listed for the **barleycorn field** in BR 16-18 are

$A = 13\ 20,\quad d = 56\ 40,\quad p = 23\ 20.$

The corresponding geometric figure is *a double 1/4-circle segment.*

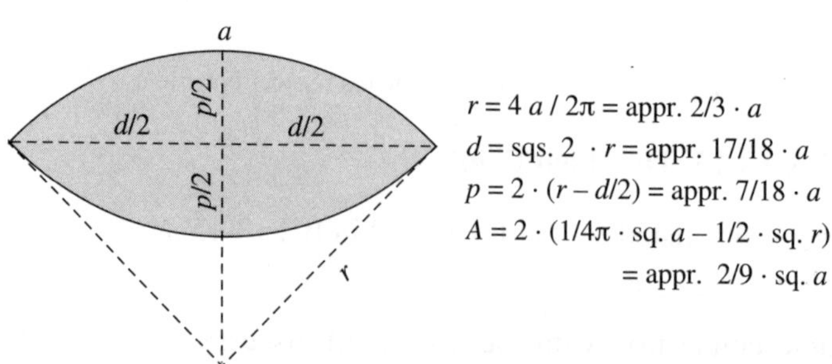

Fig. 12.3.3. BR 16-18. Parameters for the 'barleycorn field', a double 1/4-circle segment.

Indeed, if the arc of the segment is called *a*, the remaining parameters for the double circle segment can be computed as follows (Fig. 12.3.3):

the radius $r = 4a / 2\pi =$ appr. $2/3\ a$
$d =$ sqs. $2 \cdot r =$ appr. $17/12 \cdot 2/3 \cdot a = 17/18 \cdot a = ;56\ 40 \cdot a$
$p = 2 \cdot (r - d/2) =$ appr. $7/18 \cdot a = ;23\ 20 \cdot a$
$A = 2 \cdot \{1/4 \cdot 1/4\pi \cdot$ sq. $(4\ a) - 1/2 \cdot$ sq. $r\} =$ appr. $2/9 \cdot$ sq. $a = ;13\ 20 \cdot$ sq. a

12.3 d. BR 19-21. The 'ox-eye'

The constants listed for the **ox-eye** in BR 19-21 are

$A = 16\ 52\ 30, \quad d = 52\ 30, \quad p = 30.$

The corresponding geometric figure is *a double 1/3-circle segment.*

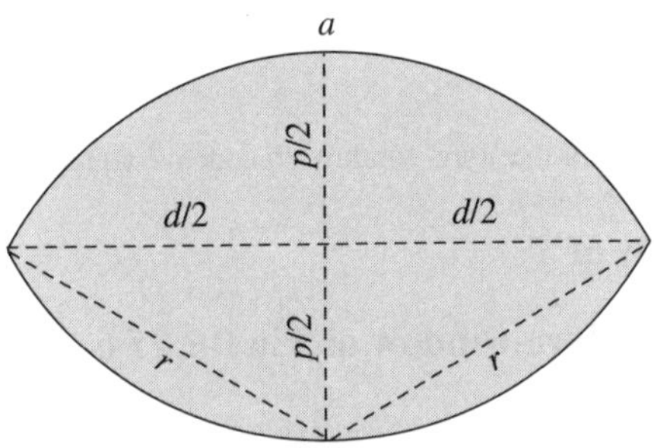

$p = r = 3\ a / 2\pi =$ appr. $1/2 \cdot a$
$d =$ sqs. $3 \cdot r =$ appr. $7/8 \cdot a$
$A = 2 \cdot (3/4\pi \cdot$ sq. $a - d/2 \cdot p/2)$
$=$ appr. $9/32 \cdot$ sq. a

Fig. 12.3.4. BR 16-18. Parameters for the 'ox-eye', a double 1/3-circle segment.

Indeed, if the arc of the segment is called *a*, the remaining parameters for the double circle segment can be computed as follows (Fig. 12.3.4):

$p = r = 3a / 2\pi =$ appr. $1/2\ a = ;30 \cdot a$
$d =$ sqs. $3 \cdot r =$ appr. $7/8 \cdot a = ;52\ 30 \cdot a$
$A = 2 \cdot \{1/3 \cdot 1/4\pi \cdot$ sq. $(3\ a) - d/2 \cdot p/2\} =$ appr. $9/32 \cdot$ sq. $a = ;16\ 52\ 30 \cdot$ sq. a

12.3 e. BR 22-24. The 'lyre-window'

The constants listed for the **lyre-window** (sound-hole) in BR 22-24 are

$A = 26\ 40, \quad d = 1\ 20, \quad p = 33\ 20.$

The corresponding geometric figure is *a concave square*. Indeed, if the arc of a concave square such as the one in Fig. 12.3.5 below is called *a*, the remaining parameters for the concave square can be computed as follows:

the radius $r = 4a / 2\pi =$ appr. $2/3\ a$

$d = 2 \cdot r =$ appr. $4/3 \cdot a =$1;20 $\cdot a$
$p =$ (sqs. $2 - 1) \cdot d =$ appr. $5/9 \cdot a =$;33 20 $\cdot a$
$A = A$(square) – A(circle) = sq. $d - 16/4\pi \cdot$ sq. $a =$ appr. 4/9 sq. $a =$;26 40 $\cdot$ sq. a

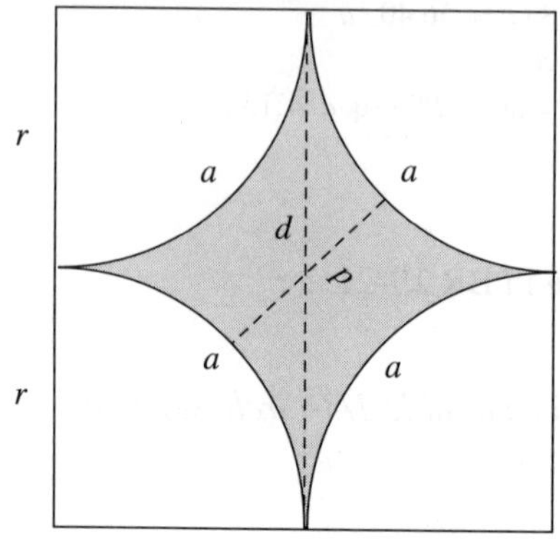

$r = 4\, a \,/\, 2\pi =$ appr. $2/3 \cdot a$
$d = 2\, r =$ appr. $4/3 \cdot a$
$p =$ (sqs. $2 - 1) \cdot d =$ appr. $5/9 \cdot a$
$A =$ sq. $d - 16/4\pi \cdot$ sq. a
= appr. $4/9 \cdot$ sq. a

Fig. 12.3.5. BR 22-24. Parameters for the 'lyre-window', a concave square.

12.3 f. BR 25. The 'lyre-window of 3'

The only constant listed for the **lyre-window of 3** in BR 25 is

$A =$15.

The corresponding geometric figure is *a concave triangle*. Indeed, if the arc of a concave triangle such as the one in Fig. 12.3.6 below is called a, the area of the concave triangle can be computed as follows:

$A = A$(triangle) – A(semi-circle) = sqs. $3 \cdot$ sq. $r - 18/4\pi \cdot$ sq. $a =$ appr. $1/4 \cdot$ sq. a

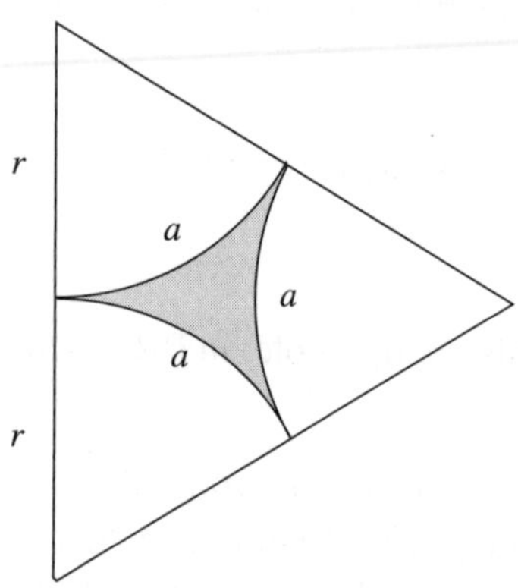

$r = 6\, a \,/\, 2\pi =$ appr. $1 \cdot a$

$A =$ sqs. $3 \cdot$ sq. $r - 18/4\pi \cdot$ sq. a
= appr. $1/4 \cdot$ sq. a

Fig. 12.3.6. BR 25. Parameters for the 'lyre-window of 3', a concave triangle.

12.4. W 23291-x § 1. A Late Babylonian Double Segment and Lune

Even if both Hippocrates' quadratures of lunes and the OB list of constants for various plane geometric figures in the table of constants BR are concerned with circle segments and computations of areas, it is not quite clear what the connection is between Hippocrates' lunes and the OB double segments. The connection is much more clear in the case of **W 23291-x § 1**, an exercise in a Late Babylonian mathematical recombination text (Friberg, *et al.*, *BaM* 21 (1990).

W 23291-x § 1, solution

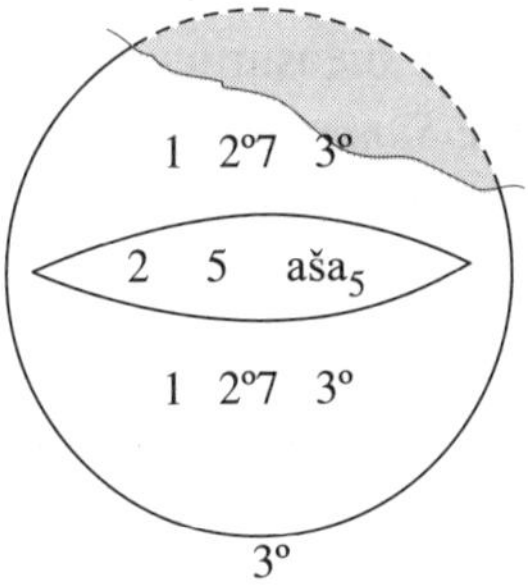

explanation

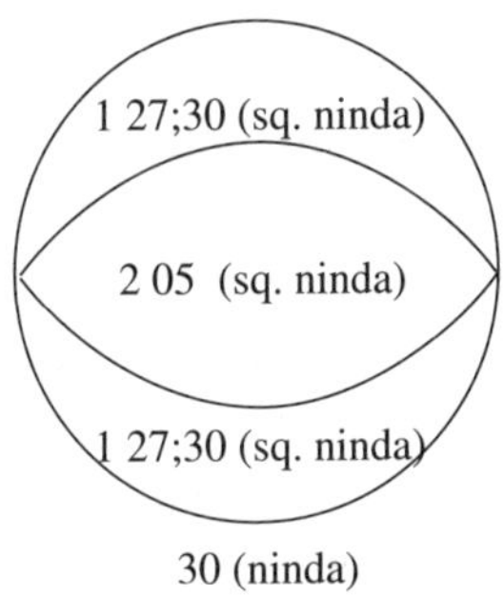

10 is my [········].	???
10 of the expansion of the heart is what?	???
20 steps of 10 is 3 20.	20 · 10 = 3 20.
Since 10 is 1/2 of *20*.	Since 10 = 1/2 · 20
7 30, 1/2 of 15	take 1/2 · ;15 = ;07 30
to 30 pair, then 37 30.	;07 30 + ;30 = ;37 30
3 20 steps of 37 30 is 2 05,	3 20 · ;37 30 = 2 05 (sq. ninda)
1 iku 25 šar, this is the field.	= 1 iku 25 šar, the area of the heart
30 of the crescent field,	One half of a crescent.
the area is what?	Its area = ?
30 steps of 30 is 15,	;30 · ;30 = ;15
5 50 go, 1 27 30,	;15 · 5 50 = 1 27;30 (sq. ninda)
1 ubu 37 1/2 šar,	= 1 ubu 37 1/2 šar
this is 1 crescent field.	= the area of 1 crescent
Steps of 2 he went,	Since there are two crescents
1 27 30 steps 2 go, then 2 55,	1 27;30 · 2 = 2 55
1 iku 1 ubu 25 šar,	= 1 iku 1 ubu 25 šar
this is 2 crescent fields.	= the area of 2 crescents.
Heap them, then	The sum of the areas of all
all of them, then 3 iku.	the three parts of the circle = 3 iku

The rather poor diagram accompanying the exercise shows a circle divided into three parts. In the text, the central part is called šà 'heart, core, inner part', while the two outer parts are called u_4.sakar 'crescent'. The area of the 'heart' is recorded in the diagram as 2 05 (sq. ninda), and the area of each crescent as 1 27;30 (sq. ninda). As noted in the last lines of the text, the area of the whole circle is

(2 05 + 2 · 1 27;30) sq. ninda = 5 00 sq. ninda = 3 iku (1 iku = 100 sq. ninda).

The use of the traditional kind of length and area measure in this exercise (the ninda and the square-ninda and its multiples) is interesting. Actually, the recombination text W 23291-x is in its entirety a collection of examples of mathematical exercises of various ways of measuring the size of fields; it starts with three exercises employing the *Sumerian/Old Babylonian area measure*, followed by a number of exercises employing instead several kinds of *Late Babylonian reed measure or seed measure, etc.* Actually, the use of OB area measure in the first three exercises is an indication that those exercises are borrowed from some OB source, although the text of the exercises has been "translated" into the Late Babylonian mathematical jargon.

The statement of the problem in W 23291 § 1 is very brief and partly destroyed. The only piece of information that can be extracted from it is that the object of the exercise is the 'extension' (dikšu) of a 'heart'. Also the first part of the solution procedure, the computation of the area of the 'heart', is quite hard to understand.

The computation of the area of the two crescents is somewhat more straight-forward. Thus, as indicated by the number '30' at the lower end of the circle in the diagram, the length of the half circle which forms the outer arc of a crescent is 30 (ninda). Therefore, the crescent is half the size of a "normalized" crescent with the outer arc equal to 1 00 (ninda), so that the area of the crescent is ;15 = 1/4 of the area of a normalized crescent. Accordingly, the area of a crescent is computed in the text as

A_{crescent} = sq. ;30 · 5 50 = ;15 · 5 50 = 1 27;30 (sq. ninda).

Here '5 50' must be *the known area of a normalized crescent.*

In a similar way, the area of the whole circle (with the circumference 2 · 30 = 1 00) could have ben computed directly as

A_{circle} = sq. 1 · 5 00 = 5 00 (sq. ninda).

The area of the central 'heart' could then have been computed as the area of the circle minus the sum of the areas of the two crescents. Instead, the text prefers to compute the area of the 'heart' directly. However, since the length of the arc bounding the 'heart' is not known, the area of the 'heart' cannot be computed with departure from the presumably known area of a normalized 'heart'.

What is known is only that the (longest) transversal of the 'heart' is approximately equal to 20, the diameter of the circle. Therefore, in order to understand the curious computation of the area of the 'heart' in W 23291-x § 1, it may be a good idea to investigate how the crossline and the area of a 'barleycorn field' or an 'ox-eye' can be computed *in terms of the length of the transversal* rather than in terms of the length of the arc.

In the case of *the barleycorn field* (see Fig. 12.3.3),

$p_{\text{barleycorn}}$ = (sqs. 2 – 1) · d = appr. 5/12 · d = ;25 · d = 8;20 when d = 20,
$A_{\text{barleycorn}}$ = (π/4 – 1/2) · sq. d = appr. 1/4 · sq. d = 1 40 when d = 20.

Similarly, in the case of *the ox-eye* (see Fig. 12.3.4),

$p_{\text{ox-eye}}$ = sqs. 3 / 3 · d = appr. 7/12 · d = ;35 · d = 11;40 when d = 20,
$A_{\text{ox-eye}}$ = 2/3 · (π/3 – sqs. 3 / 4) · sq. d = appr. 3/8 · sq. d = 2 30 when d = 20.

At the same time, the area of a *circle* with the diameter d is equal to

A_{circle} = π/4 · sq. d = appr. 3/4 sq. d = ;45 · sq. d.

Knowing this, the area C of *the crescents one either side of a barleycorn field or an ox-eye* can be computed as follows:

$C_{\text{barleycorn}}$ = appr. (3/4 – 1/4)/2 · sq. d = 1/4 · sq. d = 1 40 when d = 20,
$C_{\text{ox-eye}}$ = appr. (3/4 – 3/8)/2 · sq. d = 3/16 · sq. d = 1 15 when d = 20.

These results are listed together in Fig. 12.4.1 below.

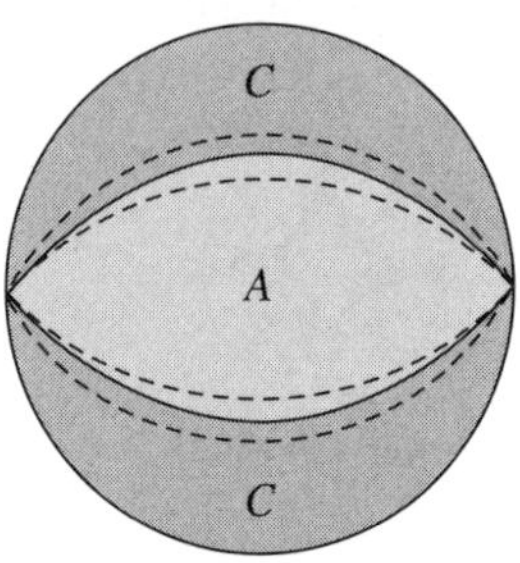

barleycorn	ox eye	heart
(1/4-circle segm.)	(1/3-circle segm.)	(half-sum)
d = 20	d = 20	d = 20
p = 8;20	p = 11;40	p = 10
A = 1 40	A = 2 30	A = 2 05
C = 1 40	C = 1 15	C = 1 27;30

Fig. 12.4.1. The barleycorn field, the ox-eye, the heart, and the associated crescents.

It is now easy to check that the area of the heart, $A = 2\ 05$, both recorded in the diagram in W 23291-x § 1 and computed in the text of that exercise, is the half-sum (mean value) of the areas of the barleycorn field and the ox-eye. The crossline p of the heart is not mentioned in the text of the exercise, but it is likely that it was thought of as the half-sum of the crosslines of the barleycorn field and the ox-eye, $(8;20 + 11;40)/2 = 10$, under the naive assumption that a double segment with the half-sum of the areas of the two known double segments would also have the half-sum of the crosslines. It is possible that the '10' mentioned obscurely in the statement of the problem is the length of the crossline of the heart.

The crescent associated with the heart is also, in a similar way, the half sum of the crescents associated with the other two double segments. It is possible that the crescent of the barley.corn field was deemed to be too thick and the crescent of the ox-eye too thin compared to the ideal image of a crescent, as it is depicted in several known Kassite *kudurrus* (boundary stones). See the example in Fig. 12.4.2 below.

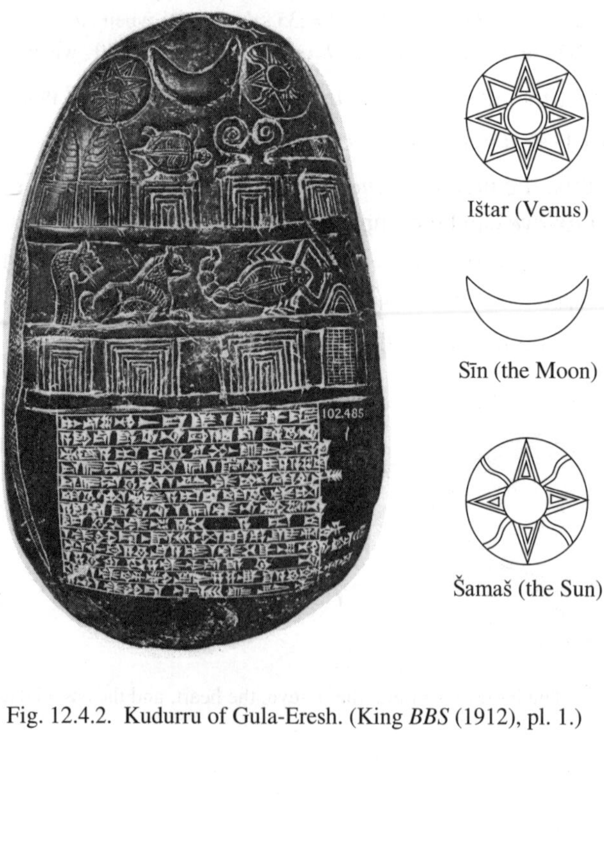

Fig. 12.4.2. Kudurru of Gula-Eresh. (King *BBS* (1912), pl. 1.)

It still remains to explain the computations in the text of W 23291-x § 1. The computation of the area of the crescent associated with the 'expansion of the heart' was probably quite simply based on a known *set of parameters for a normalized crescent*, similar to the sets of parameters for various plane geometric figures in the table of constants BR (Sec. 6.2 above):

$$C = 5\ 50, \quad d = 40, \quad p = 20.$$

The explanation for the computation of the area of the 'heart' itself is not quite so obvious. Nevertheless, the first step of the solution procedure, to compute the product 20 · 10 = 3 20, may be explained as the computation of the product $d \cdot r$, where d is the diameter and r the radius of the circumscribed circle. Next, the remark "since 10 is 1/2 of 20" may be a reference to the fact that the crossline p of the 'heart' is 1/2 of the diameter of the circle. Therefore, the area of the 'heart' was reckoned to be half-way between the areas of the barleycorn field and the ox-eye. As a result of this consideration, the area of the 'heart' was computed as

$$A_{\text{heart}} = A_{\text{barleycorn}} + 1/2 \cdot (A_{\text{ox-eye}} - A_{\text{barleycorn}}) = (;30 + 1/2 \cdot ;15) \cdot d \cdot r.$$

This explanation makes sense if the areas of the barleycorn field and the ox-eye and their crescents were known to have the values

$$A_{\text{barleycorn}} = C_{\text{barleycorn}} = \text{appr. } ;30 \cdot d \cdot r,$$
$$A_{\text{ox-eye}} = ;45 \cdot d \cdot r, \quad C_{\text{ox-eye}} = \text{appr. } ;22\ 30 \cdot d \cdot r.$$

Such equations for the areas of the double segments and their crescents can be compared with the following equation for the area of a semicircle:

$$A_{\text{semicircle}} = \text{appr. } ;45 \cdot d \cdot r.$$

(See entry 54 of the OB table of constants NSd = YBC 5022 (Neugebauer and Sachs, *MCT* (1945) text Ud).)

Note, by the way, that ;30 · $d \cdot r$ is also the area of the half-square inscribed in a semicircle of diagonal d and radius r. (Compare with the first of Hippocrates' lunes in Fig. 12.2.1, which has the same area as the half-square inscribed in the same semicircle as the lune.)

Another interesting observation is that, at least approximately, (the area of) *a circle is divided in three equal parts by an inscribed barleycorn field and its two crescents.* Similarly, at least approximately, *a circle is divided in four equal parts by the two halves of an inscribed ox-eye together with its two crescents.*

12.5. A Remark by Neugebauer Concerning BM 15285 # 33

In Neugebauer, *MKT 1* (1935), 137-142, the geometric figures on the first published fragment of BM 15285 are discussed. This is the triangular fragment containing exercises in columns iii-v and vi-viii of the reconstructed text. (See Fig. 6.2.3 above.) One of the figures caught his particular attention, namely the one in # 33 (which Neugebauer called figure XV). He interpreted the central figure in # 33 as the outer boundary of a figure composed of three partly overlapping circles, as shown in Fig. 12.5.1:

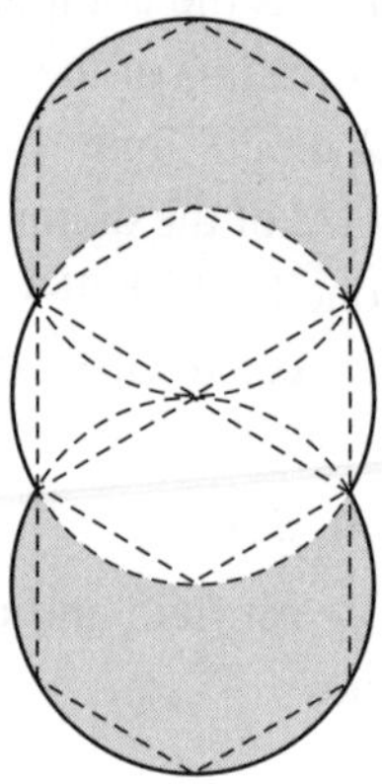

$$A_{\text{ox-eye}} = 2/6\, A_{\text{circle}} + 4/6\, (A_{\text{circle}} - A_{\text{hexagon}})$$

$$= 2/3\, A_{\text{circle}} - 1/3\, A_{\text{hexagon}}$$

$$A_{\text{x}} = 3\, A_{\text{circle}} - 2\, A_{\text{ox-eye}}$$

$$\Rightarrow$$

$$A_{\text{x}} = 5/3\, A_{\text{circle}} + 2/3\, A_{\text{hexagon}} \quad \text{or} \quad = A_{\text{circle}} + 2\, A_{\text{lune}}$$

$$\Rightarrow$$

$$A_{\text{circle}} = 3\, A_{\text{lune}} - A_{\text{hexagon}}$$

Fig. 12.5.1. Neugebauer's computation of the area of the figure in BM 15285 # 33.

Neugebauer could not know, 26 years before the publication in 1961of the table of constants BR = *TMS* 3, that Old Babylonian mathematicians called a double 1/3-circle-segment an 'ox-eye'. Nevertheless, he correctly computed the area A_{x} of the figure in BM 15285 # 3 (whose name is not mentioned in the preserved part of the text of BM 15285) as follows:

$$A_{\text{x}} = \text{(exactly) } 3\, A_{\text{circle}} - 2\, A_{\text{ox-eye}} = 5/3\, A_{\text{circle}} + 2/3\, A_{\text{hexagon}}.$$

He then made the observation that the area A_{x} is also the sum of the areas of one circle and two lunes (grey in Fig. 12.5.1), with

$$A_{\text{lune}} = 1/3\, (A_{\text{circle}} + A_{\text{hexagon}}) \quad \text{or, equivalently,} \quad A_{\text{circle}} = 3\, A_{\text{lune}} - A_{\text{hexagon}}.$$

Neugebauer then observes that this result implies that

"If this lune can be squared, then also the circle is squareable."

On the other hand, he concludes, this way of looking at the result would hardly occur to a Babylonian mathematician.

Chapter 13

Traces of Babylonian Metric Algebra in the *Arithmetica* of Diophantus

Introduction[33]

Diophantus and his work is described in the first lines of Chapter 1 of Bashmakova's *DDE* (1997) in the words

> "Diophantus represents one of the most difficult riddles in the history of science. We do not know when he lived, and we do not know his predecessors who may have worked in the same area. His works resemble a fire flashing in an impenetrable darkness"

On p. 3-4 (*ibid.*) it is stated that

> "But the most mystifying riddle is the works of Diophantus. Only six of the 13 books which make up the 'Arithmetic' have come down to us. Their style and contents differ radically form the classical ancient works on number theory and algebra whose models we know from Euclid's 'Elements' and his 'Data' and from the lemmas of Archimedes and Apollonius. The 'Arithmetic' is undoubtedly the result of numerous investigations which are completely unknown to us.We can only guess at its roots and admire the richness and beauty of its results."

Bashmakova then goes on to give a general description of the basic methods employed by Diophantus in what she interprets as his search for rational points on algebraic curves or, more precisely, for rational solutions to indeterminate equations of the second or third order. Bashmakova's approach is mathematically interesting but unhistoric.

In *GA* (1990), Chapter 3, Sesiano gives a sketch of certain "Pre-algebraic aspects in the *Arithmetica* of Diophantus". In that insightful essay,

33.The ideas discussed in this chapter were first presented at the *International Conference on the History of Mathematics and Education of Mathematics* in Baotao, China, 1991.

the basic methods of the *Arithmetica* are illustrated through a handful of well chosen examples. It will be shown below, among other things, that *quite a few of those basic examples can be explained as non-geometric reformulations of problems from Babylonian metric algebra.* It is, therefore, no longer necessary to postulate that Diophantus based his work on contributions by now forgotten predecessors to Diophantus in Athens or Alexandria. It is much more likely that Diophantus got his inspiration from some humble collection of originally Babylonian mathematical problems, perhaps inscribed on a number of Egyptian demotic or Greek-Egyptian papyrus rolls. (*Cf.* the discussion in Friberg, *UL* (2005), Sec. 3.5 b and Sec. 44 of the demotic *P.Carlsberg 30* # 2 and the Greek-Egyptian *P.Mich. 620.* See also the discussion in *op. cit.*, Sec. 2.1 b of the OB theme text YBC 4652 and the Late Babylonian fragment BM 34800.)

13.1. Determinate Problems in Book I of Diophantus' *Arithmetica*

The Babylonian influence in Book I of Diophantus' *Arithmetica* is obvious and well known. As a matter of fact, *Ar.* I *is organized in the same way as an OB mathematical theme text.* In the following partial table of contents, the notations are of the same kind as in the useful "Conspectus of Problems in the *Arithmetica*" in Sesiano, *Books IV to VII* (1982), 460 *ff.*, where the letters *a*, *b*, *c*, ··· stand for unknown magnitudes, and the letters *k*, *l*, *m*, *n*, *p*, *q*, ··· for given magnitudes. The letter D indicates that there is a *diorism* (a necessary restriction on the given magnitudes).

***Arithmetica* I, (partial) table of contents**

§ 1	1. $a + b = m, \quad a - b = n$	$m, n = 100, 40$	$a, b = 70, 30$
	2. $a + b = m, \quad a = p \cdot b$	$m, p = 60, 3$	$a, b = 45, 15$
	3. $a + b = m, \quad a = p \cdot b + l$	$m, p, l = 80, 3, 4$	$a, b = 61, 19$
	4. $a - b = n, \quad a = p \cdot b$	$n, p = 20, 5$	$a, b = 25, 5$
	5. $a + b = m, \quad 1/p \cdot a + 1/q \cdot b = n$	$m, n, p, q = 100, 30, 3, 5$ D	$a, b = 75, 25$
	6. $a + b = m, \quad 1/p \cdot a - 1/q \cdot b = n$	$m, n, p, q = 100, 20, 4, 6$ D	$a, b = 88, 12$
§ 2	7. $a - k = p \cdot (a - l)$	$k, l, p = 100, 20, 3$	$a = 140$
	8. $a + k = p \cdot (a + l)$	$k, l, p = 100, 20, 3$ D	$a = 20$
	9. $k - a = p \cdot (l - a)$	$k, l, p = 100, 20, 6$ D	$a = 4$
	10. $k + a = p \cdot (l - a)$	$k, l, p = 20, 100, 4$ D	$a = 76$
	....		
§ 4	14. $a \cdot b = p \cdot (a + b)$	$p = 3 \quad (b = 12)$ D	$a, b = 4, 12$
§ 5	15. $a + k = p \cdot (b - k), b + l = q \cdot (a - l)$	$k, l, p, q = 30, 50, 2, 3$	$a, b = 98, 94$

§ 6	16. $a + b = k, b + c = l, c + a = m$	k, l, m = 20, 30, 40 D	a, b, c = 15, 5, 25
	....		
§ 12	27. $a + b = m, \quad a \cdot b = k$	m, k = 20, 96 D P	a, b = 12, 8
	28. $a + b = m$, sq. a + sq. $b = k$	m, k = 20, 208 D P	a, b = 12, 8
	29. $a + b = m$, sq. a – sq. $b = k$	m, k = 20, 80	a, b = 12, 8
	30. $a - b = n, \quad a \cdot b = k$	n, k = 4, 96 D P	a, b = 12, 8
§ 13	31. sq. a + sq. $b = p \cdot (a + b), \quad a = q \cdot b$	p, q = 3, 5	a, b = 6, 2
	32. sq. a + sq. $b = p \cdot (a - b), \quad a = q \cdot b$	p, q = 3, 10	a, b = 6, 2
	33. sq. a – sq. $b = p \cdot (a + b), \quad a = q \cdot b$	p, q = 3, 6	a, b = 9, 3
	34. sq. a – sq. $b = p \cdot (a - b), \quad a = q \cdot b$	p, q = 3, 12	a, b = 9, 3
	....		

This could just as well have been the table of contents for an Old or Late Babylonian theme text with metric algebra problems, except for the *diorisms*, and for the fact that fractions such as, for instance, $1/p$ and $1/q$ in *Ar.* I.5-6, in a Babylonian mathematical text typically would have "non-regular sexagesimal values", most commonly 1/7, 1/11, 1/13, or 1/14.

(Also all the other books of Diophantus' *Arithmetica* are in form, although not in content, similar to OB theme texts.)

The *diorism* in *Ar.* I.5, for instance, says that

"The latter given number must be such that it lies between the numbers arising when the given fractions respectively are taken of the first given number."

Indeed,

Since $1/q \cdot (a + b) < 1/p \cdot a + 1/q \cdot b < 1/p \cdot (a + b)$ when $p < q$,
it is necessary that $1/q \cdot m < k < 1/p \cdot m$,
as in the example $m, n, p, q = 100, 30, 3, 5$ where $1/5 \cdot 100 < 30 < 1/3 \cdot 100$.

The letter P associated with the rectangular-linear or quadratic-linear systems of equations in ## 27, 28, 30, indicates that there is a supplement to the *diorism* mentioning the word *plasmatikón*, of unknown significance here. In # 27, for instance, the *diorism* says that

"The square of half the sum must exceed the product by a square number.
And this is *plasmatikón*"

The background to the *diorism* is, of course, that if $a + b = n$ and $a \cdot b = k$ are given, then $a - b$ can be computed by use of the equation

sq. $(a - b)/2$ = sq. $(a + b)/2 - a \cdot b$ = sq. $m/2 - k$.

Therefore, there exists a solution to the problem (in rational numbers) only if sq. $m/2 - k$ is a square (of a rational number). Now, since the Greek word *plásma* means ‘form, image’, *etc*., it is likely that *plasmatikón* stands for ‘representable’ and that the meaning of the mentioned obscure phrase is

> “And this can be shown in a diagram.”

The diagram in question would in this case (after transformation to metric algebra notations) be like the one in Fig. 13.1.1, left (below).

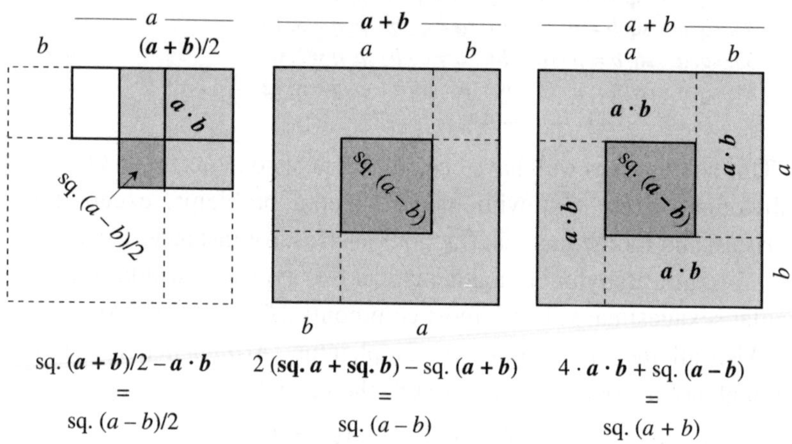

Fig. 13.1.1. Diagrams explaining the *diorisms* for Diophantus’ *Ar.* I.27, 28, 30.

In *Ar.* I.28, the *diorism* says

> “Double the sum of their squares must exceed the square of their sum by a square.”

This is because here $a - b$ can be computed by use of the equation

$$\text{sq. } (a - b)/2 = 2\ (\text{sq. } a + \text{sq. } b) - \text{sq. } (a + b) = 2\ k - \text{sq. } m.$$

The corresponding diagram is the one in Fig. 13.1.1, middle.

In *Ar.* I.30, finally, the *diorism* says that

> “Four times the product, together with the square of the difference must be a square.”

The background to the *diorism*, in this case, is that if $a - b = n$ and $a \cdot b = k$ are given, then $a + b$ can be computed by use of the equation

$$\text{sq. } (a + b) = 4 \cdot a \cdot b + \text{sq. } (a - b) = 4 \cdot k + \text{sq. } n.$$

The corresponding diagram is the one in Fig. 13.1.1, right.

Note that there is no *diorism* in *Ar.* I.29 for the simple reason that the problem in that case can be reduced to a *linear* equation.

It is remarkable that in both *Ar.* I.28 and *Ar.* I.30 *the solution procedure does not use the method suggested by the diorism*! Thus, in *Ar*. I.28, for instance, the solution procedure is like this (Thomas, *SIHGM* (1980), 537):

"Let it be required to make their sum 20 and the sum of their squares 208.
Let their difference be 2 *s*, and let the greater be *s* + 10, again adding half the sum, and the lesser 10 – *s*. Then again their sum is 20 and their difference 2 *s*.
It remains to make the sum of their squares 208. But the sum of their squares is 2 sq. *s* + 200. Therefore 2 sq. *s* + 200 = 208 and it follows that *s* = 2.
To return to the hypotheses—the greater = 12 and the lesser = 8. And these satisfy the conditions of the problem."

Here, *Diophantus does not work with two unknowns (in the Babylonian way)*, as suggested by the form of the *diorism*. Instead, *he prefers to work, as he always does, with only one unknown,* for which he has a special symbol, resembling an *s*. Setting $a - b = 2\ s$, he gets that

$$a = s + (a + b)/2 = s + 10 \quad \text{and} \quad b = (a + b)/2 - s = 10 - s.$$

Consequently,

$$\text{sq. } a + \text{sq. } b = \text{sq. } (s + 10) + \text{sq. } (10 - s) = 2 \text{ sq. } s + 200 = 208.$$

Therefore, 2 sq. $s = 8$, sq. $s = 4$, so that $s = 2$, $a = 2 + 10 = 12$, $b = 10 - 2 = 8$.

Among all the determinate problems in *Arithmetica* I, there is actually one *indeterminate* problem, namely ***Ar.* I.14**:

Find two numbers such that their product has a given ratio to their sum
One of the two numbers must be greater than the number representing the ratio.

Let the product be 3 times the sum, and let one of the numbers be *s*.
The other must be greater than 3, let it be 12.
The product is 12 *s*, the sum 12 + *s*.
Therefore 12 *s* equals 3 *s* + 36, and *s* = 4.
The two numbers are 4 and 12.

In the solution procedure, the indeterminate problem $a \cdot b = p \cdot (a + b)$ with $p = 3$ is made determinate by arbitrarily assuming that one of the numbers is 12 (greater than $p = 3$). The *diorism* is not explained, but since

$$a \cdot b = p \cdot (a + b) \quad \Rightarrow \quad (a - p/2) \cdot (b - p/2) = \text{sq. } p/2$$

it follows that one of $a - p/2$ and $b - p/2$ must be greater than $p/2$.

It is interesting to compare this quite uninteresting solution procedure with the corresponding solution procedure in the parallel OB exercise **AO 6770 # 1** (Friberg, *RC* (2007), Sec. 11.2 k).

AO 6770 # 1, literal translation	explanation
The length and the front	$u + s$
as much as the field I let be equal.	$= A = u \cdot$
You, in your doing:	Do it like this:
The step to its two you set.	Write down two copies of the unknown t
Out of it 1 you tear off.	Compute $t - 1$
The opposite you resolve.	Find $1 / (t - 1)$
With the step that you set	Take the other t
you let (them) hold each other.	and compute the product $t \cdot 1 / (t - 1)$
The front it gives you.	This is the value of s

This exercise is in several ways outstanding among OB mathematical exercises. Thus, it is the only known OB mathematical exercise where *the solution procedure is completely abstract, without an illustrating numerical example*. It is also one of very few OB exercises dealing with *an indeterminate problem*, and it is almost unique in that it has *a special term for the unknown, here called the 'step'*.

The obscure wording of the solution procedure can be explained as follows: The length is (silently) assumed to have the unknown value t. Two copies are made of this unknown value. One copy is used to form the new value $1/(t - 1)$. This new value is multiplied with the other copy of t. The result is the front s. Therefore,

$$u, s = t, t / (t - 1), \text{ for any given value of } t.$$

The procedure is based on the silent assumption that $t > 1$, which is necessary, since u and s, the sides of a rectangle, must have positive values.

For a metric algebra proof of the solution rule, see Friberg, *op. cit.*

13.2. Four Basic Examples in Book II of Diophantus' *Arithmetica*

13.2 a. *Ar.* II.8 (Sesiano, *GA* (1990), 84; Thomas, *op. cit.*, 551)

To divide a given square number into two squares.

Let it be required to divide 16 into two squares[34].

34. It is important to understand that, for want of better alternatives, Diophantus, like his Babylonian predecessors, often introduced arbitrary numerical values into the mathematical discussion of a problem, just as we would introduce symbolic values like k, l, m. Therefore, in many such cases, what seems to be only the solution to a *special* case of a given problem was certainly intended to demonstrate the *general* case.

And let first the square be sq. s
(which Diophantus writes as D^{y}. 1, an abbreviation for '1 times s in po(wer)').
Then the other will be 16 – sq. s. It is required therefore to make 16 – sq. s a square.
I take the square of any amount of 'numbers' (that is s) minus as many units as there are in the side of 16.
Let it be $2\,s - 4$, and the square itself 4 sq. $s + 16 - 16\,s$, and this equals 16 – sq. s.
Add to both sides the subtracted terms and take like from like.
Then 5 sq. s equals 16 s, and s becomes equal to 16 fifths.
One will therefore be 256/25, one 144/25, and the two together make 400/25or 16, and each is a square.

In this solution procedure, Diophantus chooses 16 as the given square and tries to set 16 – sq. s equal to, for instance, sq. $(2\,s - 4)$. Then

$$16 - \text{sq. } s = \text{sq. } (2\,s - 4) = 4 \text{ sq. } s - 16\,s + 16, \quad \text{so that} \quad 5 \text{ sq. } s = 16\,s \quad \text{and} \quad s = 16/5.$$

A *modern* interpretation of Diophantus' method in *Ar.* II.8 is illustrated in the diagram in Fig. 13.2.1 below:

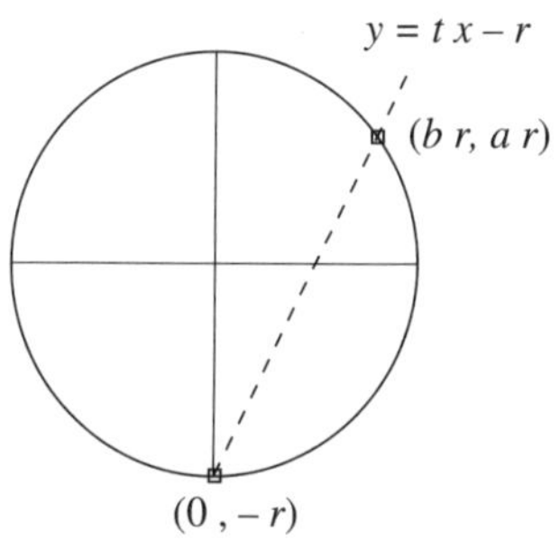

$$x^2 + y^2 = r^2, \quad y = t\,x - r$$
$$\Rightarrow$$
$$x^2 + t^2 x^2 = 2r\,t\,x$$
$$\text{or} \quad x + t^2 x = 2r\,t \quad \text{if } x \neq 0$$
$$\Rightarrow$$
$$(x, y) = (b\,r, a\,r) \quad \text{with}$$
$$b = 2\,t/(t^2 + 1), \quad a = (t^2 - 1)/(t^2 + 1$$

Fig. 13.2.1. *Ar.* II, 8. A modern interpretation in terms of the chord method.

As shown in the diagram, if r^2 is the given rational square number, then a line with rational slope t is drawn from the point $(0, -r)$ on a circle of radius r until it cuts the circle in a second point. The coordinates of this point are the solution to a pair of linear equations with rational coefficients and are therefore themselves rational. By varying t, all the rational points on the circle can be reached.

In Diophantus' example, $r, t = 4, 2$, so that the coordinates of the second point are $b\,r = 4/5 \cdot 4 = 16/5$ and $a\,r = 3/5 \cdot 4 = 12/5$. Hence, 16 = sq. 16/5 + sq. 12/5 = 256/25 + 144/25.

An alternative explanation of the solution procedure in *Ar.* II.8, *in terms of Babylonian metric algebra*, is presented in Fig. 13.2.2 below.

In this alternative interpretation, a right triangle with its short sides in the ratio $t : 1$ is inscribed in a semicircle with the radius r, with the side proportional to t along the diameter. If a radius is drawn from the center of the semicircle to the vertex of the right triangle, as in Fig. 13.2.2, a small right triangle is formed with the sides (r, p, q), where $q = t \cdot p - r$. According to the diagonal rule,

$$\text{sq. } p + \text{sq. } (t \cdot p - r) = \text{sq. } r, \quad \text{so that} \quad (\text{sq. } t + 1) \cdot \text{sq. } p = 2t \cdot p \cdot r.$$

Consequently,

$$p = \{2\,t \,/\, (\text{sq. } t + 1)\} \cdot r \quad \text{and} \quad q = t \cdot p - r = \{(\text{sq. } t - 1)/\{(\text{sq. } t + 1)\} \cdot r.$$

Therefore the small right triangle has the sides

$$r, p, t \cdot p - r = \{1, 2\,t/(\text{sq. } t + 1), (\text{sq. } t - 1)/\{(\text{sq. } t + 1)\} \cdot r.$$

A new application of the diagonal rule then shows that

$$\text{sq. } r = \text{sq. } (2\,t/(\text{sq. } t + 1) \cdot r) + \text{sq. } ((\text{sq. } t - 1)\,/(\text{sq. } t + 1) \cdot r).$$

This is the desired representation of the given square as a sum of two squares, in its most general form.

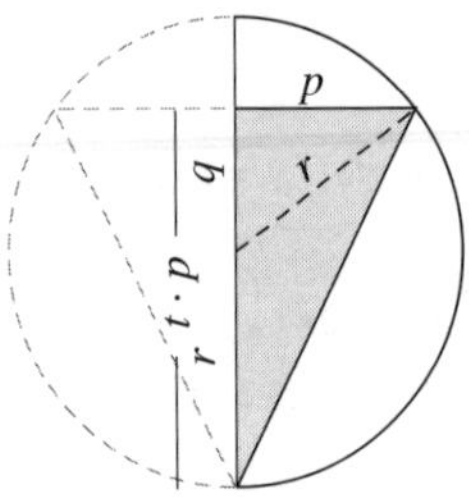

$$\text{sq. } p + \text{sq. } q = \text{sq. } r, \quad q = t \cdot p - r$$
$$\Rightarrow$$
$$(\text{sq. } t + 1) \cdot \text{sq. } p = 2\,t \cdot p \cdot r$$
$$\Rightarrow$$
$$p = \{2\,t/(\text{sq. } t + 1)\} \cdot r$$
$$q = \{(\text{sq. } t - 1)/(\text{sq. } t + 1)\} \cdot r$$

Fig. 13.2.2. *Ar.* II, 8. Interpretation in terms of Babylonian metric algebra.

The proof of Ar. II.8 can be interpreted as the derivation of a generating rule for (rational) diagonal triples. Note the similarity of this derivation with the proposed derivation of the generating rule used in the OB table text Plimpton 322 (Fig. 3.2.1, right) Note also the similarity of the diagram in Fig. 13.2.2 with (a part of) the diagram on the OB clay tablet *TMS* 1 (Fig. 1.12.4 above).

When $r = 1$ and $t = m/n$, one gets *a generating rule for diagonal triples c, b, a with c = 1. Cf.* the explanation in Sec. 11.9 of the term im.gíd.da in *TMS* 23 as a reference to a table of diagonal triples with $c = 1$.

13.2 b. *Ar.* II.9 (Sesiano, *GA* (1990), 85; Heath, *DA* (1964), 145)

To divide a given number which is the sum of two squares into two other squares

In his solution procedure, Diophantus lets the given number be 13 = 9 + 4 = sq. 3 +

sq. 2. He assumes that the two new squares are sq. $(s + 2)$ and sq. $(t\,s - 3)$, where s is unknown and t arbitrary, for instance $t = 2$. Then he gets that sq. $(s + 2)$ + sq. $(2\,s - 3) = 5$ sq. $s - 8\,s + 13$ which is required to be $= 13$.Therefore $s = 8/5$. Hence the two squares are sq. $(8/5 + 2)$ = sq. $18/5 = 324/25$ and sq. $(16/5 - 3)$ = sq. $1/5 = 1/25$.

A *modern* interpretation of Diophantus' method in *Ar.* II.9 is illustrated in Fig. 13.2.3 below:

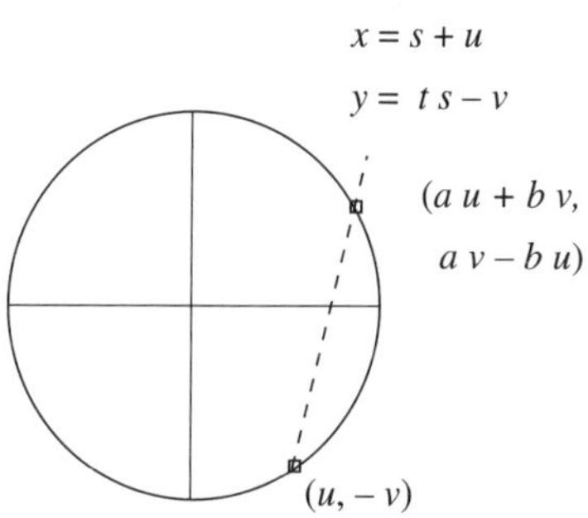

$$x^2 + y^2 = u^2 + v^2,$$
$$x = s + u, \quad y = t\,s - v$$
$$\Rightarrow$$
$$s^2 + t^2 s^2 = 2\,v\,t\,s - 2\,u\,s$$
$$\text{or} \quad s + t^2 s = 2\,v\,t - 2\,u$$
$$\Rightarrow$$
$$s = (2\,v\,t - 2\,u)/(t^2 + 1)$$
$$\Rightarrow$$
$$(x, y) = (a\,u + b\,v,\ a\,v - b\,u) \quad \text{with}$$
$$b = 2\,t/(t^2 + 1), \quad a = (t^2 - 1)/(t^2 + 1)$$

Fig. 13.2.3. *Ar.* II, 9. A modern interpretation in terms of the chord method.

An alternative explanation of the solution procedure in *Ar.* II.9, *in terms of Babylonian metric algebra*, is presented in Fig. 13.2.4 below. In this alternative interpretation, a right trapezoid with the height : (the difference of the parallel sides) = t : 1 is inscribed in a semicircle, with the height along the diameter. In addition to t, the lower parallel u and its distance v from the center of the circle are known.

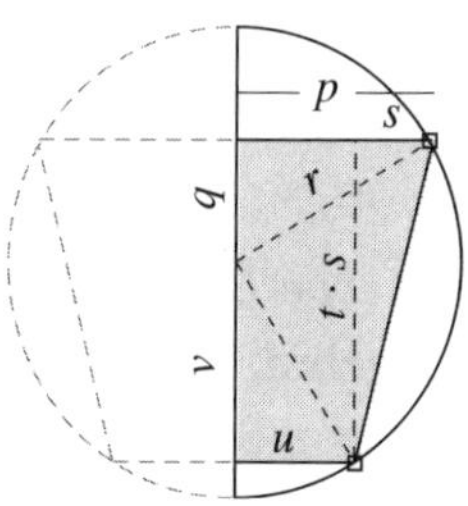

$$\text{sq. } p + \text{sq. } q = \text{sq. } u + \text{sq. } v$$
$$p = s + u, \quad q = t \cdot s - v$$
$$\Rightarrow$$
$$\text{sq. } (s + u) + \text{sq. } (t \cdot s - v) = \text{sq. } u + \text{sq. } v$$
$$\Rightarrow$$
$$s = (2\,t \cdot v - 2\,u) / (\text{sq. } t + 1)$$
$$\Rightarrow$$
$$(p, q) = (a\,u + b\,v,\ a\,v - b\,u) \quad \text{with}$$
$$b = 2\,t/(t^2 + 1), \quad a = (t^2 - 1)/(t^2 + 1)$$

Fig. 13.2.4. *Ar.* II, 9. Interpretation in terms of Babylonian metric algebra.

The upper parallel p and its distance q from the center of the circle can then be computed as follows: Let $p = s + u$. Then $q = t \cdot s - v$, and it follows from two applications of the diagonal rule that

$$\text{sq.}\ (s + u) + \text{sq.}\ (t \cdot s - v) = \text{sq.}\ p + \text{sq.}\ q = \text{sq.}\ r = \text{sq.}\ u + \text{sq.}\ v \quad (r = \text{the radius}).$$

This equation for the unknown s can be reduced to

$$\text{sq.}\ s + \text{sq.}\ t \cdot \text{sq.}\ s = 2\,t \cdot v \cdot s - 2\,u \cdot s, \quad \text{or} \quad s + \text{sq.}\ t \cdot s = 2\,t \cdot v - 2\,u.$$

Consequently,

$$s = (2\,t \cdot v - 2\,u) / (\text{sq.}\ t + 1).$$

Therefore,

$$p = s + u = a \cdot u + b \cdot v \quad \text{and} \quad q = t \cdot s - v = a \cdot v - b \cdot u,$$

where

$$b = 2\,t / (\text{sq.}\ t + 1), \quad a = (\text{sq.}\ t - 1) / (\text{sq.}\ t + 1).$$

In *Ar.* II.9, $u, v = 2, 3$ and $t = 2$, so that $b, a = 4/5, 3/5$. Compare with the related computations in the case of a *confluent quadrilateral bisection problem* in *TMS* 23 (Sec. 11.9), where b, a = ;48 (4/5), ;36 (3/5).

13.2 c. *Ar.* II.10 (Sesiano, *GA* (1990), 86; Heath, *DA* (1964), 146)

To find two square numbers having a given difference.

In his solution procedure, Diophantus lets the given difference be 60. He assumes that the sides of the two squares are s and $s + 3$, where s is unknown and 3 an arbitrarily given number, the square of which is not greater than the given difference. Then he gets that sq. $(s + 3)$ – sq. $s = 6\,s + 9$ which is required to be = 60. Therefore $s = 8\ 1/2$. Hence the two squares are sq. 11 1/2 = 132 1/4 and sq. 8 1/2 = 72 1/4.

Here Diophantus lets the *indeterminate* equation

$$\text{sq.}\ p - \text{sq.}\ q = D \quad (\text{with } D = 60)$$

be replaced by the *determinate* system of equations

$$\text{sq.}\ p - \text{sq.}\ q = D, \quad p - q = n \quad (\text{with sq.}\ n < D, \text{ for instance } n = 3).$$

This is *a quadratic-linear system of equations of type B3b* (see Sec. 1.1 above). It can easily be solved by use of metric algebra, interpreting D as the area of *a square difference* as in Fig. 1.5.2, or as the area of *a square band* as in Fig. 1.13.5. The solution is

$$p = (D/n + n)/2 = (D + \text{sq.}\ n) / 2\,n, \quad q = (D/n - n)/2 = (D - \text{sq.}\ n) / 2\,n.$$

13.2 d. *Ar.* II.19 (Sesiano, *GA* (1990), 86; Heath, *DA* (1964), 146)

To find three squares such that the difference between the greatest and the middle has to the difference between the middle and the least a given ratio.

Diophantus lets the given ratio be 3: 1. He lets the first square be sq. s and the second square sq. $(s + 1) =$ sq. $s + 2\,s + 1$. Then the third square must be sq. $s + 4 \cdot (2\,s + 1) =$ sq. $s + 8\,s + 4$. If it is also equal to sq. $(s + p)$, then, he says, either $2\,p > 8$ and sq. $p < 4$ (this cannot happen) or $2\,p < 8$ and sq. $p > 4$ (then $2 < p < 4$). Diophantus chooses $p =$ 3. It follows that sq. $s + 8\,s + 4 =$ sq. $s + 6\,s + 9$, so that $2\,s = 5$ and $s = 2\ 1/2$. Hence the three squares are sq. 5 1/2 = 30 1/4, sq. 3 1/2 = 12 1/4, and sq. 2 1/2 = 6 1/4.

In terms of Babylonian metric algebra, the problem in *Ar.* II.19 can be interpreted as finding the two fronts s_a, s_k and the transversal d in a 2-striped trapezoid, when the partial areas are to each other in a given ratio. (See above, Sec. 11.3 d-f.)

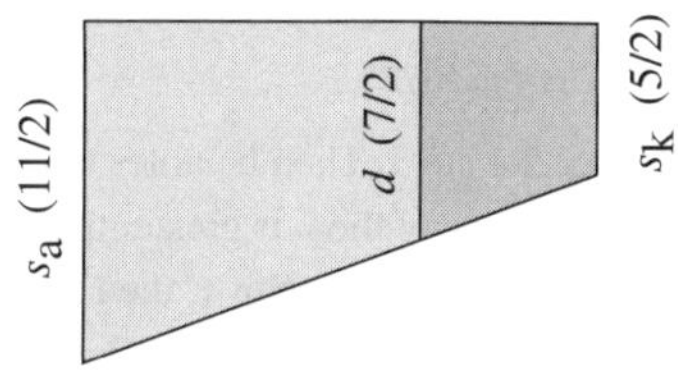

(sq. s_a – sq. d) : (sq. d – sq. s_k) = q : 1
$(s_a - s_k) : (d - s_k) = p : 1$
$(p = q = 3, \quad d - s_k = 1)$

Fig. 13.2.5. *Ar.* II.19. Interpretation in terms of Babylonian metric algebra.

As in the similar case of *Ar.* II.10, this indeterminate problem for the three unknowns s_a, s_k, and d can be made determinate through the introduction of additional (linear) equations between the unknowns. Here, these extra condition are (essentially) that

$(s_a - s_k) : (d - s_k) = p : 1$ and $d - s_k = 1$.

In other words, $(s_a, d, s_k) = (s + p \cdot t,\ s + t,\ s)$ for some value of t. Nothing essential is lost by assuming, as Diophantus does, that $t = 1$. Then, if q is the given ratio between the partial areas, it follows that

sq. $(s + p)$ – sq. $(s + 1) = q \cdot \{$sq. $(s + 1)$ – sq. $s\}$ or, after simplification,
$(2\,s + p + 1) \cdot (p - 1) = q \cdot (2\,s + 1)$.

Hence, the solution in this general case is

either $s = (q + 1 -$ sq. $p) / (2\,p - 2\,q - 2)$ or $s = ($sq. $p - q - 1) / (2\,q + 2 - 2\,p)$.

In the special case considered by Diophantus, that is when $p = q = 3$, the corresponding solution is

$s = 5/2$ so that $(s_a, d, s_k) = (s + p, s + 1, s) = (5\ 1/2, 3\ 1/2, 2\ 1/2)$.

13.3. *Ar.* "V".9. Diophantus' Method of Approximation to Limits

***Ar.* "V".9** (Sesiano, *GA* (1990), 92; Heath, *DA* (1964), 95, 206)

To divide unity into two parts so that if the same given number is added to either part the result will be a square.

Let the given number be N. It is required to find two parts of unity u, v and two numbers m, n such that

$$N + u = \text{sq. } a, \quad N + v = \text{sq. } b, \quad u + v = 1, \quad a > b.$$

Then also

$$2\,N + 1 = \text{sq. } a + \text{sq. } b, \quad \text{and} \quad 0 < \text{sq. } a - (N + 1/2) < 1/2, \quad 0 < (N + 1/2) - \text{sq. } b < 1/2.$$

Diophantus assumes that certain conditions are satisfied[35] so that $2\,N + 1$ is indeed the sum of two squares,

$$2\,N + 1 = \text{sq. } a' + \text{sq. } b'.$$

These are the essential steps of the solution procedure:

Diophantus chooses $N = 6$ so that $2\,N + 1 = 13 = \text{sq. } 2 + \text{sq. } 3$. Then he notes that it is necessary to divide 13 into two squares so that each one of them is greater than 6, and that if 13 is divided into two squares with a difference less than 1, then the problem is solved. He takes half of 13, which is 6 1/2, and then wants to find a fraction which together with 6 1/2 gives a square.

To achieve this Diophantus multiplies 6 1/2 with 4 and looks for the inverse square of a 'number' which together with 26 gives a square. He multiplies with the square of the number and gets that 26 squares of the number plus 1 unit must be a square. Setting the side of that square equal to 5 numbers plus 1, he finds that the number is 10.His conclusion is that 6 1/2 plus 1/400 equals the square of 51/20.

Diophantus now assumes that the sides of the two squares with the sum 13 are of the form $2 + 11$ 'numbers' and $3 - 9$ 'numbers'. The sum of the squares is then 202 squares of the number plus 13 units minus 10 numbers, which shall be equal to 13 units. It follows that the number is 5/101. Consequently, the sides of the two squares are 257/101 and 258/101. The squares of these sides exceed 6 units by 4843/10201 and 5358/10201, respectively.

The problem is solved.

A well known *modern* interpretation of Diophantus' method in *Ar.* "V".9 *in terms of the chord method* is illustrated in Fig. 13.3.1 below: The idea is actually quite simple. First a', b' are found such that

$$\text{sq. } a' + \text{sq. } b' = 2\,N + 1.$$

---

35. The text is partly destroyed precisely at this crucial point of the exposition.

Then c is computed as a close approximation to sqs. $(N + 1/2)$. Therefore,

sq. c + sq. c is very close to $2\,N + 1$.

In the final step, the point (a, b) is found as the intersection of the circle of radius r, where sq. $r = 2\,N + 1$, and the straight line through the two points (a', b') and (c, c). Then, since (a, b) is close to (c, c) and c is close to the square side of $N + 1/2$, it follows that also a and b are close to $N + 1/2$, as requested.

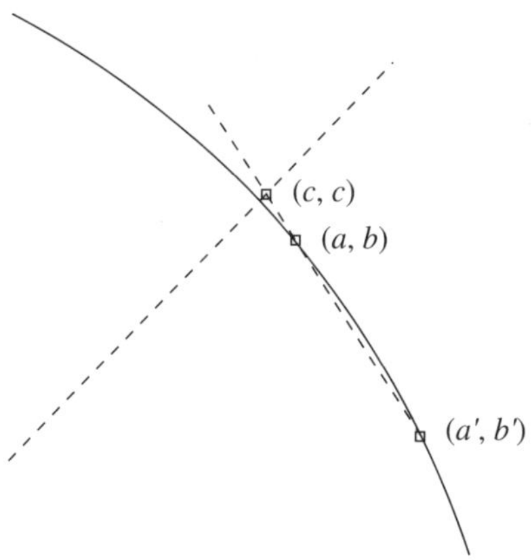

sq. a' + sq. $b' = 2\,N + 1$

sq. c very close to $N + 1/2$

$\Rightarrow$

sq. a + sq. $b = 2\,N + 1$ and

sq. a and sq. b both very close to $N + 1/2$

Fig. 13.3.1. *Ar.* "V".9. Modern interpretation in terms of the chord method.

In terms of Babylonian metric algebra, the problem in *Ar.*"V".9 can be interpreted as follows (see Fig. 13.3.2 below):

Let a trapezoid have the given upper and lower fronts s_a and s_k. Then the trapezoid is bisected (divided in two stripes of equal area) by a transversal d satisfying *the Babylonian trapezoid bisection equation*

sq. s_a + sq. s_k = 2 sq. d.

It can happen, however, that there *does not exist any rational solution to this equation*. Is it then possible to find instead *a confluent trapezoid bisection*, as in Fig. 13.3.2 below, *with the upper and lower transversals d_a and d_k very nearly equal*? If it is, then the confluent trapezoid bisection will serve as an approximate bisection of the given trapezoid.[36]

That it is possible to find such an approximate bisection of a given trapezoid will be shown below in a procedure completely parallel with the procedure in Diophantus' *Ar.* "V".9.

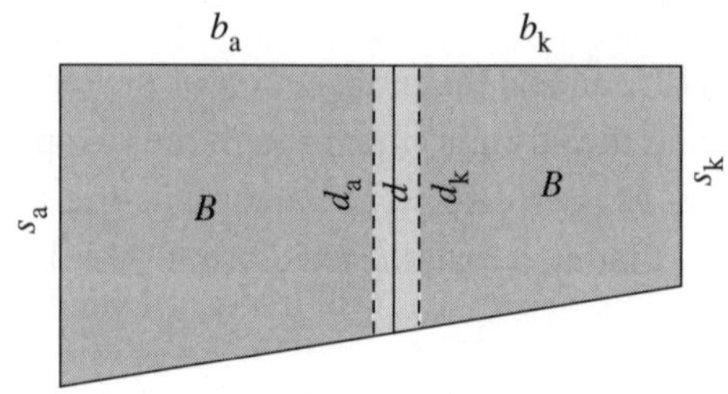

sq. s_a + sq. s_k = 2 · sq. d
d not rational
sq. s_a + sq. s_k = appr. 2 · sq. d^*
d^* rational
$b_a : b_k = (s_a - d^*) : (d^* - s_k)$

Fig. 13.3.2. *Ar.* "V".9. Interpretation in terms of Babylonian metric algebra.

Thus, let the given fronts of the trapezoid be s_a, s_k = 3, 2. Then

sq. s_a + sq. s_k = 9 + 4 = 13 = 2 sq. d so that sq. d = 6 1/2 (d not rational).

A first, obvious approximation to d is then

d = sqs. 6 1/2 = sqs. 26/4 = appr. 5/2, error: 6 1/2 – sq. 5/2 = 1/4.

A second, improved approximation is obtained by use of the OB "additive square side rule" (*cf.* the discussion in Friberg, *BaM* 28 (1997) § 8):

d^* = 5/2 + (6 1/2 – sq. 5/2) / (2 · 5/2) = 5/2 + 1/4 / 5 = 5/2 + 1/20 = 51/20.

The new error is quite small:

6 1/2 – sq. 51/20 = 26/4 – 2601/400 = 1/400 (= sq. 1/20).

This way of computing a good approximation to sqs. 6 1/2 is closely related to the method used by Diophantus. He sets sq. 2 · 6 1/2 + sq. (1/s) = sq. (5 + 1/s), or 26 sq. s + 1 = sq. (5 s + 1). This equation for s can be reduced to sq. s = 10 s. Hence, s = 10. The final result is that 6 1/2 + sq. (1/20) = sq. (51/20).
In the general case of this method, if p/q is a first approximation to sqs. N, set sq. $q \cdot N$ + sq. (1/s) = sq. (p + 1/s), or sq. (q s) · N + 1 = sq. (p s + 1). Equivalently, (sq. $q \cdot N$ – sq. p) · sq. s = 2 p s. Hence, s = 2 p / (sq. $q \cdot N$ – sq. p). The final result is that N + sq. (1/ q s) = sq. (p/q + 1/ q s), with 1/ q s = (N – sq. p/q) / 2 p/q.

---

36. *Cf.* the OB table text Plimpton 322 (Sec. 3.3 above), which can be interpreted as a recording of the result of an attempt to find a solution to the indeterminate quadratic equation sq. a +sq. b = sq. c, *with a and b very nearly equal*, thus with sq. c nearly equal to 2 sq. a. Indeed, the first entry in the table corresponds to the solution (c, b, a) = (2 49, 2 00, 1 59) = (169, 120, 119). — *Cf.* also the OB exercise AO 6484§ 7 a (Friberg, *RC* (2007), Appendix 7, in particular Fig. A7.5) which may be interpreted as the construction of a right triangle (c, b, a) = (1;00 00 16 40, 1, 0;00 44 43 20) with *c and b very nearly equal.* Here c = (t + 1/t)/2, with t = 1 21 / 1 20 = 1;00 45. It is likely that the construction was based on the observation that 1 21 and 1 20 are relatively large "regular sexagesimal twins".

Consider now a confluent trapezoid bisection as in Fig. 13.3.2, with

$b_a : b_k = (s_a - d^*) : (d^* - s_k) = (3 - 51/20) : (51/20 - 2) = 9/20 : 11/20.$

Then also, by similarity, $(s_a - d_a) : (d_k - s_k) = 9/20 : 11/20$, so that

$d_a = 3 - 9/20\ s, \quad d_k = 2 + 11/20\ s.$

The value of s is determined by the equation

sq. $(3 - 9/20\ s)$ + sq. $(2 + 11/20\ s)$ = sq. d_a + sq. $d_k = 2 \cdot 6\ 1/2 = 13.$

This equation can be reduced to

$(81 + 121) \cdot$ sq. $s = 20 \cdot (54 - 44) \cdot s$ which gives $s = 20 \cdot 10/202 = 100/101.$

Consequently,

$d_a = 3 - 9/20\ s = 3 - 45/101 = 258/101$ and $d_k = 2 + 11/20\ s = 2 + 55/101 = 257/101.$

Note that the corresponding squares are

sq. $d_a = 66564/10201 = 6\ 1/2 + 257\ 1/2\ /\ 10201 = 6 + 4358/10201$ and
sq. $d_k = 66049/10201 = 6\ 1/2 - 257\ 1/2\ /10201 = 6 + 4843/10201.$

Remark: A Babylonian mathematician could have avoided the need to divide by a non-regular sexagesimal number like 101, simply by assuming that the given fronts of the trapezoid were equal to $3 \cdot 101$ and $2 \cdot 101$.

13.4. *Ar.* III.19. A Square Number Equal to a Sum of Two Squares in Four Different Ways

***Ar.* III.19** (Heath, *DA* (1964), 166; Heath, *HGM 2* (1981), 481-483)

To find 4 numbers such the square of their sum plus or minus any one of them gives a square.

The basic idea in Diophantus' solution procedure is the following: Suppose that there exists a square number which is *equal to the sum of two squares in four different ways*. In other words, suppose that, for some number d, the equation sq. d = sq. a + sq. b has four distinct solutions (a_j, b_j), j = 1, 2, 3, 4. Then sq. $d \pm 2\ a_j\ b_j$ = sq. $(a_j \pm b_j)$ for j = 1, 2, 3, 4. Hence, it is also true that

sq. $(s \cdot d) \pm$ sq. $s \cdot 2\ a_j \cdot b_j$ = sq. $(s \cdot a_j \pm s \cdot b_j)$, j = 1, 2, 3, 4, for every s > 0.

Therefore, the four numbers sq. $s \cdot 2\ a_j \cdot b_j$, j = 1, 2, 3, 4, will solve the stated problem, provided that the value of s is chosen so that

$s \cdot d$ = sq. $s \cdot 2\ (a_1 \cdot b_1 + a_2 \cdot b_2 + a_3 \cdot b_3 + a_4 \cdot b_4).$

This will happen if $s = d\ /\ \{2\ (a_1 \cdot b_1 + a_2 \cdot b_2 + a_3 \cdot b_3 + a_4 \cdot b_4)\}.$

It remains to find a number d with the wanted property. To do this, Diophantus proceeds as follows: He starts by taking two right-angled triangles with small numbers for the sides, such as 3, 4, 5 and 5, 12, 13. The sides of each triangle are multiplied by the hypothenuse of the other triangle. The result is the two triangles **39, 52, 65** and **25, 60, 65**. These are two right-angled triangles with the same hypothenuse. However, 65 can be represented as the sum of two squares in two ways, 16 + 49 and 64 + 1. This is true, says Diophantus, *because 65 is the product of 13 and 5, each of which numbers is the sum of two squares.*

Next, Diophantus takes the square sides 7 and 4 of 49 and 16 and forms a right-angled triangles with these numbers. That is the triangle **33, 56, 65**. In the same way, 64 and 1 have the square sides 8 and 1, and Diophantus forms the triangle **16, 63, 65** with these numbers. In this way, he finds four right-angled triangles with the same hypothenuse.

With $d = 65$, Diophantus now computes the 4-fold areas of the four triangles, all multiplied by sq. s and obtains in this way sq. s multiplied by

$2 \cdot 39 \cdot 52 =$ **4056**, $2 \cdot 25 \cdot 60 =$ **3000**, $2 \cdot 33 \cdot 56 =$ **3696**, $2 \cdot 16 \cdot 63 =$ **2016**.

The sum of these four numbers is equal to on one hand sq. $s \cdot 12768$, on the other hand $s \cdot 65$. Therefore $s = 65/12768$, and the four requested numbers are

sq. $(65/12768) \cdot 4056 =$ **17136600/16302182**, *etc.*

Diophantus' arguments for finding his examples of four right triangles with the same hypothenuse are not very clear. It is possible, though, that he was inspired by some now lost Babylonian collection of mathematical exercises metric algebra exercises dealing with "cyclic quadrilaterals". (A *cyclic* quadrilateral is a quadrilateral which can be inscribed in a circle. Squares and rectangles, symmetric triangles, and symmetric trapezoids are the simplest cases of cyclic quadrilaterals.) In spite of the deplorable lack of direct evidence, this conjecture is strongly supported by extrapolation from two different directions. On one hand there are the examples discussed above of OB examples of symmetric or equilateral triangles and squares inscribed in circles (see Figs. 1.12.4 and 6.2.4-5.) On the other hand, there are the examples of more general cyclic quadrilaterals which can be found in early Indian mathematical texts, such as the commentary to the *Aryabhatiya* by Bhâskara I (522 A. D.), the *Brāhmasphuṭasiddhānta* of Brahmagupta (628), the *Ganita-sâra-samgraha* by Mahâvîra (850), or the *Lîlâvatî* by Bhâskara II (1150).[37] The discussion below is an attempt to explain the possibly Babylonian background to the Indian construction of two important classes of cyclic rational quadrilaterals, and thereby also the ideas on which Diophantus may have relied in his construction of the data for the solution to *Ar.* III, 19.

37.See Datta and Singh, *HHM 2* § 21, in particular pp. 235 *ff.*

Everywhere rational cyclic quadrilaterals

An "everywhere rational quadrilateral" (*cf.* again Datta and Singh, *HHM* 2 § 21, in particular pp. 235 *ff.*) is a quadrilateral with rational sides, heights, diagonals, and area. A cyclic quadrilateral is one that can be inscribed in a circle. *Rectangles with rational sides and rational diagonals* are the simplest example of *everywhere rational cyclic quadrilaterals*.Such rectangles can be constructed by joining two copies of a rational right triangle along a common diagonal. *Everywhere rational symmetric* (isosceles) *trapezoids* can be constructed by joining together two rational right triangles along a common side as in Fig. 13.4.1, before drawing two additional sides to complete the trapezoid.

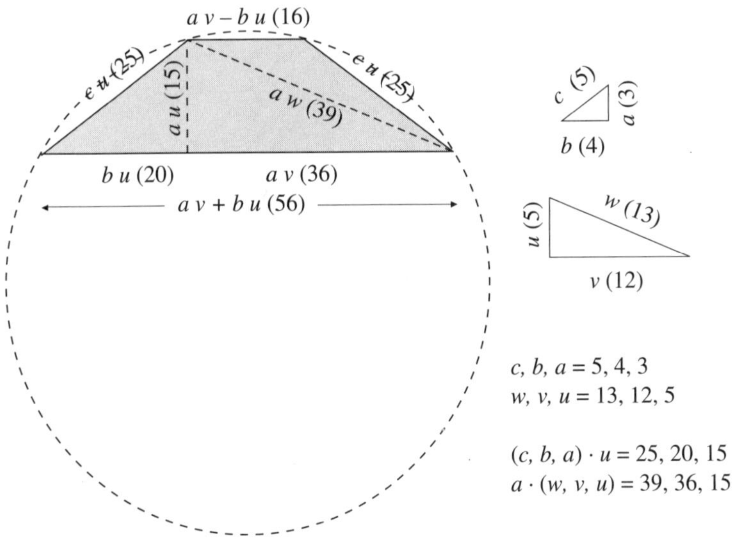

Fig. 13.4.1. An everywhere rational symmetric trapezoid.

If a circle passes through three of the four vertices of a symmetric trapezoid, it will also pass through the fourth vertex, because of the symmetry. That is why a symmetric trapezoid is a cyclic quadrilateral.

There is a single known mathematical cuneiform text showing that OB mathematicians were familiar with the trick of constructing *everywhere rational non-symmetric triangles* as joins of suitably scaled-up versions of rational right triangles. That text is VAT 7531. (See Fig. 1.12.7 above.)

A single known example of an *everywhere rational symmetric trapezoid* in a mathematical cuneiform text is the trapezoid in the Seleucid text **VAT 7848 # 4** (Neugebauer and Sachs, *MCT* Y (1945); the last third of the 1st millennium BCE). The given sides of that trapezoid are 50, 30, 14, and 30. The height, which is computed as the height of a symmetric triangle with the sides 30 and the base 50 – 14 = 36, is 24. Consequently, the diagonal d (which is not computed in the text) can be found as follows:

$$\text{sq. d} = \text{sq. } 24 + \text{sq. } (50 + 14)/2 = \text{sq. } 24 + \text{sq. } 32 = \text{sq. } 40, \quad d = 40.$$

Thus, the diagonal, the base, and one of the sides form a right triangle. Therefore, *the base of the trapezoid is also the diameter of the circumscribed circle*, as shown in Fig. 13.4.2 below.

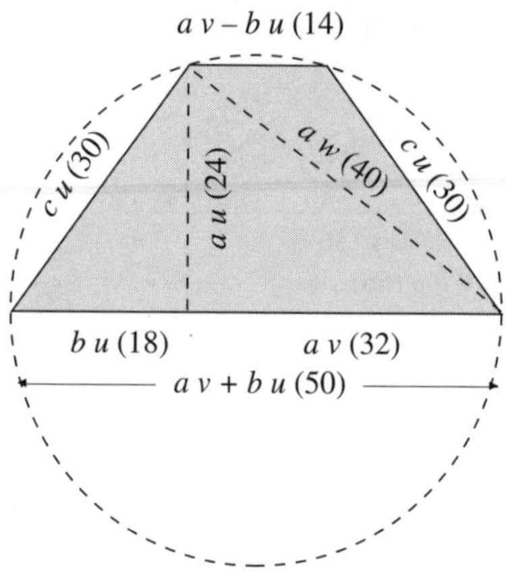

Fig. 13.4.2. VAT 7848 #4. A Seleucid everywhere rational symmetric trapezoid.

What may be called an "everywhere rational birectangle" is formed by joining along a common diagonal of length $c \cdot w$ its two "primary right sub-triangles", two rational right triangles with the sides $(c, b, a) \cdot w$ and $c \cdot (w, v, u)$, respectively. Thus, birectangles are quadrilaterals with *two opposite right angles*, and therefore obviously *cyclic* quadrilaterals. The common diagonal of the two right triangles, coinciding with the diameter of the circumscribed circle, is the "first diagonal" of the birectangle. See Fig. 13.4.3 below, where $c, b, a = 5, 4, 2$, and $w, v, u = 13, 12, 5$.

An alternative way of constructing an everywhere rational birectangle is to start with two everywhere rational non-symmetric triangles, one a join of two right triangles with the sides $(c, b, a) \cdot u$ and $a \cdot (w, v, u)$, the other a join of two right triangles with the sides $(c, b, a) \cdot v$ and $b \cdot (w, v, u)$. The two triangles will both have the base $a \cdot v + b \cdot u$. If they are joined along this common base, the result will be the same everywhere rational birectangle as the one constructed as the join of two right triangles, and the common base will be a "second diagonal" of the birectangle. As a consequence of this alternative construction, it is clear that *both the second diagonal and the two heights against the second diagonal are rational.*

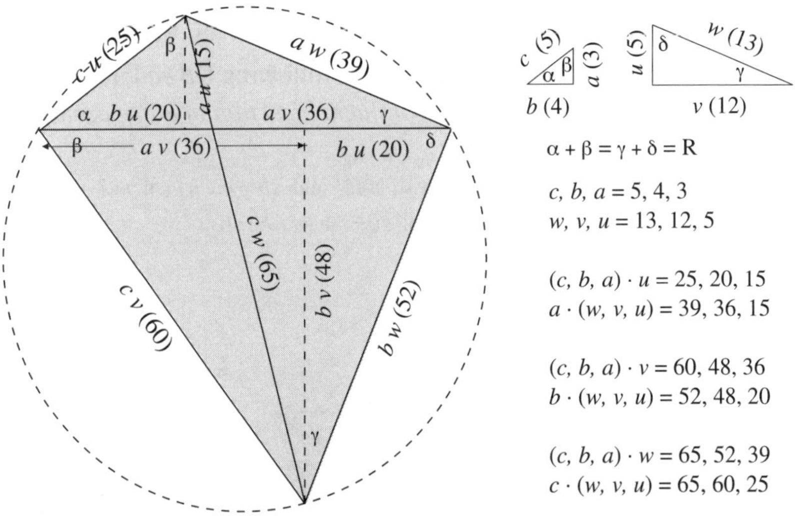

Fig. 13.4.3. An everywhere rational birectangle and its three pairs of rational sub-triangles.

Diophantus' *Ar.* III.19, Birectangles, and the OB Composition Rule

For every given everywhere rational birectangle inscribed in a circle (grey in Fig. 13.4.4 below) there is a pair of interesting "associated everywhere rational birectangles" inscribed in the same circle. One of these associated birectangles is composed of the two right triangles with the sides

$$c \cdot (w, v, u) \quad \text{and} \quad (c\,w,\, b\,v + a\,u,\, a\,v - b\,u).$$

The other one is composed of the two right triangles with the sides

$$(c, b, a) \cdot w \quad \text{and} \quad (c\,w,\, a\,v + b\,u,\, b\,v - a\,u).$$

In the example in Fig. 13.4.4, the sides of the generating triangles for the given birectangle are

$$c, b, a = 5, 4, 3 \quad \text{and} \quad w, v, u = 13, 12, 5.$$

The first associated birectangle is then composed of the two right triangles

$c \cdot (w, v, u) =$ **65, 60, 25** and $c\,w,\ b\,v + a\,u,\ a\,v - b\,u =$ **65, 63, 16**.

The second associated birectangle is composed of the two right triangles

$(c, b, a) \cdot w =$ **65, 52, 39** and $c\,w,\ a\,v + b\,u,\ b\,v - a\,u =$ **65, 56, 33**.

It follows directly from these representations of the sides of the associated right triangles for any given rational birectangle that

sq. $(b\,v + a\,u)$ + sq. $(a\,v - b\,u)$ = sq. $(c \cdot w)$ = (sq. a + sq. b) · (sq. u + sq. v), and
sq. $(a\,v + b\,u)$ + sq. $(b\,v - a\,u)$ = sq. $(c \cdot w)$ = (sq. a + sq. b) · (sq. u + sq. v).

The second identity is closely related to the following *OB equations for the upper and lower transversals in a confluent bisection of a trapezoid*:

If $d_a = a \cdot u_a + b \cdot u_k$, and $d_k = b \cdot u_a - a \cdot u_k$,
where a = [(sq. m – sq. n)/2 / (sq. m + sq. n)/2 and $b = m \cdot n$ / (sq. m + sq. n)/2,
then sq. a + sq. b = 1, and sq. d_a + sq. d_k = sq. u_a + sq. u_k.

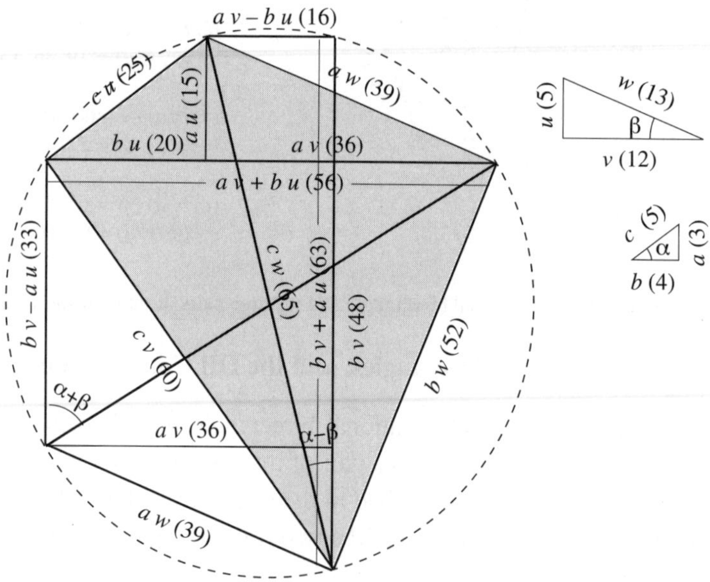

Fig. 13.4.4. A rational birectangle (grey) and its two associated rational birectangles.

These OB equations can be interpreted as saying that the "composition" of a *diagonal triple* 1, b, a with a *transversal triple* u_a, d, u_k is a new transversal triple with the same transversal d. Similarly, the equations for the sides of the birectangles associated with a birectangle say that *the compo-*

sition of a diagonal triple c, b, a with a diagonal triple w, v, u (in one of two possible ways) is a new diagonal triple with the diagonal $c \cdot w$.

Now, consider again the way in which Diophantus in *Ar.* III.19 constructs an example of four right triangles with the same hypothenuse. He starts by taking two right triangles with small numbers for the sides, such as *c, b, a* = 5, 4, 3 and *w, v, u* = 13, 12, 5, and he multiplies the sides of each triangle with the hypothenuse of the other. The result is the two triangles $(c, b, a) \cdot w$ = 65, 52, 39 and $c \cdot (w, v, u)$ = 65, 60, 25. *These are two of the right triangles making up the two associated birectangles in the example in Fig. 13.4.4.* [38] Diophantus continues by saying that 65 can be represented as a sum of two squares in two ways, as 16 + 49, and as 64 + 1, "because 65 is the product of 13 and 5, each of which is the sum of two squares". He then takes the square sides 4 and 7 in the first case and 8 and 1 in the second case and forms the right-angled triangles with the sides 65, 56, 33 and 65, 63, 16 from these numbers. *These are the other two of the right triangles making up the two associated birectangles in the example above.*

It is not immediately obvious what Diophantus means with his rather cryptic explanation, cited above. Anyway, here is a quite plausible *metric algebra explanation* of Diophantus' reasoning: Apparently, Diophantus knew that he could construct any number of everywhere rational birectangles by starting with a pair of right triangles with the sides

$$c, b, a = (\text{sq. } m + \text{sq. } n),\ 2\, m \cdot n,\ (\text{sq. } m - \text{sq. } n),$$
$$w, v, u = (\text{sq. } p + \text{sq. } q),\ 2\, p \cdot q,\ (\text{sq. } p - \text{sq. } q).$$

He would therefore think of the first diagonal in the birectangle as

$$d = c \cdot w = (\text{sq. } m + \text{sq. } n) \cdot (\text{sq. } p + \text{sq. } q).$$

Applying the OB composition rule, he would draw the conclusion that

$$d = \text{sq. } (m\, p + n\, q) + \text{sq. } (m\, q - n\, p) \quad \text{and} \quad d = \text{sq. } (n\, p + m\, q) + \text{sq. } (m\, p - n\, q).$$

With the generating pairs for the triples 5, 4, 3 and 13, 12, 5, namely

$$m, n = 2, 1 \quad \text{and} \quad p, q = 3, 2$$

this would imply that

$$d = \text{sq. } 8 + \text{sq. } 1 = 65 \ \text{ or } \ d = \text{sq. } 7 + \text{sq. } 4 = 65.$$

38.Diophantus mentions the number for the diagonal *last* in each diagonal triple, while in OB mathematics the largest number in any series of numbers is usually mentioned *first.* I follow the Babylonian tradition and mention the diagonal first in each triple.

This result would simultaneously tell him that $d = 65$ is the diagonal of the two right triangles with the short sides

$2 \cdot 8 \cdot 1$, sq. $8 -$ sq. $1 = 16, 63$ and $2 \cdot 7 \cdot 4$, sq. $7 -$ sq. $4 = 56, 33$.

An explanation like this of Diophantus' construction in *Ar.* III.19 suggests that *Diophantus may have been familiar both with diagrams like the one in Fig. 13.4.5 and with applications of the OB composition rule.*

Remark: If m, n and p, q are generating pairs for the triples c, b, a and w, v, u, then $m\,p + n\,q,\ m\,q - n\,p$ is a generating pair for the triple $c\,w,\ a\,v - b\,u,\ b\,v + a\,u$, while $n\,p + m\,q,\ m\,p - n\,q$ is a generating pair for the triple $c\,w,\ a\,v + b\,u,\ b\,v - a\,u$. It is, for instance, easy to see that

$$\text{sq. } (m\,p + n\,q) - \text{sq. } (m\,p - n\,q) = 2\,m\,n \cdot 2\,p\,q + (\text{sq. } m - \text{sq. } n) \cdot (\text{sq. } p - \text{sq. } q) = b\,v + a\,u.$$

The hypothesis that Diophantus may have been familiar with some construction like the one in Fig. 13.4.4 above is supported by the fact that birectangles are known to play an important role in at least two known Greek mathematical texts. One such text is Euclid's *Elements* II.9 (see Fig. 1.6.1 above). The other one is the Greek-Egyptian papyrus fragment ***P.Cornell 69*** (Friberg, *UL* (2005), Sec. 4.7 c), where problem # 3 is illustrated by a diagram like the one in Fig. 13.4.5 below, left.

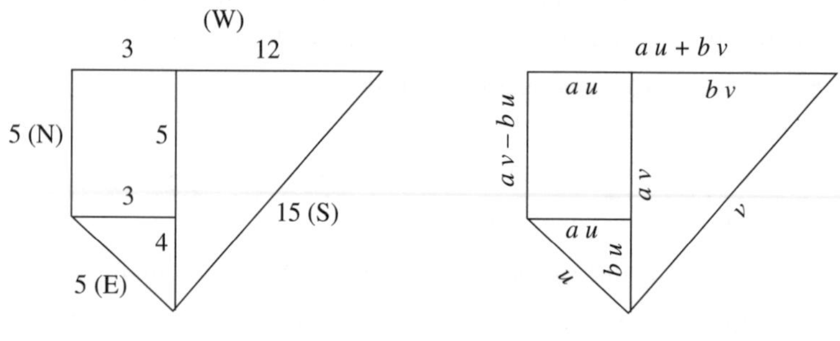

Fig. 13.4.5. *P.Cornell 69* # 3. A problem for a birectangle.

The text of the problem is almost completely destroyed, except for the computations of sq. $15 = 225$ and sq. $5 = 25$. However, this is enough for the following tentative reconstruction of the problem and its solution.

Let the sides of a birectangle be 15, 15, 5, 5. What is then the area of the birectangle? Let the sides of the birectangle be called u, v, $a\,u + b\,v$, and $a\,v - b\,u$, where $u = 5$, $v = 15$, $a\,u + b\,v = 15$, $a\,v - b\,u = 5$, and sq. a + sq. b = sq. $c = 1$. I is easy to find the values of a and b as the solution to a pair of linear equations. It turns out that $a = (2 \cdot 5 \cdot 15)/(\text{sq. } 15 + \text{sq. } 5) = 150/250 = 3/5$ and that $b = (\text{sq. } 15 - \text{sq. } 5)/(\text{sq. } 15 + \text{sq. } 5) = 200/250 = 4/5$. Consequently, $b\,v = 4/5 \cdot 15 = 12$, $a\,u = 5/5 \cdot 5 = 3$, and $b\,u = 4/5 \cdot 5 = 4$. (These values are indicated in the diagram.) The area can now be computed as, for instance, $6 \cdot 9 + 3 \cdot (9 + 5)/2 = 54 + 21 = 75$.

13.5. *Ar.* "V".30. An Applied Problem and Quadratic Inequalities

An indeterminate combined price problem

***Ar.* "V".30**, the last exercise in *Ar.* "V", appears there totally out of context. The statement of the problem, in the form of an epigram, is reproduced below in a free translation, following Czwalina, *ADA* (1952), 94.

> Someone who was obliged to do his shipmates a favor mixed together jars (of wine) at 8 drachmas and jars at 5 drachmas. As the price of all of them he gave a square, which increased by a given number gives you another square, the side of which is the number of all the jars. Consider this, my boy, and say how many jars there were at 8 drachmas and how many jars at 5 drachmas!

A related problem can be found in the OB text **YBC 4698** (Friberg, *UL* (2005), Fig. 2.1.17 and Sec. 2.1 f). Among the various "commercial problems" in YBC 4698 is the following "price and weight problem":

YBC 4698 # 4, literal translation	explanation
Its 1 30 iron, its 9 gold.	The price of iron is 1 30, the price of gold 9
1 mina of silver is given.	The combined total price is 1 mina of silver
Iron and gold,	How much iron and how much gold,
1 shekel, then buy.	if the combined weight is 1 shekel?

The text is vaguely formulated and without any known OB parallel. The question seems to be that if iron and gold are 90 (sic!) and 9 times more valuable than silver, and if 1 shekel of iron and gold together is bought for 1 mina of silver, what are then the amounts of iron and gold, respectively? The question can be reformulated (in modern terms) as a system of linear equations. If a shekels is the weight of the iron and g shekels the weight of the gold, then these equations are:

$$a + g = 1, \quad 1\ 30 \cdot a + 9 \cdot g = 1\ 00.$$

Systems of linear equations of the same type are known from the pair of OB problem texts **VAT 8389** and **VAT 8391**, solved there by use of a variant of the rule of false value, starting with a partial solution, satisfying only the first of the two given equations. (See Friberg, *RC* (2007), Sec. 11.2 m: Fig. 11.2.14 left.) In the case of YBC 4698 # 4, the first step would be to give a and g the initial false values

$a^* = g^* = ;30.$

If these values are tried in the second equation the result is that

$1\ 30 \cdot a^* + 9 \cdot g^* = 49;30,$

which gives a deficit of 10;30 compared to the wanted value 1 00.

To decrease the deficit, a^* is increased and g^* decreased by the small amount ;01. The result is that the deficit is decreased by a corresponding amount, namely

$;01 \cdot (1\ 30 - 9) = 1;21$ (a *regular* sexagesimal number with the reciprocal ;44 26 40).

Hence, the whole deficit can be eliminated if a^* is increased and g^* decreased by the larger amount

$;01 \cdot 10;30 \cdot 1/1;21 = ;10\ 30 \cdot ;44\ 26\ 40 = ;07\ 46\ 40.$

Therefore, the correct solution is that

$a = ;37\ 46\ 40,\ \ g = ;22\ 13\ 20.$

In terms of OB units of weight measure, the answer to the stated question in YBC 4698 # 4 is that the amounts of iron and gold are, respectively,

1/2 shekel 23 1/3 barleycorns of iron and 1/3 shekel 6 2/3 barleycorns of gold.

The answer is correct, since

$;37\ 46\ 40 + ;22\ 13\ 20 = 1$ and $;37\ 46\ 40 \cdot 1\ 30 + ;22\ 13\ 20 \cdot 9 = 56;40 + 3;20 = 1\ 00.$

Consider now again the question in ***Ar.* "V".30**. It can be reformulated as follows (in modern terms): Let p be the price paid for a certain number of jars at 8 drachmas each, let q be the price paid for a certain number of jars at 5 drachmas each, and let n be the total number of jars of both kinds. Suppose, as Diophantus does, that the given arbitrary number is 60! Then

$p/8 + q/5 = n,\ \ p + q = \text{sq. } n - 60 = \text{sq. } m,$ for some unspecified number m.

This is an *indeterminate* problem. What Diophantus does in his solution procedure is, essentially, that he first shows that n must lie *between certain limits*. Then he chooses arbitrarily a value for n between these limits and gets in this way a *determinate* problem for p and q, of the same type as the system of equations in YBC 4698 # 4, which he then solves.

The restrictions on the size of n are consequences of the fact that

$p/8 + q/5 < p/5 + q/5 = (p + q)/5$ and $p/8 + q/5 > p/8 + q/8 = (p + q)/8$.

Therefore,

(sq. $n - 60)/8 < n <$ (sq. $n - 60)/5$ so that sq. $n - 8\,n < 60 <$ sq. $n - 5\,n$.

These "quadratic inequalities" can be solved as follows by completion of the square although Diophantus gives only the result of the computations:

sq. $(n - 4) < 60 + 16 = 76$ $\Rightarrow$ $n - 4 <$ sqs. $76 =$ appr. 8 3/4,
sq. $(n - 2\ 1/2) > 60 + 6\ 1/4 = 66\ 1/4$ $\Rightarrow$ $n - 2\ 1/2 >$ sqs. $66\ 1/4 =$ appr. 8 9/64.

Instead of using the exact result of the computation, namely that

$10\ 41/64 < n < 12\ 3/4$ (approximately),

Diophantus mentions only the somewhat more narrow limits

$11 < n < 12$.

It remains to take care of the added restriction that sq. $n - 60 =$ sq. m. Diophantus sets $m = n - s$, for some unknown s, and finds that then

sq. $n - 60 =$ sq. $(n - s) =$ sq. $n - 2\,n \cdot s +$ sq. s so that $n =$ (sq. $s + 60) / 2\,s$.

Since he has assumed that $11 < n < 12$, it then follows that

$22\,s <$ sq. $s + 60 < 24\,s$.

This is *a new pair of quadratic inequalities*, from which follows that

$18\ 13/16 < s < 21\ 1/6$ (approximately).

Diophantus is content with saying, somewhat less exactly, that

$19 < s < 21$ so that he can choose $s = 20$.

With $s = 20$, he finds that the total number of jars is

$n =$ (sq. $s + 60) / 2\,s = 460/40 = 11\ 1/2$.

With this value for n, which is between the previously established limits, the given system of equations for p and q takes the simplified form

$p/8 + q/5 = n = 11\ 1/2$, $p + q =$ sq. $n - 60 =$ sq. $11\ 1/2 - 60 = 72\ 1/4$.

This determinate system of linear equations is then solved as follows:

Let $q/5 = t$. Then $q = 5\,t$ and $p = 92 - 8\,t$.

Insertion of these values into the second equation gives that

$92 - 3\,t = p + q = 72\ 1/4$ so that $t = 6\ 7/12 = 79/12$.

Therefore, the number of jars of each kind is

$q/5 = t = 6\ 7/12$ and $p/8 = 11\ 1/2 - 6\ 7/12 = 4\ 11/12$.

This is the answer given by Diophantus to the question in *Ar.* "V".30. It is easy to check that the result is correct. Indeed,

$p/8 + q/5 = 4\ 11/12 + 6\ 7/12 = 11\ 1/2 = n,$
$p + q = 39\ 4/12 + 32\ 11/12 = 72\ 1/4 = \text{sq. } 8\ 1/2 = \text{sq. } m,$
$\text{sq. } n - 60 = 132\ 1/4 - 60 = 72\ 1/4 = \text{sq. } m.$

13.6. *Ar.* "VI". A Theme Text with Equations for Right Triangles

The Babylonian influence in the Greek Book "VI" of Diophantus' *Arithmetica* is just as obvious as the Babylonian influence in Book I (see Sec. 13.1 above). *Ar.* "VI" is, just like *Ar.* I, *organized in the same way as an OB mathematical theme text.* This ought to be clear from the following table of contents (*cf.* Heath, *HGM 2* (1981), 507-514), where c, b, a, $P = c + b + a$, and A stand for *the unknown diagonal, the unknown sides, the unknown perimeter, and the unknown area of a right triangle*, while m, n, and r are undetermined, and k stands for an arbitrary given value.

An often used tool in *Ar.* "VI" is the application of *a suitably scaled-down version of the generating rule*

$$c = \text{sq. } p + \text{sq. } q, \quad b = 2\,p \cdot q, \quad a = \text{sq. } p - \text{sq. } q.$$

***Arithmetica* "VI"**, table of contents (sq. means *square*, cu. means *cube*)

	problem		p, q	c, b, a
1 a.	$c - a = \text{cu. } m, \quad c - b = \text{cu. } n$		10, 2	104, 96 40
b	$c + a = \text{cu. } m, \quad c + b = \text{cu. } n$		11/8, 2	377, 352, 135
2 a	$A + k = \text{sq. } m$	($k = 5$)	24/5, 5/24	(332401, *etc.*) · 1/31800
b	$A - k = \text{sq. } m$	($k = 6$)	8/3, 3/8	(4177, *etc.*) · 1/504
c	$k - A = \text{sq. } m$	($k = 10$)	80, 1/80	(40960001, *etc.*) · 1/825600
3 a	$A + a = k$	($k = 7$)		(25, 7, 24) · 1/4
b	$A - a = k$	($k = 7$)		(25, 7, 24) · 1/3
c	$A + (b + a) = k$	($k = 6$)		(53, 28, 45) · 1/18
d	$A - (b + a) = k$	($k = 6$)		(53, 28, 45) · 6/35
e	$A + (c + a) = k$	($k = 4$)	9, 5	(53, 45, 28) · 4/105
f	$A - (c + a) = k$	($k = 4$)	9, 5	(53, 45, 28) · 1/6
4 a	$A + b = \text{sq. } m, \quad A + a = \text{sq. } n$			(5, 4, 3) · 4/19
b	$A - b = \text{sq. } m, \quad A - a = \text{sq. } n$			(5, 4, 3) · 4/5
c	$A - c = \text{sq. } m, \quad A - a = \text{sq. } n$		4, 1	(17, 15, 8) · 1/
d	$A + c = \text{sq. } m, \quad A + a = \text{sq. } n$		4, 1	(17, 15, 8) · 1/77
5	To find a rational bisector of an acute angle in a right triangle			
6 a	$A + c = \text{sq. } m, \quad P = \text{cu. } n$			(629, 621, 100) · 1/50

b	$A + c =$ cu. m, $P =$ sq. n		(24153953, *etc.*) · 1/628864
c	$A + a =$ sq. m, $P =$ cu. n	4, 1	(17, 15, 8) · 1/5
d	$A + a =$ cu. m $P =$ sq. n	8, 1	(65, 63, 16) · 1/9
7 a	$P =$ sq. m, $A + P =$ cu. n	512/17, 1	(309233, *etc.*) · 1/4708
b	$P =$ cu. m, $A + P =$ sq. n	[............]	(5968, 4400, 4032) · 1/225
8 a	sq. $c =$ sq. $m + m$, sq. $c / a =$ cu. $n + n$		(5, 3, 4) · 3/4
b	$c =$ cu. $m + m$, $b =$ cu. r, $a =$ cu. $n - n$		10, 8, 6

Here is, for comparison, a table of contents for ***TMS 5*** (*cf.* Sec. 1.11 above), an OB theme text from Susa with metric algebra problems for one, two, or three *squares*. In the table of contents, A stands for the area of a square with the side s, A_1, A_2, A_3 stand for the areas of squares with the sides s_1,s_2, s_3, and A_c stands for the area of a square with the side $c \cdot s$,

TMS 5, table of contents

	equation	c (coefficient)	s, s_1, s_2, s_3 (sides of squares)
1 a	$s = k$, $c \cdot s = ?$	2, 3, 4, 2/3, 1/2, *etc.*	30, 35, 4 05 (= 5 · 7 · 7), *etc.*
b	$s + c \cdot s = k$	1/11, 2/11, *etc.*	55, 6 05 (= 5 : 11 · 11), *etc.*
c	$s - c \cdot s = k$	2/3, *etc.*, 1/7, *etc.*	30, 35, *etc.*
2 a	$s = k$, $c \cdot A = ?$	2/3, *etc.*, 1/7, *etc.*	30, 35, 4 05, *etc.*
b	$c \cdot A = k$	1/11, 2/11, *etc.*	55, 10 05 (= 5 · 11 · 11), *etc.*
3 a	$s = k$, $A_c = ?$	2, 1/3, 1/7, *etc.*	30, 35, 4 05, *etc.*
b	$A + A_c = k$	2, 3, 4, 2/3, 1/2, *etc.*	30, 35, 4 05, *etc.*
c	$A - A_c = k$	2, 3, 4, 2/3, 1/2, *etc.*	30, 35, 4 05, *etc.*
4 a	$A + c \cdot s = k$	2, 3, 4, 2/3, 1/2, *etc.*	30, 35, 4 05
b	$A - c \cdot s = k$	2, 3, 4, 2/3, 1/2, *etc.*	30, 35, 4 05
c	$c \cdot s - A = k$	1, 2, 2/3	30
d	$c \cdot s = A$	1/2	30
5	....		
6	$A - c \cdot A = k$	1/3, 1/4, 1/3 · 1/4, *etc.*	30, 35, 4 05, *etc.*
7 a	$s_1 = k$, $(s_1 - s_2)/2 = l$		(30, 20)
b	$s_2 = k$, $(s_1 - s_2)/2 = l$		(30, 20)
c	s_1, $s_2 = k, l$, $A_1 + A_2 = ?$		(30, 20)
d	$A_1 + A_2 = k$, $s_1 = l$		(30, 20)
e	$A_1 + A_2 = k$, $s_2 = l$		(30, 20)
f	$A_1 + A_2 = k$, $s_1 + s_2 = l$		(30, 20)
...	[..................]		[..................]
8 a	$s_1 = k$, $s_1 - s_2 = l$		(30, 20)
b	$A_1 - A_2 = k$, $s_1 - s_2 = l$		(30, 20)
c	$A_1 - A_2 = k$, $s_2 = c \cdot s_1$	1/7, 1/7 · 1/7	(35, 5), (4 05,5)
9 a	s_1, s_2, $s_3 = k, l, m$, $A_1 - A_2 = ?$, $A_2 - A_3 = ?$		(30, 20, 10)

b	$A_1 - A_2 = A_2 - A_3 = k,\ \ s_1 + s_2 + s_2 = l$	(30, 20, 10)
c	$A_1 - A_2 = A_2 - A_3 = k,\ \ s_1 - s_2 = s_2 - s_2 = l$	(30, 20, 10)

BM 80209 (Friberg, *JCS* (1981)) is a brief OB theme text, probably from Sippar, with metric algebra problems for squares and *circles*. Here is an abbreviated table of contents (for more details, see Sec. 1.10 above), with *s, d* standing for the side and diagonal of a square and *A, a, d* standing for the area, the circumference, and the diameter of a circle:

BM 80209, table of contents

	equation	*c* (coefficient)	*s, a*
1	$s = k$, sq. s = ?		[···]
2	$s = k,\ \ d = ?$		20
3	$s = k$, *dikšum* = ? (meaning not clear)		10
4	$A = k, a = ?$		10, 40, 50, 60
5 a	$A + c \cdot a = k$	2', 1, 1 2', *etc.*	10
b	$A - c \cdot a = k$	2', 1, 1 2', *etc.*	10
6	$A_1 - A_2 = k,\ \ a_1 - a_2 = l$		10, 30, 40, 50
7	$A + d + a = k$		10, 20, 30, 40

In addition, similar OB theme texts are known with metric algebra problems for *rectangles with the sides in a fixed ratio* or for *semicircles* (Friberg and Al-Rawi (*to be published*)).

Strictly speaking, only the problems in *Ar.* "VI" §§ 3 a-3 f (## 6-11 in the customary numbering) look like metric algebra problems in the mentioned OB theme texts, because all the other problems set various combinations of the area and the sides of a right triangle equal to undetermined squares or cubes or more complicated undetermined expressions. The appearance is deceptive, however, since *also the problems in Ar. "VI" §§ 3 a-3 f (## 6-11) are indeterminate*. Consider, for instance *Ar.* "VI" § 3 a (# 6):

***Ar.* "VI".6** (Heath, *DA* (1964), 228)

To find a right-angled triangle such that the area added to one of the perpendiculars makes a given number.

Diophantus chooses 7 as the given number and assumes that, for instance, (3*s*, 4*s*, 5*s*) are the sides of the given right-angled triangle. Then ($A + a$ =) 6 sq. s + 3 s = 7. This is a quadratic equation for *s*, which has a (rational) solution only if the square of half the coefficient of *s* plus the product of the coefficient for sq. *s* and the given number 7 is a (rational) square. However, sq. 1 1/2 + 6 · 7 (= 49 1/2) is not a square. Diophantus now replaces the right triangle with the sides (3, 4, 5) with a new with the

perpendiculars p and 1. Then $(A + a =)\ p/2 \cdot$ sq. $s + 1 \cdot s = 7$, and this equation has a (rational) solution only if sq. $1/2 + p/2 \cdot 7$ is a square, that is if $14\,p + 1$ is a square. Also, since m and 1 are the sides of a right-angled triangle, sq. $p + 1$ must be a square. Therefore, (sq. $p + 1) - (14\,p + 1) =$ sq. $p - 14\,p = p \cdot (p - 14)$ is a square difference. Since $p \cdot (p - 14) =$ sq. $\{p + (p - 14)\}/2 -$ sq. $\{p - (p - 14)\}/2$, Diophantus sets $14\,p + 1 =$ sq. $\{p - (p - 14)\}/2 =$ sq. 7, which gives $p = 24/7$. Therefore, the "auxiliary triangle" with the perpendiculars p and 1 is (24/7, 1, 25/7) or (24, 7, 25).
If then $(a, b, c) = (24\,s, 7\,s, 25\,s)$, it follows that $(A + a =)$ 84 sq. $s + 7\,s = 7$. This quadratic equation has the solution $s = 1/4$, so that $(a, b, c) = (24, 7, 25) \cdot 1/4$.

This problem and its solution, both typical for the style of Diophantus' *Arithmetica*, are potentially very important for the following reason. The series of problems in *Ar.*"VI" § 3 (## 6-11),

$A + a = k, \quad A - a = k, \quad A + (b + a) = k, \quad A - (b + a) = k,$ *etc.*
(where A and a are the area and a short side of a right triangle)

looks like a series of determinate problems of the same type as similar series of OB problems for *squares, rectangles with the sides in a given ratio, circles,* or *semicircles.* Yet they are all indeterminate. The problem in # 6, for instance, leads to a pair of indeterminate equations for p of the form

$14\,p + 1 =$ sq. m, sq. $p + 1 =$ sq. n (m and n undetermined).

And so on. What all this means is that it is not inconceivable that *some OB mathematician was inadvertently led to consider indeterminate problems of the Arithmetica type when trying to work out a series of problems of standard type for the area and the sides of a right triangle.*

This conjecture may sound quite far off, but it is to some extent corroborated by the testimony of four strangely formulated interest problems in the OB brief theme text **VAT 8521** (Neugebauer, *MKT* (1935-37) *1*, 351 *ff.*; *3*, 58). Here is the text of the first of those problems:

VAT 8521 # 1, literal translation	explanation
For 1 mina of silver 12 shekels he gave.	Interest: 12 shekels per mina.
May he give you (as) interest a square.	May the interest be a square number
Set 1 mina, set interest 12 shekels.	For 1 mina and 12 shekels in interest
Set 1 40 the 'step'	Let, for instance,
that (as) a square he gives to you.	1 40 (= 100) be the square number
12, the interest, to 1 mina lift, 12.	The interest on 1 mina is 12 shekels
The opposite of 12 resolve, 5.	1/12 = ;05
To 1 40 the step that you took lift	1 40 / 12 = 1 40 · ;05 = 8;20
8 20 the initial silver.	The initial capital was 8 minas 20 shekels

In this OB exercise, the interest is the usual 1/5 of the capital, expressed as 12 shekels on each mina (= 60 shekels). However, instead of stipulating that the interest shall be a given weight of silver, the text nonsensically says that it shall be a square number. (The meaning of the term 'step' in this context is not clear, it may be simply 'unspecified number'.) There is no other known OB mathematical text with a similar requirement

The other exercises in VAT 8521 develop the theme. Exercise # 3 is essentially identical with # 1, only with the square 36 instead of 1 40. Here is a list of the form of the interest in the 4 exercises:

# 1.	May he give you (as) interest a square.	chosen 'step': 1 40 = sq. 10
# 2.	May he give you (as) interest a cube.	chosen 'step': 7 30 (00) = cu. 30
# 3.	May he give you (as) interest a square.	chosen 'step': 36 = sq. 6
# 4.	May he give you (as) interest a 'cube minus 1'	chosen 'step': 18 = cu. 3 – sq. 3

Neugebauer (*op. cit.*) makes the following comment:

"It is likely that what we have here is a degenerate form of some other problem type, mathematically more meaningful but therefore also more difficult"

Neugebauer's hunch may have been correct. As a matter of fact, there is at least one uncanny similarity between the problems in VAT 8521 and the problems in *Ar.* "VI". Thus, in *Ar.* "VI", all the undetermined right hand sides of the equations are of one or the other of the following 5 types

a) sq. m b) cu. m c) sq. $m + m$ d) cu. $m + m$ e) cu. $m - m$.

In VAT 8521, the right sides of the equations are of the 3 types

a) sq. m b) cu. m c) cu. $m - m$.

In other words, the undetermined right hand sides in *Ar.* "VI" are

squares, cubes, quasi-squares, and quasi-cubes,

while in VAT 8521 they are

squares, cubes, and quasi-cubes.

The name "quasi-cube" was suggested in Friberg, *RC* (2007), Sec. 2.4, as a suitable notation for *various combinations of cubes, squares, and linear terms* which appear in three known OB mathematical table texts organized in the same way as the more familiar tables of cube sides.

The types of quasi-cubes that appear in the mentioned table texts are

quasi-cubes of the form cu. m + sq. $m = m \cdot m \cdot (m + 1)$ in MS 3899 and VAT 8492
quasi-cubes of the form cu. m + 3 sq. $m + 2\,m = m \cdot (m + 1) \cdot (m + 2)$ in MS 3048

Hand copies of the three table texts can be found in Friberg (*op. cit.*) Figs. 2.4.1 and 2.4.2, where they are followed by a discussion of the possible reason for the existence of such table texts.

Properly speaking, MS 3899 and MS 3048 are tables of "quasi-cube sides". Here are, for instance, the first two lines of the table on MS 3048:

6.e	1	íb.si$_8$	6	makes	1	equalsided	$(6 = 1 \cdot 2 \cdot 3)$
24.e	2	íb.si$_8$	24	makes	2	equalsided	$(24 = 2 \cdot 3 \cdot 4)$

Ar. "VI".16. A right triangle with a rational bisector

Here is the statement of *Ar.* "VI" § 5 (# 16), a problem appearing *quite out of context* in the theme text *Ar.*"VI" :

***Ar.* "VI".16** (Heath, *DA* (1964), 240; Czwalina, *ADA* (1952), 105)

To find a right-angled triangle in which the bisector of an acute angle is rational.

Diophantus sets the bisector equal to 5 s, a segment of one of the perpendiculars equal to 3 s, and the other perpendicular equal to 4 s. The whole first perpendicular is chosen as 3, so that the second segment is $3 - 3\,s$. Then the diagonal of the triangle is $4/3 \cdot (3 - 3\,s) = 4 - 4\,s$. Since the square of the diagonal is the sum of the squares of the perpendiculars, it follows that 16 sq. $s + 16 - 32\,s = 16$ sq. $s + 9$. Hence, $s = 7/32$. The rest is clear: If everything is multiplied by 32, the first perpendicular is 96, the second is 28, the hypothenuse is 100, the bisector is 35, and the segments 21 and 75.

Diophantus' method in *Ar.* "VI".16 can be explained as follows, in terms of metric algebra:

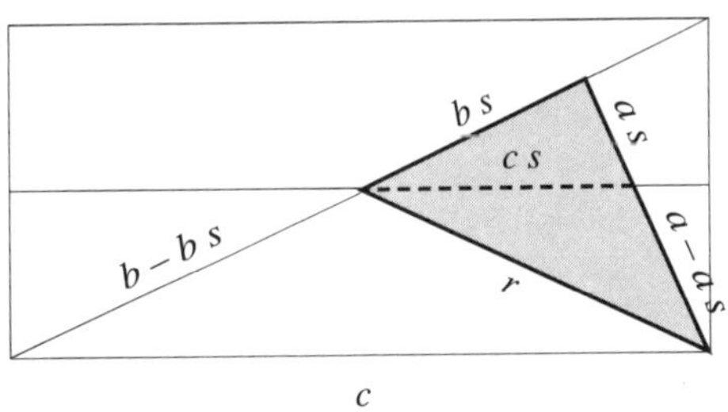

$r = b/a \cdot (a - a\,s) = b - b\,s$

$\Rightarrow$

sq. $(b - b\,s)$ = sq. a + sq. $b\,s$

$\Rightarrow$

s = (sq. b – sq. a) / (2 sq. b)

Fig. 13.6.1. Ar. "VI".16. A right triangle with a rational bisector of an acute angle.

In a right triangle with the base a the bisector of the opposite angle cuts off a right sub-triangle with the sides $c\,s$, $a\,s$, $b\,s$, where s is unknown and c, b, a a diagonal triple, for instance 5, 4, 3. Let r be the third side of the whole triangle, which then has the sides r, a, $b\,s$. It follows that (*cf. El.* VI.3)

$r : (a - a\,s) = b\,s : a\,s = b : a$, so that $r = b/a \cdot (a - a\,s) = b - b\,s$.

Consequently, by the diagonal rule,

sq. $(b - b\,s)$ = sq. a + sq. $b\,s$, so that s = (sq. b – sq. a)/(2 sq. b), *etc.*

In particular,

if $c, b, a = 5, 4, 3$, then $s = 7/32$, so that $r, a, b\,s, c\,s = (100, 96, 28, 35)\,/\,32$.

The whole solution procedure becomes exceedingly obvious if the construction is imagined to take place inside a rectangle as in Fig. 13.6.1.

13.7. *Ar.* V.7-12. A Section of a Theme Text with Cubic Problems

Of the seven lost books of Diophantus' *Arithmetica*, four (Books IV-VII) have survived in Arabic translations (Sesiano, *Books IV to VII* (1982); Rashed, *DA 3-4* (1984)). Of particular importance for the present discussion is the following brief theme text in the Arabic Book V:

***Ar.* V.7-12** (cu. means cube)		*m* (or *n*), *k*	*a, b*
V.7	$a + b = m$, *cu.* a + *cu.* $b = k$	20, 2240 D P	12, 8
V.8	$a - b = n$, *cu.* a – *cu.* $b = k$	10, 2170 D	13, 3
V.9	$a + b = m$, *cu.* a + *cu.* $b = k \cdot$ sq. $(a - b)$	20, 140 D	12, 8
V.10	$a - b = n$, *cu.* a – *cu.* $b = k \cdot$ sq. $(a + b)$	10, 8 1/8 D	15, 5
V.11	$a - b = n$, *cu.* a + *cu.* $b = k \cdot (a + b)$	4, 28 D	6, 8
V.12	$a + b = m$, *cu.* a – *cu.* $b = k \cdot (a - b)$	8, 52 D	6, 2

Note the similarity with the following theme text in *Ar.*I (Sec. 13.1 above):

***Ar.* I.27-30**		*m* (or *n*), *k*	*a, b*
I.27.	$a + b = m$, $a \cdot b = k$	$m, k = 20, 96$ D P	12, 8
I.28.	$a + b = m$, sq. a + sq. $b = k$	$m, k = 20, 208$ D P	12, 8
I.29.	$a + b = m$, sq. a – sq. $b = k$	$m, k = 20, 80$	12, 8
I.30.	$a - b = n$, $a \cdot b = k$	$n, k = 4, 96$ D P	12, 8

It is likely that *Ar.* V.7-12 and *Ar.* I.27-30 are two brief excerpts from one *common source*, a larger theme text with metric algebra problems, probably of Babylonian origin.

The *diorisms* in *Ar.* I.27-30 (necessary conditions for the existence of positive rational solutions), marked D in the catalog above, have already been mentioned. There are similar *diorisms* in V.7-12, as for instance in V.7, where the stated necessary condition is that

$(4 \cdot k$ – cu. $m)/(3 \cdot k)$ = sq. p for some undetermined p.

What is much more interesting is that there is also a remark in V.7

which, apparently, is the translation into Arabic of the Greek remarks in I.27-30, that the *diorism* is *plasmatikón* 'representable'. According to the interpretation in Sec. 13.1 above, this cryptic expression means that *the necessary condition can be explained by use of a diagram.*[39] In the case of V.7, the origin of the necessary condition is the following: As is explicitly shown in the solution procedure, Diophantus knew that

$$\text{cu. } (a + b) = \text{cu. } a + 3 \text{ sq. } a \cdot b + 3\, a \cdot \text{sq. } b + \text{cu. } b.$$

Since a and b are assumed to be solutions to the problem

$$a + b = m, \quad \text{cu. } a + \text{cu. } b = k,$$

it follows from this identity that

$$\text{cu. } m = k + 3\, m \cdot a \cdot b.$$

Hence, a and b are solutions to the *rectangular-linear system of equations*

$$a \cdot b = (\text{cu. } m - k)/(3\, m), \quad a + b = m.$$

To solve this system of equations, one may start with the observation that

$$\text{sq. } (a - b) = \text{sq. } (a + b) - 4\, a \cdot b = \text{sq. } m - 4\, (\text{cu. } m - k)/(3\, m) = (4 \cdot k - \text{cu. } m)/(3 \cdot k).$$

This identity implies that a necessary condition for the existence of a rational solution a, b is that $(4 \cdot k - \text{cu. } m)/(3 \cdot k)$ is a square. This is a necessary condition stated in *Ar.* V.7. Since the necessary condition was obtained in the process of solving a rectangular-linear system of equations, the necessity of the condition can be demonstrated *by use of a diagram* like the one in Fig. 13.1.1, right. That is why the necessary condition is *plasmatikón*.

In *Ar.* V.7, just as in *Ar.* I.27-28 and 30, Diophantus does not use the solution procedure suggested by the form of the *diorism* and the statement that the necessary condition is *plasmatikón*. Instead, he starts by setting

$$a = m/2 + s = 10 + s, \quad b = m/2 - s = 10 - s,$$

and concludes that

$$\text{cu. } a = 1000 + \text{cu. } s + 30 \text{ sq. } s + 300\, s, \quad \text{cu. } b = 1000 + 30 \text{ sq. } s - \text{cu. } s - 300\, s.$$

Therefore,

$$\text{cu. } a + \text{cu. } b = 2000 + \text{sq. } s = 2240, \quad 60 \text{ sq. } s = 240, \quad \text{sq. } s = 4, \quad \text{and} \quad s = 2.$$

39.The meaning of the obscure term *plasmatikón* and its alleged Arabic counterpart is a hotly debated issue. Conflicting interpretations can be found in, for instance, Sesiano (*op. cit.*, 192) and Rashed (*op. cit.*, 133-138). See also Christianidis, *Hist. Sci.* 6 (1995).

13.8. *Ar.* IV.17. Another Appearance of the Term 'Representable'

In addition to *Ar.* I.27-28 and 30 and *Ar.* V.7, there are just two other problems in Diophantus' *Arithmetica* where a necessary condition is qualified by the term 'representable', namely IV.17 and IV.19. Here is the context in which those two problems appear (Sesiano, *op. cit.*, 186-198):

Ar. IV.14-22 (p, q are undetermined numbers)		k, 1 or m	a, b
IV.14	$k \cdot a =$ cu. p, $l \cdot a =$ sq. q	10, 5	8/5
IV.15	$k \cdot a =$ cu. p, $l \cdot a =$ sq. p	10, 4	25/16
IV.16	$k \cdot b =$ cu. p, $k \cdot a = p$	10	3, 2700
IV.17	$k \cdot$ sq. $b =$ cu. p, $k \cdot$ sq. $a = p$, $b = m \cdot a$	5, 20 D P	2, 40
IV.18	$k \cdot$ cu. $b =$ sq. p, $k \cdot$ cu. $a = p$, $b = m \cdot a$	8, 3 D	3/2, 9/2
IV.19	$k \cdot a =$ cu. p, $l \cdot a = p$	20, 5 D P	2/5
IV.20	$k \cdot$ cu. $a =$ sq. p, $l \cdot$ cu. $a = p$	200, 5 D	2
IV.21	$k \cdot$ sq. $a =$ cu. p, $l \cdot$ sq. $a = p$	40 1/2, 2 D	3/2
IV.22	$k \cdot$ cu. $a =$ cu. p, $l \cdot$ cu. $a = p$	91 1/8, 2 D	3/2

There is no indication that this group of problems is in any way related to the problems *Ar.* I.27-28 and 30 and *Ar.* V.7, other than the *diorisms* and two cases of the term 'representable' in IV.17 and IV.19.

The necessary condition in IV.17 is of the following form:

$m \cdot k =$ sq. q with q undetermined.

It is possible that what is meant by saying that this condition is 'representable' (in a diagram) is that m, k, p are related as the two segments of the diameter and the upright in a semicircle, as in Fig. 1.7.2 above, right.

The necessary conditions in IV.17-22 are the following ones:

IV.17 $m \cdot k =$ sq. q
IV.18 $k =$ cu. q
IV.19 $k \cdot l =$ sq. q
IV.20 $k \cdot l =$ cu. q
IV.21 $k \cdot l =$ sq. sq. q
IV.22 $k \cdot l =$ sq. q, $k \cdot$ cu. $l =$ sq. cu. q

It is clear that the necessary conditions in IV.17 and IV.19 are of the same kind and can both be 'represented' in a geometric diagram like Fig. 1.7.2, right. On the other hand, the other necessary conditions cannot readily be represented geometrically.

Chapter 14

Heron's, Ptolemy's, and and Brahmagupta's Area and Diagonal Rules

14.1. *Metrica* I.8 / *Dioptra* 31. Heron's Triangle Area Rule

The area of a triangle with given sides can be found by first computing the height against one of the sides. Then the area is half the product of the height and the side. However, in his famous theorem ***Metrica* I.8**, Heron of Alexandria shows how the area of a triangle can be computed *without the need to first compute a height in the triangle*. According to Al-Bīrūnī (973-1038) in *The Book Concerning the Chords*, the proof of this theorem is due to Archimedes. The proof is reproduced below, together with an explanation in terms of metric algebra. The notations used in the metric algebra version of the proof are the ones appearing in Fig. 14.1.1, right

In the *lettered diagram* associated with the identical texts *Metrica* I.8 and *Dioptra* 31(see Høyrup, *BSSM* 17 (1997)), a circle is inscribed in the given triangle ABC (Fig. 14.1.1, left). D, E, F are the three points where the circle touches the sides of the triangle, and O is the center of the circle. BH is equal to AF, OL is orthogonal to OC, and BL is orthogonal to BC.

In the corresponding *metric algebra diagram* (Fig. 14.1.1, right), the sides a, b, c of the triangle are cut by the points where the circle touches the sides into six segments, of lengths u, u, v, v, w, w. The bisectors of the three angles of the triangle and the normals against the three sides from the center of the circle cut the triangle into three pairs of right triangles. The angles at the center of the circle of these right triangles are α, α, β, β, γ, γ. The right triangle drawn below the given triangle has the angle δ at its lower vertex and the sides $v + w$ and t. The line joining the center of the circle to the same lower vertex is orthogonal to the line from the center of the circle to the lower right corner of the given triangle and cuts the segment u in two pieces p and q. The radius of the circle is r, and half the perimeter of the triangle is s.

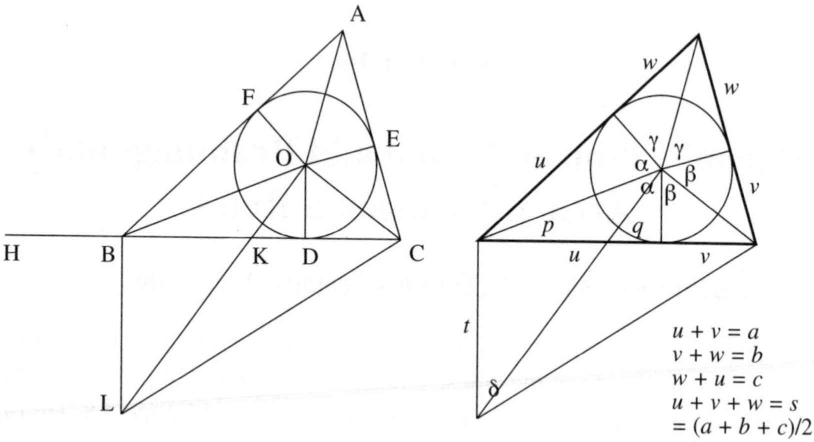

Fig. 14.1.1. Heron's *Metrica* I.8. Left: a lettered diagram. Right: a metric algebra diagram.

***Metrica* I.8** (Heath, *HGM 2* (1981), 322)	explanation in terms of metric algebra
ABC = BOC + COA + AOB	$2\,A = (u + v) \cdot r + (v + w) \cdot r + (w + u) \cdot r$
ABC = CH · OD	$A = (u + v + w) \cdot r = s \cdot r$
sq. ABC = sq. CH · sq. OD	sq. A = sq. s · sq. r
since COL and CBL are right	the lower triangle and its adjoining right
COBL is a cyclic quadrilateral	triangles form an *overlapping birectangle*
angles COB + CLB = 2 R	$\alpha + \beta + \delta$ = two right angles
angles COB + AOF = 2 R	$\alpha + \beta + \gamma$ = two right angles
angle AOF = angle CLB	$\gamma = \delta$
AOF and CLB are similar triangles	therefore, by similarity,
BC : BL = AF : FO = BH : OD	$(u + v) : t = w : r$ and
CB : BH = BL : OD = BK : KD	$(u + v) : w = t : r = p : q$
CH : HB = BD : DK	$s : w = (u + v + w) : w = (p + q) : q = u : q$
sq. CH : CH · HB = BD · DC : CD · DK	sq. $s : s \cdot w = u \cdot v : v \cdot q$
= BD · DC : sq. OD	$= u \cdot v$: sq. r
sq. ABC = sq. CH · sq. OD	sq. A = sq. s · sq. r
= CH · HB · BD · DC	$= (s \cdot w) \cdot (u \cdot v) = s \cdot (s - a) \cdot (s - b) \cdot (s - c)$

It is evident from this metric algebra explanation of *Metrica* I.8 that the basic ideas in Archimedes' proof of the triangle area rule are:

1) $A = s \cdot r$ where $s = u + v + w = (a + b + c)/2$
2) $\text{sq. } r = u \cdot v \cdot w \,/\, (u + v + w) = (s - a) \cdot (s - b) \cdot (s - c) \,/\, s$

In other words, *the essential part of Archimedes' proof is the computation of the radius of the inscribed circle in terms of the segments u, v, w cut off by the inscribed circle*. This realization leaves us with two equally plausible explanations of how Archimedes can have found his proof of the triangle area rule. One possible situation is that he first computed the radius of the inscribed circle and then saw that he could obtain the triangle area rule as an easy corollary. Another possibility is that he already knew the rule and found a new proof for it in terms of the radius of the inscribed circle.

Assume that, as in the second alternative, Archimedes already knew the triangle area rule in the form sq. $A = s \cdot (s - a) \cdot (s - b) \cdot (s - c)$, but that he also knew the more obvious rule $A = s \cdot r$. He can then have understood that he could obtain the former rule from the latter if he could prove that

$$\text{sq. } r = (s - a) \cdot (s - b) \cdot (s - c) \,/\, s \quad \text{or} \quad \text{sq. } r = u \cdot v \cdot w \,/\, (u + v + w).$$

He can also have seen that it would be easier to prove that

$$(u + v + w) : w = u \cdot v : \text{sq. } r.$$

This observation may have been a decisive first step towards the proof in the form that we know it from *Metrica* I.8.

14.2. Two Simple Metric Algebra Proofs of the Triangle Area Rule

One simple proof of the triangle area rule is the one explicitly dismissed by Heron in the first few lines of *Metrica* I.8, namely the rather obvious proof in terms of the height against one of the sides of the given triangle.

Let, as in Fig. 1.8.1, right, above, p, q be the segments into which the side b of the given triangle is divided by the height against b. Then

$$\text{sq. } c - \text{sq. } p = \text{sq. } h = \text{sq. } a - \text{sq. } q.$$

Therefore, p and q are solutions to the quadratic-linear system of equations

$$p + q = b, \quad \text{sq. } p - \text{sq. } q = \text{sq. } c - \text{sq. } a.$$

This was a familiar fact already in OB mathematics. See the discussion of VAT 7531 in Sec. 1.1, in particular Fig. 1.12.7. Consequently,

$$p = b/2 + (\text{sq. } c - \text{sq. } a) \,/\, 2\,b = (\text{sq. } b + \text{sq. } c - \text{sq. } a) \,/\, 2\,b, \quad \textit{etc.}$$

When p is known in this form, sq. h can be expressed as follows

$$\text{sq. } h = \text{sq. } c - \text{sq. } p = (c + p) \cdot (c - p)$$
$$= (2\, b \cdot c + \text{sq. } b + \text{sq. } c - \text{sq. } a) \cdot (2\, b \cdot c - \text{sq. } b - \text{sq. } c + \text{sq. } a) \,/\, \text{sq. } (2\, b)$$
$$= \{\text{sq. } (c + b) - \text{sq. } a\} \cdot \{\text{sq. } a - \text{sq. } (c - b)\} \,/\, \text{sq. } (2\, b).$$

The triangle area rule then follows, in the form

$$\text{sq. } (4\, A) = \text{sq. } (2\, b \cdot h) = \text{sq. } (2\, b) \cdot \text{sq. } h = \{\text{sq. } (c + b) - \text{sq. } a\} \cdot \{\text{sq. } a - \text{sq. } (c - b)\}.$$

It is easy to get from there to the rule in the more symmetric form

$$\text{sq. } A = s \cdot (s - a) \cdot (s - b) \cdot (s - c) \quad \text{with} \quad s = (a + b + c)/2.$$

Another metric algebra proof, reproduced by Id and E. S. Kennedy in King and M. S. Kennedy (*eds.*), *SIES* (1969), 492-494, is given by **Al-Shannī** in the medieval manuscript **MS 223**, **Bibliothèque Orientale, Université Saint Joseph, Beirut**. The basic idea in Al-Shannī's proof is to inscribe the given triangle in a right triangle (essentially the lower right half of the rectangle in the diagram in Fig. 14.2.1 below) in a suitable way, and then make two applications of Ptolemy's diagonal rule (Heath, *HGM* 2 (1981), 278), which states that *in any cyclic quadrilateral the sum of the products of the two pairs of opposite sides equals the product of the diagonals*. Below, Al-Shannī's proof is replaced by a simplified version.

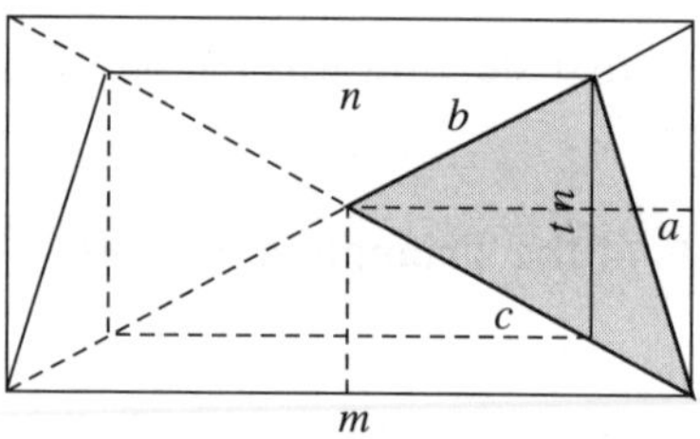

$$2\, A = m/2 \cdot t\, n$$
$$\text{sq. } (c + b) - \text{sq. } a = m \cdot n$$
$$\text{sq. } a - \text{sq. } (c - b) = t\, m \cdot t\, n$$
$$\Rightarrow$$
$$\text{sq. } (4\, A) = \text{sq. } (m \cdot t\, n)$$
$$= (m \cdot n) \cdot (t\, m \cdot t\, n)$$
$$= \{\text{sq. } (c + b) - \text{sq. } a\} \cdot \{\text{sq. } a - \text{sq. } (c - b)\}$$

Fig. 14.2.1. A simplified version of Al-Shannī's proof of the triangle area rule.

Consider a given triangle with the sides c, b, a. Construct two concentric, parallel, and similar rectangles with the sides m, $t\, m$ and n, $t\, n$, respectively, $n < m$, and such that the side c of the given triangle coincides with the half-diagonal of the larger rectangle, while the side b coincides with the half-diagonal of the smaller rectangle, as in Fig. 14.2.1. Then $c + b$ is the diagonal in a symmetric trapezoid with the sides m, a, n, a. Therefore,

$$\text{sq. } (c + b) - \text{sq. } a = m \cdot n,$$

in view of Ptolemy's diagonal rule. Similarly, a is the diagonal in a symmetric trapezoid with the sides $t\,m$, $c - b$, $t\,n$, $c - b$. Therefore,

$$\text{sq. } a - \text{sq. } (c - b) = t\,m \cdot t\,n.$$

Combining the two identities, one finds that

$$\{\text{sq. } (c + b) - \text{sq. } a\} \cdot \{\text{sq. } a - \text{sq. } (c - b)\} = (m \cdot n) \cdot (t\,m \cdot t\,n).$$

On the other hand, the given triangle can be formed by joining together two triangles with the common base $t\,n$ and with the sum of their heights against this base equal to $m/2$. (See again Fig. 14.2.1.) Therefore, twice the the area A of the given triangle can be computed as

$$2\,A = m/2 \cdot t\,n.$$

Consequently, the triangle area rule is obtained in the form

$$\text{sq. } (4\,A) = \text{sq. } (m \cdot t\,n) = (m \cdot n) \cdot (t\,m \cdot t\,n) = \{\text{sq. } (c + b) - \text{sq. } a\} \cdot \{\text{sq. } a - \text{sq. } (c - b)\}.$$

14.3. Simple Proofs of Special Cases of Brahmagupta's Area Rule

In his ***Brāhmasphuṭasiddhānta* XII.21** (Colebrooke, *AAMS* (1973), 295), the Indian astronomer Brahmagupta (628) formulated as follows a rule for the area of a triangle or quadrilateral in terms of the sides:

> "The product of half the sides and countersides is the inexact area of a triangle or quadrilateral. Half the sum of the sides set down four times, and in turn lessened by the sides and multiplied together, the product is the exact area."

The *inexact area rule* mentioned here by Brahmagupta is clearly the proto-Sumerian/Sumerian/Babylonian "false area rule" for triangles and quadrilaterals. (See, for instance, Sec. 11. 3 a above.)

The *exact* area rule is stated here in the same form for triangles and 'quadrilaterals'. Brahmagupta gives no information about which kind of quadrilaterals he has in mind. However, the area rule is *correct only for cyclic quadrilaterals.*[40]

According to Al-Bīrūnī, Brahmagupta's area rule was first found by some anonymous Indian mathematician (Tropfke, *GE 4* (1940), 155). However, Tropfke himself (*op. cit.*, 154) was of the opinion that

---

40. For a fairly exhaustive account of the interesting history of cyclic quadrilaterals in ancient Indian and Islamic or more recent Western mathematical works, the reader is advised to consult Tropfke, *GE 4* (1940), 150-169. In Europe, the first *algebraic* proof of the area rule for general cyclic quadrilaterals was given by P. Naudé in 1727 (*op. cit.*, 166). A simple proof by use of trigonometry was found by N. Fuss in 1797 (*op. cit.*, 167).

> "It is hardly likely that an Indian mathematician could derive the area rule algebraically, and even less could Indian mathematicians accomplish it by geometric means."

This verdict may be much too harsh, since *correct derivations of the area rule for the most interesting special kinds of (cyclic) quadrilaterals* are so easy to find that they may have been known to Indian mathematicians, or even to their Babylonian predecessors. This will be shown below.

What has to be shown is that if p, s, q, t (in this order) are the sides of a quadrilateral, then the area of the quadrilateral is given by the equation

$$\text{sq. } A = (p+q+s-t)/2 \cdot (p+q-s+t)/2 \cdot (s+t+p-q)/2 \cdot (s+t-p+q)/2,$$

or simply

$$\text{sq. } (4\,A) = (p+q+s-t) \cdot (p+q-s+t) \cdot (s+t+p-q) \cdot (s+t-p+q).$$

Now, this rule is *correct for triangles*, because

if $t = 0$, then the rule is reduced to Heron's triangle area rule.

The rule is trivially *correct for squares and rectangles*. Indeed,

if $p = q$ and $s = t$, then the rule is reduced to $\text{sq. } A = p \cdot p \cdot s \cdot s = \text{sq. } (p \cdot s)$.

The rule is also *correct for symmetric trapezoids*, because

if p and q are parallel, $p > q$, and $s = t$, then the rule is reduced to

$$\text{sq. } A = (p+q)/2 \cdot (p+q)/2 \cdot (2\,s+p-q)/2 \cdot (2\,s-p+q)/2$$
$$= \text{sq. } (p+q)/2 \cdot (\text{sq. } s - \text{sq. } (p-q)/2) = \text{sq. } \{(p+q)/2 \cdot h\},$$

where h is the height of the trapezoid.

That the rule is *correct for birectangles* can be shown as follows:

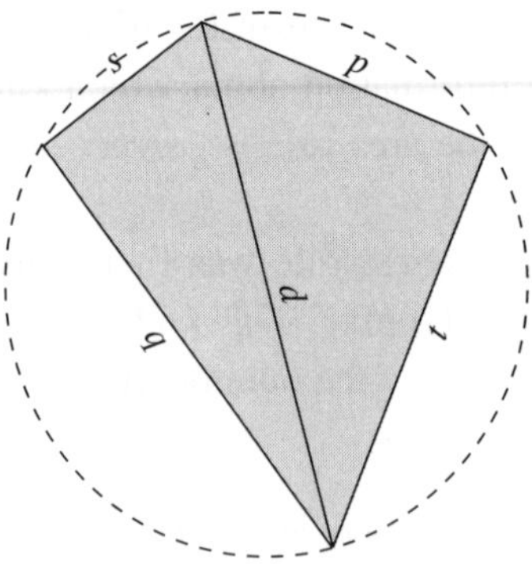

$\text{sq. } p + \text{sq. } t = \text{sq. } q + \text{sq. } s$
and $2\,A = p \cdot t + q \cdot s$
$\Rightarrow$
$4\,A = 2\,p \cdot t + 2\,q \cdot s$
$= \text{sq. } (p+t) - \text{sq. } (q-s)$
$= \text{sq. } (q+s) - \text{sq. } (t-p)$
$\Rightarrow$
$\text{sq. } (4\,A)$
$= \{\text{sq. } (p+t) - \text{sq. } (q-s)\}$
$\cdot \{\text{sq. } (q+s) - \text{sq. } (t-p)\}$

Fig. 14.3.1. An easy metric algebra derivation of the area rule in the case of a birectangle.

Let p, s, q, t (in this order) be the given sides of a birectangle, let d be the first diagonal, and let A be the area, as in Fig. 14.3.1. Then, clearly,

$$\text{sq. } p + \text{sq. } t = \text{sq. } d = \text{sq. } q + \text{sq. } s \quad \text{and} \quad 2\,A = p \cdot t + q \cdot s.$$

Consequently,

$$4\,A = 2\,p \cdot q + 2\,s \cdot t = \text{sq. } (p + q) - \text{sq. } (t - s) \qquad (\text{if } t > s, \text{ for instance}),$$

but also

$$4\,A = 2\,p \cdot q + 2\,s \cdot t = \text{sq. } (s + t) - \text{sq. } (q - p) \quad (\text{if } q > p, \text{ for instance}).$$

Through multiplication of the two alternative results, one arrives at

$$\text{sq. } (4\,A) = \{\text{sq. } (p + q) - \text{sq. } (t - s)\} \cdot \{\text{sq. } (s + t) - \text{sq. } (q - p)\}.$$

After factorization, this expression becomes

$$\text{sq. } (4\,A) = (p + q + s - t) \cdot (p + q - s + t) \cdot (s + t + p - q) \cdot (s + t - p + q).$$

Finally, it is easy to show that Brahmagupta's area rule is *correct for cyclic quadrilaterals with orthogonal diagonals*, which here, for simplicity, will be referred to as "cyclic orthodiagonals". As is well known, there is a simple relation between birectangles and cyclic orthodiagonals, which, incidentally, shows that *since Brahmagupta's area rule holds for birectangles, it holds also for cyclic orthodiagonals*.

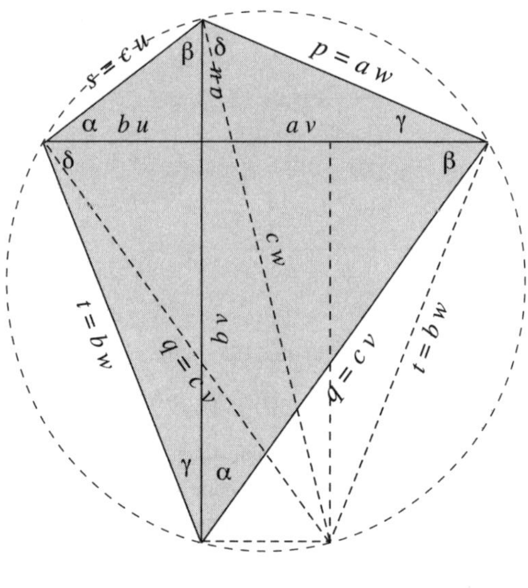

Fig. 14.3.2. A simple relation between cyclic orthodiagonals and birectangles.

Indeed, all that is required to pass from a birectangle to a cyclic orthodiagonal is to join together the two pairs of right sub-triangles of a birectangle in a different way. See Figs. 13.4.3, 14.3.2. Evidently, the transformation does not change the area of the quadrilateral, and it changes only the order of the sides, from p, s, q, t to p, s, t, q.

14.4. Simple Proofs of Special Cases of Ptolemy's Diagonal Rule

As mentioned above, in Sec. 14.2, Ptolemy's diagonal rule says that

In any cyclic quadrilateral the sum of the products of the two pairs of opposite sides equals the product of the diagonals.

Ptolemy's well known proof is simple enough, but it is interesting that it is possible to find alternative proofs in special cases by use of metric algebra.

Thus, in the case of a *rectangle* with the sides u, s, u, s and diagonals d, Ptolemy's diagonal rule is identical with the *OB diagonal rule*:

$$\text{sq. } u + \text{sq. } s = \text{sq. } d.$$

In the case of *a symmetric trapezoid* with the sides m, s, n, s and diagonals d, Ptolemy's diagonal rule is identical with the *OB trapezoid diagonal rule* discussed in Appendix 1:

$$m \cdot n + \text{sq. } s = \text{sq. } d.$$

A simple metric algebra proof is given in Fig. 14.4.1 below.

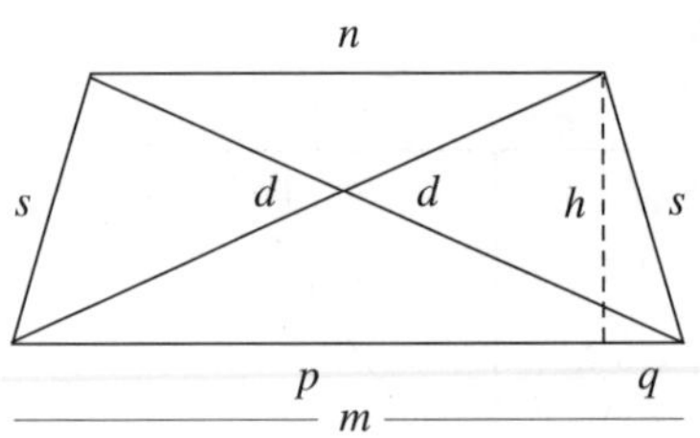

$$\begin{aligned}&\text{sq. } d - \text{sq. } p\\&= \text{sq. } h\\&= \text{sq. } s - \text{sq. } q\\&\Rightarrow\\&\text{sq. } d - \text{sq. } s\\&= \text{sq. } p - \text{sq. } q\\&= (p+q)\cdot(p-q)\\&= m \cdot n\end{aligned}$$

Fig. 14.4.1. A proof of Ptolemy's diagonal rule in the case of a symmetric trapezoid.

In the case of a *birectangle* (see Fig. 14.3.2),

$p = a\,w$ and $q = c\,v$ is one pair of opposite sides,
$s = c\,u$ and $t = b\,w$ is another pair of opposite sides,
$d = c\,w$ and $e = b\,u + a\,v$ are the two diagonals.

Then Ptolemy's diagonal rule follows easily, because

$$p \cdot q + s \cdot t = a\,w \cdot c\,v + c\,u \cdot b\,w = c\,w \cdot (a\,v + b\,u) = d \cdot e.$$

Similarly, in the case of a *cyclic orthodiagonal* (see again Fig. 14.3.2)

$p = a\,w$ and $t = b\,w$ is one pair of opposite sides,

$s = c\,u$ and $q = c\,v$ is another pair of opposite sides,
$d = a\,u + b\,v$ and $e = b\,u + a\,v$ are the two diagonals.

Here, too, Ptolemy's diagonal rule follows easily, because

$$p \cdot t + s \cdot q = a\,w \cdot b\,w + c\,u \cdot c\,v = a\,b \cdot \text{sq.}\,w + \text{sq.}\,c \cdot u\,v$$
$$= a\,b \cdot (\text{sq.}\,u + \text{sq.}\,v) + (\text{sq.}\,a + \text{sq.}\,b) \cdot u\,v = (a\,u + b\,v) \cdot (b\,u + a\,v) = d \cdot e.$$

A simple derivation of Ptolemy's diagonal rule can be given also in the case of an *overlapping birectangle*, by which is meant a cyclic quadrilateral formed by two partly overlapping right triangles with a common diagonal, as in Fig. 14.4.2 below.

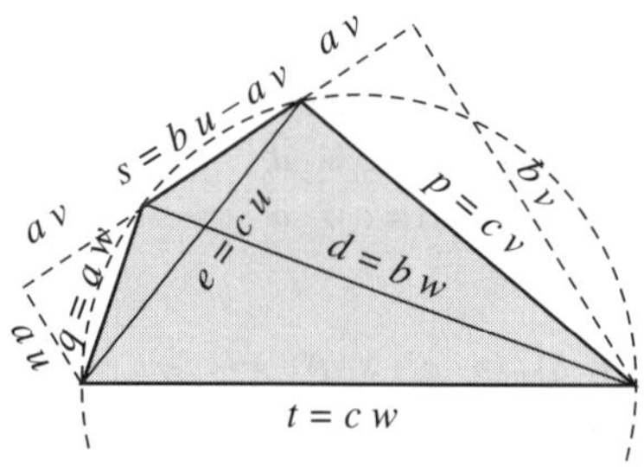

$$p\,q + s\,t$$
$$= c\,v \cdot a\,w + (b\,u - a\,v) \cdot c\,w$$
$$= b\,u \cdot c\,w = b\,w \cdot c\,u$$
$$= d \cdot e$$

Fig. 14.4.2. Ptolemy's diagonal rule in the case of an overlapping birectangle.

Note that the diagram in Fig. 14.4.2 is essentially identical with the diagram used in Ptolemy's proof of a subtraction rule for chords. See Heath, *HGM 2* (1981), 279. (Note also the occurrence of an overlapping birectangle in Archimedes' proof of the triangle area rule.)

Although the proof of Ptolemy's diagonal rule for an overlapping bi rectangle as in Fig. 14.4.2 is exceedingly simple, *there seems to be no simple way of proving Brahmagupta's area rule for an overlapping birectangle in the same way as the area rule was proved above for a birectangle or a cyclic orthodiagonal!*

It is not difficult to find a first expression for the area A of an overlapping birectangle, for instance as follows: It is obvious from the diagram in Fig. 14.4.2 that A is equal to the area of a trapezoid with the parallel sides $b\,v$, $a\,u$ and the height $b\,u + a\,v$, minus the areas of two right triangles with the sides $b\,v$, $a\,v$ and $a\,v$, $a\,u$. Therefore,

$$2\,A = (b\,v + a\,u) \cdot (b\,u + a\,v) - (a\,v \cdot b\,v + a\,u \cdot a\,v) = b\,v \cdot b\,u + a\,u \cdot b\,u.$$

The trouble is that it seems to be difficult to identify this simple equation for the area with the one given by Brahmagupta in terms of the sides.

14.5. Simple Proofs of Special Cases of Brahmagupta's Diagonal Rule

Brahmagupta's rule for the lengths of the diagonals of a quadrilateral in terms of the sides is formulated as follows in ***Bss*** **XII.28** (Colebrooke, *AAMS* (1973), 300):

> "The sums of the products of the sides about both the diagonals being divided by each other, multiply the quotients by the sum of the products of opposite sides; the square-roots of the results are the diagonals in a quadrilateral with unequal sides"

The rule is *correct for a cyclic orthodiagonal*. Indeed, if p, s, t, q are the sides and d, e the diagonals of a cyclic orthodiagonal, then with the notations in Fig. 14.3.2 above

$p \cdot t + s \cdot q = d \cdot e$ (as in Ptolemy's diagonal rule; see above),
$p \cdot s + t \cdot q = a\,w \cdot c\,u + b\,w \cdot c\,v = c\,w \cdot (a\,u + b\,v) = c\,w \cdot d,$
$p \cdot q + s \cdot t = a\,w \cdot c\,v + c\,u \cdot b\,w = c\,w \cdot (a\,v + b\,u) = c\,w \cdot e.$

Combining these results, one finds that

$d \cdot e = p \cdot t + s \cdot q$ and $d / e = (p \cdot s + t \cdot q) / (p \cdot q + s \cdot t) \Rightarrow$
sq. $d = (p \cdot t + s \cdot q) \cdot (p \cdot s + t \cdot q) / (p \cdot q + s \cdot t)$ and
sq. $e = (p \cdot t + s \cdot q) \cdot (p \cdot q + s \cdot t) / (p \cdot s + t \cdot q).$

The rule is also *correct for a birectangle*. Indeed, if p, s, q, t are the sides and d, e the diagonals of a birectangle, then with the notations in Fig. 14.3.2 above

$p \cdot q + s \cdot t = d \cdot e$ (as in Ptolemy's diagonal rule; see above),
$p \cdot s + q \cdot t = a\,w \cdot c\,u + c\,v \cdot b\,w = c\,w \cdot (a\,u + b\,v) = d \cdot (a\,u + b\,v),$
$p \cdot t + q \cdot s = a\,w \cdot b\,w + c\,v \cdot c\,u = a\,b \cdot$ sq. w + sq. $c \cdot u\,v =$
$= a\,b \cdot$ (sq. u + sq. v) + (sq. a + sq. b) $\cdot\, u\,v = (a\,u + b\,v) \cdot (a\,v + b\,u) = (a\,u + b\,v) \cdot e.$

And so on, as in the case of a cyclic orthodiagonal.

14.6. A Proof of Brahmagupta's Diagonal Rule in the General Case

A simple proof of Brahmagupta's area rule in *Yuktibhāṣā* and *Kriyākramarī* (India, 16th century; see Amma, *GAMI* (1979), Sec. 5.5.13) is actually a straightforward modification of the first of the proofs of Heron's area rule mentioned in Sec. 14.2 above. In metric algebra notations, as in Fig. 14.6.1 below, the proof can be explained as follows:

Let the sides of a (concave) quadrilateral be p, s, q, t and let the two diagonals be d and e. Let h and k be the heights in the triangles with the sides p, s, e and q, t, e, respectively. Then the area of the quadrilateral is

$A = e/2 \cdot h + e/2 \cdot k$, so that $4\,A = 2\,e \cdot (h + k)$.

Finally, let f and g be the segments of e to the left of h and k, respectively. Then d is the diagonal in a rectangle with the length $h + k$ and the width $g - f$ (if, say, $g > f$). Therefore,

$$\text{sq.}\ (h + k) = \text{sq.}\ d - \text{sq.}\ (g - f).$$

The lengths of the segments g and f are known to be

$$g = (\text{sq.}\ e + \text{sq.}\ q - \text{sq.}\ t)\ /\ (2\,e) \quad \text{and} \quad f = (\text{sq.}\ e + \text{sq.}\ s - \text{sq.}\ p)\ /\ (2\,e).$$

Consequently,

$$2\,e \cdot (g - f) = (\text{sq.}\ q + \text{sq.}\ p) - (\text{sq.}\ s + \text{sq.}\ t).$$

Hence,

$$\begin{aligned}\text{sq.}\ (4\,A) &= \text{sq.}\ (2\,e) \cdot \text{sq.}\ (h + k) = \text{sq.}\ (2\,e) \cdot \{\text{sq.}\ d - \text{sq.}\ (g - f)\} \\ &= \text{sq.}\ (2\,d \cdot e) - \text{sq.}\ \{(\text{sq.}\ q + \text{sq.}\ p) - (\text{sq.}\ s + \text{sq.}\ t)\} \\ &= \{2\,d \cdot e + (\text{sq.}\ q + \text{sq.}\ p) - (\text{sq.}\ s + \text{sq.}\ t)\} \cdot \{2\,d \cdot e - (\text{sq.}\ q + \text{sq.}\ p) + (\text{sq.}\ s + \text{sq.}\ t)\}.\end{aligned}$$

In the next step of the procedure *it must be assumed that the quadrilateral is cyclic so that Ptolemy's diagonal rule can be applied.* Then

$$2\,d \cdot e = p \cdot q + s \cdot t$$

and it follows from two completions of squares that (if, say, $t > s$, $q > p$)

$$\text{sq.}\ (4\,A) = \{\text{sq.}\ (q + p) - \text{sq.}\ (t - s)\} \cdot \{\text{sq.}\ (t + s) - \text{sq.}\ (q - p)\}.$$

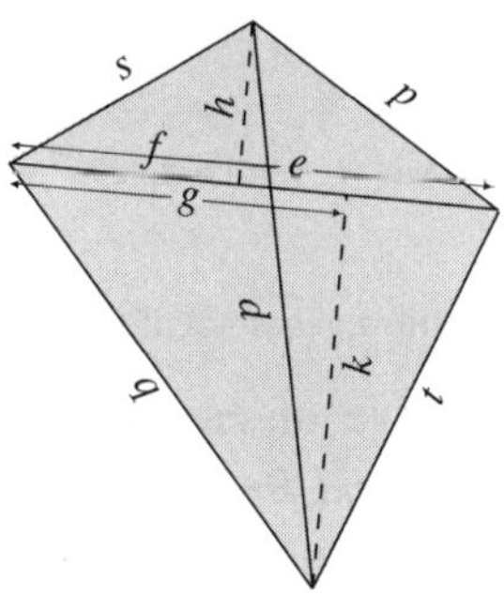

$$4\,A = 2\,e \cdot (h + k)$$
$$\text{sq.}\ (h + k) = \text{sq.}\ d - \text{sq.}\ (g - f)$$
$$g = (\text{sq.}\ e + \text{sq.}\ q - \text{sq.}\ t)\ /\ (2\,e)$$
$$f = (\text{sq.}\ e + \text{sq.}\ s - \text{sq.}\ p)\ /\ (2\,e)$$
$$2\,e \cdot (g - f) = (\text{sq.}\ q + \text{sq.}\ p) - (\text{sq.}\ s + \text{sq.}\ t)$$
$$\Rightarrow \text{ if } \boldsymbol{d \cdot e = p \cdot q + s \cdot t} \text{ then } \text{sq.}\ (4\,A) =$$
$$\text{sq.}\ (2\,d \cdot e) - \text{sq.}\ \{(\text{sq.}\ q + \text{sq.}\ p) - (\text{sq.}\ s + \text{sq.}\ t)\} =$$
$$\{\text{sq.}\ (q + p) - \text{sq.}\ (t - s)\} \cdot \{\text{sq.}\ (t + s) - \text{sq.}\ (q - p)\}$$

Fig. 14.6.1. A simple proof of Brahmagupta's area rule in the general case.

The proof is then complete. Note that this means that *the problem of proving Brahmagupta's area rule has been reduced by a straightforward use of metric algebra to the problem of proving Ptolemy's diagonal rule.*

Unfortunately, there does not seem to be any simple way of proving Ptolemy's diagonal rule *in the general case* by use of metric algebra.

A quick look at Ptolemy's proof of his diagonal rule (Heath, *HGM 2* (1981), 278) will show where the difficulty lies. Expressed in terms of metric algebra notations, Ptolemy's proof proceeds as follows:

Let a cyclic quadrilateral have the sides p, s, q, t and the diagonals d, e (as in Fig. 14.6.2). Then according to ***Elements* III.21**: *in a circle angles on the same arc are equal*, the angle (α) between s and e equals the angle between d and t, and the angle (β) between p and e equals the angle between d and q. Draw the line v cutting e into e' and e'' so that also the angle (ε) between s and v equals the angle between d and p. Then the triangles with the sides s, v, e' and d, p, t are similar, and it follows that $s : e' = d : t$, so that $s \cdot t = d \cdot e'$. For a similar reason, $p \cdot q = d \cdot e''$. Consequently,

$$s \cdot t + p \cdot q = d \cdot (e' + e'') = d \cdot e.$$

Note that the proposition El. III.21 which is used in this seemingly simple proof is totally beyond the scope of Babylonian-type metric algebra!

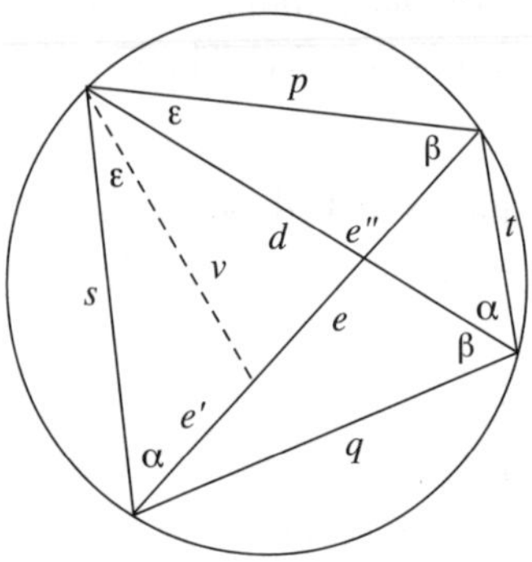

The triangles with the sides s, v, e' and d, p, t are similar $\Rightarrow s : e' = d : t \Rightarrow s \cdot t = d \cdot e'$.

The triangles with the sides p, v, e'' and d, s, q are similar $\Rightarrow p : e'' = d : q \Rightarrow p \cdot q = d \cdot e''$.

Therefore, $\boldsymbol{s \cdot t + p \cdot q} = d \cdot (e' + e'') = \boldsymbol{d \cdot e}$.

Fig. 14.6.2. The proof of Ptolemy's diagonal rule in metric algebra notations.

Conclusion. The fairly detailed discussion above of *Heron's and Brahmagupta's area rules*, as well as of *Ptolemy's and Brahmagupta's diagonal rules* can be summarized in the following way: All those rules can be derived in a simple and straightforward way by use of metric algebra, at least as long as no other cyclic figures are considered than *triangles, rectangles, symmetric trapezoids, birectangles, and cyclic orthodiagonals*. What that means is that it is not unlikely that all those rules were first discovered (in these special cases) either by Babylonian mathematicians (although there is no direct evidence for that), or by Greek or Indian mathematicians working in the Babylonian tradition.

Chapter 15

Theon of Smyrna's Side and Diagonal Numbers and Ascending Infinite Chains of Birectangles

The Greek "side and diagonal numbers algorithm" has been discussed extensively in many previous studies of the topic, such as, for instance, Heath, *HGM 1* (1981 (1921)), 91-93, Knorr, *EEE* (1975), Chapter 2, and Fowler, *MPA* (1987), Secs. 2.4(e), 3.6(b).

The key reference is a passage from Theon of Smyrna's ***Expositio Rerum Mathematicarum ad Legendum Platonem Utilium***, below in a translation borrowed from Fowler, *op. cit.*, Sec. 2.4(e):

Theon of Smyrna, *ERMLPU*, 42-5.

"Just as numbers potentially contain triangular, square, and pentagonal ratios, and ones corresponding to the remaining figures, so also we can find side and diagonal ratios appearing in numbers in accordance with the generative principles; for it is from these that the figures acquire balance. Therefore since the unit, according to the supreme generative principle, is the starting point of all the figures, so also in the unit will be found the ratio of the diagonal to the side.

For instance, two units are set out, of which we set one to be a diagonal and the other a side, since the unit, as the beginning of all things, must have it in its capacity to be both side and diagonal. Now there are added to the side a diagonal and to the diagonal two sides, for as great as is the square on the side, taken twice, the square on the diagonal [is] taken once. The diagonal therefore became the greater and the side became the less. Now in the case of the first side and diagonal, the square on the unit diagonal will be less by a unit than twice the square on the unit side; for units are equal, and 1 is less by a unit than twice 1.

Let us add to the side a diagonal, that is, to the unit let us add a unit; therefore the side will be two units. To the diagonal let us now add two sides, that is, to the unit let us add two units; the diagonal will therefore be three units. Now the square on the side of two units will be 4, while the square on the diagonal of three units will be 9; and 9 is greater by a unit than twice the square on the side 2.

Again, let us add to the side 2 the diagonal of three units; the side will be 5. To the diagonal of three units let us add two sides, that is, twice 2; there will be 7. Now the square from the side 5 will be 25, while that from the diagonal 7 will be 49; and 49 is less by a unit than twice 25.

Again, if you add to the side 5 the diagonal 7, there will be 12. And if to the diagonal 7 you add twice the side 5, there will be 17. And the square on 17 is greater by a unit than twice the square of 12.

When the addition goes on in the same way in sequence, the proportion will alternate; the square on the diagonal will be now greater by a unit, now less by a unit, than twice the square on the side; and such sides and diagonals are both expressible.

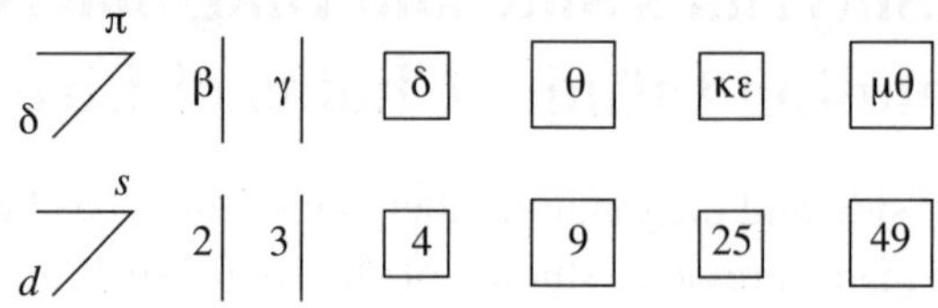

The squares on the diagonals, alternating one by one, are now greater by a unit than double the squares on the sides, now less than double by a unit, and the alternation is regular. All the squares on the diagonals will therefore become double the squares on the sides, equality being produced by the alternation of excess and deficiency by the same unit, regularly distributed among them; with the result that in their totality they do not fall short of nor exceed the double. For what falls short in the square on the preceding diagonal exceeds in the next one.

The *algebraic* meaning of this passage is clear. A double sequence of 'sides' and 'diameters' s_n, d_n is formed in a *recursive* procedure starting with a pair of units. The steps of the recursive procedure are the following:

$$d_1, s_1 = 1, 1 \quad \text{and} \quad d_{n+1} = d_n + 2\, s_n, \quad s_{n+1} = d_n + s_n \quad \text{for} \quad n = 1, 2, 3, \ldots$$

In this way are formed the pairs

$$d_1, s_1 = 1, 1, \quad d_2, s_2 = 3, 2, \quad d_3, s_3 = 7, 5, \quad d_4, s_4 = 17, 12, \quad \text{and so on.}$$

It is observed that

$$\text{sq. } 1 = \text{sq. } 1 \cdot 2 - 1, \quad \text{sq. } 3 = \text{sq. } 2 \cdot 2 + 1, \quad \text{sq. } 7 = \text{sq. } 5 \cdot 2 - 1, \quad \text{sq. } 17 = \text{sq. } 12 \cdot 2 + 1.$$

From this observation is inferred the conclusion that

$$\text{sq. } d_n = \text{sq. } s_n \cdot 2 - 1 \quad \text{for all odd } n, \quad \text{and} \quad \text{sq. } d_n = \text{sq. } s_n \cdot 2 + 1 \quad \text{for all even } n.$$

Therefore, *in average*, sq. d_n = sq. $s_n \cdot 2$.

Note the diagram in the text, headed by a figure with two straight lines forming an angle, one horizontal and tagged with a π for πλευραὶ 'sides', the other slanting and tagged with a δ for διάμετροι 'diagonals'.

15.1. The Greek Side and Diagonal Numbers Algorithm[41]

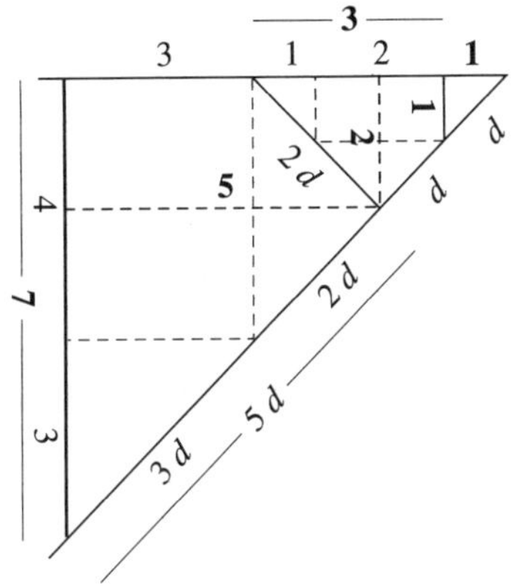

sq. $d = 2$

$d_1, s_1 = \mathbf{1, 1}$
$d_2, s_2 = 1 + 2 \cdot 1, 1 + 1 = \mathbf{3, 2}$
$d_3, s_3 = 3 + 2 \cdot 2, 3 + 2 = \mathbf{7, 5}$
$d_4, s_4 = 7 + 2 \cdot 5, 7 + 5 = \mathbf{17, 12}$
$d_5, s_5 = 17 + 2 \cdot 12, 17 + 12 = \mathbf{41, 29}$
etc.

Fig. 15.1.1. The first few steps of the side and diagonal numbers algorithm.

In the diagram in Fig. 15.1.1, the initial step of a geometric algorithm producing the side and diagonal numbers is a half-square with the side 1 and the diagonal *d*. These are the 'units' for the sides and the diagonals, respectively. One of the sides is extended indefinitely to the left and the diagonal is extended indefinitely to the left and downwards. In this way a diagram is formed which closely resembles the leftmost figure in the small diagram illustrating Theon's cited passage.

In the second step of the algorithm, a birectangle with two given sides, 1 and *d*, is joined to the initial half-square, so that a new half-square is formed, one with the side 2 *d*, and therefore with the diagonal 4. Consequently, the third and fourth sides of the birectangle are 3 and 2 *d*. The side 2 *d* of the birectangle is the diagonal of a half-square with the side 2. Now, a crucial observation is that the two new sides 3 and 2 *d* of the birectangle are *nearly equal*. Therefore, 2 *d* is the *inexpressible diameter of* 2, while 3 can be interpreted as an *expressible diameter of* 2. The closeness of the approximation is demonstrated by the equation

41.The ideas discussed in this chapter were first presented at a mathematical meeting at Niagara Falls in the summer of 1996.
The proposed explanation in Sec. 15.1 of the side and diagonal numbers in terms of a chain of birectangles is related to a similar proposal in Hofmann, *Centaurus* 5 (1956).

sq. 3 = 2 · sq. 2 + 1.

In the third step of the algorithm, a second birectangle, with two given sides 2 d and 3, is joined to the first birectangle in such a position that a half-square is formed with the side 7 and therefore with the diagonal 7 d. Therefore the two other sides of the birectangle are 7 and 5 d, with 7 and 5 d even more nearly equal than 3 and 2 d. Here 5 d is the geometric, irrational diagonal of 5, and 7 is the rational diagonal of 5. This time,

sq. 7 = 2 sq. 5 – 1.

And so on. The general step in the algorithm is described by Proclus (410-85), as follows:

Proclus, *Comm. on Plato's Republic, ii.27.11-22* (Thomas, *SIHGM 1* (1939), 137)

"The Pythagoreans proposed this elegant theorem about the diameters and sides, that when the diameter receives the side of which it is the diameter it becomes a side, while the side, added to itself and receiving its diameter, becomes a diameter. And this is proved graphically in the second book of the *Elements* by him (Euclid): If a straight line is bisected and a straight line is added to it, the square on the whole line including the added straight line and the square on the latter by itself are together double of the square on the half and of the square on the straight line made up of the half and the added straight line."

It has generally been taken for granted that the theorem in the *Elements* referred to in this passage is *El.* II.10, since the statement of the proposition in *El.* II.10 is very close to the cited statement in Proclus' commentary. However, as will be shown below, it is equally possible that the proposition in *El.* II referred to by Proclus is *El.* II.9, the proposition proved by use of a birectangle. (*Cf.* Fig. 1.6.1 above.)

In step (n + 1) of the algorithm, a birectangle with the given sides d_n and $s_n \cdot d$ is joined to the n-th birectangle along a vertical side of length d_n or a slanting side of length $s_n \cdot d$. As shown by the diagram in Fig. 15.1.2, the other sides of the new birectangle are then d_{n+1} and $s_{n+1} \cdot d$, where

$$d_{n+1} = d_n + 2\, s_n \quad \text{and} \quad s_{n+1} = d_n + s_n.$$

These are, evidently, the rules for the formation of the successive side and diagonal numbers described by Theon of Smyrna and Proclus in the cited passages. Now, since the birectangle can be divided into two right triangles joined along a common diagonal, it follows that

$$\text{sq. } (2\, s_n + d_n) + \text{sq. } d_n = \text{sq. } (s_n \cdot d) + \text{sq. } \{(d_n + s_n) \cdot d\} = 2\, \{\text{sq. } s_n + \text{sq. } (d_n + s_n)\}.$$

This is precisely *El.* II.9, with 2 s_n as the bisected straight line and with d_n as the straight line added to it. The result can be rephrased as

sq. d_{n+1} + sq. d_n = 2 sq. s_n + 2 sq. s_{n+1}.

Equivalently,

sq. d_{n+1} – 2 sq.s_{n+1} = 2 sq. s_n – sq. d_n.

This is a recursion formula with the known initial value

2 sq. s_1 – sq. d_1 = 2 sq. 1 – sq. 1 = 1.

Therefore,

2 sq. s_n – sq. d_n = 1 for all odd n, and sq. d_n – 2 sq. s_n = 1 for all even n.

In other words,

2 sq. s_n = sq. d_n + 1 for all odd n, and 2 sq. s_n = sq. d_n – 1 for all even n.

All this agrees perfectly with the description given by Theon of Smyrna of the properties of the side and diagonal numbers.

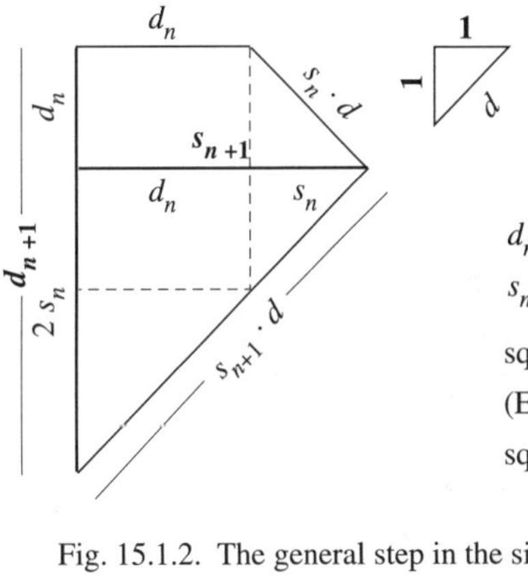

sq. d = 2

$d_{n+1} = d_n + 2\,s_n$ $\quad$ $d_n = 2\,s_{n+1} - d_{n+1}$

$s_{n+1} = d_n + s_n$ $\quad$ $s_n = d_{n+1} - s_{n+1}$

sq. d_{n+1} + sq. d_n = 2 {sq. s_{n+1} + sq. s_n}

(El. II.9) ⇒

sq. d_{n+1} – 2 sq. s_{n+1} = sq. (s_n d) – 2 sq. s_n

Fig. 15.1.2. The general step in the side and diagonal numbers algorithm.

The result must have intrigued the ancient Greek mathematicians not only because it is an "elegant theorem", but also because it leads to *a series of increasingly improved approximations to the square side of* 2. Indeed, it can be interpreted, in modern terms, as saying that

sq. (d_n /s_n) = 2 ± sq. (1/s_n) where s_n > the (n - 1)th power of 2.

(The estimate for s_n follows from the recursion formula for s_n.)

15.2. MLC 2078. The Old Babylonian Spiral Chain Algorithm

There is no known direct parallel to the Greek side and diagonal numbers algorithm in Babylonian mathematics. However, there is nothing in the proposed explanation of the algorithm that would have been beyond the capabilities of Old Babylonian mathematicians, and the idea of a "geometric algorithm" producing a series of rational sides seems to have been well known. Two OB examples have been mentioned already.

One such example is IM 55357 (Sec. 4.3 above), where a right triangle with the rational sides 1 15, 1 00, 45 is cut into *a chain of increasingly small rational right sub-triangles* by a series of heights, alternatingly against the diagonal and against the long side of the given triangle.

A second example is provided by the *chains of increasingly larger rational bisected trapezoids* considered in the problem text AO 17264, with three consecutive bisected trapezoids (Fig. 11.6.1 above).

A third example may possibly be provided by the well known but previously never adequately explained *algorithm table* **MLC 2078** (Neugebauer and Sachs, *MTC* (1945), 35), referred to in *MCT* as a table of exponents and logarithms.

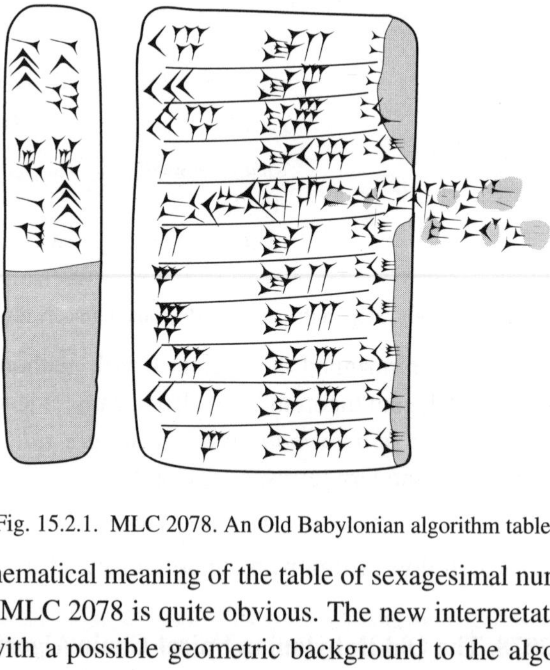

Fig. 15.2.1. MLC 2078. An Old Babylonian algorithm table.

The mathematical meaning of the table of sexagesimal numbers in two columns on MLC 2078 is quite obvious. The new interpretation below is concerned with a possible geometric background to the algorithm table.

Here is a copy and a tentative translation of the table in the text:

MLC 2078, transliteration			tentative translation			
15.e	2	íb.si$_8$	;15	makes	2	a square side
30.e	4	íb.si$_8$	;30	makes	4	a square side
45.e	8	íb.si$_8$	;45	makes	8	a square side
1.e	16	íb.si$_8$	1	makes	16	a square side
ga-mi-ru-um 4 · · · · · · · · · ··			? ? ? ? ?			
2.e	1	íb.si$_8$	2	is the	1st	square side
4.e	2	íb.si$_8$	4	is the	2nd	square side
8.e	3	íb.si$_8$	8	is the	3rd	square side
16.e	4	íb.si$_8$	16	is the	4th	square side
32.e	5	íb.si$_8$	32	is the	5th	square side
1 04.e	6	íb.si$_8$	1 04	is the	6th	square side
1.15.e	32	íb.si$_8$	1;15	makes	32	a square side
1 30.e	1 04	íb.si$_8$	1;30	makes	1 04	a square side

The table can be divided into two or three parts. The first part consists of the first four lines and is (possibly) concluded with the text in the fifth line. Unfortunately, that text is badly preserved and without a known parallel in OB mathematics. No translation of it was offered in the original publication of MLC 2078. The word *ga-mi-ru-um* may be some derived form of the verb *gamārum* 'to complete' (although that derived form is not in the dictionary). The two lines on the edge of the clay tablet are, obviously, a continuation of the four lines in the first part of the table.

The second part of the table consists of the 6 lines after the line of text.

The Sumerian word íb.si$_8$ literally means something like 'it is equal'. It is used most often in tables of square sides, where a phrase like 4.e 2 íb.si$_8$ can be translated loosely as '4 makes 2 a square side', or simply '4 has the square side 2'. The corresponding phrase 8.e 2 íb.si$_8$ in a table of cube sides (less common) means '8 makes 2 a cube side', or simply '8 has the cube side 2'. The same kind of phrase appears also in the table of "quasi-cube sides" MS 3048 (see Sec. 13.6), and in Plimpton 322 (Sec. 3.3). The tentative interpretation of the table text MLC 2078 proposed here is based on the assumption that it is related to the spiraling chain of half-squares and right trapezoids shown in Fig. 15.2.2 below.

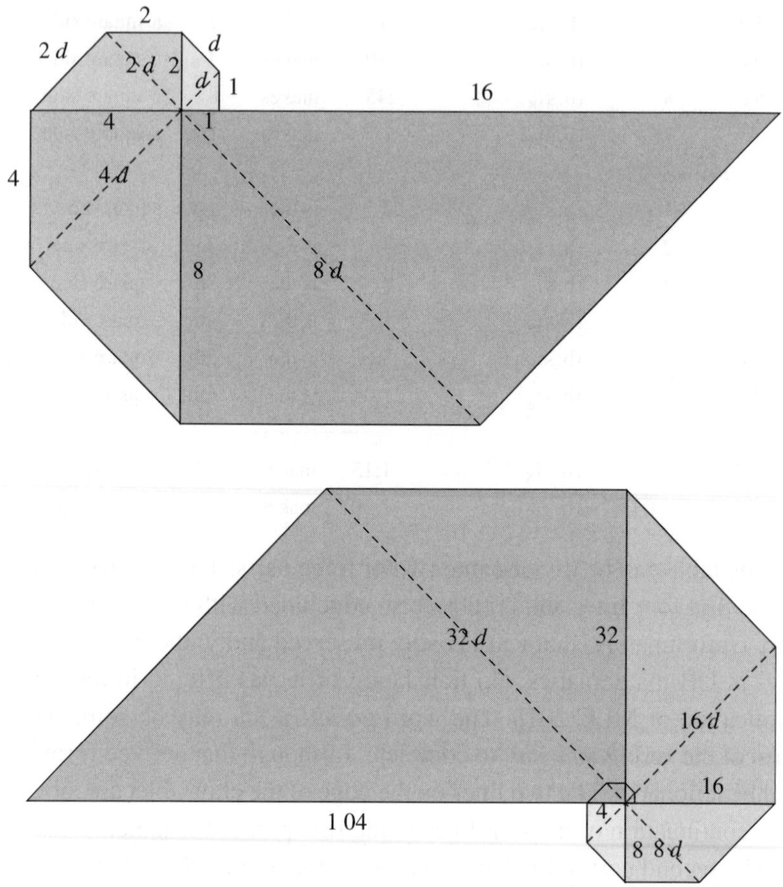

Fig. 15.2.2. A chain of similar right trapezoids, made up of pairs of half-squares.

According to this interpretation, this OB "spiral chain algorithm" begins, just like the Greek side and diagonal numbers algorithm interpreted as in Fig. 15.1.1 above, with a half-square with the sides 1, 1, d, where sq. $d = 2$. This half-square and a larger half-square with the sides d, d, 2 are joined along the diagonal of the first half-square, forming a right(-angled) trapezoid with the sides 1, 1, d, 2. This right trapezoid is then, in its turn, joined along its side of length 2 with a right trapezoid of the same form, but twice as large, that is with the sides 2, 2, 2 *d*, 4. And so on.

In terms of this spiral chain algorithm, the table text MLC 2058 can be explained as follows. The first four lines say that

After ;15 = 1/4 turn of the spiral, the square side 2 is reached,
after ;30 = 1/2 turn of the spiral, the square side 4 is reached,
after ;45 = 3/4 turns of the spiral, the square side 8 is reached,
after 1 full turn of the spiral, the square side 16 is reached.

The enigmatic text line then probably says something like

After 4 quarter-turns, the spiral is complete.

Indeed, after the fourth quarter-turn, the spiral begins to overlap itself. See the upper diagram in Fig. 15.2.2.

The second part of the table text is a kind of *inverse of the first part*, apparently saying that

The square side 2 is reached after 1 quarter-turn,
the square side 4 is reached after 2 quarter-turns,
the square side 8 is reached after 3 quarter-turns,
the square side 16 is reached after 4 quarter-turns,
the square side 32 is reached after 5 quarter-turns,
the square side 1 04 (= 64) is reached after 6 quarter-turns.

Thus, in this second part of the table the spiral continues, overlapping itself, as in the lower diagram in Fig. 15.2.2.

Finally, for the sake of symmetry, the first part of the table is continued on the edge of the clay tablet, with two additional lines saying that

After 1;15 = 1 1/4 turns of the spiral, the square side 32 is reached,
after 1;30 = 1 1/2 turns of the spiral, the square side 1 04 (= 64) is reached,

15.3. Side and Diagonal Numbers When Sq. *p* = Sq. *q* · *D* – 1

The proposed interpretation of the Greek side and diagonal numbers algorithm in terms of a chain of birectangles can easily be extended to the case when the initial half-square with the sides *d*, 1, 1, where sq. *d* = 2, is replaced by *a right triangle* with the sides ***q* · *d*, *p*, 1**, where sq. ***d* = *D*** and **sq. *p* = sq. *q* · *D* – 1**. Examples of such triangles are

d, 2, 1 with *D* = sq. *d* = 5, and 5 *d*, 18, 1 with *D* = sq. *d* = 13.

Application of the algorithm in the mentioned cases will yield rational approximations to the square sides of 5 and 13, in the same way as Theon's algorithm yields approximations to the square side of 2.

The general step in the side and diagonal numbers algorithm in the case when sq. p = sq. $q \cdot D - 1$ is shown in Fig. 15.3.1 below. Note that the algorithm can be run through backwards: If d_{n+1} and s_{n+1} are known, then d_n and s_n can be obtained algebraically as the solutions to a system of linear equations, or geometrically through inspection of the diagram.

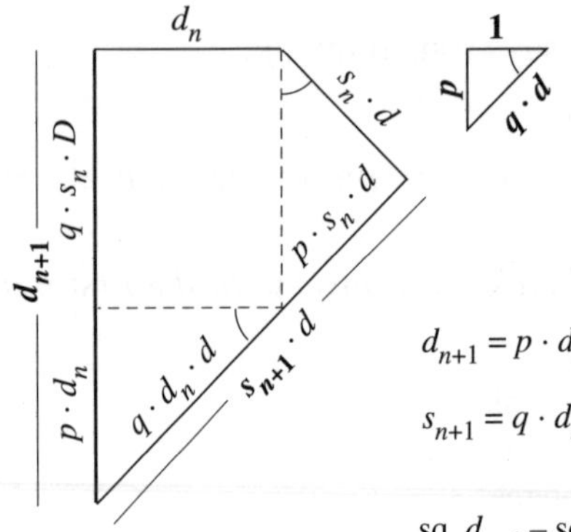

sq. p = sq. $q \cdot D - 1$

D = sq. d

$d_{n+1} = p \cdot d_n + q \cdot s_n \cdot D$ $\qquad$ $d_n = q \cdot s_{n+1} \cdot D - p \cdot d_{n+1}$

$s_{n+1} = q \cdot d_n + p \cdot s_n$ $\qquad$ $s_n = q \cdot d_{n+1} - p \cdot s_{n+1}$

sq. d_{n+1} – sq. $s_{n+1} \cdot D$ = sq. $s_n \cdot D$ – sq. d_n

Fig. 15.3.1. The general step in the case when sq. p = sq. $q \cdot D - 1$.

In the case when $D = 13$, for instance,

$d_1 = p = 18$, $s_1 = q = 5$, sq. $s_1 \cdot D$ – sq. d_1 = sq. $5 \cdot 13$ – sq. $8 = 325 - 324 = 1$,
$d_2 = 18 \cdot 18 + 5 \cdot 5 \cdot 13$ = sq. 18 + sq. $5 \cdot 13 = 324 + 325 = 649$,
$s_2 = 5 \cdot 18 + 18 \cdot 5 = 90 + 90 = 180$,
sq. d_2 – sq.$s_2 \cdot D$ = sq. 649 – sq. $180 \cdot 13 = 421201 - 421200 = 1$, *etc.*

15.4. Side and Diagonal Numbers When Sq. p = Sq. $q \cdot D + 1$

With the necessary modifications, the side and diagonal numbers algorithm can be made to work also in the case when the initial right triangle has the sides $q, p \cdot d$, 1, where sq. $d = D$ and sq. p = sq. $q \cdot D + 1$. Examples of such triangles are

2, d, 1 with sq. $d = 3$, and 5, 2 d, 1 with sq. $d = 6$.

The diagram in Fig. 15.4.1 for the case when sq. $d = 3$ starts with an initial right triangle with the sides 2, d, 1. Here d is the *inexpressible* and 2 an *expressible* diagonal of 1, and

sq. 2 – sq. $1 \cdot 3 = 1$.

To the initial right triangle is joined the first birectangle, with the given sides 2 and *d*. The second pair of sides in the birectangle can then be shown to be 7 and 4 *d*. Here 4 *d* is the *inexpressible diagonal* corresponding to the side 4, while 7 is the *expressible diagonal*. Note that 7 and 4 *d*, the two long sides of the birectangle are nearly equal. Indeed,

sq. 7 – sq. 4 · 3 = 49 – 16 · 3 = 1.

In the second birectangle, 15 *d* and 26 are the inexpressible and expressible diagonals of 15, and

sq. 26 – sq. 15 · 3 = 676 – 225 · 3 = 1.

And so on.

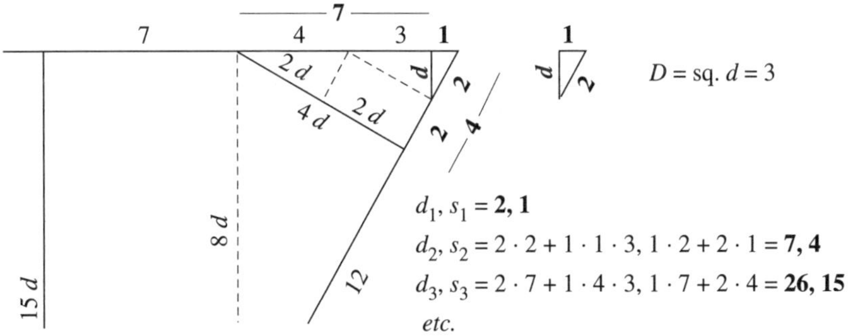

Fig. 15.4.1. The first few steps of the algorithm when $D = \text{sq. } d = 3$.

The general step of the side and diagonal numbers algorithm in the case when sq. p = sq. q · D + 1 is shown in Fig. 15.4.2 below:

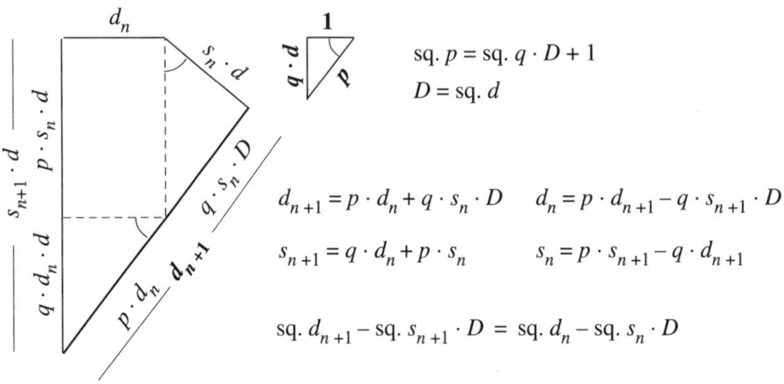

Fig. 15.4.2. The general step in the case when sq. p = sq. $q \cdot D + 1$.

Chapter 16

Greek and Babylonian Square Side Approximations

16.1. *Metrica* I.8 b. Heron's Square Side Rule

An explicit rule for the approximation of square sides (square roots) is given in Heron of Alexandria's *Metrica* I.8 (in connection with an example of the application of Heron's triangle area rule):

***Metrica* I.8 b** (Bruins, *CCPV 3* (1964), 189; Schöne, *HAVD* (1903), 19)

"But as 720 does not have a rational side, we shall with the smallest difference take the side like this:

Since the closest square to 720 is 729 and it has the side 27, divide 720 by 27, it becomes 26 and two thirds.

Add 27, it becomes 53 and two thirds. Of this the half, it becomes 26 1/2 1/3.

Thus, the side of 720 is very close to 26 1/2 1/3.

For 26 1/2 1/3 on itself becomes 720 1/36.

Thus, the difference is the 36th part of the unit.

However, if we want the difference to be smaller than the 36th part, we shall replace 729 by 720 1/36 that we now have found, and when we do the same again, we shall find that the difference becomes much smaller than1/36,"

This is an explicit example of "Heron's square side rule", a general rule for the computation of an approximate square side to a given area number. The rule can be formulated as follows:

Let D be a given area number and let a be a known approximation to sqs. D.

Set $r = (a + D/a)/2$. Then r is an *improved* approximation to sqs. D.

To get an even more accurate approximation, repeat the process.

An alternative, more precise, formulation of the rule is as follows:

> Let D be a given area number and let u be an approximation *from above* to sqs. D. Set $r = (u + D/u)/2$. Then r is an *improved* approximation *from above* to sqs. D.
>
> If, instead, s is an approximation *from below* to sqs. D, set $r = (s + D/s)/2$. Then r is an *improved* approximation to sqs. D, but *from above*.

In Heron's example in *Metrica* I.8,

$D = 720$, $u = 27$, $D/u = 720/27 = 26\ 18/27 = 26\ 2/3$,
$r = (27 + 26\ 2/3)/2 = 26\ 1/2\ 1/3$,
sq. $u - D = 729 - 720 = 9$, sq. $r - D = 720\ 1/36 - 720 = 1/36$.

A simple geometric proof of Heron's area rule is given in Fig. 16.1.1 below. (*Cf. El.* II.14, Figs. 1.7.1-2.)

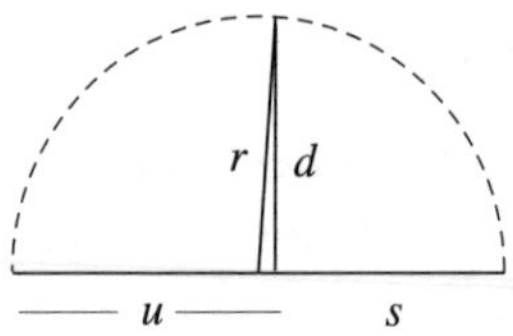

$u \cdot s = \text{sq. } d = D$, $u > s$
$s = D/u < d < u$ or $u = D/s > d > s$
$r = (u + D/u)/2 > d$ or $r = (s + D/s)/2 > d$
sq. $r - D$ = sq. $(u - D/u)/2$ = sq. $(D/s - s)/2$

Fig. 16.1.1. Geometric explanation of Heron's square side rule in *Metrica* I.8.

16.2. Heronic Square Side Approximations

A complete and detailed survey of all examples of square side approximations in Heron's collected works is presented in Hofmann, *JDMv* 43 (1934). In the first 3 examples, the first approximation is *from above*:

1	$D = 720$	$u = 27$	$s = 26\ 2/3$	$r = (u + s)/2 = 26\ 1/2\ 1/3$
2	$D = 63$	$u = 8$	$s = 8 - 1/8$	$r = (u + s)/2 = 7\ 1/2\ 1/4\ 1/8\ 1/16$
3	$D = 250$	$u = 16$	$s = 15\ 5/8$	$r = (u + s)/2 = 15\ 13/16$

In the next 12 examples, the first approximation is *from below*:

4	$D = 72$	$u = 8$	$s = 9$	$r = (u + s)/2 = 8\ 1/2$
5	$D = 96$	$u = 9$	$s = 10\ 1/3$	$r = (u + s)/2 = 9\ 1/2\ 1/3$
6	$D = 150$	$u = 12$	$s = 12\ 1/2$	$r = (u + s)/2 = 12\ 1/4$

etc.

In all but 2 of these examples, the first approximation is an integer.

In 7 further examples, an integer p giving a good approximation to the square side is found directly:

18 $D = 288$ = sq. $12 \cdot 2$ = sq. $17 - 1$ $p = 17$ (sqs. 34 = appr. 35/6)

	(p is here the approximate diagonal of a square with the side 12)		
19	$D = 675 = \text{sq. } 15 \cdot 3 = \text{sq. } 26 - 1$	$p = 26$	(sqs. 3 = appr. 26/15)
	(p is here the approximate height of an equilateral triangle with the side 30)		
20	$D = 1224 = \text{sq. } 6 \cdot 34 = \text{sq. } 35 - 1$	$p = 35$	
21	$D = 144\ 1/2 = \text{sq. } 12 + 1/2$	$p = 12$	
22	$D = 195\ 1/2\ 1/4\ 1/8\ 1/16 = \text{sq. } 14 - 1/16$	$p = 14$	
23	$D = 75600 = \text{sq. } 60 \cdot 21 = 25 \cdot (\text{sq. } 55 - 1)$	$p = 275$	(sqs. 21 = appr. 55/12)
24	$D = 67500 = \text{sq. } 150 \cdot 3 = \text{sq. } 10 \cdot (\text{sq. } 26 - 1)$	$p = 259$	(sqs. 3 = appr. 26/5)

Sixteen further examples are of the same types as the ones already mentioned, but with various round-offs.

The five remaining examples in Hofmann's survey are especially interesting. According to Hofmann's interpretation, they are all examples of applications of *a second, more accurate square side rule*, namely

$$\sqrt{(x^2 - 1)} = \text{appr. } x - 1/(2\,x - 1) + 1/\{(2\,x - 1)\,(2\,x + 1)\} = x - 1/\{2\,x - 1/(2\,x)\}.$$

The five special examples are:

41	$D = 3$	$= \text{sq. } 2 - 1$	$p =$ **2 – 1/3 + 1/15**
42	$D = 135$	$= 9 \cdot (\text{sq. } 4 - 1)$	$p = 3 \cdot$ (**4 – 1/7 + 1/63**) = 11 1/2 1/14 1/21
43	$D = 216$	$= 9 \cdot (\text{sq. } 5 - 1)$	$p = 3 \cdot$ (**5 – 1/9 + 1/99**) = 14 2/3 1/33
44	$D = 1575$	$= 25 \cdot (\text{sq. } 8 - 1)$	$p = 5 \cdot$ (**8 – 1/15 + 1/255**) = 39 2/3 1/51
45	$D = 6300$	$= 100 \cdot (\text{sq. } 8 - 1)$	$p = 10 \cdot$(**8 – 1/15 + 1/255**) = 79 1/3 1/34 1/102

16.3. A New Explanation of Heron's Accurate Square Side Rule

The side and diagonal numbers algorithm in the case when

$$\text{sq. } a = \text{sq. } c \cdot D - 1, \quad D = \text{sq. } d$$

(see Fig. 15.3.1 above) is governed by the *recursive equations*

$$d_1, s_1 = a, c, \quad d_{n+1}, s_{n+1} = a\,d_n + c\,s_n \cdot D,\ c\,d_n + a\,s_n \quad \text{for} \quad n = 1, 2, 3, \ldots$$

It is, in certain situations, convenient to express these equations more concisely in terms of a *formal multiplication* of pairs of numbers:[42]

$$(d_1, s_1) = (a, c), \quad (d_{n+1}, s_{n+1}) = (a, c) \cdot (d_n, s_n) \quad \text{for} \quad n = 1, 2, 3, \ldots$$

The result of iterated birectangular compositions of the initial pair (a, c) with the factor (a, c) can then be interpreted as *formal powers* of (a, c):

42. The oldest documented use of this kind of formal multiplication can be found in Brahmagupta, ***Bss*** **XVIII.65-66** (Colebrooke, *AAMS* (1973), 363). See also, for instance, Weil, *NTATH* (1984), § IX.)

$(d_n, s_n) = (a, c)^n, \quad n = 1, 2, \cdots.$

Take, for instance, the case when ***D* = 2.** Then

$(a, c) = (1, 1), \quad (1, 1)^2 = (1, 1) \cdot (1, 1) = (3, 2), \quad (1, 1)^3 = (1, 1) \cdot (3, 2) = (7, 5),$ *etc.*

In this case, it follows from the equation

sq. d_{n+1} – sq. $s_{n+1} \cdot 2$ = sq. $s_n \cdot 2$ – sq. d_n for all n

that (in a deliberately anachronistic notation)

sq. d_n – sq. $s_n \cdot 2 = r_n$ where $r_n = -1$ when n is odd, but $r_n = +1$ when n is even.

The result can be expressed in a concise way by expanding the number pairs (d_n, s_n) to number triples $(d_n, s_n; r_n)$. Thus, when ***D* = 2**:

$(d_1, s_1; r_1) = (1, 1; -1), \quad (1, 1; -1)^2 = (3, 2; 1), \quad (1, 1; -1)^3 = (7, 5; -1),$ *etc.*

The situation in the case when, as in Fig. 15.4.2 above,

sq. c = sq. $a \cdot D + 1$, D = sq. d

is perfectly parallel, except that c and a change place, and

sq. d_{n+1} – sq. $s_{n+1} \cdot D$ = sq. d_n – sq. $s_n \cdot D$ so that $r_n = 1$ for all n.

Therefore, in this case,

$(d_1, s_1; r_1) = (c, a; 1), \quad (d_{n+1}, s_{n+1}; 1) = (c, a; 1) \cdot (d_n, s_n; 1)$ for all n,

and consequently

$(d_n, s_n; 1) = (c, a; 1)^n, \quad n = 1, 2, \cdots.$

When ***D* = 3**, for instance, (Fig. 15.4.1 above):

$(d_1, s_1; r_1) = (2, 1; 1), \quad (2, 1; 1)^2 = (7, 4; 1),$ $\mathbf{(2, 1; 1)^3 = (26, 15; 1)}$, *etc.*

Therefore, the "first", "second", and "third" approximations to **sqs. 3** are

2/1 = 2, 7/4 = 1 1/2 1/4, and 26/15 = **1 2/3 1/15**.

The corresponding errors are

sq. 2 – 3 = 1, sq. 7/4 – 3 = sq. 1/4, sq. 26/15 – 3 = sq. 1/15.

The case when ***D* = 6** is only slightly more complicated:

$(d_1, s_1; r_1) = (5, 2; 1), \quad (5, 2; 1)^2 = (49, 20; 1),$ $\mathbf{(5, 2; 1)^3 = (485, 198; 1)}$, *etc.*

Thus, the first, second, and third approximations to **sqs. 6** are

5/2 = 2 1/2, 49/20 = 2 1/3 1/10 1/60, and 485/198 = **2 1/3 1/9 1/198**.

The corresponding errors are

sq. 5/2 – 6 = sq. 1/2, sq. 49/20 – 6 = sq. 1/20, sq. 485/198 – 6 = sq. 1/198.

Similarly in the case when ***D* = 7**:

$(d_1, s_1; r_1) = (8, 3; 1)$, $(8, 3; 1)^2 = (127, 48; 1)$, $\mathbf{(8, 3; 1)^3 = (2024, 765; 1)}$, *etc.*

Therefore, the first, second, and third approximations to **sqs. 7** are

8/3 = 2 2/3, 127/48 = 2 1/2 1/8 1/48, 2024/765 = **2 1/2 1/9 1/30 1/765**.

It should be obvious by now what the respective errors are in this case.

The case when ***D*** **= 15** is just as simple as the case when $D = 3$:

$(d_1, s_1; r_1) = (4, 1; 1)$, $(4, 1; 1)^2 = (31, 8; 1)$, $\mathbf{(4, 1; 1)^3 = (244, 63; 1)}$, *etc.*

Therefore, the first, second, and third approximations to **sqs. 15** are

4/1 = 4, 31/8 = 3 1/2 1/4 1/8, and **244/63 = 3 5/6 1/42 1/63**.

Note that all the mentioned approximations to sqs. 3, sqs. 6, sqs. 7, and sqs. 15 are *from above*.

The special examples 41-45 of Heronic square side approximations considered by Hofmann (*op. cit.*), can now be explained as follows, all in terms of *third approximations*:

41	**sqs. 3 =**	**1 2/3 1/15** (= 26/15)		
42	sqs. 135 =	3 · **sqs. 15** = 3 · **3 5/6 1/42 1/63** =	11 1/2 1/14 1/21	*Geom.* 15.4
43	sqs. 216 =	6 · **sqs. 6** = 6 · **2 1/3 1/9 1/198** =	14 2/3 1/33	*Geom.* 16.34
44	sqs. 1575 =	15 · **sqs. 7** = 15 · **2 1/2 1/9 1/30 1/765** =	39 2/3 1/51	*Geom.* 15.11
45	sqs. 6300 =	30 · **sqs. 7** = 30 · **2 1/2 1/9 1/30 1/765** =	79 1/3 1/34 1/102	*Geom.* 15.10

Note that in Heron's work accurate expressions for sqs. 3 appear on three occasions in connection with *equilateral triangles*, namely

$A_{\text{equil. tr.}}$ = **1/3 1/10** · sq. s	(1/4 · sqs. 3 = 1/4 · **26/15** = 13/30)	*Geom.* 10.1
$h_{\text{equil. tr}}$ = $\boldsymbol{s}$ **– 1/10 1/30** · s	(1/2 sqs. 3 = 1/2 · **26/15** = 26/30)	*Geom.* 10.3
when $s = 30$: $h_{\text{equil. tr}}$ = sqs. (900 – 900/4) = sqs. 675 = 15 · **26/15** = 26		*Geom.* 10.12

Hofmann's explanation of the five special cases 41-45 is related in the following way to the explanation above in terms of *third approximations*:

$D = \text{sq. } c - 1$

$\Rightarrow (d_1, s_1; r_1) = (c, 1; 1),$

$(d_2, s_2; r_2) = (c, 1; 1)^2 = (\text{sq. } c + D, 2\,c; 1) = (2 \text{ sq. } c - 1, 2\,c; 1)$

$(d_3, s_3; r_3) = (c, 1; 1)^3 = (c \cdot (2 \text{ sq. } c - 1) + 2\,c \cdot D, 1 \cdot (2 \text{ sq. } c - 1) + c \cdot 2\,c; 1)$

$= (4 \text{ cu. } c - 3\,c, 4 \text{ sq. } c - 1; 1).$

Consequently, if $D = \text{sq. } c - 1$, the first, second and third approximations to sqs. D are, in agreement with Hofmann's explanation,

$\boldsymbol{d_1/s_1} = c,$

$\boldsymbol{d_2/s_2}$ = (2 sq. c – 1)/(2 c) = **$\boldsymbol{c}$ – 1/(2 $\boldsymbol{c}$)**,
$\boldsymbol{d_3/s_3}$ = (4 cu. c – 3 c)/(4 sq. c – 1) = **$\boldsymbol{c}$ – 1/{2 $\boldsymbol{c}$ – 1/(2 $\boldsymbol{c}$)}**.

Note, however, that in order to be able to make use of his explanation, Hofmann did not consider the simple cases $D = 6$ and $D = 7$ in his special examples 43-45, but instead $D = 4 \cdot 6 = 24$ and $D = 9 \cdot 7 = 63$.

16.4. Third Approximations in Ptolemy's *Syntaxis* I.10

The preliminaries to the Table of Chords in **Book I.10 of Ptolemy's *Syntaxis*** or the *Almagest* (150) include the computation of the side of a regular polygon inscribed in a circle, expressed as a multiple of the 120th part of the diameter of the circle, when the regular polygon in question has 10, 5, 6, 4, or 3 sides. (This is equivalent to computing the chords of 36°, 72°, 60°, 90°, and 120°. See Heath, *HGM* 2 (1921), 276-278.)

In all these cases, except the case of the hexagon, a crucial step in the procedure is to find *a very accurate approximation* to the square side of a small integer, namely 5 in the case of the decagon and the pentagon, 2 in the case of the square, and 3 in the case of the equilateral triangle. A close look at the approximations to these square sides actually appearing in *Syntaxis* I:10 reveals that they were probably computed, just like Heron's accurate square side approximations, as *third approximations,* however *with departure from good first approximations*. The approximations mentioned by Ptolemy are:

1 sqs. 7200 = **60 sqs. 2 = 84;51 10**
2 sqs. 10800 = **60 sqs. 3 = 103;55 23**
3 sqs. 4500 = **30 sqs. 5 = 67;04 55**

These approximations can be explained as follows:

sqs. 2: $(17, 12; 1)^2$ = (sq. 17 + sq. 12 · 2, 2 · 17 · 12; 1 · 1) = (577, 408; 1),
$(17, 12; 1)^3$ = (577 · 17 + 408 · 12 · 2, 577 · 12 + 408 · 17; 1 · 1) = **(19601, 13860; 1)**
60 · sqs. 2 = 19601/231 = 84 197/231 = 84 51/60 13/4620 = appr. **84;51 10**

sqs. 3: $(7, 4; 1)^2$ = (sq. 7 + sq. 4 · 3, 2 · 7 · 4; 1 · 1) = (97, 56; 1)
$(7, 4; 1)^3$ = (97 · 7 + 56 · 4 · 3, 97 · 4 + 56 · 7; 1 · 1) = **(1351, 780; 1)**
60 · sqs. 3 = 1351/13 = 103 12/13 = 103 55/60 1/156 = appr. **103;55 23**

sqs. 5: $(9, 4; 1)^2$ = (sq. 9 + sq. 4 · 5, 2 · 9 · 4; 1 · 1) = (161, 72; 1)
$(9, 4; 1)^3$ = (161 · 9 + 72 · 4 · 5, 161 · 4 + 72 · 9; 1 · 1) = **(2889, 1292; 1)**

60 · sqs. 5 = 67 53/646 = 67 4/60 149/9690 = appr. **67;04 55**

Remark. Ptolemy's computation of sqs. 4500 is explained in the commentary to the *Almagest* by Theon of Alexandria (380) by use of a geometric diagram (see Heath, *HGM* 2, 60-63). Theon's method is based on sexagesimal place value notation and iteration. It is interesting that a close parallel to this method may have been used in scribe schools in Mesopotamia in the *Old Akkadian* period, 500 years *before* the time of the OB mathematical cuneiform texts. See Friberg, *CDLJ* (2005/2), Fig. 14.

16.5. The General Case of Formal Multiplications

In the preceding section, formal multiplications were considered with factors of the form $(p, q; -1)$ or $(p, q; 1)$. It was shown that these two kinds of formal multiplications can be used to construct sequences of increasingly accurate approximations to the square side of D, when either sq. p = sq. $q \cdot D - 1$ (Fig. 15.3.1) or sq. p = sq. $q \cdot D + 1$ (Fig. 15.4.2).

It is possible to modify the diagrams in Figs. 15.3.1 and 15.3.2 so that they cover also the more general cases of formal multiplications with factors of the form $(p, q; - r)$ or $(p, q; + r)$. This is done, in the first of the two cases, in Fig. 16.5.1 below.

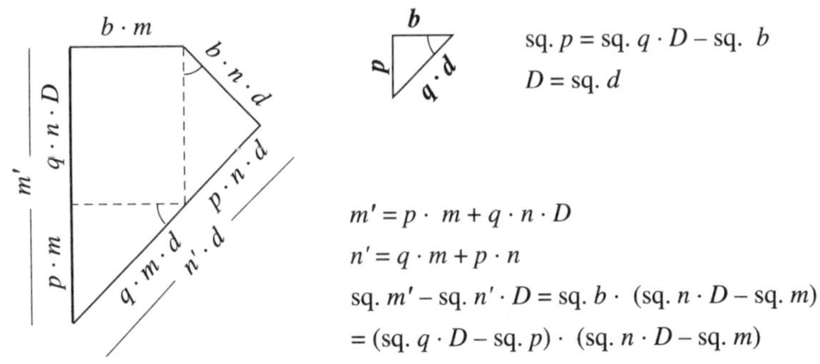

Fig. 16.5.1. Formal multiplication in a more general case.

The result can be formulated as the rule[43] that

$(p, q; - r) \cdot (m, n; s) = (p\, m + q\, n \cdot D,\ q\, m + p\, n;\ - r\, s)$.

43. Actually *Brahmagupta's rule* (*Bss* **XVIII.65-66** (Colebrooke, *AAMS* (1973), 363)).

In the second case, the following rule can be verified in a similar way:

$$(p, q; r) \cdot (m, n; s) = (p\,m + q\,n \cdot D,\, q\,m + p\,n;\, r\,s).$$

It follows from this second rule that, in particular,

$$(m, n; s)^2 = (\text{sq. } m + \text{sq. } n \cdot D,\, 2\,m\,n;\, \text{sq. } s).$$

Therefore, *there is a simple connection between Heron's square side rule in Metrica I. 8 and formal multiplication with regard to D*. Indeed,

$$(m/n + D \cdot n/m)/2 = (\text{sq. } m + \text{sq. } n \cdot D)/(2\,m\,n).$$

This means that, for any given initial approximation m/n to sqs. D the improved approximation to sqs. D given by Heron's square side rule coincides with the "second" approximation yielded by formal multiplication. Note, by the way than an iterated application of Heron's square side rule yields not the "third" but the "fourth" approximation.

Note also that it is easy to see that

$$\text{sq. } (\text{sq. } m + \text{sq. } n \cdot D) - \text{sq. } (2\,m\,n) \cdot D = \text{sq. } (\text{sq. } m - \text{sq. } n \cdot D),$$

which is another way of proving that $(m, n; s)^2 = (\text{sq. } m + \text{sq. } n \cdot D, 2\,m\,n; \text{sq. } s)$.

16.6. A New Explanation of the Archimedian Estimates for Sqs. 3

In the proof of Proposition 3 in Archimedes' *Measurement of the Circle*, the arguments are based on the following *precise lower and upper bounds for sqs. 3*, or, more precisely, for the ratio of the diameter of a circle to the side of a circumscribed hexagon:

$$265/153 < \text{sqs. } 3 < 1351/780.$$

Archimedes says nothing about the origin of these very accurate estimates. Many different explanations have been proposed by various authors in the course of more than a century. The difficulty is to reach a consensus about which of these explanations is the one that has the greatest chance of being historically correct. In *HGM* 2(1921), 51, Heath writes:

> "How did Archimedes arrive at these particular approximations? No puzzle has exercised more fascination upon writers interested in the history of mathematics."

More than fifty years later the matter was still not settled. In *AHES* 15 (1975/76), Knorr writes:

> "These values have stimulated a massive scholarly commentary."

The answer to the question proposed below, a further development of the answer proposed by Knorr (*op. cit.*), has the advantage of linking Archimedes' estimates to Heron's second square side rule, as well as to Ptolemy's accurate approximations to the square sides of 2, 3, and 5.

Knorr begins his discussion of Archimedes' estimates with the observation that a side and diagonal numbers sequence for the square side of 3 is generated by the formula

$$d_{n+1} = d_n + 3\, s_n,\quad s_{n+1} = d_n + s_n,\quad d_1 = 2,\quad s_1 = 1.$$

From the terms of this sequence can be obtained the approximations

sqs. 3 = 2/1, 5/3, 14/8 = 7/4, 19/11, 52/30 = 26/15, 71/41, 194/112 = 97/56, **265/153**, *etc.*

The *eighth* term of the sequence is Archimedes' lower estimate, and the *eleventh* term is his upper estimate. This does not sound right, so Knorr proposes a revised sequence with fewer terms. His point of departure is that 5/3 is a *lower estimate* for sqs. 3 and that 3/(5/3) = 9/5 is a corresponding *upper estimate*. Therefore, he introduces the following modified side and diagonal numbers sequence, making use of a *weighted average*:

$$d_{n+1}= 9\, s_n + 5\, d_n,\quad s_{n+1} = 5\, s_n + 3\, d_n,\quad d_1 = 5,\quad s_1 = 3.$$

The resulting sequence of lower and upper estimates is

5/3, 52/30 = 26/15, **265/153**, 2702/1560 = **1351/780**.

Thus, in this revised sequence, the *third* and *fourth* terms coincide with Archimedes' estimates, which sounds convincing.

In terms of formal multiplications with regard to $D = 3$, Knorr's analysis can be explained as follows: Knorr's first sequence actually starts with the lower estimate 1/1, so that, more correctly, Archimedes' estimates are the *ninth* and *twelfth* terms of the sequence, namely

$$(1, 1; -2)^9 \text{ and } (1,1; -2)^{12}.$$

The terms of Knorr's modified side and diagonal numbers sequence are

$$(1, 1; -2)^{3n} = (5, 3; -2)^n,\quad n = 1, 2, \cdots.$$

Note that here, for instance,

$$(5, 3; -2)^2 = (5 \cdot 5 + 3 \cdot 3 \cdot 3, 2 \cdot 5 \cdot 3; 4) = (52, 30; 4) = (26, 15; 1).$$

The transformation of (52, 30; 4) into (26, 15; 1) is a way of exploiting the circumstance that, through elimination of the square factor 4,

$$\text{sq. } 52 - \text{sq. } 30 \cdot 3 = 4 \;\Rightarrow\; \text{sq. } 26 - \text{sq. } 15 \cdot 3 = 1.$$

A similar transformation is needed for every second term of the sequence.

Knorr's explanation can be refined in the following way (Friberg, *BaM* 28 (1997), Sec. 9 d): Assume that Archimedes started with *the OB standard approximation to the square side of 3, which was 7/4* (= 1;45). He could then easily obtain the following improved estimates:

$(7, 4; 1)^2 = (7 \cdot 7 + 4 \cdot 4 \cdot 4, 2 \cdot 7 \cdot 4; 1) = (97, 56; 1)$, and
$\mathbf{(7, 4; 1)^3} = (97 \cdot 7 + 56 \cdot 4 \cdot 3, 97 \cdot 4 + 56 \cdot 7; 1) = \mathbf{(1351, 780; 1)}$.

Therefore, Archimedes can have been obtained his *upper estimate* of sqs. 3 as the *third approximation* when starting with the *upper estimate 7/4*.

To get an accurate lower estimate, Archimedes would have to start with a relatively accurate *lower* estimate. As such he could choose 5/3, from which he could derive the following improved estimates:

$(5, 3; -2)^2 = (5 \cdot 5 + 3 \cdot 3 \cdot 3, 2 \cdot 5 \cdot 3; 4) = (52, 30; 4) = (26, 15; 1)$, and
$\mathbf{(5, 3; -2)^3} = (26 \cdot 5 + 15 \cdot 3 \cdot 3, 26 \cdot 3 + 15 \cdot 5; 1 \cdot - 2) = \mathbf{(265, 153; -2)}$.

Therefore, Archimedes can have been obtained his *lower estimate* of sqs. 3 as the *third approximation* when starting with the *lower estimate* 5/3.

Remember, by the way, that one or several applications of Heron's (first) square side rule will only lead to *upper* estimates of the square side. Note also that Archimedes' upper estimate coincides with the accurate approximation to sqs. 3 used in Ptolemy's *Syntaxis* I.10! (Sec. 16.4 above.)

16.7. Examples of Babylonian Square Side Approximations

The additive and subtractive square side rules

The Demotic mathematical papyrus ***P.BM 10520*** (Early Roman?) contains a couple of exercises with explicit examples of the application of a method for the approximate computation of square sides. (*Cf.* Friberg, *UL* (2005), Sec. 3.3 f.) One of them is reproduced below.

***P.BM 10520* § 6 a** (Parker, *DMP* (1972) # 62).

Let 10 be reduced to its square side.
You shall count 3 3 times, result 9, remainder 1. (Its) 2', result 2'.
You let 2' make part of 3, result 6'.
You add 6' to 3, result 3 6'. It is the square side.
Let it be known, namely:
You shall count 3 6', 3 6' times, result 10 $\overline{36}$.
Its difference of square side $\overline{36}$.

The computations in this exercise can be explained as follows:

1) sqs. 10 = sqs. (sq. 3 + 1) = appr. 3 + (1/2)/3 = 3 1/6.
2) Check: sq. 3 1/6 = 10 1/36. Error: 1/36.

Apparently, the rule in its general form was

sqs. (sq. $s + R$) = appr. $s + R/(2\ s)$.

This is Heron's square side rule (Sec. 16.1) in another form. Indeed,

sqs. D = appr. s, D = sq. $s + R$ $(R > 0)$
$\Rightarrow$ $(s + D/s)/2$ = (sq. $s + D)/(2\ s)$ = (2 sq. $s + R)/(2\ s) = s + R/(2\ s)$.

The rule in this form presumes that the given approximation s is a *lower* estimate for sqs. D. It may be called the "additive square side rule". An independent, metric algebra derivation of this rule is shown in Fig. 16.7.1:

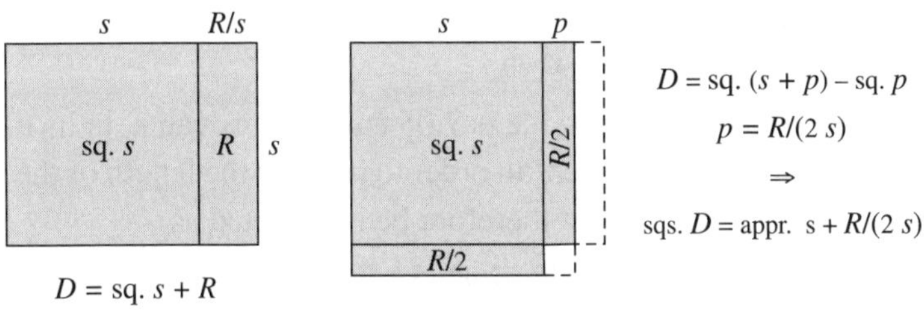

Fig. 16.7.1. A metric algebra proof of the additive square side rule.

The diagram shows that the improved approximation is an *upper* estimate for the square side.

There is also a corresponding "subtractive square side rule" with a similar metric algebra proof. This rule, another reformulation of Heron' square side rule, presumes that the given approximation u is an *upper* estimate for sqs. D. Note that

sqs. D = appr. u, D = sq. $u - R$ $(R > 0)$
$\Rightarrow$ $(u + D/u)/2$ = (sq. $u + D)/(2\ u)$ = (2 sq. $u - R)/(2\ u) = u - R/(2\ u)$.

There is no known Babylonian mathematical text with an explicit formulation of a square side rule. However, since computations with squares and rectangles are much more common in Babylonian mathematics than computations with circles, it is likely that the square side rule used in Babylonian mathematics was the additive/subtractive square side rule (Fig. 16.7.1) rather than Heron's square side rule (Fig. 16.1.1).

Late and Old Babylonian approximations to sqs. 2

A moderately good approximation to sqs. 2 appears in the following isolated exercise in the *Late Babylonian* mixed problem text AO 6484 (Neugebauer, *MKT* 1 (1935), 98; Thureau-Dangin, *TMB* (1938) 158):

AO 6484 § 8, literal translation	explanation
The great divider of an equalside, 10 cubits.	The diagonal d of a square is 10 c.
The length of the equalside is what?	The side s = ?
10 · 42 30 *go, it is* 7 05, the length.	d · ;42 30 = s
7 05 · 1 25 go, it is 10 25, the great divider.	s · 1;25 = d

In this exercise, the length of the diagonal of a square is given, d = 10 cubits. The side of the square is computed as '42 30' times the diagonal. The constant has to be interpreted as

1/(sqs. 2) = 1/2 · sqs. 2 = appr. ;42 30.

The obtained result is that the side is 7;05 cubits. This value, in its turn, is multiplied by the constant '1 25' in order to retrieve the length of the diagonal. The constant '1 25' must therefore be interpreted as

sqs. 2 = appr. 1;25 with **sq. 1;25 = 2;00 25** (= 2 1/144).

An approximation like this may, of course, have been obtained by trial and error, but it is also possible that it was obtained through a combined application of the additive and subtractive square side rules. Indeed,

1) sqs. 2 = sqs. (sq.1 + 1) = appr. 1 + 1/2 = 1;30 (= 3/2), with sq.1;30 = 2;15
2) sqs. 2 = sqs. (sq.1;30 – ;15) = appr. 1;30 – ;15/3 = 1;30 – ;05 = 1;25 (= 17/12)

A famous example of an *Old Babylonian* text with an accurate square side approximation is **YBC 7289** (Neugebauer and Sachs, *MCT* (1945), 42), a round hand tablet showing a square with the side 30. Along the diagonal of the square are inscribed the numbers '1 24 51 10' and '42 25 35'. These numbers can be interpreted as

sqs. 2 = 1;24 51 10 with sq. 1;24 51 10 = 1;59 59 59 38 01 40

and

1/(sqs. 2) = 1/2 · sqs. 2 = appr. 1;24 51 10/2 = ;42 25 35.

The two approximations 1;25 and 1;24 51 10 to sqs. 2 are mentioned in two OB tables of constants (**BR = *TMS* 3, NSe = YBC 7243**), in the following way:

1 25	constant of the diagonal of a square	BR 31
1 24 51 10	the diagonal of an equalside	NSe 10

It is interesting that 1;24 51 10 is the same accurate approximation to sqs. 2 as the one used in Ptolemy's *Syntaxis* I.10! Indeed, it was shown in Sec. 16. 4 above that Ptolemy's approximation can be obtained as follows:

$(17, 12;1)^3 = (19601, 13860; 1)$ which gives the third approximation
$60 \cdot$ sqs, $2 = 60 \cdot 19601/13860 = 19601/231 =$ appr. 84;51 10.

The corresponding accurate approximation to sqs. 2 is

sqs. 2 = 84;51 10 / 60 = 1;24 51 10.

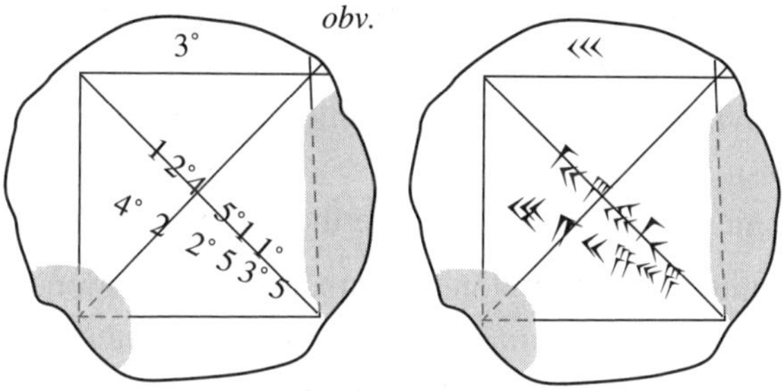

Fig. 16.7.2. YBC 7289. An OB hand tablet with an accurate approximation to sqs. 2.

Late and Old Babylonian approximations to sqs. 3

In the *Late Babylonian* mathematical text **W 23291 § 4 b** (see Sec. 7.7 above), a computation rule for equilateral triangles is stated as follows:

1 peg-head-field, equilateral, the one with the 8th torn out.
Stroke steps of ditto, and steps of *26 15* go.

This abstract rule is accompanied by a numerical example and a diagram showing a triangle with the sides given as '1' and the height as '52 30'.

What the rule means is that in an equilateral triangle with the side s, the height h and the area A can be computed as follows:

$h = s - 1/8 \cdot s = (1 - 1/8) \cdot s = ;52\ 30 \cdot s$ and
$A = h \cdot s/2 = ;26\ 15 \cdot$ sq. s.

Evidently, this rule is based on the following series of approximations:

sqs. 3 = 1;45 (= 7/4), 1/2 · sqs. 3 = ;52 30 (= 7/8), 1/4 · sqs. 3 = ;26 15 (= 7/16).

In the next exercise, **W 23291 § 4 c**, a *more accurate* computation rule for equilateral triangles is stated in the following form:

1 peg-head-field, equilateral, that with a 10th and a 30th torn off.
Stroke steps of ditto, and steps of 26 go.

Here, obviously,

$h = s - 1/10\ 1/30 \cdot s = (1 - 1/10\ 1/30) \cdot s = (1 - ;08) \cdot s = ;52 \cdot s$ and
$A = h \cdot s/2 = ;26 \cdot \text{sq. } s$.

The rule is based on the following series of *accurate approximations*:

sqs. 3 = 1;45 (= 26/15), 1/2 · sqs. 3 = ;52 (= 26/30), 1/4 · sqs. 3 = ;26 (= 26/60).

These are the same accurate approximations as the ones used in the pseudo-Heronic *Geometrica*, which can be obtained as *third approximations*, starting with 2/1! What is particularly surprising is that this is the only known example of the use in a Babylonian text of *a sum of parts* such as 1/10 1/30. What is even more surprising is that precisely the same rather weird expression for the height, $h = s - 1/10\ 1/30 \cdot s$, reappears in the pseudo-Heronic *Geometrica* 10:3! (See the end of Sec. 16.3 above.)

The continuity of the Babylonian mathematical tradition is demonstrated by the fact that the following constants for an equilateral triangle appear in an *OB* table of constants (**G = IM 52916**; Robson, *MMTC* (1999), 40):

A peg-head, the one with the eighth torn out, 26 15 its constant — G rev. 7
The transversal of the triangle, 52 30 its constant — G rev. 8

Thus here, just as in W 23291 § 4 b, the height of an equilateral triangle is given as $h = (1 - 1/8) \cdot s$, expressed with an almost identical phrase!

The same rule for the computation of the height of an equilateral triangle appears also in the *Kassite* (post-OB) mathematical text MS 3876, in which the weight is computed of an icosahedron built of 20 equilateral triangles (called 'gaming-pieces') made of 1 finger thick copper sheets. In that text, the area of each one of the 20 equilateral triangles is computed as follows, in preparation for the computation of its volume:

MS 3876 # 3 (Friberg, *RC* (2007), Sec. 11.3)

If 3 cubits each a gaming-piece is equalsided, the volume (is) what?
Half (of) 15, the front, break, then 7 30.
7 30 steps of 15, the second front, 1 52 30, halved. 14 03 45, its eighth tear off, then
1 38 26 15 (is) the ground (of) one gaming-piece-field that you see.

Here, the area of an equilateral triangle with the side 3 cubits = ;15 ninda is computed in the following steps:

1 $s/2 \cdot s$ = ;07 30 n. · ;15 n. = ;01 52 30 sq. n.
2 $1/8 \cdot s/2 \cdot s$ = 1/8 · ;01 52 30 = ;00 14 03 45 sq. n
3 $s/2 \cdot s - 1/8 \cdot s/2 \cdot s$ = ;01 52 30 sq. n – ;00 14 03 45 sq. n. = ;01 38 26 15 sq. n.

A Late Babylonian approximation to sqs. 5

W 23291§ 4 a in the same Late Babylonian text where §§ 4 b and 4 c are based on two different approximations to sqs. 3, is devoted to the computation of the area of a symmetric triangle with the base and the height both equal to '1' (actually 1 00 ninda). The exercise is illustrated by a diagram showing a symmetric triangle, with the length of each sloping side given as 1 07 ninda 2', that is as 1 07;30 ninda.This length can have been computed in the following way:

sqs. (sq. 1 00 + sq. 30) = 30 · sqs. (sq. 2 + sq. 1) = 30 · sqs. 5 = 30 · 2;15 = 1 07;30.

The approximation to sqs. 5 apparently used in this computation is

sqs. 5 = 2;15 (= 9/4),

a value resulting from an application of the additive square side rule.

Note that the Babylonian approximations to the square sides of 2, 3, and 5, namely 1;25 = 12/17, 1;45 = 7/4, and 2;15 = 9/4 were the points of departure for Ptolemy's accurate approximations $(17, 12; 1)^3$, $(7, 4; 1)^3$, and $(9, 4; 1)^3$, while the OB accurate approximation to sqs. 2 was the same as Ptolemy's accurate approximation. (See Sec. 16.4 above.)

Late and Old Babylonian exact computations of square sides

In the Seleucid (late Late Babylonian) series of igi-igi.bi problems **AO 6484 § 7 a-d** (see Sec. 1.13 above), the following square sides of many-place sexagesimal numbers are mentioned:

sqs. 33 20 04 37 46 40	= 44 43 20	§ 7 a
sqs. 3 02 15	= 13 30	§ 7 b
sqs. 5 34 04 37 46 40	= 18 16 40	§ 7 c
sqs. 15 00 56 15	= 3 52 30	§ 7 d

These square sides can have been computed as shown below (in relative values), by use of the OB additive/subtractive square side rule:

33 20 = appr. 33 45 = sq. 45, 33 20 = sq. 45 – 25 § 7 a
sqs. 33 20 = appr. 45 – 25/(2 · 45) = 45 – 5/18 = 45 – 16 40 = 44 43 20
sq. 44 43 20 = 33 20 04 37 46 40 (the exact answer)

5 34 = appr. sq. 18 = 5 24, 5 34 = sq. 18 + 10 § 7 c
sqs. 5 34 = appr. 18 + 10/(2 · 18) = 18 + 5/18 = 18 + 16 40 = 18 16 40
sq. 18 16 40 = 5 34 04 37 46 40 (the exact answer)

The two examples show how seemingly difficult computations of square sides of given many-place sexagesimal numbers *are actually simpler than expected when the given numbers are perfect squares.*

It is easy to find similar examples of computations of square sides of many-place sexagesimal numbers that are actually perfect squares in *OB* mathematical texts. One such example can be found in ***TMS* 20** (Bruins and Rutten (1961)). In *TMS* 20, a quadratic equation is set up for the area A, the length a, and the divider (transversal) d of a 'lyre-window' (*cf.* Fig. 6.2.6 above). The equation is

$A + a + d = B =$ 1 16 40.

Inserting the constants for the lyre-window, as in Fig. 6.2.6, one gets:

$c_A \cdot$ sq. $a + a + c_d \cdot a =$ 26 40 · sq. a + (1 + 1 20) · $a = B =$ 1 16 40.

In the solution procedure for this quadratic equation, the need arises to find the square side of

$c_A \cdot B$ + sq. $(1 + c_d)/2$ = 1 55 44 26 40.

The indicated square side 1 23 20 of 1 55 44 26 40 can have been computed in the following way, using sexagesimal relative place value notation:

1 55 44 26 40 = 20 · 20 · 17 21 40, 17 21 40 = sqs. 4 + 1 21 40
sqs. 17 21 40 = appr. 4(00) + 1 21 40/8 (00) = appr. 4 10, sq. 4 10 = 17 21 40
20 · 4 10 = 1 23 20 (the exact answer).

The example shows how the computation of square sides of given many-place sexagesimal numbers *is considerably simplified when it is possible to find by inspection square factors of the given number.*

The idea of first eliminating any obvious square factors from a given many-place sexagesimal number before attempting to find the square side is demonstrated *explicitly* by use of the example

sqs. 26 00 15 = 30 · sqs. 1 44 01 = 30 · 1 19 = 39 30.

in the OB exercise **IM 54472**. See Muroi, *HSc* 9 (1999). The idea was apparently routinely applied in OB mathematics. This is shown by 35 examples from 22 OB mathematical texts cited by Muroi (*op. cit.*).

The idea is demonstrated *explicitly* also by an example inscribed on the round hand tablet ***UET 6/2 222*** from Ur (Robson, *MMTC* (1999), 252; Friberg, *RA* 94 (2000), 108). The text on the reverse of *UET 6/2 222* is organized as a table with seven rows and three columns:

	1 03 45	
	1 03 45	
15	**1 07 44 03 45**	16
15	18 03 45	16
17	4 49	
	3 45	
	1 03 45	

In the first three rows of *UET 6/2 222 rev.* are inscribed two copies of the number n = 1 03 45, followed by their product, the square of 1 03 45:

sq. 1 03 45 = 1 07 44 03 45.

In the next four lines of the table, the square side of 1 07 44 03 45 is computed by use of a factorization algorithm based on the properties of sexagesimal numbers in Babylonian relative (floating) place value notation.

rev.

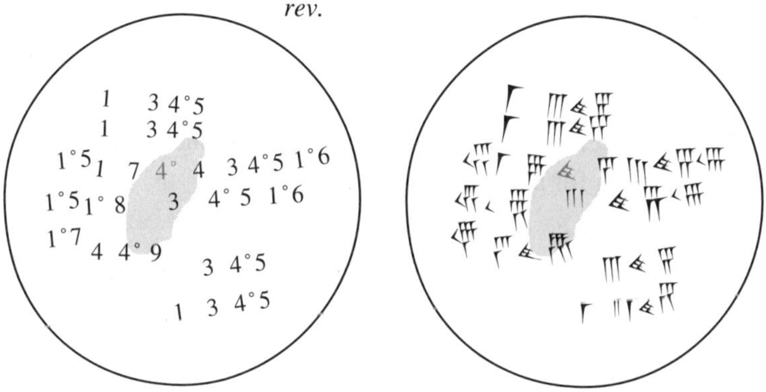

Fig. 16.7.3. *UET 6/2* 222 *rev*. A square side algorithm using elimination of square factors.

The first step of the algorithm exploits the fact that the "trailing part" of the given number 1 07 44 03 45 is 3 45 = sq. 15, which is the reciprocal of 16 = sq. 4, in the sense that 3 45 · 16 = 1 (times some power of 60). Now, if a sexagesimal number ends with 03 45, the whole number contains 3 45 as a factor. In the present case, for instance,

1 07 44 03 45 / 3 45 = (1 07 44 · 1 00 00 + 3 45) / 3 45 = 1 07 44 · 16 + 1
= 18 03 44 + 1 = 18 03 45.

Therefore, the meaning of line 3 of the table is that if 1 07 44 03 45 is multiplied by 16 (written to the right) then the factor 3 45 is eliminated, and simultaneously the factor 15 (written to the left) is eliminated from the square side of 1 07 44 03 45. The number 18 03 45 remaining after the elimination of the factor 3 45 is written in the middle of line 4 of the table. The process is repeated, and the result after the removal of another factor 3 45 is the number 4 49, which is written in the middle of line 5. Since the square side of 4 49 is 17, the number 17 is written to the left in line 5.

Now, finally, sqs. 1 07 44 03 45 is computed in lines 5-6 as follows:

$$15 \cdot 15 = 3\ 45, \quad 3\ 45 \cdot 17 = 1\ 03\ 45.$$

A beautiful example of the probable application of a factorization method of this kind is offered by the amazing OB problem text ***TMS* 19b** (Høyrup, *LWS* (2002), 194), where the sides u, s and the diagonal d of a rectangle are required to satisfy the following system of equations:

$$u \cdot s = A = 20, \quad \text{sq. } u \cdot u \cdot d = B = 14\ 48\ 53\ 20.$$

The solution procedure in the text for this highly unusual system of equations can be explained as follows:

$$\text{sq. } B = \text{sq. } (\text{sq. } u \cdot u \cdot d) = \text{sq. sq. } u \cdot \text{sq. } u \cdot \text{sq. } d$$
$$= \text{sq. sq. } u \cdot \text{sq. } u \cdot (\text{sq. } u + \text{sq. } s) = \text{sq. sq. } u \cdot (\text{sq. sq. } u + \text{sq. } (u \cdot s)).$$

Therefore, the given system of equations can be reduced to the following quadratic equation for a new unknown a:

$$\text{sq. } a + \text{sq. } A \cdot a = \text{sq. } B, \quad a = \text{sq. sq. } u.$$

This quadratic equation is solved in the usual way, beginning with[44]

$$\text{sq. } (a + 1/2 \cdot \text{sq. } A) = \text{sq. } B + \text{sq. } (1/2 \cdot \text{sq. } A) = 3\ 39\ 28\ 43\ 27\ 24\ 26\ 40 + 11\ 06\ 40$$
$$= 3\ 50\ 35\ 23\ 27\ 24\ 26\ 40.$$

The square side of the 8-place sexagesimal number 3 50 35 23 27 24 46 40 is given in the text as 15 11 06 40, without any indication of how this result was obtained. It is likely, however, that the square side was found without trouble through an application of the same factorization technique as the one in *UET 6/2* 222. Indeed, 15 11 06 40 = 20 · 20 · 20 · 10 · 41, so that

$$3\ 50\ 35\ 23\ 27\ 24\ 26\ 40 = \text{sq. } 20 \cdot \text{sq. } 20 \cdot \text{sq. } 20 \cdot \text{sq. } 10 \cdot \text{sq. } 41.$$

Another beautiful example of the application of the same factorization

44. The reader can find a discussion of the many interesting copying or calculation errors in this text in Høyrup (*op. cit.*), footnotes 222-235. The errors are corrected here.

technique is given by the round hand tablet **Ist. Si. 428** from Sippar (Friberg, *RlA* 7 (1990), Sec. 5.3 a). There the square side of 2 02 02 02 05 05 04 is computed by use of the factorization method, as follows:

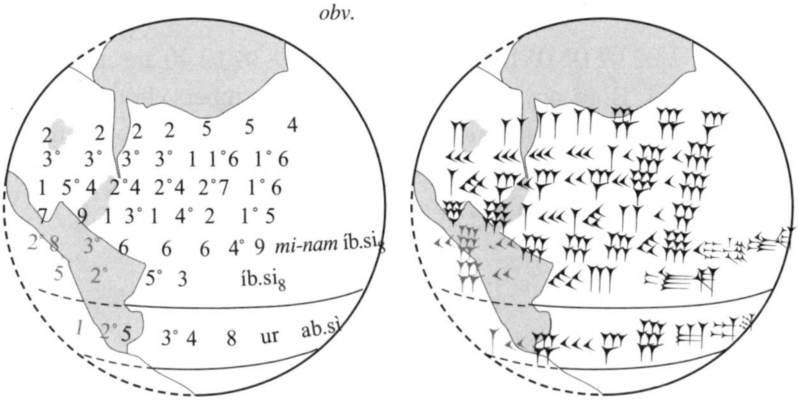

Fig. 16.7.4. Ist. Si. 428. Computation of the square side of 2 02 02 02 05 05 04.

1. 2 02 02 02 05 05 04 = 4 · 30 30 30 31 16 16
2. 30 30 30 31 16 16 = 16 · 1 54 24 24 27 16
3. 1 54 24 24 27 16 = 16 · 7 09 01 31 42 15
4. 7 09 01 31 42 15 = 15 · 28 36 06 06 49
5. sqs. 28 36 06 = sqs. (sq. 5 + 3 36 06) = appr. 5 + 21 = 5 21
6. sqs. 28 36 06 06 49 = sqs. (sq. 5 21 – 1 15 53 11) = appr. 5 21 – 7 = 5 20 53
7. sq. 5 20 53 = 28 36 06 06 49 (exactly)
8. sqs. 2 02 02 02 05 05 04 = 2 · 4 · 4 · 30 · 5 20 53 = 1 25 34 08 (exactly)

Steps 5-6 are reconstructed here, since there is no explicit indication in the text of how sqs. 28 36 06 06 49 was computed.

How did the author of this problem find out in the first place that the funny number 2 02 02 02 05 05 04 is a square number? He probably started with the even funnier 7-place sexagesimal number 2 02 02 02 02 02 02 and found by computation, in some way, that its 4-place square side is

sqs. 2 02 02 02 02 02 02 = 1 25 34 08 (approximately).

Therefore, the author of Ist.Si. 428 may have found 1 25 34 08 as an accurate *4-place approximation* to sqs. 2 02 02 02 02 02 02, and then used this prior knowledge in order to construct for his students a surprisingly elegant problem with a funny number and an *exact* answer!

It is interesting to note that the construction of this exercise may have been inspired by the observation that an accurate approximation to sqs. 2 is 1 24 51 10, where sq. 1;24 51 10 = 1;59 59 59 38 01 40. Therefore,

sqs. 1 59 59 59 38 01 40 = 1 24 51 10.

Obviously, 2 02 02 02 05 05 04 and 1 59 59 59 38 10 40 are numbers of the same kind. Both are 7-place sexagesimal numbers, both are exact squares, and both are "funny numbers" in the sense that one of them contains four sexagesimal places 02 in a row, while the other contains three sexagesimal places 59 in a row.

Chapter 17

Theodorus of Cyrene's Irrationality Proof and Descending Infinite Chains of Birectangles

17.1. *Theaetetus* 147 C-D. Theodorus' Metric Algebra Lesson

Below is reproduced a brief excerpt from Plato's well known account in the dialogue *Theaetetus* of a lesson given by Theodorus of Cyrene. *Cf.* Knorr's translation in *EEE* (1975), 62, which is followed by a detailed discussion of several difficult words in the text.

"*Theaet.*: Theodorus drew something for us about powers,
about the one of three feet and the one of five feet,
demonstrating that these are not commensurable in length with the foot,
and selecting in this way each one separately up to the one of seventeen feet.
But in this, in some way, he ran into trouble.
Now, this is what occurred to us:
Since we recognized the powers to be unlimited in number,
we might try to collect them under a single name,
by which we would designate all these powers."

Here 'power' is a term for (geometric) square, a power of 3, 5, or 17 feet is a square with the area 3, 5, or 17 square feet, and the statement that 'these (powers) are not commensurable in length with the foot' means that the sides of these squares are not commensurable with a length of 1 foot.

So, apparently, Theodorus gave a lesson in mathematics for Theaetetus and his fellow students, demonstrating by use of diagrams that the sides of squares with areas ranging from 3, and 5, all the way up to 17, square feet (avoiding, of course, the trivial cases, 4, 9, and 16 square feet) are not commensurable with the unit length of 1 foot. However, at 17 feet he could not continue, for some unspecified reason.

In the last part of the cited passage (and its continuation, which is omitted here), Theaetetus seems to claim that he and some others managed to do what Theodorus had been unable to do, namely to take care of the general case of squares of N square feet for *all* non-square positive integers N.

All this is relatively easy to understand. There is only one obscure point, although an extremely important one for any proposed interpretation of the meaning of the whole passage. Did Theodorus have trouble with the case of 17 square feet, or was he unable to continue *beyond* this case?

17.2. A Number-Theoretical Explanation of Theodorus' Method

Knorr (*EEE* (1975), Sec. 3.5) is definitely of the opinion that the text says that Theodorus *could not* handle the case of 17 square feet, and he has found a quite ingenious explanation of how Theodorus may have reasoned.[45] Knorr's basic idea is to let the square side of N be represented by the upright in a right triangle with the hypothenuse $(N + 1)/2$ and the base $(N - 1)/2$, if N is odd, and by half the upright in a right triangle with the hypothenuse $N + 1$ and the base $N - 1$, if N is even. Knorr also proves the following crucial lemma of his own design (*op. cit.*, 158):

> If in a right triangle triple of integers the hypothenuse is even, then also the upright and the base are even.

Four cases are considered by Knorr, *op. cit.*, Chapter 6. The first case is:

> $\boldsymbol{N = 4\,n + 3}$: Then $(N + 1)/2 = 2\,n + 2$ and $(N - 1)/2 = 2\,n + 1$. *Assume that* the square side of N (square units) is commensurable with the unit, the rational ratio between the two being, in its lowest terms, $a : b$. Then a is the upright of a right triangle with the hypothenuse $(2\,n + 2) \cdot b$ and the base $(2\,n + 1) \cdot b$. Here the hypothenuse is even, so that also the upright a and the base $(2\,n + 1) \cdot b$ are even. Hence both a and b are even, which is a contradiction, since the ratio $a : b$ is supposed to be in its lowest terms. This means that the assumption was false, and therefore the square side of N is *not* commensurable with the unit.

The remaining three cases are:

> $\boldsymbol{N = 8\,n + 5}$, $\boldsymbol{N = 4\,n + 2}$, and $\boldsymbol{N = 4\,n}$.

Also in these cases, the assumption that the square side and the unit length

45. In *EEE*, Chapter 4 Knorr has written a critical review of a series of earlier attempts to explain Theodorus' method. His conclusion is that they all fail to satisfy one or another of several basic criteria he has set up for a successful explanation of the method.

have a rational ratio leads to a contradiction.

In this way, it can be proved that the square side is irrational when

$N = 3, 7, 11, 15, 19, \cdots$	$(\boldsymbol{N = 4\,n + 3})$
$N = 5, 13, 21, \cdots$	$(\boldsymbol{N = 8\,n + 5})$
$N = 6, 10, 14, 18, \cdots$	$(\boldsymbol{N = 4\,n + 2})$
$N = 8, 12, 20, \cdots$	$(\boldsymbol{N = 4\,n})$

This takes care of all non-square N strictly between 2 and 17. However, *in the case N = 17, the method proposed by Knorr does not work.* This result seems to confirm Knorr's hypothesis that the method proposed by him is also the method used by Theodorus and that Theodorus stopped at $N = 17$, because he could not find a proof in that particular case.

17.3. An Anthyphairetic Explanation of Theodorus' Method

Another way of explaining Theodorus' method to prove the irrationality of square sides is discussed by Knorr in *EEE* (1975), Sec. 4.3, following in the steps of Zeuthen, *OKDVSF* (1910), Heath, *HGM 1* (1981), 206-8, and Heller, *Centaurus* 5 (1956). Here follows an account of Knorr's explanation, somewhat simplified and improved.

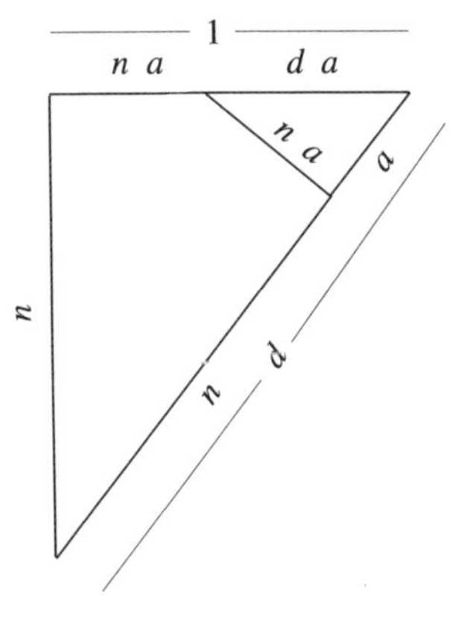

sq. $n + 1$ = sq. $d = D$
(for instance $D = 2, n = 1$ or $D = 5,\ n = 2$)

$d = n + a, \quad 1 = n\,a + d\,a$

$\Rightarrow$

$d - n \cdot 1 = a, \quad 1 - n \cdot a = d\,a$

$\Rightarrow$

$1/a = n + d, \quad d = n + 1/(n + d)$

$\Rightarrow$

$d = [n, \overline{2\,n}]$

Fig. 17.3.1. An anthyphairesis proof of the irrationality of sqs. (sq. n + 1).

Consider first the case when d = sqs. (sq. n + 1) for a positive integer n. Draw a right triangle with the sides d, n, 1, and divide the right triangle into a *symmetric birectangle* with the sides n, n, $n\ a$, $n\ a$ and a right triangle with the sides $(d, n, 1) \cdot a$, as in Fig. 17.3.1 above. Apply the Euclidean division algorithm (*anthyphairesis*) to the pair d, 1. As is evident from the diagram in Fig. 17.3.1, the first couple of steps of the algorithm are

$$d - n \cdot 1 = a, \quad 1 - n \cdot a = d\, a.$$

The next couple of steps are, similarly,

$$d \cdot a - n \cdot a = \text{sq. } a, \quad n - n \cdot \text{sq. } a = d \text{ sq. } a.$$

And so on, forever. Therefore, d and 1 are *incommensurable* (*El.* X.2).

The argument is, of course, closely connected to the expansion of d into a *continued fraction*. This can be shown as follows:

$$d = n \cdot 1 + a, \quad 1 = n \cdot a + d\, a \quad \Rightarrow \quad d = n + a, \quad 1/a = n + d \quad \Rightarrow \quad d = n + 1/(n + d).$$

(The last equation above can also be proved algebraically, since it is a consequence of an application of the conjugate rule to the left hand side of the equation sq. d – sq. n = 1.) Now by substitution of the expression for d into the expression itself it follows that

$$d = n + 1/(n + d) = n + 1/(2\, n + 1/(n + d)) = \cdots .$$

The substitution can be repeated any number of times. Therefore,

$$d = n + 1/(2\, n + 1/(2\, n + 1/(2\, n + \cdots))) = [n, \overline{2\, n}].$$

The case when d = sqs. (sq. n – 1) for a positive integer n can be explained similarly, although it is somewhat more complicated:

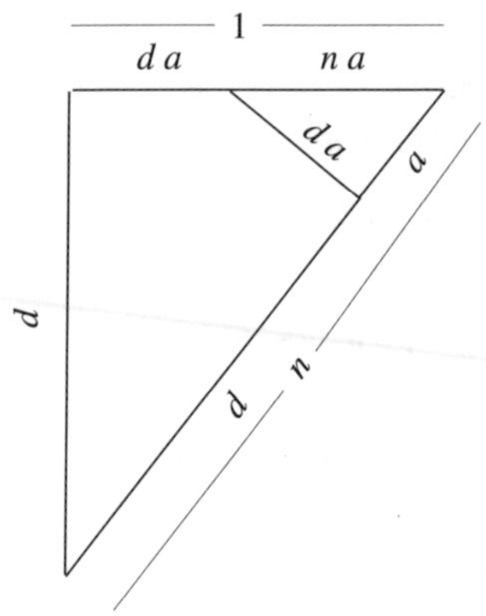

sq. $n - 1$ = sq. $d = D$

(for instance $D = 3$ and $n = 2$)

$n = d + a, \quad 1 = n\, a + d\, a \quad \Rightarrow$

$d = (n - 1) \cdot 1 + (1 - a), \quad 1 = 1 \cdot (1 - a) + a,$

$(1 - a) = (n - 1) \cdot a + d\, a \quad \Rightarrow$

$1/(1 - a) = 1 + a/(1 - a), \quad (1 - a)/a = (n - 1) + d,$

$d = (n - 1) + 1/(1 + a/(1 - a))$

$\quad = (n - 1) + 1/(1 + 1/((n - 1) + d))$

$\Rightarrow \quad d = [(n - 1), \overline{1, 2\,(n - 1)}]$

Fig. 17.3.2. An anthyphairesis proof of the irrationality of sqs. (sq. n – 1).

Draw a right triangle with the sides n, d, 1, and divide the right triangle into a *symmetric birectangle* with the sides d, d, $d\,a$, $d\,a$ and a right triangle with the sides (n, d, 1) · a, as in Fig. 17.3.2 above. Apply the Euclidean division algorithm (*anthyphairesis*) to the pair d, 1. As is evident from the diagram in Fig. 17.3.2,

$$n = d + a \quad \text{and} \quad 1 = n\,a + d\,a.$$

Therefore, the first three steps of the algorithm are

$$d = (n-1)\cdot 1 + (1-a), \quad 1 = 1\cdot(1-a) + a, \quad (1-a) = (n-1)\cdot a + d\,a.$$

(The middle step is trivially true.) The next three steps are similar. And so on, forever. Therefore, d and 1 are *incommensurable* (*El.* X.2).

The argument is, in this case, too, closely connected to the expansion of d into a *continued fraction*, which can be shown as follows:

$$d = (n-1)\cdot 1 + (1-a), \quad 1 = 1\cdot(1-a) + a, \quad (1-a) = (n-1)\cdot a + d\,a$$
$$\Rightarrow \; d = (n-1) + (1-a), \quad 1/(1-a) = 1 + a/(1-a), \quad (1-a)/a = (n-1) + d$$
$$\Rightarrow \; d = (n-1) + 1/(1 + a/(1-a)) = (n-1) + 1/(1 + 1/((n-1) + d)).$$

Now by substitution of the expression for d into the expression itself it follows that

$$d = (n-1) + 1/(1 + 1/(2\,(n-1) + 1/(1 + 1/((n-1) + d)))).$$

The substitution can be repeated any number of times. Therefore,

$$d = (n-1) + 1/(1 + 1/(2\,(n-1) + 1/(1 + 1/((2\,(n-1) + \cdots))))) = [(n-1), \overline{1, 2\,(n-1)}].$$

This anthyphairetic explanation of Theodorus' method is completed as follows (Knorr, *op. cit.*, 124):

> "For any non-square integer M, integers p and q are to be found such that $p^2\,M = q^2 \pm 1$ (a relation frequently known as the Fermat-, or Pell-, equation). Having specified p and q for a given M, the irrationality of $\sqrt{q^2 \pm 1}$ (whichever pertains) has been established from the previous constructions. Hence, $\sqrt{p^2\,M}$ is irrational; as this equals $p\sqrt{M}$, the irrationality of M follows."

17.4. A Metric Algebra Explanation of Theodorus' Method

Ingenious as it is, Knorr's number-theoretical/geometrical explanation (Sec. 17.2 above) of the method used by Theodorus for his incommensurability proof is not quite satisfactory, for the following reasons. First, a method of this kind would be *an isolated phenomenon* in the corpus of Greek mathematics. There are no known parallel or related methods. Secondly, the mention of *feet* as units for area and length places Theodorus' method squarely in the same tradition as the pseudo-Heronic *Geometrica* (ms SV) and *Stereometrica* (see Sec. 18.1 below). Since both *Geometrica* and *Stereometrica* show clear signs of having been influenced by the Babylonian mathematical tradition, it is likely that also Theodorus' irrationality proof was influenced by that tradition.

Indeed, Plato's account in *Theaetetus* 147 C-D may have been intended to be a demonstration of how Greek-style mathematics, exemplified by Theaetetus' general proof of the irrationality of square sides (of non-squares), was superior to Babylonian-style metric algebra, exemplified by Theodorus' attempted proof, which ran into difficulties.

It will be shown below how an alternative explanation of the method used by Theodorus may be phrased in terms of *a descending infinite chains of birectangles*, which means that Theodorus' method of proof may have been closely related to Theon of Smyrna's side and diagonal numbers algorithm, in the form in which it was explained in Secs. 15.3-4 above.

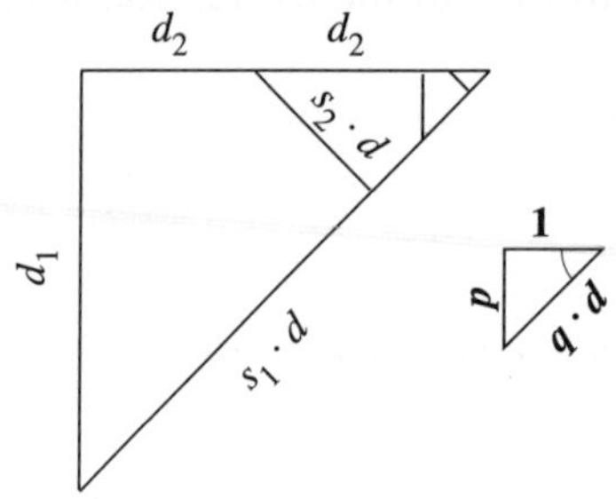

$$\text{sq. } p = \text{sq. } q \cdot D - 1, \quad D = \text{sq. } d$$

$$d_{n+1} = q \cdot s_n \cdot D - p \cdot d_n$$
$$s_{n+1} = q \cdot d_n - p \cdot s_n$$

$$\text{sq. } s_1 \cdot D - \text{sq. } d_1 = 0$$
$$\Rightarrow$$
$$\text{sq. } d_n - \text{sq. } s_n \cdot D = 0 \quad \text{for} \quad n = 1, 2, \cdots$$

Fig. 17.4.1. A metric algebra proof of the irrationality of sqs. D, when sq. p = sq. $q \cdot D - 1$.

Consider the case when sq. p = sq. $q \cdot D - 1$, D = sq. d, for some given pair of positive integers p, q, and *assume that the pair d, 1 is commensurable*. Then there exists another pair of integers d_1, s_1 such that

$$d_1 = s_1 \cdot d.$$

Now, construct a birectangle as in Fig. 15.3.1 above. Let the longer sides of the birectangle be d_1, $s_1 \cdot d$, and let the sides of a *descending* chain of birectangles be determined by the recursive equations

$$d_{n+1} = q \cdot s_n \cdot D - p \cdot d_n, \quad s_{n+1} = q \cdot d_n - p \cdot s_n \quad \text{for} \quad n = 1, 2, \cdots.$$

See Fig. 17.4.1 above. Then

$$\text{sq. } d_1 - \text{sq. } s_1 \cdot D = 0 \;\Rightarrow\; \text{sq. } d_2 - \text{sq. } s_2 \cdot D = 0 \;\Rightarrow\; \text{sq. } d_3 - \text{sq. } s_3 \cdot D = 0 \;\Rightarrow\; \cdots$$

Consequently,

$$d_n = s_n \cdot d \quad \text{for} \quad n = 1, 2, \cdots.$$

This means that all the birectangles in the chain are *symmetric birectangles*, similar to the first birectangle in the chain. Therefore, the chain goes on forever, and the sides d_n of the birectangles form *an infinite strictly decreasing sequence of positive integers.* Since this is impossible, the initial assumption was incorrect. Therefore *the pair d, 1 is not commensurable.*

A similar argument holds also in the case when sq. p = sq. $q \cdot D + 1$, D = sq. d, for some given pair of positive integers c, a.

In the discussion below, it will be convenient to say that

p/q is an "optimal approximation" to sqs. D when sq. p = sq. $q \cdot D \pm 1$.

The result obtained above can then be expressed in the following way:

sqs. D (D non-square) is irrational if there exists an *optimal* approximation p/q to sqs. D.

Since, for instance,

2 = sq. $1 + 1$, 5 = sq. $2 + 1$, 10 = sq. $3 + 1$, 17 = sq. $4 + 1$,

and

3 = sq. $2 - 1$, 8 = sq. $3 - 1$, 15 = sq. $4 - 1$,

it follows immediately that

the square sides of 2, 5, 10, 17 and 3, 8, 15 are irrational,

and that the corresponding optimal approximations are 1/1, 2/1, 3/1, 4/1 and 2/1, 3/1, 4/1, respectively.

It is also clear that

the square sides of 12 and 18 are irrational,

since sqs. 12 = 2 · sqs 3 and sqs 18 = 3 · sqs. 2.

Another group of cases can be taken care of as follows: Take, for instance, $D = 6$. It is clear that sq. 2 = sq. 1 · 6 – 2. Now, in view of the discussion in Sec. 16.5 above,

$$(2, 1; -2)^2 = (4 + 6, 2 \cdot 2; 4) = (10, 4; 4) = (5, 2; 1) \quad \text{when} \quad D = 6.$$

Hence, sq. 5 = sq. 2 · 6 + 1, and 5/2 is an optimal approximation to sqs. 6.

Generally, when sq. p = sq. $q \cdot D \pm 2$, so that therefore sq. p + sq. $q \cdot D$ = 2 sq. $q \cdot D \pm 2$ is even, a similar argument shows that

$$(p, q; \pm 2)^2 = (\text{sq. } p + \text{sq. } q \cdot D, 2\, p \cdot q; 4) = (\text{sq. } q \cdot D \pm 1, p \cdot q; 1).$$

For instance, since sq. 3 = sq. 1 · 7 + 2, it follows that

$$(3, 1: 2)^2 = (9 + 7, 2 \cdot 3; 4) = (16, 6; 4) = (8, 3; 1) \quad \text{when} \quad D = 7.$$

Hence, sq. 8 = sq. 3 · 7 + 1 and 8/3 is an optimal approximation to sqs. 7.

Therefore, it is convenient to say that

p/q is a "pre-optimal approximation" to sqs. D, when sq. p = sq. $q \cdot D \pm 2$.

Then,

sqs. D is irrational if there exists a *pre-optimal* approximation p/q to sqs. D.

In particular,

the square sides of 6, 7, 11, 13, 14, 18 are irrational.

Now, consider instead the case $D = 13$, with sq. 3 = 13 – 4. Then,

$(3, 1; -4)^2 = (9 + 13, 2 \cdot 3; 16) = (22, 6; 16) = (11, 3; 4)$ and

$(3, 1; -4)^3 = (33 + 3 \cdot 13, 9 + 11; -16) = (72, 20; -16) = (36, 10; -4) = (18, 5; -1)$.

Therefore *a couple of formal multiplications* is enough to show that

sq. 3 = 13 – 4 ⇒ sq. 11 = sq. 3 · 13 – 4 ⇒ sq. 18 = sq. 5 · 13 – 1.

This means that in this case the "third approximation" is optimal.

Although it is not necessary, it can be shown also that

$(3, 1; -4)^6 = (18, 5; -1)^2 = (324 + 25 \cdot 13, 2 \cdot 18 \cdot 5; 1) = (649, 180; 1)$.

Compare the example above and Brahmagupta's ***Bss* XVIII.69** (Colebrooke, *AAMS* (1973), 365) is a rule saying, without a proof, that

if sq. p = sq. $q \cdot D + 4$, then sq. p^* = sq. $q^* \cdot D + 1$,

where

p^* = (sq. p – 3)/2 · p and q^* = (sq. p – 1)/2 · q.

The rule is applied in ***Bss* XVIII.70** to the case when $p = 4$, $q = 2$, $D = 3$, in which case $p^* = 26$, $q^* = 15$, and sq. 26 – sq. 15 · 3 = 1.[46] (Cf. the explanation in Sec. 16.3 above of Hofmann's example 41: sqs. 3 = 26/15.)

Actually, the proof of the rule is easy, since,

$(p^*, q^*; 1) = (p, q; 4)^3$.

A similar rule in ***Bss* XVIII.71** says, without proof, that

if sq. p – sq. $q \cdot D = -4$, then sq. p^* – sq. $q^* \cdot D = 1$,

where

p^* = [(sq. p + 3) · (sq. p + 1)/2 – 1] · (sq. p + 2) and
q^* = (sq. p + 3) · (sq. p + 1)/2 · p · p.

---

46. For a brief but informative account of Brahmagupta's treatment of what he called the "square-nature" equation sq. p = sq. $q \cdot D + r$, see Weil, *NT* (1984), 19-22.

The rule is applied in ***Bss* XVIII.72** to the case when $p = 3$, $q = 1$, $D = 13$, in which case $p^* = 649$, $q^* = 180$, and sq. 649 – sq. 180 · 13 = 1.

The rule can be proved by observing that, in this case,

$(p^*, q^*; 1) = (p, q; -4)^6$.

In view of these rules, it is convenient to say that

p/q is a "pre-pre-optimal approximation" to D when sq. p = sq. $q \cdot D \pm 4$.

If Theodorus used the metric algebra method outlined in this section (or the related anthyphairetic/continued fraction method outlined in Sec. 17.3) he apparently omitted the case $D = 2$, because it was already well known, and he explained in some detail the two model cases $D = 3$ and $D = 5$ ("the one of three feet and the one of five feet"), where sq. 2 = sq. 1 · 3 + 1, while sq. 2 = sq. 1 · 5 – 1. Having done that, all that remained for him to do was to show that he could find optimal approximations to the square sides of *all* squares of D (square) feet, where D is any non-square positive integer. He did not manage to do that, but it is possible that he showed how far he had come by writing out for his students a tabular survey of (essentially) the following kind:

D	**equation**	**(*p*, *q*; *r*)**	**appr. value of *d***	**type of appr.**
2	sq. 1 = 2 – 1	(1, 1; – 1)	1	optimal
3	sq. 2 = 3 +1	(2, 1; 1)	2	optimal
5	sq. 2 = 5 – 1	(2, 1; – 1)	2	optimal
6	sq. 2 = 6 – 2	(2, 1; – 2)	2	pre-optimal
	sq. 5 = sq. 2 · 6 + 1	$(2, 1; -2)^2 = (5, 2; 1)$	5/2 = 2 1/2	optimal
7	sq. 3 = 7 + 2	(3, 1; 2)	3	pre-optimal
	sq. 8 = sq. 3 · 7 + 1	$(3, 1; 2)^2 = (8, 3; 1)$	8/3 = 2 2/3	optimal
8	sqs. 8 = 2 · sqs. 2			—
10	sq. 3 = 10 – 1	(3, 1; – 1)	3	optimal
11	sq. 3 = 11 – 2	(3, 1; – 2)	3	pre-optimal
	sq. 10 = sq. 3 · 11 + 1	$(3, 1; -2)^2 = (10, 3; 1)$	10/3 = 3 1/3	optimal
12	sqs. 12 = 2 · sqs. 3			—
13	sq. 3 = 13 – 4	(3, 1; – 4)	3	pre-pre-optimal
	sq. 11 = sq. 3 · 13 + 4	$(3, 1; -4)^2 = (11, 3; 4)$	11/3 = 3 2/3	pre-pre-optimal
	sq. 18 = sq. 5 · 13 – 1	$(3, 1; -4)^3 = (18, 5; -1)$	18/5 = 3 3/5	optimal
14	sq. 4 = 14 + 2	(4, 1; 2)	4	pre-optimal
	sq. 15 = sq. 4 · 14 + 1	$(4, 1; 2)^2 = (15, 4; 1)$	15/4 = 3 3/4	optimal
15	sq. 4 = 15 + 1	(4, 1; 1)	4	optimal
17	sq. 4 = 17 – 1	(4, 1; – 1)	4	optimal
18	sqs. 18 = 3 · sqs. 2			—

Theodorus could not continue past $D = 17$ (not counting the trivial case $D = 18$) for the reason that *he could not find an optimal, pre-optimal, or pre-pre-optimal approximation to the square side of 19.* Note that the best approximations by integers to the square side of 19 are 4 and 5, where

sq. 4 = 19 – **3** and sq. 5 = 19 + **6**.

It is interesting to note that in the applications of Heron's improved square side rule by use of "third approximations" in examples 41-45 (see Sec. 16.3 above) the improved square side approximations are based on the *optimal* first approximations 2/1 for sqs. 3, 4/1 for sqs 15, 5/2 for sqs. 6, and 8/3 for sqs. 7, the same optimal approximations as in the survey above!

Remark 1. A way of solving the equation sq. p = sq. $q \cdot D + 1$ in the general case was presented by the Indian mathematician Jayadeva in the eleventh century, however with no indication of a strict proof.[47] (A totally different solution method, with a complete proof, was published much later (in 1767) by Lagrange.) Applying Jayadeva's "cyclic process" one can show that when $D = 19$ the equation has the solution $p, q = 170, 39$. These values are clearly so large that Theodorus must have given up long before he could find them through trial and error.

Remark 2. In the tabular survey above (and elsewhere in this chapter) the use of what looks like common fractions has been allowed only for the sake of convenience. It is unlikely that Theodorus would have written, for instance, the optimal approximation to sqs.13 in the form 18/5 = 3 3/5 as here in the survey. (The corresponding *sum of parts* of the standard Greek/Egyptian kind is 3 1/2 1/10.) For this reason, the proper way to understand 18/5, for instance, is not as a fraction but as '18 divided by 5'. (The earliest documented use of common fractions, in a very explicit form, is in the early Roman Greek-Egyptian mathematical papyrus ***P.BM 10520* § 5** (Friberg, *UL* (2005), Sec. 3.3 e). An even earlier example of the use of (a kind of) common fractions is in the Egyptian demotic mathematical papyrus (Ptolemaic, 3rd c. BCE) ***P.Cairo* § 1** (Friberg, *op. cit.*, Sec. 3.1 c).

47. See Weil, *NT* (1984), 22-24. See also Srinivasiengar, *HAIM* (1967), Chapter 10, or Datta and Singh, *HHM* (1962 (1938)), Chs. 16-17.

Chapter 18

The Pseudo-Heronic *Geometrica*

18.1. *Geometrica*, a Compilation of Various Sources

Høyrup's informative paper in *ANwR 7* (1997) contains a detailed comparison of the contents of the Heronic ***Metrica*** and the "pseudo-Heronic" ***Geometrica*** with related issues in the Near Eastern "practical tradition" and in various Arabic or Western medieval mathematical problem collections. He begins the comparison with the following summary of what Heiberg himself wrote in Latin in the preface of *HAOO 5* (1914) about the sources he had used for his compilated version of *Geometrica*:

"(1) *Geometrica* 'was not made by Hero, nor can a Heronian work be reconstructed by removing a larger or smaller number of interpolations' (p. xxi).

(2) Mss AC represent a book which, with additions, changes and omissions, only reached the present shape in Byzantine times; it was not meant to serve field mensuration directly but was for use 'in [commercial and legal] life' and in general education (p. xxi).

(3) Manuscript S, with the closely related ms V, was intended to serve youth studying 'architecture, mechanics and field mensuration' in the 'University of Constantinople' and thus 'more familiar with theoretical mathematics' – a use which in Heiberg's view agrees with the presence of Heron's (more or less) genuine *Metrica* in the same manuscript (p. xxiii).

(4) Both versions of the work merge (in their own ways) 'various problem collections together with Heronian and Euclidean excerpts' (p. xxiv). · · · As things are, only a very careful observation or reading of the Latin preface to *volume V* of the *Opera omnia* will reveal that a work contained in volume IV is a modern conglomerate of two (indeed more) ancient conglomerates. · · · He (Heiberg) also seems to

have known, but does not say it too directly, that the origin of the 'problem collections' was neither Heronian nor Euclidean."

The following brief comparison of *Geometrica* with *Metrica* in Heath, *HGM 2* (1981 (1921)), 318, is also interesting:

"The mensuration in the *Geometry* has reference almost entirely to the same figures as those measured in Book I of the *Metrica*, the difference being that in the *Geometry*
(1) the rules are not explained but merely applied to examples,
(2) a large number of numerical illustrations are given for each figure,
(3) the Egyptian way of writing fractions as the sum of submultiples is followed,
(4) lengths and areas are given in terms of particular measures, and the calculations are lengthened by a considerable amount of conversion from one measure into another."

To these important remarks about the nature of *Geometrica* can be added here the following observations:

(1) In *Geometrica* there are not, as in *Metrica*, any lettered diagrams. Nor are there any discussions of chains of "givens", in the style of Euclid's *Data*.
(2) In *Metrica,* lengths are measured in 'units' or abstract numbers. The "particular measures" used in *Geometrica*, are in mss AC 'ropes' (*schoinía*) and 'fathoms' (*órgyiai*), but in mss SV 'feet' (as also in *Stereometrica* (Heiberg, *HAOO 5* (1914)).
(3) Høyrup chose to neglect the testimony of Greek-Egyptian mathematical papyrus fragments from the Ptolemaic and Roman periods in Egypt (see Friberg, *UL* (2005), Chapter 4) in his otherwise quite extensive comparison of geometric themes and terminology in *Metrica*, *Geometrica*, and medieval mathematical problem collections.

Høyrup (*op cit.*) ends as follows his scrutiny of, among other things, the terminology used in *Geometrica*, mss AC, mss SV, and the independent source ms S, Chapter 24:

"·If we think of Moritz Cantor's old metaphor, according to which the development of mathematics is to be likened to a river landscape, the river that had sprung from Near Eastern geometrical practice had dissolved itself in later antiquity into a delta, in a multitude of independent streams now running together, now splitting apart. Hero knew some of them and used them – at times literally – in *Metrica*; *Geometrica*/AC collected others, *Geometrica*/S and S:24 still others. Further studies of terminology and style may help us sort out more details; given the complexity of the situation and the paucity of sources for precisely the practitioners' level of mathematical activity, however, we are not very likely to get very far."

18.2. *Geometrica* mss AC

Near the beginning of ms A, after a long metrological introduction, there is a brief note mentioning that

> One must also know that a *módios* of seed is 40 *lítra*,
> and each *lítra* seeds 5 fathoms of land.

What this means is that here, just as in Late Babylonian field plans and mathematical texts (see Friberg, *BaM* 28 (1997), Chs. 1-3), the size of a field is measured in "seed measure" rather than area measure. The *fixed conversion rate* is

> 1 *lítra* in seed measure = 5 (square) fathoms in area measure, or
> 1 *módios* in seed measure (= 40 *lítra*) = 200 (square) fathoms = 2 (square) ropes.

After the mentioned brief note follow two *metrological conversion tables*. No similar Late Babylonian tables are known (*cf.* Friberg, *GMS* 3 (1993)).

For width and length (meaning area measure) of	5 fathoms make	1 *lítra*
width and length of	10 fathoms make	2 *lítra*
width and length of	15 fathoms make	3 *lítra*
....	....	....
width and length of	100 fathoms make	20 *lítra*
width and length of	200 fathoms make	40 *lítra*
....	....	....
width and length of	1000 fathoms make	200 *lítra*
width and length of	2000 fathoms make	400 *lítra*
....	....	....
width and length of	10000 fathoms make	2000 *lítra*
200 fathoms	are land for	1 *módios*
300 fathoms	are land for	1 1/2 *módios*
....	....	....
1000 fathoms	are land for	5 *módios*
2000 fathoms	are land for	10 *módios*
....	....	....
10000 fathoms	are land for	50 *módios*

The metrological tables are followed by what is, essentially, a *hand book in mensuration* of the kind that one finds in medieval Arabic and Western texts. (See Høyrup (*op cit.*), 74-78.) It begins with exercises for *squares, rectangles, trapezoids, and right-angled triangles*. In *Geom.* 7.1-3/AC, for instance, the base and upright of a right-angled triangle are 4 ropes = 40 fathoms and 3 ropes = 30 fathoms, respectively. The size of the triangle is then computed first as 6 (square) ropes, with the corresponding seed measure equal to $1/2 \cdot 6 = 3$ *módios*, and then as 600 (square) fathoms, with the corresponding seed measure $1/200 \cdot 600 = 3$ *módios*.

Then follows what must be an *interpolation* in mss AC, because only *abstract* numbers are used for lengths and surfaces. In ***Geom.* 8.1**, it is shown how to construct a right-angled triangle by "the method of Pythagoras", when an *odd* number is given. (*Cf.* the discussion in Sec. 3.1 above.) The method is demonstrated only by an example, in the case when the upright 5 of the triangle is given. Then the base is $(5 \cdot 5 - 1)/2 = 12$, and the hypothenuse is the base + 1 = 13. In modern notations:

$c, b, a = (\text{sq. } a + 1)/2, (\text{sq. } a - 1)/2, a$ for any odd number a.

In ***Geom.* 9.1** it is then shown by an example how to construct a right-angled triangle by the "method of Plato" when an *even* number is given. Let the upright be 8. Then the base is $8/2 \cdot 8/2 - 1 = 15$, and the hypothenuse is the base + 2 = 17. In modern notations:

$c, b, a = \text{sq. } a/2 + 1, \text{sq. } a/2 - 1, a$ for any even number a.

The derivations of these rules by use of (metric) algebra is obvious. In the first case (the "method of Pythagoras"), c and b are obtained as the solutions to the quadratic-linear system of equations

$\text{sq. } c - \text{sq. } b = \text{sq. } a, \quad c - b = 1$ where a is a given odd number.

In the second case (the "method of Plato"), c and b are obtained as the solutions to the quadratic-linear system of equations

$\text{sq. } c - \text{sq. } b = \text{sq. } a, \quad c - b = 2$ where a is a given even number.

After this brief interruption, the hand book continues with exercises for *equilateral triangles*. In ***Geom* 10.1-5**, the following rules are stated:

$A = 1/3\ 1/10 \cdot \text{sq. } s, \quad h = s - 1/10\ 1/30 \cdot s \qquad (1/3\ 1/10 = 26/60, \quad 1 - 1/10\ 1/30 = 52/60).$

(*Cf.* Sec. 16.3 above.) If, for instance, $s = 10$ ropes, then

$A = (33\ 1/3 + 10)$ (square) ropes = 43 1/3 (square) ropes = 21 *módios* 26 2/3 *lítra*.

In ***Geom.* 11** the topic is *isosceles (symmetric) triangles*, and in ***Geom.* 12** *scalene (non-symmetric) triangles*. In ***Geom.*12.1-14**, for instance, the example is the triangle with the sides 13, 14, and 15 ropes. The rule used for the computation of the segment q of the base can be explained as follows, with the same notations as in Fig. 1.8.1 above, right:

$q = \{(\text{sq. } b + \text{sq. } a - \text{sq. } c)/2\}/b.$

This means, apparently, that the computation of the segment q in *Geom.*12.14 is based on Euclid's *El.* II.13. The examples in *Geom.* 12.15, 12.23, 12.28, and 12.29 are similar.

In ***Geom.* 12.30**, the area of the triangle with the sides 13, 14, 15 ropes is computed by use of *Heron's triangle area rule* (*cf.* Sec.14.1).

In ***Geom.* 12.33**, an *obtuse-angled* triangle with the sides 17, 9, 10 ropes is considered. The rule for the computation of the extension q of the base is the following, with the same notations as in Fig. 1.8.1 above, left:

$q = \{(\text{sq. } c - \text{sq. } b - \text{sq. } a)/2\}/b$ $\qquad$ $\{(289 - 81 - 100)/2\}/9 = 54/9 = 6$ (ropes).

Apparently, therefore, the computation of the segment q in *Geom.*12.33 is based on Euclid's *El.* II.12.

Another rule for the computation of the extension q of the base of the same triangle is used in ***Geom.* 12.38**. There q is computed as follows:

$q = \{(\text{sq. } c - \text{sq. } a)/b + b)/2 - b$ $\qquad$ $\{(289 - 100)/9 + 9\}/2 - 9 = 15 - 9 = 6$ (ropes).

This alternative rule can be found by use of *metric algebra*, by looking for solutions to the quadratic-linear system of equations (*cf.* Fig. 1.8.1, left):

$\text{sq. } p - \text{sq. } q = \text{sq. } c - \text{sq. } a, \quad p - q = b.$

It is interesting to compare the rules for scalene triangles in *Geom.* 12.1, 12.30 and 12.38 with the rules used in the Greek-Egyptian mathematical papyrus fragments *P.Chicago litt. 3* and *P.Cornell* 69 (Friberg, *UL* (2005) Secs. 4.7 b-c). In the exercise ***P.Chicago litt. 3* # 2**, a non-symmetric trapezoid is reduced, by removal of a central rectangle, to a non-symmetric triangle with the sides 13, 14, 15 ropes. The segment q of the base (*op. cit.*, Fig. 4.7.1) is computed as follows:

$q = \{b - (\text{sq. } c - \text{sq. } a)/b)/2 - b$ $\qquad$ $\{14 - (225 - 169)/14\}/2 = 5$ (ropes).

In ***P.Chicago litt. 3* # 3**, a non-symmetric trapezoid is reduced, by removal of a central parallelogram, to a non-symmetric obtuse-angled triangle with the sides 15, 4, 13 ropes. Then the extension q of the base is found as:

$q = \{(\text{sq. } c - \text{sq. } a)/b - b)/2 - b$ $\qquad$ $\{(225 - 169)/4 - 4\}/2 = 5$ (ropes).

***P.Cornell 69* # 2** is a similar exercise, where a given non-symmetric trapezoid is reduced to the same triangle with the sides 15, 4, 3 *báion* (1/16 rope). Here, the sum p of the base and the extension of the base is found as

$p = \{(\text{sq. } c - \text{sq. } a)/b + b)/2$ $\qquad$ $\{(225 - 169)/4 + 4\}/2 = 9$ (ropes).

It is clear that the computation rules in *Geom.* 12.38, *P.Chicago litt. 3* ## 2-3 and *P.Cornell 69* # 2 are closely related. Therefore, all of them were probably obtained in the same way by use of *metric algebra.*

The topic dealt with in ***Geom.* 12.41-71** is *isosceles triangles with inscribed squares.* The treatise on mensuration continues in ***Geom.* 13-16** with exercises for various kinds of *quadrilaterals*, and ends in ***Geom.* 17-20** with exercises for *circles and circle segments.* (See the discussion in Høyrup, *ANwR 7* (1997) 87-90.)

18.3. *Geometrica* ms S 24

While *Geometrica* mss A is throughout a rather dull hand book in mensuration, the independent source **"ms S 24" = *Geom.* 24.1-51** is a composite of (a) another series of dull exercises concerned with triangles with inscribed or circumscribed circles or squares, and (b) an interesting collection of *cleverly designed problems* (what Høyrup likes to call "riddles").[48]

Take, for instance, ***Geom.* 24.1**. The problem in that exercise is *an indeterminate pair of equations for four unknowns*:

> To find two fourangled fields [such the perimeter of the second is the threefold of that of the first and] such the area of the first is the threefold of that of the second.

The following solution is given in the manuscript:

> I make it so: Cube 3, result 27. This twice, result 54. Now take away 1 unit,
> the rest is 53. Thus one side shall be 53 feet, the other side 54 feet.
> And [the sides] of the other field like this: set together 53 and 54, result 107 feet.
> Make this times 3 [and take away 3 units]: result 318 feet.
> Thus that of the first side shall be 318 feet, the second side 3 feet.
> The area of one becomes 954 feet, and of the other 2862 feet.

Note that here, as everywhere else in ms S, lengths are measured in feet and the size of fields in (square) feet, that is in *area measure.*[49]

The explanations for the solution procedure in *Geom.* 24.1 suggested.

48. For reproductions of the pages of *Geom.* 24, see Bruins, *CCPV 1* (1964), I:52-72.
49. Compare with the fact that, as mentioned, in the known Greek-Egyptian mathematical papyrus fragments the sizes of fields are measured in area measure. Thus, in *P.Vindob. G. 26740* (Friberg, *UL* (2005), Sec. 4.3) and in *P.Chic. litt. 3* (*op. cit.*, Sec. 4.7 b), fields are measured in sq. *schoinía* (*arouras*). In *P.Cornell 69* (*op. cit.*, Sec. 4.7 c), fields are measured in sq. *baia*, and in *P.Genève 259* (*op. cit.*, Sec. 4.7 a) they are measured in (sq.) feet.

by Heath, *HGM 2* (1981 (1921)), 442, and Bruins (*op. cit.*), III:83, are unnecessarily complicated. There is, indeed, a much easier explanation:

Given that (1) $u + v = p \cdot (a + b)$, (2) $a \cdot b = p \cdot (u \cdot v)$ (in the text $p = 3$).
Set $u = 2\,p \cdot a$.
Then (2) $\Rightarrow$ $a \cdot b = p \cdot (2\,p \cdot a) \cdot v$ $\Rightarrow$ $b = 2$ sq. $p \cdot v$.
Then (1) $\Rightarrow$ $2\,p \cdot a + v = p \cdot (a + 2 \text{ sq.} p \cdot v)$ $\Rightarrow$ $(2 \text{ cu.} p - 1) \cdot v = p \cdot a$.
This equation is satisfied when, for instance, $\boldsymbol{v = p}$ and $\boldsymbol{a = 2}$ **cu.** $\boldsymbol{p - 1}$.
It follows that $\boldsymbol{b = 2}$ **cu.** $\boldsymbol{p}$ and $\boldsymbol{u = 2\,p \cdot (2}$ **cu.** $\boldsymbol{p - 1)}$.

The solution procedure given in the text can now described as follows:

1 2 cu. $p - 1 = 54 - 1 = 53 = a$
2 2 cu. $p = 54 = b$
3 $a + b = 107$, $p \cdot (a + b) = 321 = u + v$
4 $v = p = 3$, $u = (u + v) - v = 318$
5 $u \cdot v = 318 \cdot 3 = 954$ (feet), $a \cdot b = 53 \cdot 54 = 2862$ (feet)

The extreme briefness of this solution procedure suggests that this particular exercise originally was one of several in a theme text, where full information was given only in the solution procedure for the initial exercise. Single exercises appearing like this without the needed explaining context are commonplace in Babylonian mathematical recombination texts.

A simple way of explaining the solution procedure works also in the case of ***Geom.* 24.2**, as in the case of *Geom.* 24.1:

To find a field in perimeter equal to a field and with the area the fourfold of the area.

The following solution is given in the manuscript:

I make it so: Cube 4 with itself, result 64 feet. Take away 1 unit, result: the rest is 63.
So much is each one of the perimeters of the 2 parallel-sides. Now separate the sides.
I make it so: Set 4. Take away 1 unit: 3 remains. Then one side is 3 feet.
The other side like this: From 63 take away 3, the rest is 60 feet.
[For the sides] of the other field make it so: 4 on itself, result 16 feet.
From this take away 1 unit, the result is the rest 15 feet. So much is the first side, 15 feet.
The second side like this: Take away 15 from 63, result: the rest is 48 feet.
The other side shall be 48 feet. Then the area of one is 720 feet, and of the other 180 feet.

This solution procedure can be explained as follows (with $p = 3$):

Given that (1) $u + v = a + b$, (2) $a \cdot b = p \cdot (u \cdot v)$ (in the text $p = 4$).
Set $u = p \cdot a$. Then
(2) $\Rightarrow$ $a \cdot b = p \cdot (p \cdot a) \cdot v$ $\Rightarrow$ $b = \text{sq.}\,p \cdot v$.
(1) $\Rightarrow$ $p \cdot a + v = a + \text{sq.}\,p \cdot v$ $\Rightarrow$ $(\text{sq.}\,p - 1) \cdot v = (p - 1) \cdot a$.
This equation is satisfied when, for instance, $\boldsymbol{v = p - 1}$ and $\boldsymbol{a =}$ **sq.** $\boldsymbol{p - 1}$.

It follows that $\boldsymbol{b}$ = sq. $p \cdot (p-1)$ = **cu.** $\boldsymbol{p}$ **– sq.** $\boldsymbol{p}$ and $\boldsymbol{u} = p \cdot$ (sq. $p - 1$) = **cu.** $\boldsymbol{p} - \boldsymbol{p}$.
Therefore $a + b = u + v =$ *cu.* $p - 1$.

The solution procedure given in the text is, accordingly (with $p = 4$):

1 cu. $p - 1 = 64 - 1 = 63 = a + b = u + v$
2 $p - 1 = 3 = v$
3 $(u + v) - v = 63 - 3 = 60 = u$
4 sq. $p - 1 = 16 - 1 = 15 = a$
5 $(a + b) - a = 63 - 15 = 48 = b$
6 $a \cdot b = 15 \cdot 48 = 720$ (feet), $u \cdot v = 60 \cdot 3 = 180$ (feet)

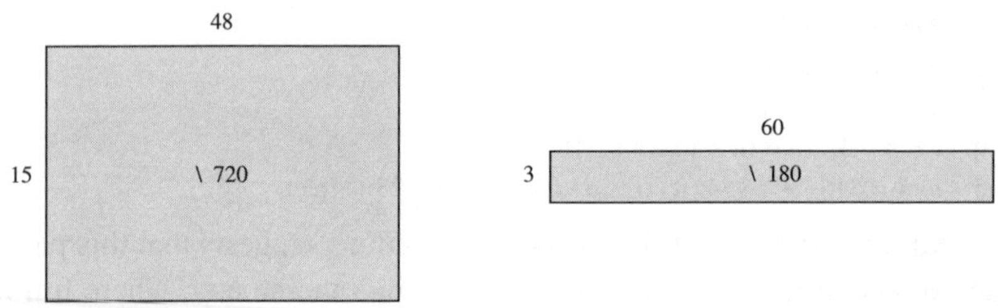

Fig. 18.3.1. *Geom.* 24.2. An indeterminate problem for two rectangles.

Nothing like the indeterminate problems in Geom. 24.1-2 appears in Diophantus' *Arithmetica*. There is also nothing like these problems in the known corpus of Babylonian mathematics. On the other hand, both the formulation of the problems and the solution procedures are so simple and straightforward that a Babylonian origin cannot be excluded. (In favor of a Babylonian origin speaks also the circumstance that in *Geom.* 24. 2 the areas of the two rectangles are multiples of 60.)

In ***Geom.* 24.4**, the problem is stated as follows:

The sum $a + b + c$ of the sides of a right-angled triangle is 50 feet. Find the sides.

It is assumed that the sides are proportional to 3, 4, 5, the "first" triple of sides that can be constructed by use of the "method of Pythagoras". Since $3 + 4 + 5 = 12$, it follows directly that $a = 3 \cdot 50/12 = 12\ 1/2$ feet, *etc*.

In ***Geom.* 24.5**, the problem is:

The area of a right-angled triangle is 5 feet. Find the sides.

The brief solution procedure can be explained as follows: According to the "method of Pythagoras", the sides can be assumed to be n, (sq. $n - 1$)/2, (sq. $n + 1$)/2, where n is odd. Then the area is

$A = n/2 \cdot (\text{sq. } n - 1)/2 = (n - 1) \cdot n \cdot (n + 1)/4$, with n odd.

Since n is odd one of the numbers is a multiple of 4, another a multiple of 2, and a third a multiple of 3. Therefore A must be a multiple of 6. Let, for instance, $A = 5 \cdot \text{sq. } 6 = 180$. Then n is a solution to the *cubic equation*

$(n - 1) \cdot n \cdot (n + 1) = 720.$

The value of n can be obtained by trial and error or from a "quasi-cube table" like the OB table text MS 3048 (Sec. 13.6 above). It is found to be $n = 9$, so that $a, b, c = 9, 40, 41$.

In ***Geom.* 24.7**,

The upright a of a right-angled triangle is 12 feet (and the area is 96 feet).

The base b and the hypothenuse c are then found as

$b = a + a/3 = 16$ feet, $c = b + b/4 = 20$ feet.

The embarrassingly simple-minded method can be explained as follows:

if a, b, c are proportional to 3, 4, 5, then $b = 4/3 \cdot a$ and $c = 5/4 \cdot b$.

Similarly in ***Geom.* 24.8-9**, where the corresponding equations are

$a = b - b/4\ (= 3/4 \cdot b)$, $c = b + b/4\ (= 5/4 \cdot b)$ and
$b = c - c/5\ (= 4/5 \cdot c)$, $a = c - c/4\ (= 3/4 \cdot c)$.

An OB problem text of a similar kind is **MS 3971 § 4** (Friberg, *RC* (2007), Sec. 10.1 d), where it is stated that

The diagonal $c = 7$ of a right triangle is given.

The remaining sides of the triangle are then computed as

$b = 7 \cdot 4/5 = 7 \cdot ;48 = 5;36$ and $a = 7 \cdot 3/5 = 7 \cdot ;36 = 4;12$.

A LB (Seleucid) exercise of a somewhat similar kind is **BM 34568 # 1** (Neugebauer, *MKT 3* (1937), 20):

*4 the length, 3 the f*ront, what is the great divider?
Since you do not know it: 2' of your length *to* your front add on, that is it.
4 the length · 30 go, then 2, *2* to 3 add on, 5. 5 is the great divider.
The third of your front to your length add on, that is the great divider.
3 the front steps of 20 go, 1, *1* to 4 add on, 5. 5 is the great divider.

Although awkwardly formulated, the text of this exercise probably wants to say that for the right triangle with the sides $c, b, a = 5, 4, 3$, the diagonal c can be expressed in these two ways as a *linear* combination of b and a:

$c = b/2 + a$ and $c = b + a/3$.

This makes little sense, but BM 34568 is a large theme text concerned with increasingly difficult systems of equations for the sides of right triangle. (An application of the diagonal rule appears already in exercise # 2.) In a theme text of this kind, the introductory first exercise should play an important role, rather than being nearly meaningless. Perhaps like this:

Let c, b, a be the sides of a right triangle. Suppose that $c = b/p + a$, for some (regular sexagesimal) number p. Then the pair c, a satisfies the following quadratic-linear system of equations:

$$\text{sq. } c - \text{sq. } a = \text{sq. } b, \quad c - a = b/p.$$

Consequently,

$$c + a = \text{sq. } b \,/\, (b/p) = p \cdot b.$$

Evidently, then, the solution to the system of equations for c and a is

$$c = (p + 1/p)/2 \cdot b, \quad a = (p - 1/p)/2 \cdot b.$$

This means that

$$c = b/p + a \;\Rightarrow\; c, b, a = \{(p + 1/p)/2, 1, (p - 1/p)/2\} \cdot b, \quad \text{and, similarly,}$$

$$c = b + a/q \;\Rightarrow\; c, b, a = \{(q + 1/q)/2, (q - 1/q)/2, 1\} \cdot a.$$

Therefore, it is possible that a teacher could use the seemingly trivial exercise # 1in BM 34568 as an introduction to an *oral* presentation of a method to find *two generating pairs of numbers* p, 1 and q, 1 for every given right triangle with rational sides. (*Cf.* the discussion of OB igi-igi.bi problems in Secs. 3.2-3 above.) In the given example, for instance,

$$5 = 4/2 + 3 \;\Rightarrow\; 5, 4, 3 = \{(2 + 1/2)/2, 1, (2 - 1/2)/2\} \cdot 4 \quad \text{and}$$

$$5 = 4 + 3/3 \;\Rightarrow\; 5, 4, 3 = \{(3 + 1/3)/2, (3 - 1/3)/2, 1\} \cdot 3.$$

A particularly interesting problem in *Geom.* 24 is ***Geom.* 24.10**, where

> The area plus the perimeter of a right-angled triangle is 280 feet.

This looks like one of the indeterminate problems for right-angled triangles in Diophantus' *Arithmetica* "VI".6-11. Nevertheless, the solution method here is totally different. It is stated, quite cryptically, like this:

> Always look for factors. Factorize by 2, 140, by 4, 70, by 5, 56, by 7, 40,
> by 8, 35, by 10, 28, by 14, 20. I find that 8 and 35 meet the condition.
> 1/8 of 280, result 35 feet. Always take 2 away from 8, the rest is 6 feet.
> Now 35 and 6 together, result 41 feet. These on themselves, result 1681 feet.
> And 35 on 6, result 210 feet. These always on 8, result 1680 feet.
> Take these away from 1681, the rest is 1. Of which the square side, result 1.

Next set 41 and take away 1 unit, the rest is 40. Of which 1/2, result 20.
This is the upright, 20 feet. And set again 41 and add 1, result 42 feet.
Of which 1/2, result 21 feet. The base shall be 21 feet.
And set 35 and take away 6, the rest is 29.

The solution method is (silently) based on the observation that if a, b, c are the sides and A the area of a right-angled triangle, then

$$(a + b + c) \cdot (a + b - c) = \text{sq. } (a + b) - \text{sq. } c = 2\, a \cdot b = 4\, A.$$

Therefore, the stated problem can be reformulated in the following way:

$$A + (a + b + c) = (a + b + c)/2 \cdot \{(a + b - c)/2 + 2\} = B = 280 \text{ (feet)}.$$

Now, if B is factorized as $B = 35 \cdot 8$, it is possible to begin by setting

$$(a + b + c)/2 = 35, \quad (a + b - c)/2 + 2 = 8, \quad \text{so that} \quad (a + b - c)/2 = 8 - 2 = 6.$$

Then

$$a + b = (a + b + c)/2 + (a + b - c)/2 = 35 + 6 = 41 \quad \text{and}$$
$$a \cdot b\,/\,2 = (a + b + c)/2 \cdot (a + b - c)/2 = 35 \cdot 6 = 210.$$

This *rectangular-linear system* of type B1a is solved as follows:

$$\text{sq. } (b - a) = \text{sq. } (a + b) - 4\, a \cdot b = \text{sq. } 41 - 8 \cdot 210 = 1681 - 1680 = 1, \quad \text{and} \quad b - a = 1.$$

Therefore,

$$b = \{(a + b) + (b - a)\}/2 = (41 + 1)/2 = 21, \;\; a = \{(a + b) - (b - a)\}/2 = (41 - 1)/2 = 20.$$

Finally, the remaining side c can be computed as

$$c = (a + b + c)/2 - (a + b - c)/2 = 35 - 6 = 29.$$

Hence, the solution to the problem is the diagonal triple $a, b, c = 20, 21, 29$.

The text of *Geom.* 24.10 ends with a verification:

$$A = 210 \text{ feet}, \quad a + b + c = 70 \quad \Rightarrow \quad A + (a + b + c) = 210 + 70 = 280.$$

Now, recall that the solution procedure in *Geom.* 24.10 starts with the factorizations

$$280 = 2 \cdot 140 = 4 \cdot 70 = 5 \cdot 56 = 7 \cdot 40 = 8 \cdot 35 = 10 \cdot 28 = 14 \cdot 20,$$

after which it is stated that “8 and 35 meet the condition”. What this means can be found out by looking at a couple of alternative factorizations of the given number 280. If, for instance, the chosen factors are 7 and 40, then

$$(a + b + c)/2 = 40 \quad \text{and} \quad (a + b - c)/2 = 7 - 2 = 5 \quad \Rightarrow$$
$$a + b = 40 + 5 = 45 \quad \text{and} \quad a \cdot b\,/\,2 = 40 \cdot 5 = 200 \quad \Rightarrow$$
$$b - a = \text{sqs. } (\text{sq. } 45 - 8 \cdot 200) = \text{sqs. } (2025 - 1600) = \text{sqs. } 425 = 5 \cdot \text{sqs. } 17.$$

Similarly, if the chosen factors are 4 and 70, say, then

$(a + b + c)/2 = 70$ and $(a + b - c)/2 = 4 - 2 = 2 \Rightarrow$
$a + b = 70 + 2 = 72$ and $a \cdot b / 2 = 70 \cdot 2 = 140 \Rightarrow$
$b - a =$ sqs. (sq.72 − 8 · 140) = sqs. (5184 − 1120) = sqs. 4064 = 4 · sqs. 254.

Thus, apparently, the factorization 280 = 35 · 8 was chosen because it was the only one that would lead to a *solution in integers.*

The proposed interpretation of *Geom.* 24.10 is supported by the testimony of ***Geom.* 24.26**, where the question is:

> A circle is inscribed in a right-angled triangle with the sides 6, 8, and 10 feet. Find the diameter of the circle.

The diameter d of the circle is computed in the following two ways:

1 $d = a + b - c = 6 + 8 - 10$ feet = 4 feet.
2 $d = (4\,A)/\,(a + b + c) = (4 \cdot 6 \cdot 4)/(6 + 8 + 10)$ feet = 96/24 feet = 4 feet.

The given triangle can be divided into three sub-triangles, each with half the diameter as height and one of the sides as base. Therefore,

$$2\,A = 2\,A_1 + 2\,A_2 + 2\,A_3 = d/2 \cdot a + d/2 \cdot b + d/2 \cdot c = d/2 \cdot (a + b + c).$$

The second computation rule in *Geom.* 24.26 follows immediately from this observation. The first computation rule then follows from the second, in view of the identity $(a + b + c) \cdot (a + b - c) = 4\,A$.

The suggested interpretation of *Geom.* 24.10 is important, because:

a) This is *the only known example of a Greek mathematical problem text containing an explicit solution procedure for a rectangular-linear system of equations.*
b) The form of the entire solution procedure looks like a typical OB solution procedure, with an algorithmic series of instructions without theoretical motivations, and with each step of a general procedure illustrated by an explicit numerical computation.
c) One crucial step of the solution procedure, the factorization of $B = 280$, would be meaningless if it was not known beforehand that *the problem is constructed so that it will have a solution in integers.* This is a typically Babylonian feature.
d) The use of the identity $(a + b + c) \cdot (a + b - c) = 2\,A$ (twice the area of a *rectangle*) is known from the Seleucid mathematical theme text BM 34568 ## 17-18.

Here is the text of **BM 34568 ## 17-18** (*cf.* Neugebauer, *MKT 3*, 14 *ff.*):

BM 34568 # 17

The length, front, and great divider add, then 12, and the field 12.
What as the length, front, and great divider?
Since you do not know:
12 · 12 2 24. 12 · 2 24. 24 from 2 24 lift, then remaining is 2.

2 · 30 go, then 1. 12 · what shall I go so that 1? 12 · 5 1. 5 is the great divider.

BM 34568 # 18

The length, front, and great divider add, then 1, 5 the field.
The length, front, and great divider · the length, front, and great divider go.
The field · 2 go, from <···> the great divider lift. What is remaining · the half go.
The length, front, and great divider · what[!] as your step do you set?
The great divider is your step.

The text of # 17 is condensed, with the elimination of words that are not really necessary, and with the use of the cuneiform sign GAM for 'times', here represented by the symbol '·'. Note that while # 17 gives a numerical example, # 18 gives the general rule (with an omission in line 3).

The questions in ## 17-18 are both of the type

a, b, c, and A are the sides, the diagonal, and the area of a rectangle.
$b + a + c$ and A are given. Find b, a, c.

The value of the diagonal c is computed by use of the identity

$$c \cdot (b + a + c) = \{\text{sq.}\ (b + a + c) - 2\,A\}/2.$$

This identity can be proved in several ways. See Neugebauer, *op. cit.*, 21, and Høyrup, *LWS* (2002), 397. The simplest proof is probably this:

$$c = \{(b + a + c) - (b + a - c)\}/2 \;\Rightarrow$$
$$c \cdot (b + a + c) = \{\text{sq.}\ (b + a + c) - (b + a + c) \cdot (b + a - c)\}/2 = \{\text{sq.}\ (b + a + c) - 2\,A\}/2.$$

Actually, the computation of the diagonal c is only the first step of the solution procedure. After c has been computed, b and a can be found without trouble as solutions to the following problem

The diagonal c and the sum $b + a$ of the sides of a rectangle are known. Find b and a.

The way to solve a problem of this kind is shown in **BM 34568 # 10.**

Another particularly interesting problem in *Geom* 24 is ***Geom.* 24.21**:

In a right-angled triangle, the upright is 15, the base 20, and the hypothenuse 25 feet. Another triangle is circumscribed at a distance of 2 feet. I look for its sides.

The problem is interesting for two reasons. First, *figures extended a certain fractional distance in all directions* are known from both OB and LB (Seleucid) mathematical texts. See the discussion of the term *dakāšu* 'to push outwards' in Friberg, *et al.*, *BagM* 21 (1990), 488, and in Friberg, *RlA* 7 (1990), Sec. 5.4 *l.* The term is used for outwards extended *squares*, *circles*, and *double circle-segments*. Secondly, the problem is interesting because the way it may have been solved (not explicitly indicated in the text) is loosely related to the proof of Heron's triangle area rule in *Metrica* I.8 (Chapter 14 above).

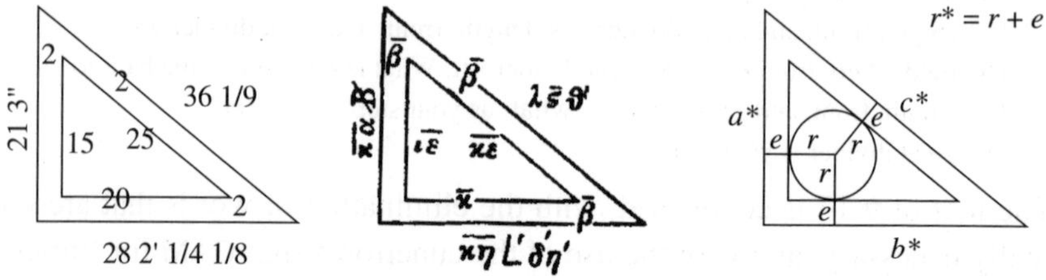

Fig. 18.3.2. Left and middle: A diagram illustrating the extension problem *Geom.* 24.21.

In *Geom.* 24.21, the following answer is given without explanation:

1 $a^* = 21\ 2/3$ feet, $b^* = 28\ 1/2\ 1/4\ 1/8$ feet, $c^* = 36\ 1/9$ feet
2 $a^* = a + 1/3\ 1/9 \cdot a$, $b^* = b + 1/3\ 1/9 \cdot b$, $c^* = c + 1/3\ 1/9 \cdot c$.

Here is a proposed explanation of how the sides of the extended triangle may have been computed: Let *a*, *b*, *c* and *a**, *b**, *c** be the sides of the given and the extended triangle, let *e* be the distance between the sides of the two triangles, and let $r = d/2$ and $r^* = d^*/2$ be the half diameters of the inscribed circles (Fig. 18.3.2, right). With respect to the common center of the inscribed circles, the two triangles are concentric, parallel, and similar. Therefore, one is a scaled-up version of the other, with the scale factor *t* given by the equations (cf. *Geom.* 24.26 above)

$t \cdot r = r^* = r + e$ so that $t - 1 = e\,/\,r = 2\,e\,/\,d = 2\,e \cdot (a + b + c)/(4\,A)$.

With the given values $e = 2$ and $a, b, c = 15, 20, 25$ (feet) it follows that

$t = 1 + 4 \cdot 60/(2 \cdot 15 \cdot 20) = 1 + 4 \cdot 1/10 = 1\ 1/3\ 1/15$.

The answer given in the text is $t = 1\ 1/3\ 1/9 = 1 + 4 \cdot 1/9$. The error is easy to explain if the author of the problem looked up the value of $4 \cdot 1/10$ in a table of fractions such as *P.Akhmîm* (Friberg, *UL* (2005), Sec. 4.5 a). He can then have chosen the wrong column in the table and found incorrectly the value of $4 \cdot 19$ as 'of $4\ \overline{3}\ \overline{9}$' instead of correctly the value of $4 \cdot 1/10$ as 'of $4\ \overline{3}\ \overline{15}$' (Baillet, *PMA* (1892) 27-28: columns 7-8).

***Geom.* 24.46-47** are well known because they are the only known examples of the *explicit solution of a quadratic equation* in a Greek mathematical text. The question in *Geom* 24.46 is

The sum $d + a + A$ of the diameter, circumference, and area of a circle is 212 feet.

Separate the three from each other.

An OB parallel to this question (without solution) can be found in **BM 80209**, a "catalog text" with equations for circles (see Sec. 1.10 above).

The solution to the problem in *Geom.* 24.46 proceeds like this:

212 · 154 = 32648, 32648 + 841 = 33489 = sq. 183 feet, 183 – 29 = 154,
1/11 · 154 = 14 feet = the diameter.

Actually, with the Archimedian approximation area of circle / square of radius of circle = appr. 3 1/7 = 22 · 1/7, the equation for d becomes

$d + a + A = d + 22 \cdot 1/7 \cdot d + 11 \cdot 1/14 \cdot \text{sq. } d = 112.$

Through multiplication with 14 · 11 = 154, this equation is reduced to

$2 \cdot 29 \cdot 11 \cdot d + \text{sq. } 11 \cdot \text{sq. } d = 154 \cdot 112 = 32648.$

This equation, in its turn, is reduced through completion of the square to

$\text{sq. } (11\, d + 29) = 32648 + \text{sq. } 29 = 32648 + 841 = 33489 = \text{sq. } 183.$

Therefore, $11\, d + 29 = 183$, and $d = (183 - 29)/11 = 14$ (feet).

18.4. *Metrica* III.4. A Division of Figures Problem

The difference in style between exercises in Heron's *Metrica* and in the pseudo-Heronic "*Geometrica*" was mentioned above (Sec. 18.1). Here is an example of the style in *Metrica* (Schöne *HA* (1903), 149), where use is made of a *lettered diagram*:

***Metrica* 3.4**

Given the triangle ABC, take away from it the triangle DEZ, given in magnitude, so that the remaining triangles ADE, BDZ, CEZ are equal to each other. If now <the sides> are divided so that AD is to DB as BZ to ZC and as CE to EA, then the triangles ADE, BDZ, and ZCE shall be equal to each other. Let now AZ be joined. Since then as BZ is to ZC, so is CE to EA, and compounding, as BC is to CZ, so is CA to AE. And therefore, as the triangle ABC is to AZC, so is AZC to AZE, and subtracting, as the triangle ABC is to ABZ, so AZC is to ECZ, which is given. And also ABC is given. · · ·

The solution procedure continues in this manner, showing first that AZC is given, and that, if AH is the height against BC, then the square of AH times the product of BZ and ZC is given. Since the square of AH is given, also the product of BZ and ZC is given, and since BC is given, Z is given, and similarly E and D are given. Therefore, DE, EZ, and ZD are given.

After this demonstration *in the style of Euclid's Data* follows *a numerical example*, with AB, BC, CA = 13, 14, 15 units, and DEZ = 24 units.

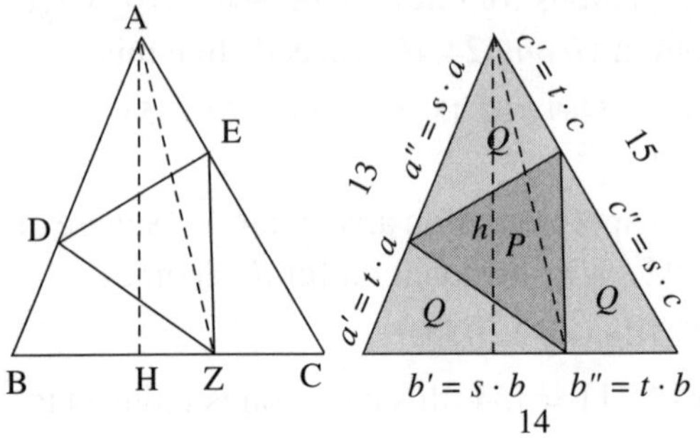

$a, b, c = 13, 14, 15$

$h = 12, \quad A = 84, \quad P = 24 \Rightarrow$

$Q = (84 - 24)/3 = 20,$

$4\,A \cdot Q = 6720$

$4\,A \cdot Q = \text{sq. } h \cdot b' \cdot b'' \Rightarrow$

$b' \cdot b'' = 6720/144 = 46$ <2/3>

$b' + b'' = 14.$ Hence

$b', b'' = 7 \pm$ sqs. 2 1/3

= appr. 7 ± 1 1/2 (1/36)

Fig. 18.4.1. *Metr.* 3.4. An elegant division of figures problem.

If the three sides of the given triangle are divided in the same ratio $s : t$, with $s + t = 1$, and if A and Q are the areas of the given triangle and the corner triangles, then the procedure shows that $Q : A = s \cdot t$. Therefore, $b' = s \cdot b$ and $b'' = t \cdot b$ are solutions to the following *rectangular-linear system of equations of type B1a:*

$b' \cdot b'' = 4\,A \cdot Q$ / sq. $h = 4 \cdot 84 \cdot 20$ / sq. $12 = 46$ 2/3, $\quad b' +\cdot\, b'' = b = 14.$

The solution given in the text is $b' = 8$ <1/2>, $b'' = 5$ 1/2.

Appendix 1

A Chain of Trapezoids with Fixed Diagonals

by Jöran Friberg and Joachim Marzahn

A.1.1. VAT 8393. A New Old Babylonian Single Problem Text

VAT 8393 is a well preserved and unusually interesting Old Babylonian mathematical cuneiform text from the Near Eastern Museum (Vorderasiatisches Museum) in Berlin, with a single metric algebra problem, illustrated by an intriguing diagram.

The diagram on the clay tablet is drawn in the usual OB way with little regard to the true proportions of various parts of the figure. The diagram is reproduced below, with corrected proportions:

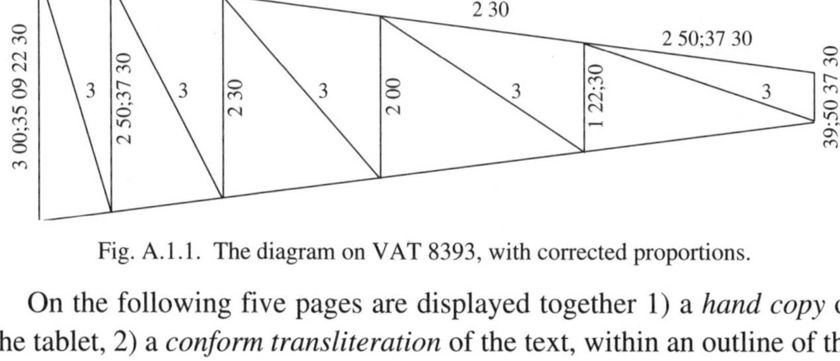

Fig. A.1.1. The diagram on VAT 8393, with corrected proportions.

On the following five pages are displayed together 1) a *hand copy* of the tablet, 2) a *conform transliteration* of the text, within an outline of the clay tablet, 3) a *transliteration sentence by sentence*, 4) a corresponding *translation* of the text, and 5) photos of the clay tablet.

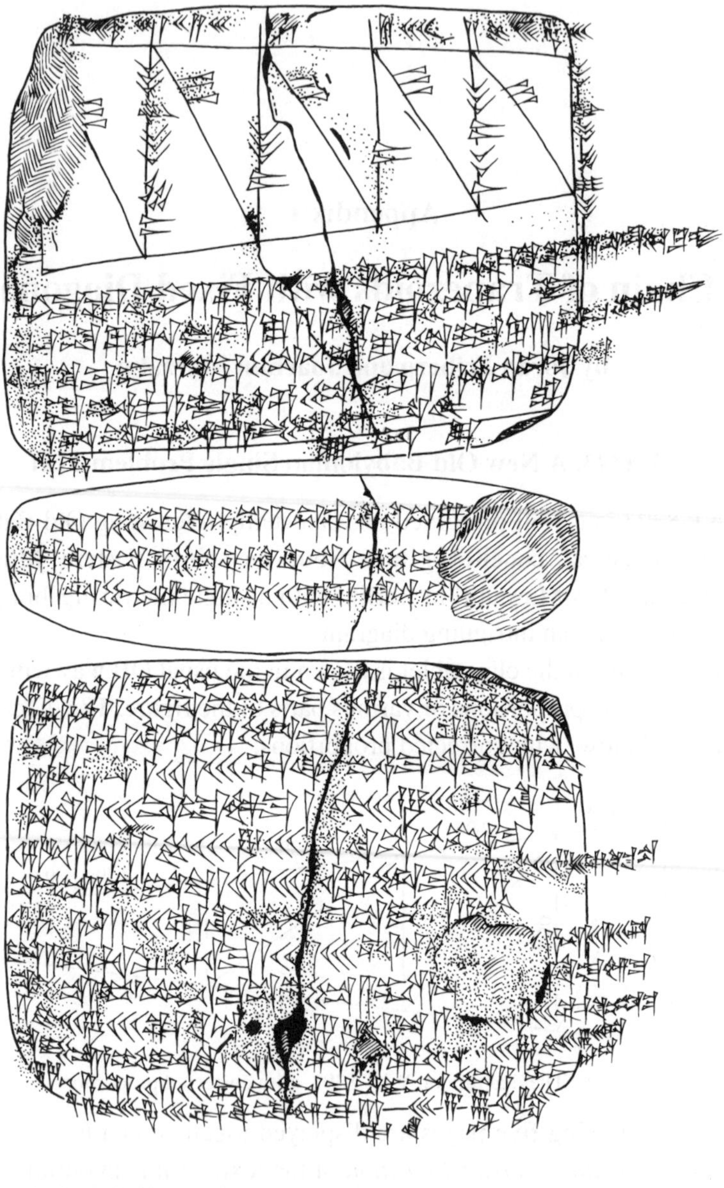

Fig. A.1.2. VAT 8393. (Hand copy by J. Marzahn, curator of the collection of clay tablets in the Near Eastern Museum, Berlin.)

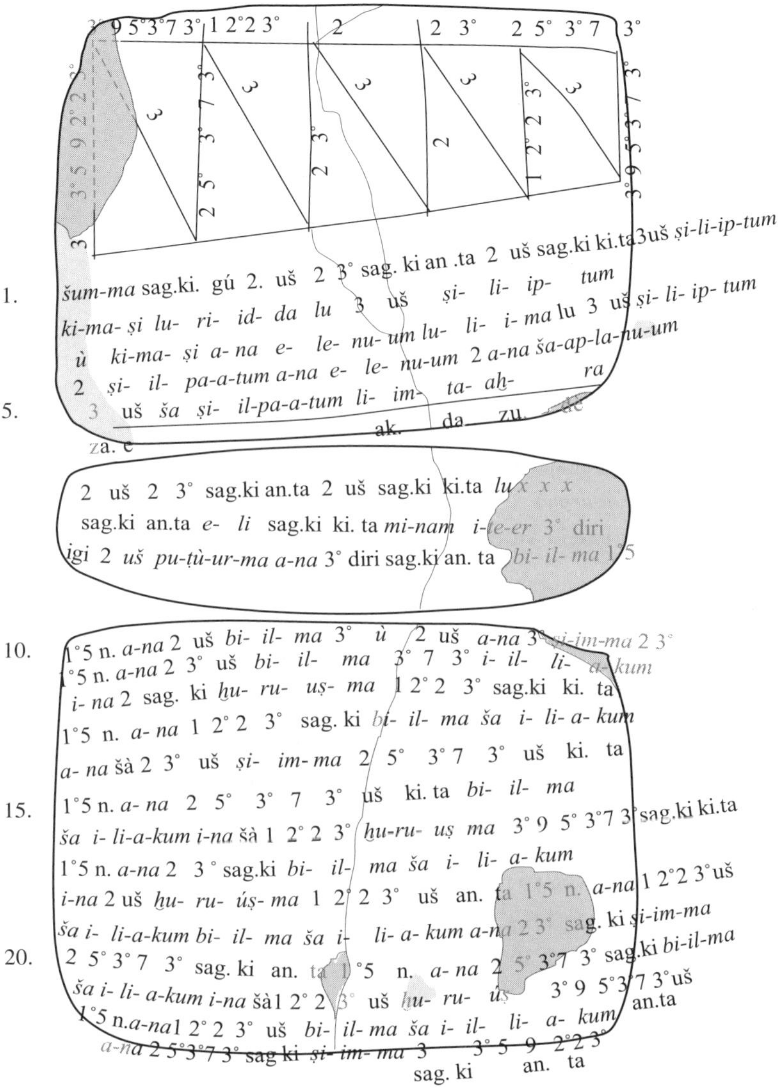

Fig. A.1.3. VAT 8393. A conform transliteration within an outline of the clay tablet. (Transliteration by J. Friberg and J. Marzahn, with the assistance of M. Krebernik.)

VAT 8393, transliteration sentence by sentence

obv.

Fig.

1 *šum-ma* sag.ki.gù 2 uš 2 30 sag.ki an.ta 2 uš sag.ki ki.ta
3 *ṣi-li-ip-tum* /
2 *ki ma-ṣi lu-ri-id-da! lu!* 3 uš *ṣi-li-ip-tum* /
3 *ù ki ma-ṣi a-na e-le-nu-um lu-li-i-ma lu* 3 uš *ṣi-li-ip-tum* /
4 2 *ṣi-il-pa-a-tum a-na e-le-nu-um* 2 *a-na ša-ap-la-nu-um* /
5 [3] uš *ša ṣi-il-pa-a-tum li-im-ta-aḫ-ra* /
6 [z]a.e ak.da.zu.[dè] /

lower edge

7 2 uš 2 30 sag.ki an.ta sag.ki ki.ta *lu*-[*x-x-x*] /
8 sag.ki an.ta *e-li* sag.ki ki.ta *mi-nam i*-[*te-er* 30 diri] /
9 igi 2 uš *pu-ṭù-ur-ma*
a-na 30 diri sag.ki an.ta [*bi-il-ma* 15] /

rev.

10 15 ninda-*nu a-na* 2 uš *bi-il-ma* 30
ù 2 uš *a-na* [30 *ṣi-im-ma* 2 30 uš ki.ta] /
11 15 ninda-*nu a-na* 2 30 uš *bi-il-ma* 37 30 *i-il-li*-[*a-kum*] /
12 *i-na* 2 sag.ki *ḫu-ru-uṣ-ma* 1 22 30 sag.ki ki.ta /
13 15 ninda-*nu a-na* 1 22 30 sag.ki [*b*]*i-il-ma ša i-li-a-kum* /
14 *a-na* šà 2 30 uš *ṣi-im-ma* 2 50 37 30 uš ki.ta /
15 15 ninda-*nu a-na* 2 50 37 30 uš ki.ta *bi-il-ma* /
16 *ša i-li-a-kum i-na šà* 1 22 30 *ḫu-ru-uṣ-ma*
39 50 37 30 sag.ki ki.ta /
17 15 ninda-*nu a-na* 2 30 sag.ki *bi-il-ma ša i-li-a-kum* /
18 *i-na* 2 uš *ḫu-ru-uṣ-ma* 1 22 30 uš an.t[a]
[15 ninda-*nu*] *a-na* 1 22 30 uš /
19 *ša i-li-a-kum bi-il-ma ša i-li-a-kum*
20 *a-n*[*a* 2 30 sa]g.ki *ṣi-im-ma* / 2 50 37 30 sag.ki an.[ta]
[1]5 ninda-*nu a-na* 2 [50] 37 30 sag.ki *bi-il-ma* / *ša i-li-a-kum*
21 *i-na* šà 1 22 [30] uš [*ḫ*]*u-ru-u*[*ṣ*] 39 50 37 30 uš anta /
22 15 ninda-*nu a-na* 1 22 30 uš *bi-il-ma ša i-il-li-a-kum* /
23 [*a-n*]*a* 2 50 37 30 sag.ki *ṣi-im-ma*
3 35 09 22 30 sag.ki an.ta

VAT 8393, translation

obv.

Fig.

1 If a trapezoid, 2 (00) the length, 2 30 the upper front, 2 sixties the lower front,
3 (00) the diagonal. /

2 How much shall I go down so that 3 (00) the diagonal? /

3 And how much above shall I go up so that 3 (00) the diagonal? /

4 *2* diagonals above, 2 below. /

5 *3* sixties what the diagonals may be equal. /

6 *Y*ou in your doing: /

lower edge

7 2 (00) the length, 2 30 the upper front, 2 (00) the lower front may *x x x*. /

8 The upper front over the lower front what does it exc*eed? 30 the excess.* /

9 The reciprocal of 2 resolve (= 1 / 2 00 = 0;00 30).
To 30 the excess of the upper front *bring (= multiply) then ;15.* /

rev.

10 ;15 the ninda (= growth rate) to 2 (00) the length bring, then 30.
And 2 (00) the length to *30 double (= add), then 2 30, the lower length.* /

11 ;15 the ninda to 2 30 the length bring, 37;30 comes up *for you.* /

12 From 2 (00) the front tear off (= subtract), then 1 22;30 the lower front. /

13 ;15 the ninda to 1 22;30 the front *br*ing. What comes up for you /

14 onto 2 30 the length double, then 2 50;37 30 the lower length. /

15 ;15 the ninda to 2 50;37 30 the lower length bring, then /

16 what comes up for you out from 1 22;30 tear off, then
39;50 37 30 the lower front. /

17 ;15 the ninda to 2 30 the front bring. What comes up for you /

18 from 2 (00) the length tear off, then 122;30 the upp*er* length.
;15 the ninda to 1 22;30 the length /

19 that comes up for you bring. What comes up for you

20 t*o 2 30 the fro*nt double, then / 2 50;37 30 the up*per* front.

21 *;1*5 the ninda to 2 *50*;37 30 the front bring, then / what comes up for you
out from 1 22;*30* the length *te*ar of*f,* 39;50 37 30 the upper length. /

22 ;15 the ninda to 1 22;30 (error!) the length bring, then what comes up for you /
to 2 50;37 30 the front double, then

23 3 00;35 09 22 30 the upper front.

Fig. A.1.4. VAT 8393. (From color photos courteously provided by Olaf M. Teßmer.)

In the metric algebra explanation below of VAT 8393, the notations used are the ones displayed in Fig. A.1.5.

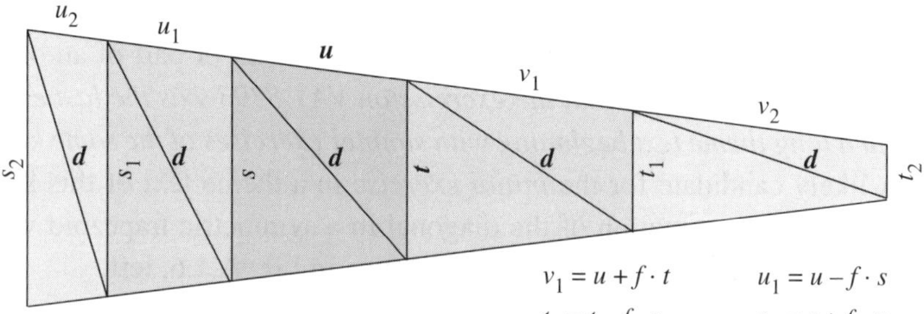

$v_1 = u + f \cdot t$ $u_1 = u - f \cdot s$

$t_1 = t - f \cdot v_1$ $s_1 = s + f \cdot u_1$

$v_2 = v_1 + f \cdot t_1$ $u_2 = u_1 - f \cdot s_1$

$t_2 = t_1 - f \cdot v_2$ $s_2 = s_1 + f \cdot u_2$

Fig. A.1.5. Metric algebra notations for the parameters of the trapezoids on VAT 8393.

Here u is the length, s and t the "upper" and "lower" fronts, and d one of the diagonals of a symmetric trapezoid, situated in the middle of *a chain of five symmetric trapezoids*. The two trapezoids "below" and the two trapezoids "above" have diagonals of the same length as the trapezoid in the middle. Their lengths are denoted v_1, v_2, and u_1, u_2, respectively, and their fronts t_1, t_2, and s_1, s_2, respectively. The question and the successive steps of the solution procedure can then be described briefly as follows:

Question:

Given a symmetric trapezoid. The length u = 2 00 (120), the upper front s = 2 30 (150), the lower front t = 2 00 (120), the diagonal d = 3 00 (180), all measured in ninda = 6 m. Find 2 symmetric trapezoids below with the diagonal d = 3 00 and 2 symmetric trapezoids above with d = 3 00 so that all the trapezoids form a chain of trapezoids.

Procedure:

1 $f = (s - t)/u$ = 30 / 2 00 = ;15 (the growth rate *nindanu* in ninda /ninda) lower edge

2 $v_1 = u + f \cdot t$ = 2 00 + ;15 · 2 00 (= 30) = 2 30 *rev.* 1

3 $t_1 = t - f \cdot v_1$ = 2 00 – ;15 · 2 30 (= 37;30) = 1 22;30 *rev.* 2-3

4 $v_2 = v_1 + f \cdot t_1$ = 2 30 + ;15 · 1 22;30 (= 20;37 30) = 2 50;37 30 *rev.* 4-5

5 $t_2 = t_1 - f \cdot v_2$ = 1 22,30 – ;15 · 2 50;37 30 (= 42;39 22 30) = 39;50 37 30 *rev.* 6-7

6 $u_1 = u - f \cdot s$ = 2 00 – ;15 · 2 30 (= 37;30) = 1 22;30 *rev.* 8-9a

7 $s_1 = s + f \cdot u_1$ = 2 30 + ;15 · 1 22;30 (= 20;37 30) = 2 50;37 30 *rev.* 9b-11a

8 $u_2 = u_1 - f \cdot s_1$ = 1 22;30 – ;15 · 2 50;37 30 (= 42;39 22 30) = 39;50 37 30 *rev.* 11b-12

9 $s_2 = s_1 + f \cdot u_2$ = 2 50;37 30 + ;15 · 39;50 37 30 (= 9;57 39 22 30) = 3 00;35 09 22 30

The numbers given or computed in the text agree with the ones displayed in the diagram (see Fig. A.1.1 above). The error in *rev.* 13, with 1 22 30 instead of 39 50 37 30, does not influence the computation of s_2. This circumstance proves that the text is a copy of all or part of an older text. Actually, *it is likely that the exercise on VAT 8393 was the last exercise in a long theme text beginning with simpler exercises of the same kind.*

A likely candidate for the *initial exercise* in a theme text of this kind would be the computation of the diagonal in a symmetric trapezoid with given sides. How that could be done is shown in Fig. A.1.6, left.

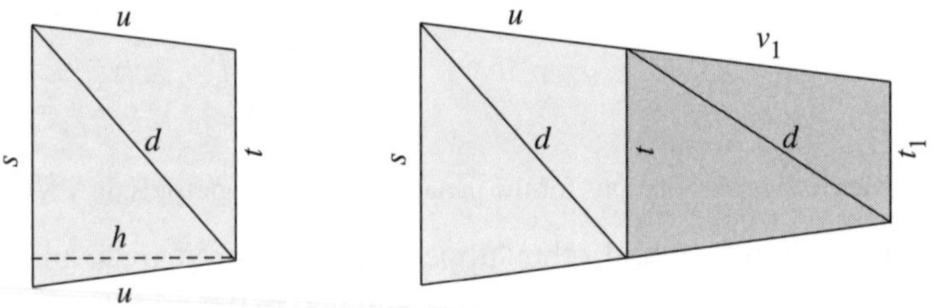

sq. u = sq. h + sq. $(s - t)/2$
sq. d = sq. h + sq. $(s + t)/2$
$\Rightarrow$
sq.*d* = sq. *u* + *s* · *t*

a) $s - t = f \cdot u$
$\Rightarrow$ $\boldsymbol{t - t_1 = f \cdot v_1}$ and $s - t_1 = f \cdot (v_1 +$
b) sq. $u + s \cdot t$ = sq. d = sq. $v_1 + t_1 \cdot t$
$\Rightarrow$ sq. v_1 − sq. $u = (s - t_1) \cdot t = f \cdot (v_1 + u)$
$\Rightarrow$ $\boldsymbol{v_1 - u = f \cdot t}$

Fig. A.1.6. A metric algebra proof of the OB fixed trapezoid diagonal rule.

Let u be the (slanting) side, s and t the fronts, d the diagonal, and h the distance between the fronts in a symmetric trapezoid. Then two applications of the *OB (rectangle) diagonal rule* show that

sq. u = sq. h + sq. $(s - t)/2$ and sq. d = sq. h + sq. $(s + t)/2$.

Through a combination of these two identities one finds that

sq. d = sq. u + sq. $(s + t)/2$ − sq. $(s - t)/2$, so that **sq. *d* = sq. *u* + *s* · *t*.**

Since the author of the exercise VAT 8393 must have been familiar with this rule, it is motivated to call it "the OB trapezoid diagonal rule". The rule is, of course, identical with Ptolemy's diagonal rule in the case of a symmetric trapezoid. See Fig. 14.4.1 above.

A likely candidate for a *second exercise* in a theme text of the kind mentioned above is an exercise where the set task is to extend a given symmetric trapezoid from below (or from above) into a brief chain of two symmetric trapezoids, both with diagonals of the same length. Such a chain may be called a "chain of symmetric trapezoids with fixed diagonals". The situation is illustrated by the example in Fig. A.1.6, right, where a symmetric trapezoid with the slanting side v_1, the fronts t, t_1, and the diagonal d is an extension from below of the given trapezoid.

Since t and d are known, the only new parameters are v_1 and t_1. Their values can be determined as the solutions to a system of two equations, one a similarity equation, the other an equation for the diagonal of the added trapezoid. The *similarity condition* says that if f is the *growth rate*, then

$$s - t = f \cdot u \quad \Rightarrow \quad t - t_1 = f \cdot v_1 \quad \text{or} \quad s - t_1 = f \cdot (v_1 + u).$$

The *fixed diagonal condition* says, in view of the OB trapezoid diagonal rule, that

$$\text{sq. } u + s \cdot t = (\text{sq. } d =) \text{ sq. } v_1 + t_1 \cdot t.$$

If these equations are combined, one finds that

$$\text{sq. } v_1 - \text{sq. } u = (s - t_1) \cdot t = f \cdot (v_1 + u) \cdot t.$$

Since sq. v_1 – sq. $u = (v_1 + u) \cdot (v_1 - u)$, it follows that

$$v_1 - u = f \cdot t.$$

This means that the equations for v_1 and t_1 can be reduced to the pair

$$\mathbf{v_1 = u + f \cdot t \quad \text{and} \quad t_1 = t - f \cdot v_1.}$$

This pair of equations my be called the "OB fixed trapezoid diagonal rule".

A corresponding pair of equations determines how the given symmetric trapezoid can be extended from above. It is also clear that the process can be repeated, so that the given trapezoid can be extended several times in either direction in a *recursive* procedure. In this way will be formed what may be called "descending or ascending chains of symmetric trapezoids with fixed diagonals". *Examples of such descending or ascending chains of trapezoids* may have been the object of successive exercises in an OB mathematical theme text of the kind mentioned above.

It was conjectured above that the exercise in VAT 8393 may originally have been the *last* exercise in a theme text of this kind. This conjecture is based on the following observation: Consider the diagrams in Figs. A.1.1 and A.1.5. If one tries to continue the descending chain of trapezoids one step further, the next slanting side will have the length

$$v_3 = v_2 + f \cdot t_2 = 2\ 50;37\ 30 + ;15 \cdot 39;50\ 37\ 30\ (= 9;57\ 39\ 22\ 30).$$

This means that $v_3 > 3 = d$, which is geometrically impossible. Therefore, the *descending* chain of trapezoids with fixed diagonals is *self-terminating*. Similarly it can be shown that the *ascending* chain is self-terminating, since, with the given parameter values,

$$u_3 = u_2 - f \cdot s_2 = 39;50\ 37\ 30 - ;15 \cdot 3\ 00;35\ 09\ 22\ 30\ (= 45;08\ 47\ 20\ 37\ 30).$$

This means that $u_3 < 0$, which is geometrically impossible.

A surprising feature of the diagrams in Figs. A.1.1 and A.1.5 is the conspicuous *lopsided symmetry of the data*, namely that

$$t = 2\ 00 = u, \quad s = 2\ 30 = v_1, \quad t_1 = 1\ 22;30 = u_1,$$
$$s_1 = 2\ 50;37\ 30 = v_2, \quad \text{and} \quad t_2 = 39;50\ 37\ 30 = u_2.$$

These unexpected relations between the values of the parameters can be explained as follows by use of a recursive argument:

$$t = u \quad \Rightarrow \quad s = t + f \cdot u = u + f \cdot t = v_1,$$
$$t = u \ \text{ and } \ s = v_1 \quad \Rightarrow \quad t_1 = t - f \cdot v_1 = u - f \cdot s = u_1,$$
$$s = v_1 \ \text{ and } \ t_1 = u_1 \quad \Rightarrow \quad s_1 = s + f \cdot u_1 = v_1 + f \cdot t_1 = v_2,$$
$$t_1 = u_1 \ \text{ and } \ s_1 = v_2 \quad \Rightarrow \quad t_2 = t_1 - f \cdot v_2 = u_1 - f \cdot s_1 = u_2.$$

Hence, the lopsided symmetry of the parameter values is an automatic consequence of the initial relation $t = u$. It is, of course, impossible to know if the author of the problem was aware of the fact that he would obtain this lopsided symmetry of the parameter values by choosing $t = u$ (= 2 00).

It is interesting to note that in certain ways the descending/ascending chain of *symmetric trapezoids with fixed diagonal* in Fig. A.1.1 is similar to the descending/ascending chain of *birectangles* in Fig. 15.1.1 above. However, only the descending chain of birectangles is self-terminating.

Strictly speaking, the idea to construct such chains of trapezoids with a fixed diagonal is completely new and unexpected. No similar constructions appear to be known from any other mathematical documents, Greek, or Islamic, or whatever.

A.1.2. VAT 8393. About the Clay Tablet

The clay tablet VAT 8393 was acquired by the Near Eastern Museum in Berlin from the art dealer David in Paris at the beginning of the 20th century, together with several hundred other clay tablets, of varied content, linguistically as well as with regard to subject matter. Unfortunately, as far as is known today, all acquisition documents were lost in the war, so that nothing more precise can be said about when the museum's acquisition of the clay tablets in the lot was made, or about from whom the dealer David had purchased them, and therefore about the provenance of the clay tablets. However, to judge from the inventory lists of the museum, the years 1913 or 1914 appear to be the most likely acquisition dates. All other texts from the same lot that are mathematical, like VAT 8393, were published long ago. In particular, VAT 8389, 8390, and 8391, with nearby catalog numbers, were published in Neugebauer's *Mathematische Keilschrift-Texte 1* (1935), pp. 317, 395, and 317. Only VAT 8393 has remained unpublished, why is not known. *Incidentally, the mathematical terminology in VAT 8393 is quite different from the terminology used in all other mathematical texts of the lot, so a common provenance can be excluded.*

For this and other reasons, the authors of this appendix are happy to be able to present here this extraordinarily interesting mathematical cuneiform text, which is an important testimony of the surprisingly high level of Old Babylonian mathematics (in some instances). The difficult reading of the cuneiform text was accomplished by the two authors in a mutual giving and taking, where J. Friberg, who initially knew the text only from less than perfect photos, had to rely on the exactness of the readings made by J. Marzahn, while the latter, whose understanding of the mathematical content was limited, had to rely on the former's explanations of the text. Gradually, the increasing understanding of the text made possible a far-reaching, error free cleaning of the text, which in its turn made it possible to finally read large parts of the text that had been obscured by dirt ever since the excavation of the clay tablet. When, eventually, in this way, the meaning of the text was revealed, both authors were pleasantly surprised.

The clay tablet is 6.9 cm high, 8.8 cm wide, and 3.2 cm thick. It consists of two large fragments glued together, with the addition of a small superficial flake on the obverse. It is no longer possible to say if the pieces were glued before the acquisition or in the course of some previous preservation procedure. Although the clay tablet remains unbaked, not much of it has been lost, so that almost the whole text can be read after the cleaning. Also the diagram which accompanies and explains the text is relatively well preserved, even if the proportions in it are far from correct. In a few places, the beginnings of straight lines in the diagram show that the same stylus was used for the drawing of the diagram as for the writing of the text.

The text is undated, but the hand writing and the inventory of cuneiform signs used in it, as well as the form of the imprint of the tip of the stylus, show that it is from the Old Babylonian period, most likely from between the 18th and 16th centuries BC. A more precise dating is probably not possible. The size of the text and the diagram relative to the size of the clay tablet, and also the sureness of the hand, indicate that the author of the text was a well educated scribe with a considerable routine. On the other hand, there are several peculiarities in the cuneiform text. The signs ŠA and TA, normally easily distinguishable, can in VAT 8393 hardly be separated from each other, they are almost completely identical. In other cases, variant writings can be found inside the text. Thus, the sign KI is written in lines 3, 7, 8, and 9 with a clear vertical wedge on the left side, a wedge that is missing in other places where it cannot be found even in a microscope and after cleaning. The case is the same with the sign LI, which only in lines 2, 3, and 8 displays its characteristic vertical wedge in the middle of the sign, while the wedge does not appear elsewhere.

It is, in view of such variations in the way of writing the signs, hard to believe that VAT 8393 is a text written by a school boy. It is more likely that it was produced by a well advanced student. This conclusion is supported by the observation above that VAT 8393 probably is an excerpt from a large and systematically organized theme text.

VAT 8393 belongs to no known group of OB mathematical texts. Characteristic terms in it are *ṣi-ip* 'double' (= add), *ḫu-ru-uṣ* 'tear off' (= subtract), *bi-il* 'bring' (= multiply), igi *pu-ṭù-ur* 'resolve the opposite' (= compute the reciprocal), *ša i-li-a-kum* 'what comes up for you' (= the result), and ninda-*nu* 'ninda' (= the growth rate). The terms *ḫu-ru-uṣ*, *bi-il*, igi *pu-ṭù-ur*, and *ša i-li-a-kum* appear also in **CBS 19761**, a mathematical fragment from Nippur (Robson, *Sciamvs* 1 (2000)), 36). Therefore, VAT 8393, too, is probably a text from Nippur. In addition, the terms *bi-il, ša i-li-a-kum*, and (the inflected form) *ni-in-da-nam* appear in the brief text **YBC 10522** (*MCT* (1945), text Uc), while the term *(n)in-da-nu* alone appears repeatedly in the two texts **MS 3052** and **MS 2792**, both from Uruk (Friberg, *RC* (2007), Chapter 10).

Appendix 2

A Catalog of Babylonian Geometric Figures

A large number of Babylonian geometric figures have been mentioned in this book, in various connections. On the following three pages, an effort is made to enumerate *all plane or solid geometric figures* appearing in cuneiform mathematical texts.

The following notations will be used:

c	the most likely form of a figure mentioned in a mathematical *table of constants*
d	the most likely intended form of a figure shown in a geometric *diagram*
p	the most likely form of a figure mentioned in a mathematical *problem text*
t	the most likely form of a figure related to entries in a mathematical *table text.*

If nothing else is said the geometric figure appears, in one way or another, in an *Old Babylonian* mathematical text. Otherwise, the following notations are used

(LB)	the figure appears *only* in a Late Babylonian or Seleucid text
(+LB)	the figure appears *also* in a Late Babylonian or Seleucid text
(+OAkk)	the figure appears *also* in an *Old Akkadian* text
(+OSum)	the figure appears *also* in an *Old Sumerian (ED III)* text
(+pr-Sum)	the figure appears *also* in a *proto-Sumerian* text.

The figures are generally of three kinds. In illustrations 1 a - 1 h are shown a basic set of plane geometric figures, divided by diagonals, transversals, *etc*. In 2 a - 2 g are shown more complicated plane geometric figures, such as figures within figures, concentric figures, repeatedly divided figures, and rings or chains of figures. In 3 a - 3 g, finally, are shown a number of solid figures, and in 3 h examples of the not very successful attempts of Babylonian mathematicians to depict such solid figures.

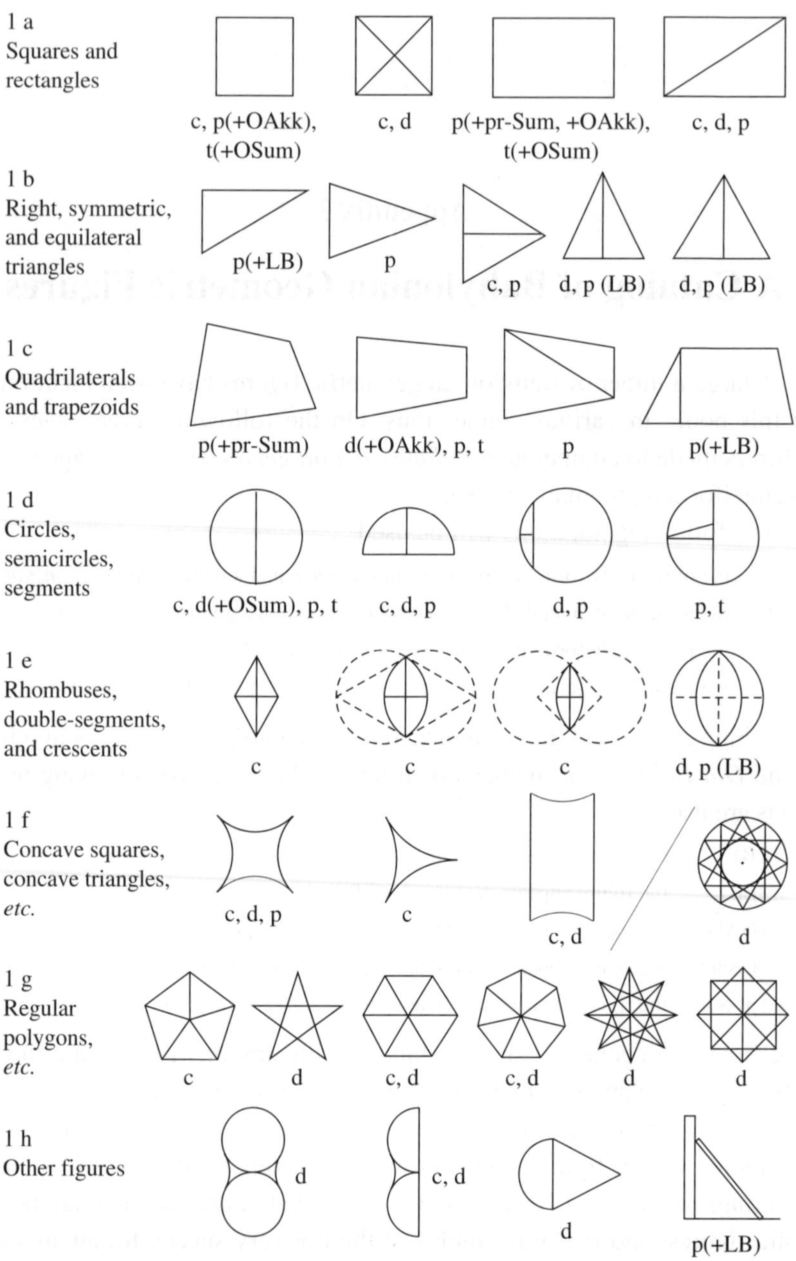
1 a
Squares and rectangles
c, p(+OAkk), t(+OSum)
c, d
p(+pr-Sum, +OAkk), t(+OSum)
c, d, p
1 b
Right, symmetric, and equilateral triangles
p(+LB)
p
c, p
d, p (LB)
d, p (LB)
1 c
Quadrilaterals and trapezoids
p(+pr-Sum)
d(+OAkk), p, t
p
p(+LB)
1 d
Circles, semicircles, segments
c, d(+OSum), p, t
c, d, p
d, p
p, t
1 e
Rhombuses, double-segments, and crescents
c
c
c
d, p (LB)
1 f
Concave squares, concave triangles, *etc.*
c, d, p
c
c, d
d
1 g
Regular polygons, *etc.*
c
d
c, d
c, d
d
d
1 h
Other figures
d
c, d
d
p(+LB)

2 a
Basic
metric algebra
figures

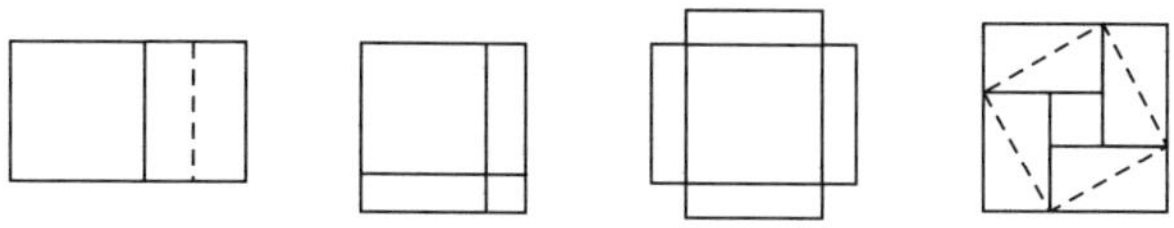

2 b.
Concentric
figures

2 c
Figures within
figures

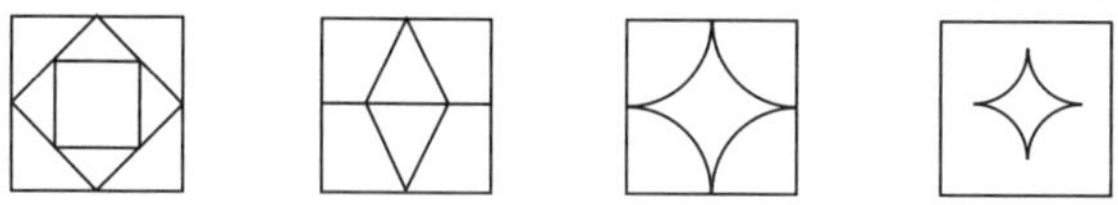

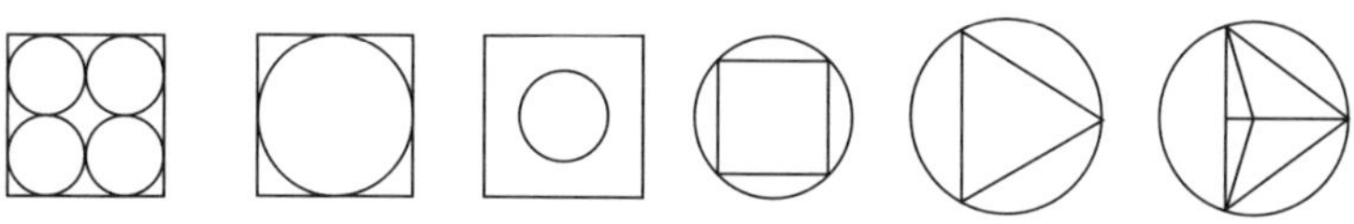

2 d
Striped triangles
and trapezoids

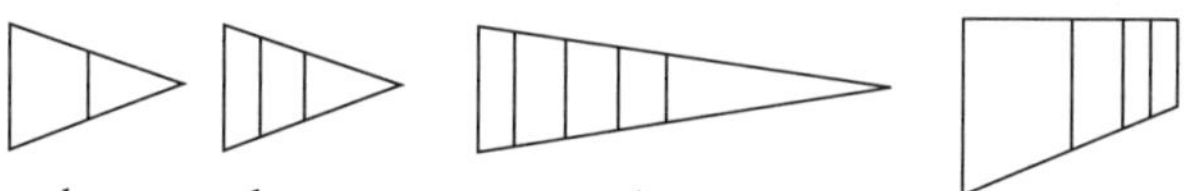

2 e
Bisected
trapezoids

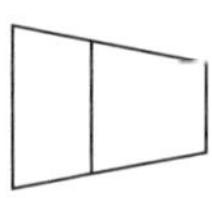

Confluent
bisections

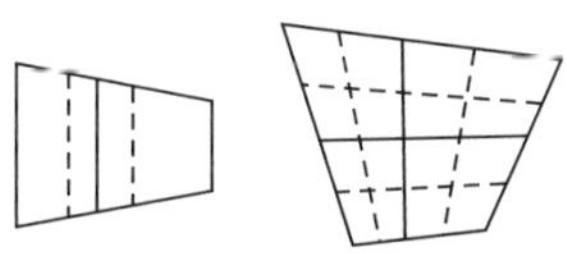

2 f
Rings of
figures

2 g
Chains of
figures

2 h
Similar
sub-triangles

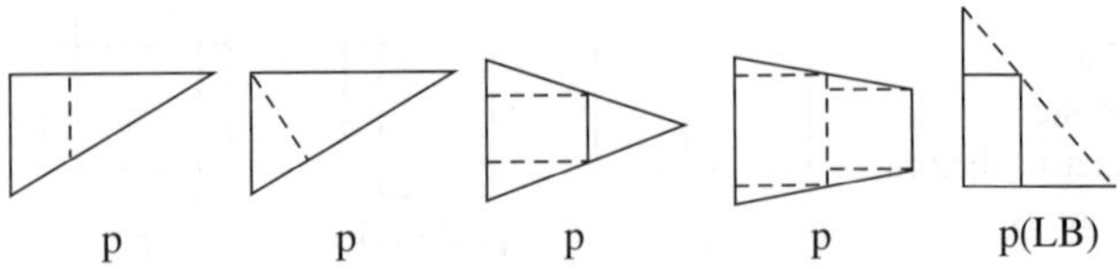

3 a
Walls or canals
with rectangular,
triangular, or
trapezoidal
cross sections

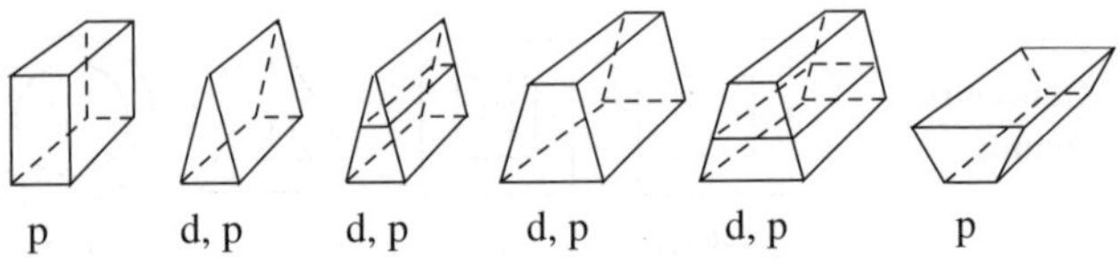

3 b
Cubes, whole
and truncated
square pyramids,
step pyramids

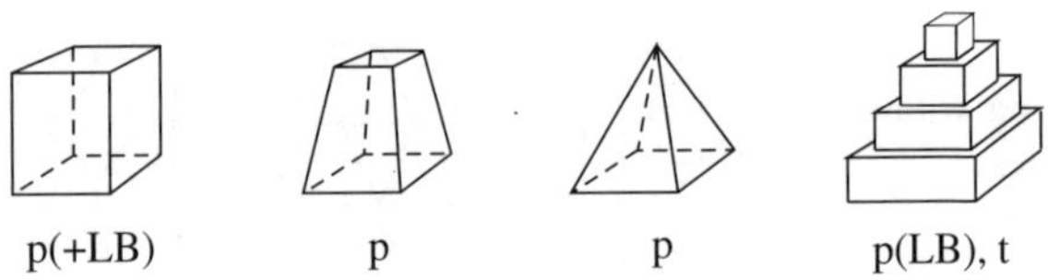

3 c
Cylinders, whole
and truncated
circular cones,
and ring-cones

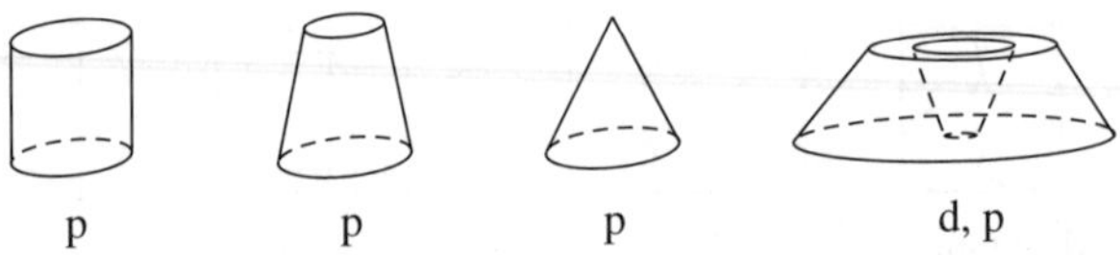

3 d
Whole and
truncated
crest pyramids

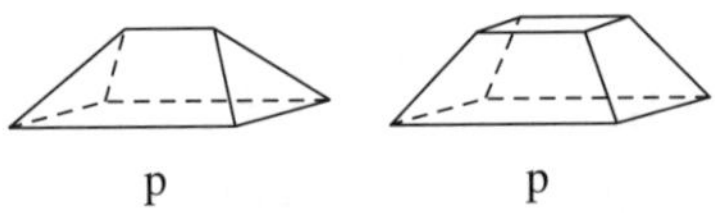

3 e
Gate with
interior
diagonal

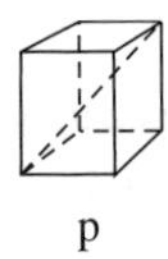

3 f
Ramps with
straight or
sloping sides

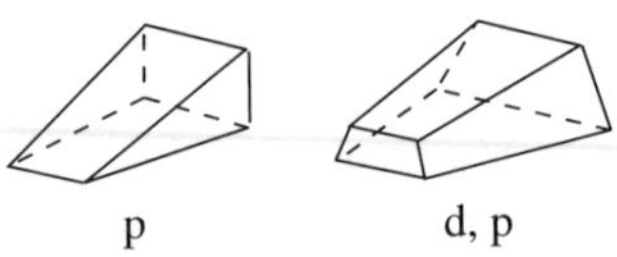

3 g
Horn figure
(icosahedron)

3 h
Drawings of
solid figures

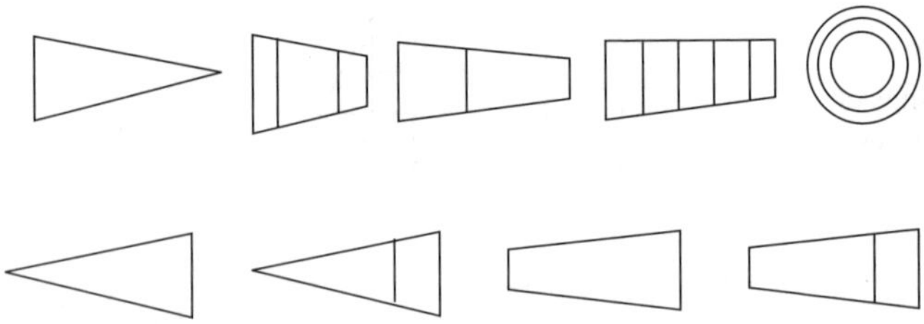

Index of Texts, Propositions, Lemmas

BM 96954+, § 1 f	9.3	9.3.3	a ridge pyramid truncated at mid-height	OB
BM 96954+ § 4	9.3		problems for cones and truncated cones	OB
BR 2-19	6.2		parameters for the circle, the semicircle, *etc.*	OB
BR 10-12	12.3a	12.3.1	parameters for the 'bow field'	OB
BR 13-15	12.3b	12.3.2	parameters for the 'boat field'	OB
BR 16-18	12.3c	12.3.3	parameters for the 'barleycorn field'	OB
BR 19-21	12.3d	12.3.4	parameters for the 'ox-eye'	OB
BR 22-24	12.3e	12.3.5	parameters for the 'lyre-window'	OB
BR 25	12.3f	12.3.6	parameters for the 'lyre-window of 3'	OB
BR 26-28	7.8		parameters for the 5-, 6-, and 7-fronts	OB
BR 31	16.7		an approximation to sqs. 2	OB
Bss XII.21	14.3	14.3.1	Brahmagupta's area rule for quadrilaterals	Ind
Bss XII.28	14.5-6	14.6.1	Brahmagupta's diagonal rule	Ind
Bss XVIII.50-51	9.4		volumes of conical piles of grain	Ind
Bss XVIII.65-66	16.5	16.5.1	formal multiplication of number pairs	Ind
Bss XVIII.69-72	17.4		Brahmagupta's solution rules	Ind
CBS 19761	App.1		a mathematical fragment from Nippur	OB
Collections IV.1, see Pappus				Gr
CUNES 50-08-001	11.3a		a combined table of areas of squares	ED III
Data 54-55	10.4	10.4.3	figures given in form and magnitude	Gr
Data 57	10.1	10.1.3	parabolic applications of parallelograms	Gr
Data 58	10.2	10.2.2	elliptic applications of parallelograms	Gr
Data 59	10.3	10.3.1	hyperbolic applications of parallelograms	Gr
Data 66	11.1		a rule for the area of a triangle	Gr
Data 84	10.5	10.5.1	an equivalence rule for quadratic equations	Gr
Data 86	10.6	10.6.1-2	a rectangular-linear system of type B5	Gr
DPA 37	5.3	5.3.2	computation of the area of a square	OAkk
DPA 36-37	1.14	1.14.1	computations of the areas of squares	OAkk
DPA 39	10.1	10.1.2	a metric division exercise	OAkk
El. I.41. 43	5.3	5.3.1	relation to the OB similarity rule	76
El. I.43	10.1	10.1.1	equal complements about the diameter	Gr
El. I.44	10.1	10.1.1	parabolic applications of parallelograms	Gr
El. II.2-3	1.2	1.2.1-2	systems of equations *vs.* quadratic equations	Gr
El. II.4, II.7	1.3	1.3.1-2	quadratic-linear *vs.* rectangular-linear systems	Gr
El. II.5-6	1.4	1.4.1-2	rectangular-linear systems of equations	Gr
El. II.8	1.5	1.5.1-2	subtractive quadratic-linear systems of equations	Gr
El. II.9-10	1.6	1.6.1	constructive counterparts to *El.* II.4, II.7	Gr
El. II.11*, II.14*	1.7	1.7.1-2	constructive counterparts to *El.* II.5-6	Gr
El. II.12-13	1.8	1.8.1	constructive counterparts to *El.* II.8	Gr
El. III.32	12.2		proposition about the chord-tangent angle	Gr
El. IV	6.1		outline of contents	Gr
El. IV.10-11	6.1	6.1.1	preliminaries to the construction of a pentagon	Gr
El. VI.19	10.4	10.4.2	similar triangles in duplicate ratio of sides	Gr
El. VI.24	10.1		parallelograms about the diameter	Gr
El. VI.25	10.4	10.4.1	a figure of given shape and size	Gr
El. VI.28	10.2	10.2.1	elliptic applications of parallelograms	Gr

Geom. 24.7-9	18.3		relations between the sides of a right triangle	Gr
Geom. 24.10	18.3		given the area plus perimeter of a right triangle	Gr
Geom. 24.21	18.3		a pair of parallel right triangles	Gr
Geom. 24.26	18.3		the diameter of a circle inscribed in a right triangle	Gr
Geom. 24.46	18.3		given the sum of diameter, arc, and area of a circle	Gr
Hippocrates	12.1	12,1,1	quadrature of lunes according to Alexander	Gr
Hippocrates	12.2	12,2.1-2	quadrature of lunes according to Eudemus	Gr
IM 31248	11.4	11.4.2	a problem for a 5-striped trapezoid	OB
IM 43996	11.2j	11.2.11	a problem for a 3-striped triangle	OB
IM 51979	7.8	7.8.2	a drawing of an octagram	OB
IM 52916 = G	7.7		a table of constants	OB
IM 54472	16.7		elimination of square factors	OB
IM 55357	4.3	4.3.1-2	a chain of right sub.triangles in a right triangle	OB
IM 58045	11.3a	11.3.1	a drawing of a bisected trapezoid	OAkk
IM 67118	5.4	5.4.2	a quadratic-rectangular system of type B5	OB
IM 121613, 1	10.4		a form and magnitude problem	OB
Ist. Si. 269	11.3f	11.3.5	diagrams of six 2-striped trapezoids	OB
Ist. Si. 428	16.7	16.7.4	the square side of 2 02 02 02 05 05 04	OB
JZSS V.2-16	9.4		outline of contents, with diagrams	Chin
Kudurru	12.4	12.4.2	images of the Sun, the Moon, and Venus	Kass
Practica Geom.	11.0		Leonardo Pisano's classical work	Eur
Liber Mahameleth	1.13		a Latin manuscript based on Islamic sources	Eur
Liu Hui	9.3		commentary to *JZSS* V	Chin
MAH 16055	11.2i	11.2.10	outline of the clay tablet with its 10 diagrams	OB
MAH 16055	11.2i	11.2.9	a problem for a 3-striped triangle	OB
Meno, 82 B - 85 B	6.1		Socrates and the slave boy	Gr
Metrica I.8	14.1	14.1.1	Heron's triangle area rule	Gr
Metrica I.8 b	16.1	16.1.1	Heron's square side rule	Gr
Metrica II	9.3		volumes of pyramids and cones	Gr
Metrica III.4	18.4		a triangle divided into four sub-triangles	Gr
MLC 1950	11.2f	11.2.6	a problem for a 2-striped triangle	OB
MLC 2078	15.2	15.2.1-2	an OB algorithm table for a chain of trapezoids	OB
MS 1938/2	6.2	6.2.9	a circle in the middle of a hexagon	OB
MS 2192	2.4	2.4.2	a ring of three trapezoids	OB
MS 2985	6.2	6.2.8	a circle in the middle of a square	OB
MS 3049 § 1 a	1.12b	1.12.5	a problem for a chord in a circle	OB
MS 3049 § 5	8.2	8.2.1	the inner diagonal of a gate	OB
MS 3050	6.2	6.2.4	a square with diagonals inscribed in a circle	OB
MS 3051	6.2	6.2.5	an equilateral triangle inscribed in a circle	OB
MS 3052 § 1 c	11.2		a problem for a triangular clay wall	OB
MS 3876, 3	8.3	8.3.1-2	the weight of a copper icosahedron	Kass
MS 3908	11.4	11.4.1	a drawing of a 3-striped trapezoid	OB
MS 3971 § 2	5.4	5.4.2	a quadratic-rectangular system of type B5	OB
MS 3971 § 3	3.2	3.2.1	five igi-igi.bi problems	OB
MS 3971 § 4	3.3		a scaling problem for diagonal triples	OB
MS 4632	9.2	9.2.2	a spherical envelope with number tokens	pre-lit

TMS 3 = BR	6.2		a table of constants	OB
TMS 5	1.11		a catalog text with metric algebra problems	OB
TMS 5	13.6		a theme text with problems for squares	OB
TMS 5 § 7-8	2.4		problems for concentric and parallel squares	OB
TMS 14	9.3	9.3.1	a problem for a ridge pyramid	OB
TMS 18	11.2e	11.2.5	a clever problem for a 2-striped triangle	OB
TMS 19 b	16.7		an amazingly elegant metric algebra problem	OB
TMS 20	16.7		an exact square side computation	OB
TMS 21	6.2	6.2.6	a concave square in the middle of a square	OB
TMS 23	11.9	11.9.1-2	confluent bisections in two directions	OB
TSS 77	6.2	6.2.1	four circles inscribed in a square	ED III
UE 3, 78	7.9	7.9.1	drawing of a conjugate pair of hyperbolas	pr-Sum
UE 3, 286	7.9	7.9.3	two entangled acrobats (a concave square)	pr-Sum
UE 3, 393	7.9	7.9.4	four entangled acrobats	pr-Sum
UE 3, 518	7.9	7.9.6	four entangled armed men (a concave square)	ED III
UET 6/2 222	16.7	16.7.3	a square side algorithm by use of factorization	OB
VA 5953	7.9	7.9.7	a pentagram of entangled bearded men	OB
VAT 7351	1.12c	1.12.7	cross-wise striped trapezoids	OB
VAT 7531	11.8	11.8.1	cross-wise striped trapezoids	OB
VAT 7621	11.7		a $2 \cdot 9$-striped trapezoid	OB
VAT 7848, 4	1.12c		a problem for a rational symmetric trapezoid	Sel
VAT 7848, 4	13.4	13.4.2	a rational symmetric trapezoid	Sel
VAT 8389	13.5		systems of linear equations	OB
VAT 8393	App.1	A.1.1-6	a chain of trapezoids with fixed diagonals	OB
VAT 8512	11.2g	11.2.7	a clever problem for a 2-striped trapezoid	OB
VAT 8512	11.3b	11.3.3	Gandz' and Huber's interpretation	OB
VAT 8521	13.6		indeterminate interest problems	OB
VAT 12593	11.3a		table of areas of squares	ED III
W 23291 § 1 a-g	1.13	1.13.1-6	metric algebra problems in terms of seed measure	LB
W 23291 § 4 a-b	7.7		two rules for the area of an equilateral triangle	LB
W 23291 § 4 a	16.7		a standard approximation to sqs. 5	LB
W 23291 § 4 b	16.7		a standard approximation to sqs 3	LB
W 23291 § 4 c	16.7		an accurate approximation to sqs. 3	LB
W 23291-x § 1	12.4	12.4.1	double circle segment and lunes	LB
W 23291-x § 2	1.13	1.13.7	a problem for five concentric circles	LB
W 23291-x § 4	1.13		metric algebra problems in terms of area measure	LB
YBC 10522	App.1		a brief text related to VAT 8393	OB
YBC 4608	11.3d		a problem for a 2-striped trapezoid	OB
YBC 4675	11.3c		a problem for a bisected quadrilateral	OB
YBC 4696	11.2h	11.2.8	outline of the clay tablet	OB
YBC 4696	11.2h		a series of problems for a 2-striped triangle	OB
YBC 4698, 4	13.5		a price and weight problem for gold and iron	OB
YBC 4709	10.8		systems of equations of types B5 and B6	Kass
YBC 7289	16.7	16.7.2	an accurate approximation to sqs. 2	OB
YBC 7359	6.2	6.2.7	a square in the middle of a square	OB

Index of Subjects

Bibliography

Chapter 1. *Elements* II and Babylonian Metric Algebra

Artmann, B. (1991) Euclid's Elements and its prehistory, in I. Mueller (ed.) *Peri Ton Mathematon. Apeiron* 24, 1-47.

Berggren, J. L. (1984) History of Greek mathematics: A survey of recent research. *Historia Mathematica* 11, 394-410.

Bruins, E. M. and M. Rutten (1961) *Textes mathématiques de Suse.* Paris: Paul Geuthner.

Christianidis, J. (ed.) (2004) *Classics in the History of Greek Mathematics.* Dordrecht/Boston/London: Kluwer Academic Publishers.

Freudenthal, H. (1977) What is algebra and what has been its history? *Archive for History of Exact Sciences* 16, 189-200.

Friberg, J. (1981) Plimpton 322, Pythagorean triples,and the Babylonian triangle parameter equations. *Historia Mathematica* 8, 277-318.

——— (1981) Methods and traditions of Babylonian mathematics, 2. An Old Babylonian catalogue text with equations for squares and circles. *Journal of Cuneiform Studies* 33, 57-64.

——— (1990) Mathematik (in English), in *Reallexikon der Assyriologie und Vorderasiatischen Archäologie, vol. 7.* Berlin & New York: de Gruyter, 531-585.

———, H. Hunger, F. Al-Rawi (1990) 'Seed and Reeds', a metro-mathematical topic text from Late Babylonian Uruk. *Baghdader Mitteilungen* 21, 483-557, pl. 46-48.

——— (1993) On the structure of cuneiform metrological table texts from the -1st millennium. *Grazer Morgenländische Studien* 3, 383-405.

——— (1997) 'Seed and Reeds Continued'. Another metro-mathematical topic text from Late Babylonian Uruk. *Baghdader Mitteilungen* 28, 251-365, pl. 45-46.

——— (1997/98) Round and almost round numbers in proto-literate metro-mathematical field texts. *Archiv für Orientforschung* 44/45, 1-58.

——— (2005) On the alleged counting with sexagesimal place value numbers in mathematical cuneiform texts from the third millennium B.C. *Cuneiform Digital Library Journal* 2005:2. http://cdli.ucla.edu/pubs/cdlj/2005/cdlj2005_002.html

——— (2005) *Unexpected Links between Egyptian and Babylonian Mathematics.*

Singapore: World Scientific.

—— (2007) *A Remarkable Collection of Babylonian Mathematical Texts.* New York: Springer .

Heath, T. L. (1921; Dover edition 1981) *A History of Greek Mathematics, 1-2.* Oxford: Clarendon Press.

—— (1926; Dover edition 1956) *Euclid. The Thirteen Books of The Elements, I-III.* New York: Dover Publications.

Høyrup, J. (1990) Algebra and naive geometry. An investigation of some basic aspects of Old Babylonian mathematical thought. *Altorientalische Forschungen* 17, 27-69, 262-354.

—— (2002) *Lengths, Widths, Surfaces. A portrait of Old Babylonian algebra and its kin.* New York, Berlin: Springer.

Joyce, D. E. *A quick trip through the Elements.* http://aleph0.clarku.edu/~djoyce/java/elements/elements.html

Melville, D. (2004) Poles and walls in Mesopotamia and Egypt. *Historia Mathematica* 34, 148-162.

Muroi, K. (1991) Two Babylonian mathematical problems concerning a timber placed against a wall. (English summary.). *Kagakushi Kenkyu* 30, 23-27.

Neugebauer, O. (1934-1936) Zur geometrischen Algebra (Studien zur Geschichte der antiken Algebra III). *Quellen und Studien B 3*, 245-259.

—— (1935-37 (reprinted 1973)) *Mathematische Keilschrift-Texte, I-III.* Berlin: Springer Verlag.

Parker, R. A. (1972) *Demotic Mathematical Papyri.* Providence, R. I. & London: Brown University Press.

Sesiano, J. (1987) Survivance médiévale en Hispanie d'un problème né en Mésopotamie. *Centaurus* 30, 18-61.

Szabó, A. (1969) *The Beginnings of Greek Mathematics.* Dordrecht: Reidel.

Tannery, P. (1882) De la solution géometrique des problèmes du second degré avant Euclide. *Mémoires de la société des sciences physiques et naturelles de Bourdeaux 2e Série* 4, 395-416.

Tropfke, J., K. (Vogel, *et al.*) (1980) *Geschichte der Elementarmathematik, 4. Auflage, Band 1: Arithmetik und Algebra.* Berlin/New York: Walter de Gruyter.

Unguru, S. (1975) On the need to rewrite the history of Greek mathematics. *Archive for History of Exact Sciences* 15, 67-115.

—— (1979) History of ancient mathematics: Some reflections on the ancient state of the art. *Isis* 70, 555-565.

—— and D. E. Rowe (1981, 1982) Does the quadratic equation have Greek roots? A study of 'geometric algebra', 'application of areas', and related problems. *Libertas Mathematica* 1, 1-49; 2, 1-62.

van der Waerden, B. L. (1976) Defence of a "shocking" point of view. *Archive for History of Exact Sciences* 15, 199-210.

Vitrac, B. (1990) *Euclide d'Alexandrie. Les Elements, 1. Introduction générale. Livres I à IV.* Paris: Presses Universitaires de France.

Weil, A. (1978) Who betrayed Euclid? *Archive for History of Exact Sciences* 19, 91-93.

Zeuthen, H. G. (1896) *Die Lehre von den Kegelschnitten im Altertum.* Copenhagen: Fischer-Benzon.

Chapter 2. *El.* I.47 and the Old Babylonian Diagonal Rule

Bruins, E. M. and M. Rutten (1961) *Textes mathématiques de Suse.* Paris: Paul Geuthner.

Friberg, J. (2007) *A Remarkable Collection of Babylonian Mathematical Texts.* New York: Springer .

Heath, T. L. (1926; Dover edition 1956) *Euclid. The Thirteen Books of The Elements, I-III.* New York: Dover Publications.

Høyrup, J. (2002) *Lengths, Widths, Surfaces. A portrait of Old Babylonian algebra and its kin.* New York, Berlin: Springer.

Loomis, E. S. (1968) *The Pythagorean Proposition. Yearbook of the National Council of Teachers of Mathematics.* New York.

Chapter 3. The Lemma *El.* X.28/29 1a, Plimpton 322, and Babylonian igi-igi.bi problems

Friberg, J. (1981) Plimpton 322, Pythagorean triples,and the Babylonian triangle parameter equations. *Historia Mathematica* 8, 277-318.

—— (2007) *A Remarkable Collection of Babylonian Mathematical Texts.* New York: Springer .

Heath, T. L. (1921; Dover edition 1981) *A History of Greek Mathematics, 1-2.* Oxford: Clarendon Press.

—— (1926; Dover edition 1956) *Euclid. The Thirteen Books of The Elements, I-III.* New York: Dover Publications.

Chapter 4. The Lemma *El.* X.32/33 and an OB Geometric Progression

Baqir, T. (1950) An important mathematical problem text from Tell Harmal. *Sumer* 6, 39-54.

Heath, T. L. (1926; Dover edition 1956) *Euclid. The Thirteen Books of The Elements, I-III.* New York: Dover Publications.

Chapter 5. *Elements* X and Babylonian Metric Algebra

Friberg, J. (2005) On the alleged counting with sexagesimal place value numbers in mathematical cuneiform texts from the third millennium B.C. *Cuneiform Digital Library Journal* 2005/2. http://cdli.ucla.edu/pubs/cdlj/2005/cdlj2005_002.html

—— (2007) *A Remarkable Collection of Babylonian Mathematical Texts.* New York: Springer .

Heath, T. L. (1926; Dover edition 1956) *Euclid. The Thirteen Books of The Elements, I-III.* New York: Dover Publications.

Høyrup, J. (2002) *Lengths, Widths, Surfaces. A portrait of Old Babylonian algebra and its kin.* New York, Berlin: Springer.

Neugebauer, O. (1935-37 (reprinted 1973)) *Mathematische Keilschrift-Texte, I-III.* Berlin: Springer Verlag.

Taisbak, C. M. (1982) *Coloured Quadrangles. A Guide to the Tenth Book of Euclid's Elements.* Copenhagen: Museum Tusculanum Press.

Knorr, W. (1983) 'La croix des mathématiciens': The Euclidean theory of irrational lines. *Bulletin of the American Mathematical Society* 9, 41-69.

Chapter 6. *Elements* IV and Babylonian Figures Within Figures

Bruins, E. M. and M. Rutten (1961) *Textes mathématiques de Suse.* Paris: Paul Geuthner.

Friberg, J. (2005) *Unexpected Links between Egyptian and Babylonian Mathematics.* Singapore: World Scientific.

—— (2007) *A Remarkable Collection of Babylonian Mathematical Texts.* New York: Springer .

Heath, T. L. (1921; Dover edition 1981) *A History of Greek Mathematics, 1-2.* Oxford: Clarendon Press.

—— (1926; Dover edition 1956) *Euclid. The Thirteen Books of The Elements, I-III.* New York: Dover Publications.

Jestin, R. R. (1937) *Tablettes sumériennes de Shuruppak au Musée de Stmboul.* Paris.: E. de Boccard.

Muroi, K. (2000) Quadratic equations in the Susa mathematical text no. 21. *Sciamvs* 1: 3-10.

Neugebauer, O. (1935-37 (reprinted 1973)) *Mathematische Keilschrift-Texte, I-III.* Berlin: Springer Verlag.

Robson, E. (1999) *Mesopotamian Mathematics 2100-1600 BC. Technical Constants in Bureaucracy and Education.* Oxford: Clarendon Press.

Chapter 7. *El.* VI.30, XIII.1-12 and Regular Polygons in Babylonian Mathematics

Andrae, W. (1937) *Berichte aus den Preußischen Kunstsammlungen* 58, 34-35.

Bruins, E. M. and M. Rutten (1961) *Textes mathématiques de Suse.* Paris: Paul Geuthner.

Friberg, J. (1997) 'Seed and Reeds Continued'. Another metro-mathematical topic text from Late Babylonian Uruk. *Baghdader Mitteilungen* 28, 251-365, pl. 45-46.

—— (2007) *A Remarkable Collection of Babylonian Mathematical Texts.* New York: Springer .

Goetze, A. (1951) A mathematical compendium from Tell Harmal. *Sumer* 7, 126-155.

Heath, T. L. (1926; Dover edition 1956) *Euclid. The Thirteen Books of The Elements, I-III.* New York: Dover Publications.

Knorr, W. (1983) 'La croix des mathématiciens': The Euclidean theory of irrational lines. *Bulletin of the American Mathematical Society* 9, 41-69.

Legrain, L. (1936) *Archaic Seal-Impressions* (Ur Excavations, 3). Oxford: Oxford University Press.

Robson, E. (1999) *Mesopotamian Mathematics 2100-1600 BC: Technical Constants in Bureaucracy and Education.* Oxford: Clarendon Press.

Rochberg-Halton, F. (1987) TCL 16, 3: Mixed tradition in Late Babylonian astrology. *Zeitschrift für Assyriologie* 77, 207-222.

Thureau-Dangin, F. (1922) *Tablettes d'Uruk à l'usage des prêtres du temple d'Anu au temps des Séleucides.* Paris: P. Geuthner.

Vitrac, B. (1998) *Euclide. Les Elements, 3.* Livre X. Paris: Presses Universitaires de France.

Chapter 8. *El.* XIII.13-18 and Polyhedrons in Babylonian Mathematics

Friberg, J. (1994) Preliterate counting and accounting in the Middle East. A constructively critical review of Schmandt-Besserat's Before Writing. *Orientalistische Literaturzeitung* 89, 477-502.

—— (1996) Pyramids and cones in ancient mathematical texts. New hints of a common tradition. *Proceedings of the Cultural History of Mathematics* 6, 80-95.

—— (2007) *A Remarkable Collection of Babylonian Mathematical Texts.* New York: Springer .

Heath, T. L. (1926; Dover edition 1956) *Euclid. The Thirteen Books of The Elements, I-III.* New York: Dover Publications.

Chapter 9. Elements XII and Pyramids and Cones in Babylonian Mathematics

Cullen, C. (2004) *The Suan Shu Shu – Writings On Reckoning.* http://www.nri.org.uk/Suan Shushu.html

Dehn, M. (1902. Ueber den Rauminhalt. *Mathematische Annalen* 55, 465-478.

Friberg, J. (1996) Pyramids and cones in ancient mathematical texts. New hints of a common tradition. *Proceedings of the Cultural History of Mathematics* 6, 80-95.

—— (2005) *Unexpected Links between Egyptian and Babylonian Mathematics.* Singapore: World Scientific.

—— (2007) *A Remarkable Collection of Babylonian Mathematical Texts.* New York: Springer .

Heath, T. L. (1921; Dover edition 1981) *A History of Greek Mathematics, 1-2.* Oxford: Clarendon Press.

—— (1926; Dover edition 1956) *Euclid. The Thirteen Books of The Elements, I-III.* New York: Dover Publications.

Jessen, B. (1939) Om Polyedres Rumfang. *Matematisk Tidskrift, A*: 35-44.

Robson, E. (1999) *Mesopotamian Mathematics 2100-1600 BC: Technical Constants in Bureaucracy and Education.* Oxford: Clarendon Press.

Schmandt-Besserat, D. (1992) *Before Writing, Vol. 1: From Counting to Cuneiform..* Austin TX, University of Texas Press.

Shen, K., Crossley, J. N., and Lun, A W.-C. (1999) *The Nine Chapters on the Mathematical Art – Companion and Commentary.* Oxford: Oxford University Press & Beijing: Science Press.

Vogel, K. (1968) *Chiu Chang Suan Shu, Neun Bücher Arithmetischer Technik.* Braunschweig: Friedr. Vieweg & Sohn.

Wagner, D. B. (1975) *Proof in Ancient Chinese Mathematics. Liu Hui on the Volumes of Rectilinear Solids* (Specialopgave i forbindelse med hovedfagseksamen i kinesisk), Københavns Universitet.

—— (1979) An early Chinese derivation of the volume of a pyramid: Liu Hui, Third century A. D. *Historia Mathematica* 6, 164-188.

Chapter 10. *El.* I.43-44, *El.* VI.24-29, *Data* 57-59, 84-86, and Metric Algebra

Friberg, J. (1990) Mathematik (in English), in *Reallexikon der Assyriologie und Vorderasiatischen Archäologie, vol. 7.* Berlin & New York: de Gruyter, 531-585.

—— (2005) On the alleged counting with sexagesimal place value numbers in mathematical cuneiform texts from the third millennium B.C. *Cuneiform Digital Library Journal* 2005/2. http://cdli.ucla.edu/pubs/cdlj/2005/cdlj2005_002.html

—— (2005) *Unexpected Links between Egyptian and Babylonian Mathematics.* Singapore: World Scientific.

Heath, T. L. (1926; Dover edition 1956) *Euclid. The Thirteen Books of The Elements, I-III.* New York: Dover Publications.

Knorr, W. (1986 (Dover edition 1993)) *The Ancient Tradition of Geometric Problems.* Boston, Birkhäuser.

Neugebauer, O. (1934-36) Serientexte in der babylonischen Mathematik. *Quellen und Studien B 3*, 106-114.

——— (1935-37 (reprinted 1973)) *Mathematische Keilschrift-Texte, I-III*. Berlin: Springer Verlag.

Saito, K. (1985) Book II of Euclid's Elements in the light of the theory of conic sections. *Historia Scientiarum* 28, 31-60.

——— (2004) Book II of Euclid's Elements in the light of the theory of conic sections. *Classics in the History of Greek Mathematics.* (ed. J. Christianidis), Dordrecht/Boston/london: Kluwer A. P., 139-167.

Taisbak, C. M. (2003) *Euclid's Data. The Importance of Being Given.* Copenhagen: Museum Tusculanum Press.

——— (2003) Exceeding and falling short: Elliptical and hyperbolical application of area. *Science in Context* 16, 299-318.

Tannery, P. (1882) De la solution géometrique des problèmes du second degré avant Euclide. *Mémoires de la société des sciences physiques et naturelles de Bordeaux* (2e Série) 4, 395-416.

Zeuthen (1917) Sur la reforme qu'a subie la mathématique de Platon à Euclide, et grâce à laquelle elle est devenue science raisonnée. (French summary in *Hvorledes Mathematiken i tiden fra Platon til Euklid blev rationel videnskab*, Copenhagen: A. F. Høst & søn. (Available online at the *Digital Mathematics Library.*)

Chapter 11. Euclid's *On Divisions* and Babylonian Striped Triangles or Trapezoids

Archibald, R. C. (1915) *Euclid's Book On Divisions of Figures*. Cambridge: University Press.

Boncompagni, B. (1862) *Scritti di Leonardo Pisano 2. Leonardi Pisano Practica Geometriae ed opuscoli*. Rome: Scienze matematiche e fisiche.

Bruins, E. M. (1951) *Nouvelles découvertes sur les mathématiques babyloniennes.* Paris.

——— (1953) Three geometrical problems. *Sumer* 9, 255-259.

——— and M. Rutten (1961) *Textes mathématiques de Suse.* Paris: Paul Geuthner.

——— (1964) *Codex Constantinopolitanus Palatii Veteris No. 1*. Leiden: E. J. Brill.

Friberg, J. (1990) Mathematik (in English), in *Reallexikon der Assyriologie und Vorderasiatischen Archäologie, vol. 7*. Berlin & New York: de Gruyter, 531-585.

——— (2005) *Unexpected Links between Egyptian and Babylonian Mathematics.* Singapore: World Scientific.

——— (2005) On the alleged counting with sexagesimal place value numbers in mathematical cuneiform texts from the third millennium B.C. *Cuneiform Digital Library Journal* 2005/2. http://cdli.ucla.edu/pubs/cdlj/2005/cdlj2005_002.html

—— (2007) *A Remarkable Collection of Babylonian Mathematical Texts.* New York: Springer .

Heath, T. L. (1921; Dover edition 1981) *A History of Greek Mathematics, 1-2.* Oxford: Clarendon Press.

Hogendijk, J. P. (1993) The Arabic text of Euclid's On Divisions. In M. Folkerts and J. P. Hogendijk (eds.) *Vestigia Mathematica.* Amsterdam/Atlanta, GA: Rodopi B. V.

Høyrup, J. (2002) *Lengths, Widths, Surfaces. A portrait of Old Babylonian algebra and its kin.* New York, Berlin: Springer.

Neugebauer, O. (1935-37 (reprinted 1973)) *Mathematische Keilschrift-Texte, I-III.* Berlin: Springer Verlag.

—— and A. Sachs (1945) *Mathematical Cuneiform Texts.* New Haven, CT: American Oriental Society.

Nissen, Hans J., Peter Damerow, and Robert K. Englund (1993) *Archaic Bookkeeping, Early Writing and Techniques of Economic Administration in the Ancient Near East,* Chicago, London: U. of Chicago Press.

Parker, R. A. (1975) A mathematical exercise – P. Dem. Heidelberg 663. *Journal of Egyptian Archaeology* 61, 189-196.

Robson, E. (1999) *Mesopotamian Mathematics 2100-1600 BC. Technical Constants in Bureaucracy and Education.* Oxford: Clarendon Press.

Thomas, I. (1939 (reprinted 1980)) *Selections Illustrating the History of Greek Mathematics. I. Thales to Euklid.* Cambridge, MA.:Harvard University Press; London: William Heinemann

Thureau-Dangin, F. (1938) *Textes mathématiques babyloniens.* Leiden: E. J. Brill.

Vaiman, A. A. (1955) Ermitazhnaya klinopisnaya matematicheskaya tablichka No. 015189. (The cuneiform mathematical table text Erm. 015189.) *Epigrafika Vostoka* 10: 71-83, pl. 1-2.

—— (1961) *Shumero-vavilonskaya matematika III - I tysyacheletiya do n. e.* (*Sumero-Babylonian mathematics in the third to first millennia BCE.*) Moskva: Izdatel'stvo vostochnoi literatury.

Chapter 12. Hippocrates' Lunes and Babylonian Figures with Curved Boundaries

Bruins, E. M. and M. Rutten (1961) *Textes mathématiques de Suse.* Paris: Paul Geuthner.

Friberg, J., H. Hunger, F. Al-Rawi (1990) 'Seed and Reeds', a metro-mathematical topic text from Late Babylonian Uruk. *Baghdader Mitteilungen* 21, 483-557, pl. 46-48.

King, L. W. (1912) *Babylonian Boundary-Stones and Memorial Tablets in the British Museum.* London: Oxford University Press.

Knorr, W. (1986 (Dover edition 1993)) *The Ancient Tradition of Geometric Problems.*

Boston: Birkhäuser.

Neugebauer, O. (1935-37 (reprinted 1973)) *Mathematische Keilschrift-Texte, I-III*. Berlin: Springer Verlag.

—— and A. Sachs (1945) *Mathematical Cuneiform Texts*. New Haven, CT: American Oriental Society.

Thomas, I. (1939 (reprinted 1980)) *Selections Illustrating the History of Greek Mathematics. I. Thales to Euklid*. Cambridge, MA.:Harvard University Press; London: William Heinemann

Vaiman, A. A. (1963) Istolkovanie geometriceskih postoyannyh iz suzskogo klinopisnogo spiska I (Suzy) (Interpretation of the geometric constants in the cuneiform table text TMS I from Susa). *Vestnik drevni istorii* I :83, 75-86.

van der Waerden, B. L. (1975) *Science Awakening*. Groningen: Nordhoff.

Chapter 13. Traces of Babylonian Metric Algebra in the *Arithmetica* of Diophantus

Bashmakova, I. G. (1997) *Diophantus and Diophantine Equations*. The Mathematical Association of America.

Christianidis, J. (1995) Maxime Planude sur le sens du terme diophantien *plasmatikón*. *Historia Scientiarum* 6, 37-41

Friberg, J. (2005) *Unexpected Links between Egyptian and Babylonian Mathematics*. Singapore: World Scientific.

Heath, T. L. (1885 (reprinted 1964)) *Diophantus of Alexandria; A Study in the History of Greek Algebra, 2nd edition*. New York: Dover.

Rashed, R. (1984) *Diophante. Les Arithmétiques. Vol. 3: Livre IV; Vol. 4: Livres V, VI, VII*. Paris: Les Belles Lettres.

Sesiano, J. (1982) *Books IV to VII of Diophantus' Arithmetica in the Arabic Translation*. New York/Heidelberg/Berli:, Springer-Verlag.

—— (1990) Frühalgebraische Aspekte in der "Arithmetica" Diophants. In *Geschichte der Algebra* (ed. E. Scholz). Mannheim: Wissenschaftsverlag.

Chapter 14. Heron's and Brahmagupta's Area Rules

Amma, S. (1979) *Geometry in Ancient and Medieval India*. Delhi: Motilal Banarsidass.

Colebrooke, H. T. (1817 (reprinted 1973)) *Algebra with Arithmetic and Mensuration from the Sanskrit of Brahmagupta and Bhaskara*. Walluf bei Wiesbaden, Sändig.

Heath, T. L. (1921; Dover edition 1981) *A History of Greek Mathematics, 1-2*. Oxford: Clarendon Press.

Høyrup, J. (1997) Sulla posizione della "formula di Erone" nei Metrica (con una nota platonica). *Bolletino di Storia delle Scienze Matematiche* 17, 3-11.

King, D. A. and M. H. Kennedy (eds.) (1983) *Studies in the Islamic Sciences by E. S. Kennedy, Colleagues and Former Students.* Beirut: American University of Beirut.

Schöne, H. (1903) *Heronis Alexandrini opera quae supersunt omnia. Vol. III: Rationes dimetiendi et commentatio dioptrica. (Vermessungslehre und Dioptra.).* Leipzig: Teubner.

Tropfke, J. (1940) *Geschichte der Elementar-Mathematik, Bd 4: Ebene Geometrie.* Berlin: Gruyter.

Chapter 15. Theon of Smyrna's Side and Diagonal Numbers and Ascending Infinite Chains of Birectangles

Fowler, D. H. (1987) *The Mathematics of Plato's Academy. A New Reconstruction.* Oxford: Oxford Science Publications.

Heath, T. L. (1921; Dover edition 1981) *A History of Greek Mathematics, 1-2.* Oxford: Clarendon Press.

Hofmann, J. E. (1956). Ergänzende Bemerkungen zum "geometrischen" Irrationalitätsbeweis der alten Griechen. *Centaurus* 5, 59-72.

Knorr, W. R. (1975) *The Evolution of the Euclidean Elements. A Study of the Theory of Incommensurable Magnitudes and Its Significance for Early Greek Geometry.* Dordrecht/Boston: D. Reidel Publishing Company.

Neugebauer, O. and A. Sachs (1945) *Mathematical Cuneiform Texts.* New Haven, CT: American Oriental Society.

Thomas, I. (1939 (reprinted 1980)) *Selections Illustrating the History of Greek Mathematics. I. Thales to Euklid.* Cambridge, MA.:Harvard University Press; London: William Heinemann

Chapter 16. Greek and Babylonian Square Side Approximations

Bruins, E. M. and M. Rutten (1961) *Textes mathématiques de Suse.* Paris: Paul Geuthner.

—— (1964) *Codex Constantinopolitanus Palatii Veteris No. 1.* Leiden: E. J. Brill.

Colebrooke, H. T. (1817 (reprinted 1973)) *Algebra with Arithmetic and Mensuration from the Sanskrit of Brahmagupta and Bhaskara.* Walluf bei Wiesbaden, Sändig.

Friberg, J. (1990) Mathematik (in English). In *Reallexikon der Assyriologie und Vorderasiatischen Archäologie.* Berlin & New York: de Gruyter.Vol. 7: 531-585.

—— (1997) 'Seed and Reeds continued'. Another metro-mathematical topic text from Late Babylonian Uruk. *Baghdader Mitteilungen* 28, 251-365; pl. 45-46.

—— (2000) Mathematics at Ur in the Old Babylonian period. *Revue d'assyriologie* 94, 97-188.

—— (2005) On the alleged counting with sexagesimal place value numbers in mathematical cuneiform texts from the third millennium BCE. *Cuneiform Digital Library*

Journal 2005/2. http://cdli.ucla.edu/Pubs/CDLJ/2005/002.html

——— (2005) *Unexpected Links between Egyptian and Babylonian Mathematics.* Singapore: World Scientific.

——— (2007) *A Remarkable Collection of Babylonian Mathematical Texts.* New York: Springer .

Heath, T. L. (1921; Dover edition 1981) *A History of Greek Mathematics, 1-2.* Oxford: Clarendon Press.

Hofmann, J. E. (1934) Über die Annäherung von Quadratwurzeln bei Archimedes und Heron. *Jahresbericht der Deutschen Matematikervereinigung* 43: 187-210.

Høyrup, J. (2002) *Lengths, Widths, Surfaces. A portrait of Old Babylonian algebra and its kin.* New York, Berlin: Springer.

Knorr, W. R. (1975/76) Archimedes and the Measurement of the Circle: A new interpretation. *Archive for History of Exact Sciences* 15, 115-140.

Muroi, K. (1999) Extraction of square roots in Babylonian mathematics. *Historia Scientiarum* 9, 127-133.

Neugebauer, O. (1935-37 (reprinted 1973)) *Mathematische Keilschrift-Texte, I-III.* Berlin: Springer Verlag.

——— and A. Sachs (1945) *Mathematical Cuneiform Texts.* New Haven, CT: American Oriental Society.

Parker, R. A. (1972) *Demotic Mathematical Papyri.* Providence, RI & London: Brown University Press.

Robson, E. (1999) *Mesopotamian Mathematics 2100-1600 BC: Technical Constants in Bureaucracy and Education.* Oxford: Clarendon Press.

Schöne, H. (1903) *Heronis Alexandrini opera quae supersunt omnia, 3. Rationes dimetiendi et commentatio dioptrica.* (*Vermessungslehre und Dioptra.*) Leipzig, Teubner..

Thureau-Dangin, F. (1938) *Textes mathématiques babyloniens.* Leiden, E. J. Brill.

Weil, A. (1984) *Number Theory, An Approach Through History. From Hammurapi to Legendre.* Boston/Basel/Stuttgart: Birkhäuser.

Chapter 17. Theodorus of Cyrene's Irrationality Proof and Descending Infinite Chains of Birectangles

Colebrooke, H. T. (1817 (reprinted 1973)) *Algebra with Arithmetic and Mensuration from the Sanskrit of Brahmagupta and Bhaskara.* Walluf bei Wiesbaden, Sändig.

Datta, B. and A. N. Singh (1938 (reprinted 1962)) History of Hindu Mathematics. A Source Book. Part 2: Algebra. Bombay/Calcutta, *etc.* : Asia Publishing House.

Friberg, J. (2005) *Unexpected Links between Egyptian and Babylonian Mathematics.* Singapore: World Scientific.

Heath, T. L. (1921; Dover edition 1981) *A History of Greek Mathematics, 1-2.* Oxford:

Clarendon Press.

Heller, S. (1956-58) Ein Beitrag zur Deutung der Theodoros-Stelle in Platons Dialog Theaetet, *Centaurus* 5, 1-58.

Hofmann, J. E. (1956). Ergänzende Bemerkungen zum "geometrischen" Irrationalitätsbeweis der alten Griechen. *Centaurus* 5, 59-72.

Knorr, W. R. (1975) *The Evolution of the Euclidean Elements. A Study of the Theory of Incommensurable Magnitudes and Its Significance for Early Greek Geometry.* Dordrecht/Boston: D. Reidel Publishing Company.

Srinivasiengar, C. N. (1967) *The History of Ancient Indian Mathematics.* Calcutta: World Press Private.

Weil, A. (1984) *Number Theory, An Approach Through History. From Hammurapi to Legendre.* Boston/Basel/Stuttgart: Birkhäuser.

Zeuthen, H. G. (1910) Sur la constitution des livres arithmétiques des Eléments d'Euclide et leur rapport à la question de l'irrationalité. *Oversigt over det Kgl. Danske Videnskabernes Selskabs Forhandlinger,* 431-473.

Chapter 18. The Pseudo-Heronic *Geometrica*

Baillet, J. (1892) *Le papyrus mathématique d'Akhmîm* (Mém. de la Mission Archéologique Française au Caire, 9).Paris: E. Leroux.

Bruins, E. M. (1964) *Codex Constantinopolitanus Palatii Veteris No. 1.* Leiden.

Friberg, J. (1981) "Methods and traditions of Babylonian mathematics, 2. An Old Babylonian catalogue text with equations for squares and circles." *JCS* 33: 57-64.

—— (1990) Mathematik (in English), in *Reallexikon der Assyriologie und Vorderasiatischen Archäologie, vol. 7.* Berlin & New York: de Gruyter, 531-585.

——, H. Hunger, F. Al-Rawi (1990) 'Seed and Reeds', a metro-mathematical topic text from Late Babylonian Uruk. *Baghdader Mitteilungen* 21, 483-557, pl. 46-48.

—— (1993) On the structure of cuneiform metrological table texts from the -1st millennium.. *Grazer Morgenländische Studien* 3, 383-405.

—— (1997) 'Seed and Reeds continued'. Another metro-mathematical topic text from Late Babylonian Uruk. *Baghdader Mitteilungen* 28, 251-365; pl. 45-46.

—— (2005) *Unexpected Links between Egyptian and Babylonian Mathematics.* Singapore.: World Scientific.

—— (2007) *A Remarkable Collection of Babylonian Mathematical Texts.* New York: Springer .

Heath, T. L. (1921; Dover edition 1981) *A History of Greek Mathematics, 1-2.* Oxford: Clarendon Press.

Heiberg, J. L. (1912) *Heronis Alexandrini opera quae supersunt omnia, 4. Heronis Definitiones cum variis collectionibus. Heronis quae feruntur Geometrica.* Leipzig:

Teubner.

——— (1914) *Heronis Alexandrini opera quae supersunt omnia, 5. Heronis quae feruntur Stereometrica et De Mensuris.* Leipzig: Teubner.

Høyrup, J. (1997) Hero, Ps-Hero, and Near Eastern practical geometry. An investigation of Metrica, Geometrica, and other treatises. In K. Döring, B. Herzhoff and G. Wöhrle (eds.) *Antike Naturwissenschaft und ihre Rezeption, vol.* 7, 67-93. Trier: Wissenschaftlicher Verlag Trier. .

——— (1997) Sulla posizione della 'formula di Erone' nei Metrica (con una nota platonica). *Bolletino di Storia delle Scienze Matematiche* 17, 3-11.

——— (2002) *Lengths, Widths, Surfaces. A portrait of Old Babylonian algebra and its kin.* New York, Berlin: Springer.

Neugebauer, O. (1935-37 (reprinted 1973)) *Mathematische Keilschrift-Texte, I-III.* Berlin: Springer Verlag.

Schöne, H. (1903) *Heronis Alexandrini opera quae supersunt omnia, 3. Rationes dimetiendi et commentatio dioptrica.* (*Vermessungslehre und Dioptra.*) Leipzig, Teubner..

App. 1. VAT 8393. A Chain of Trapezoids with Fixed Diagonals

Friberg, J. (2007) *A Remarkable Collection of Babylonian Mathematical Texts.* New York: Springer .

Neugebauer, O. and A. Sachs (1945) *Mathematical Cuneiform Texts.* New Haven, CT: American Oriental Society.

Robson, E. (2000). Mathematical cuneiform texts in Philadelphia, Part I: Problems and calculations. *SCIAMVS – Sources and Commentaries in Exact Sciences* 1, 11-48.

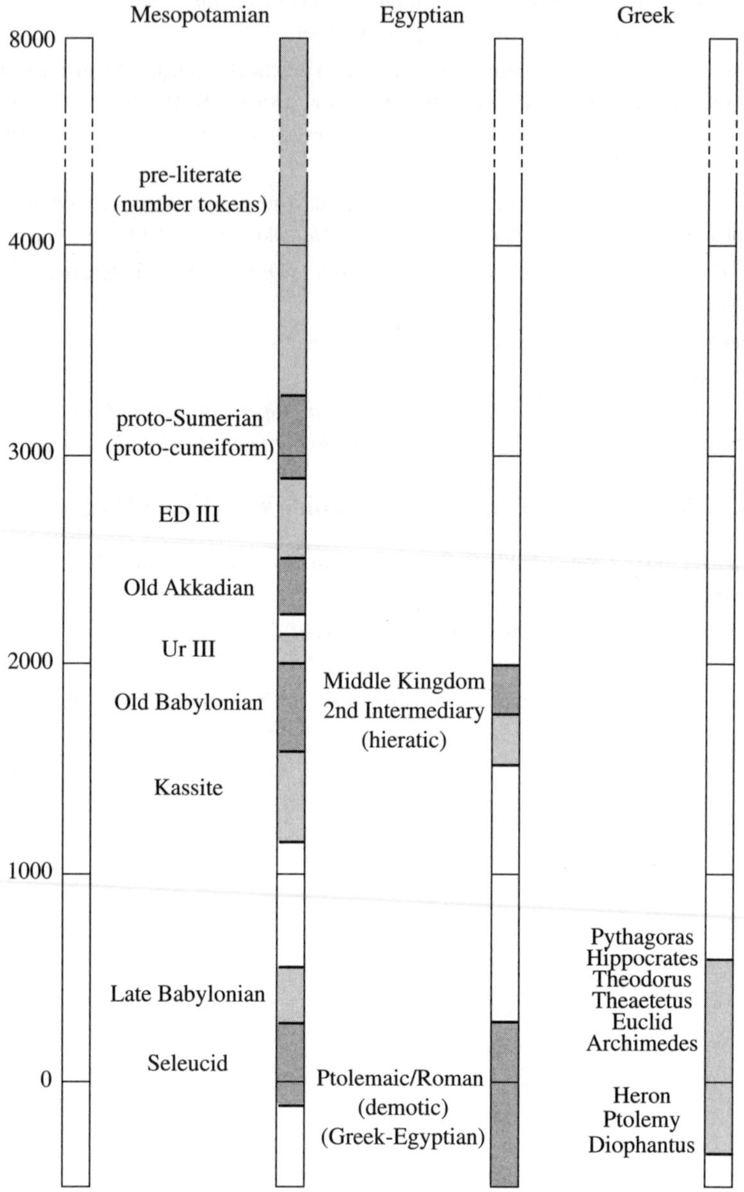

Comparative Timelines, Showing Periods of Documented Mathematical Activities.